Two-Dimensional Transition-Metal Dichalcogenides

Two-Dimensional Transition-Metal Dichalcogenides

Phase Engineering and Applications in Electronics and Optoelectronics

Edited by Chi Sin Tang, Xinmao Yin, and Andrew T. S. Wee

WILEY-VCH

Editors

Dr. Chi Sin Tang
National University of Singapore
Singapore Synchrotron Light Source
5 Research Link
117603 Singapore
Singapore

Prof. Xinmao Yin
Shanghai University
Physics Department
200444 Shanghai
China

Prof. Andrew T. S. Wee
National University of Singapore
Department of Physics
2 Science Drive 3
Department of Physics
117551 Singapore

Cover Image: © oxygen/Getty Images

Library of Congress Card No.: applied for

British Library Cataloguing-in-Publication Data
A catalogue record for this book is available from the British Library.

Bibliographic information published by the Deutsche Nationalbibliothek
The Deutsche Nationalbibliothek lists this publication in the Deutsche Nationalbibliografie; detailed bibliographic data are available on the Internet at <http://dnb.d-nb.de>.

Print ISBN: 978-3-527-35064-3
ePDF ISBN: 978-3-527-83874-5
ePub ISBN: 978-3-527-83876-9
oBook ISBN: 978-3-527-83875-2

Typesetting Straive, Chennai, India
Druck und Bindung CPI Group (UK) Ltd, Croydon, CR0 4YY

C9783527350643_101123

Contents

Preface

Two-dimensional transition-metal dichalcogenides (2D-TMDs) have undergone intense research and detailed scrutiny over the past decade. The years of intensive research yielded profound insights into their optical, electronic, mechanical, and magnetic properties and capabilities. The agglomeration of extensive scientific knowledge towards 2D-TMDs has in turn opened new frontiers and created greater potential for applications in a diverse range of disciplines and over multiple domains, such as light emitters, photo-sensors, catalysts, and clean-energy media, and the list could possibly grow longer with time.

Unlike previous publications related to 2D-TMDs, this new publication provides an in-depth yet comprehensive review concerning the polymorphic phases of 2D-TMDs, with emphasis on the phase engineering strategies and a diverse range of arising applications. We gathered experimental and theoretical experts of the respective sub-domains in the multifaceted discipline of 2D-TMDs who share valuable insights in areas including thermoelectricity, theoretical modeling, magnetic 2D-TMDs, material preparation, and the extent to which 2D-TMDs have been utilized in multiple areas of application. The book concludes by giving readers an idea how the rapid research and development of 2D-TMDs could possibly address the major issues facing humanity today.

By elegantly piecing together various aspects of 2D-TMD research and development in this publication, this reference text covers a broad range of topics that encompass the rapid scientific and technological development of polymorphic 2D-TMDs. It serves as an ideal reference for physicists, chemists, materials engineers, and technologists to better comprehend the challenges, motivate them to address the utilization of TMD-based applications, and propel this exciting field forward.

9 December 2022

Chi Sin Tang
Singapore Synchrotron Light Source
National University of Singapore
5 Research Link
117603 Singapore

Xinmao Yin
Shanghai Key Laboratory of High
Temperature Superconductors
Physics Department
Shanghai University
Shanghai, 200444, China

Andrew T. S. Wee
Department of Physics
National University of Singapore
2 Science Drive 3, 117551, Singapore

1

Two-dimensional Transition Metal Dichalcogenides: A General Overview

Chi Sin Tang[1,2] *and Xinmao Yin*[2]

[1] *National University of Singapore, Singapore Synchrotron Light Source (SSLS), 5 Research Link, Singapore, 117603, Singapore*
[2] *Shanghai University, Shanghai Key Laboratory of High Temperature Superconductors, Shanghai Frontiers Science Center of Quantum and Superconducting Matter States, Physics Department, 99 Shangda Road, Shanghai 200444, China*

1.1 Introduction to 2D-TMDs

Since graphene was first exfoliated from graphite using the mechanical cleavage method [1], 2D materials have garnered widespread interest. Atomically thin 2D-TMDs, with a formula of MX_2 (M: transition metal atom; X: chalcogen atom), form a diverse class of 2D materials with about 60 members. While one can trace back the extensive studies on bulk and multilayer TMD materials to more than half a century ago [2], it was only the groundbreaking emergence of graphene of single-atom thickness [1, 3] that led to the tremendous progress of monolayer van der Waals systems within the last two decades. With unique optoelectronic properties and robust mechanical features, 2D-TMD is a class of low-dimensional materials ideal for multiple applications in areas such as electronics, optoelectronics, and valleytronics [4–6]. They surpass graphene in terms of their functionality due to the combination of their non-zero bandgap electronic structures and their pristine yet robust layered surface properties. 2D-TMD is also a favorable class of materials in practical applications related to field-effect-transistor (FET) based systems. Hence, extensive research studies have taken center stage over the past decade to uncover both the fundamental physical properties and to unleash new frontiers for possible 2D-TMD-based device applications.

At the molecular level, diverse variations to the chemical bonding and crystal configurations of the transition metal atom component in 2D-TMDs have led to multiple structural phases that possess unique electronic properties. The semiconducting 1H phase is a quintessential example. Structural changes to one of the chalcogen planes will result in the metallic 1T phase. In addition, a unique quasi-metallic 1T'phase arises due to its distorted sandwich structure, where an array of one-dimensional zigzag transition metal chains are formed [7, 8].

Two-Dimensional Transition-Metal Dichalcogenides: Phase Engineering and Applications in Electronics and Optoelectronics, First Edition. Edited by Chi Sin Tang, Xinmao Yin, and Andrew T. S. Wee.

1.2 Crystal Structures of 2D-TMDs in Different Phases

TMDs present various structural polymorphs which have attracted huge interest in the last decade, both as an ideal platform for the fundamental study of layered quantum systems and their potential for multiple applications. Structurally, a unit layer is made up of a transition metal layer sandwiched between two chalcogen layers. Interestingly, TMDs, whether in the mono or multi-layer form, manifest themselves in different structural phases arising due to different configurations of the transition metal atom component. The common polymorphs of 2D-TMDs are the trigonal prismatic 1H phase and the octahedral 1T phase. In the case of the octahedral 1T phase structure, it has been experimentally and theoretically shown that it is dynamically unstable under free-standing conditions [7–9]. Consequently, similar to the Peierls distortion, the 1T phase will relax and buckle spontaneously into a thermodynamically more stable distorted structure known as the 1T′ phase [7–9]. Hence, 1T phase 2D-TMDs can further stabilize under favorable chemical, thermal and mechanical conditions [10], particularly into the 1T′phase.

To better understand the diverse structural properties of 2D-TMDs, the respective structural phases can be visualized by the stacking configurations of the three atomic planes (i.e. the X-M-X structure). The 1H phase corresponds to an ABA stacking configuration where the chalcogen atoms at the top and bottom atomic planes are in the same vertical position and are located on top of each other in a direction perpendicular to the layer (Figure 1.1a). In contrast, the 1T structural phase has an ABC stacking configuration displayed in Figure 1.1b. Since the 1T phase structure is unstable under freestanding conditions, it will buckle and distort into the 1T′ structure where the transition metal atoms sandwiched between the upper and lower atomic layers distort. Consequently, it forms a period doubling 2×1 structure. As viewed from the top, this structural phase consists of an array of 1D zigzag transition metal chains (Figure 1.1c). Indeed, recent investigations related to symmetry-reducing CDW properties [12] in metallic 1T phase 2D-TMDs have created new opportunities for integrated low-dimensional material-based applications, including transistor systems, nanoscale charge channels and gate switching devices [13–15]. Besides, 1T′ phase 2D-TMDs are known to possess anisotropic electronic and optical features. An in-depth understanding of its unique structure can bring new insights to its characteristics, which can then be exploited for directionally regulated charge or photon channel applications in optoelectronics and electronics.

As discussed thereafter, electronic structure calculations have indicated that while the 1T phase 2D-TMD is metallic [10, 16], the 1T′ phase counterpart possesses a unique quasi-metallic electronic structure that will be discussed later. To clearly distinguish between these two structural phases, note that while low-temperature charge density wave (CDW) phases are typically observed in 1T phase TMD systems (e.g. T_{CDW} ~120 K for $TaSe_2$ and T_{CDW}~35 K for $NbSe_2$ [17, 18]), a CDW-like lattice distortion in the form of periodic 1D zigzag chain structure unique to the 1T′phase 2D-TMD can be observed even at room temperature [7, 8]. Morphologically, while the low-temperature commensurates CDW in 1T phase, TMDs can be distinguished

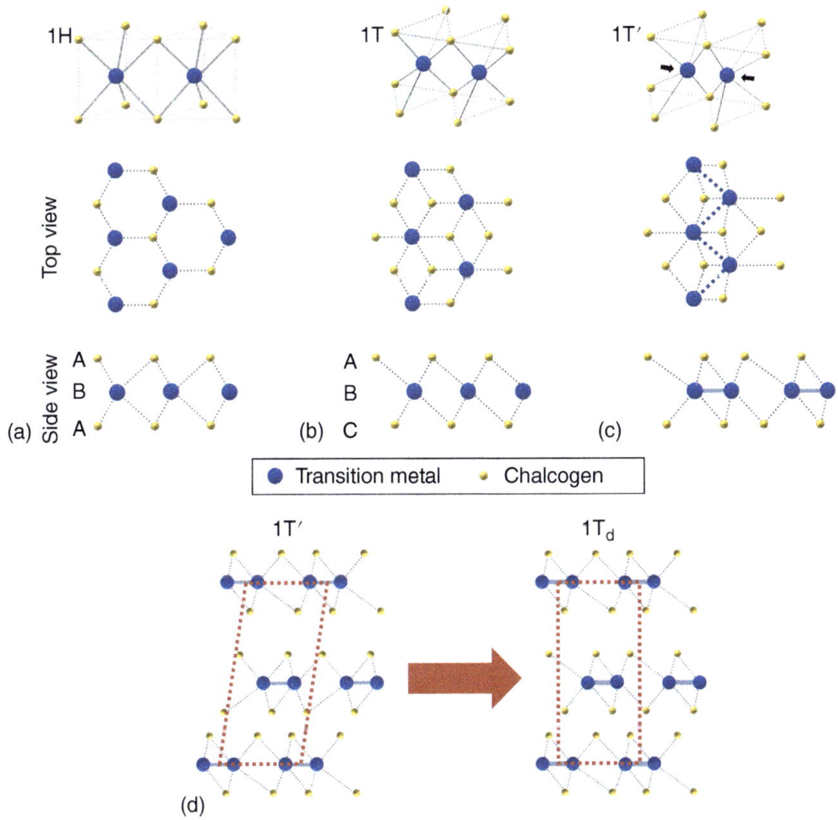

Figure 1.1 Lattice structures of 2D-TMDs in the (a) trigonal prismatic (1H), (b) octahedral (1T), and (c) distorted (1T′) phases. Stacking configuration of the atomic planes have been indicated. (d) The stacking orders that distinguish between the 1T′ and the $1T_d$ phases. The red dashed boxes serve as visual guides. Source: Tang et al. [11]/With permission of AIP Publishing.

by a unique star-of-David superlattice [19–21] typically characterized using scanning tunneling microscopy. Conversely, the 1D zigzag chains in the 1T′ phase are clearly distinguished via high-resolution transmission electron microscopy.

1.2.1 Other Structural Phases

Apart from the three common structural phases, TMDs are also present in other structural phases each with their unique optical and electronic properties. For example, $MoTe_2$ and WTe_2 would undergo a first-order phase transition from the monoclinic 1T′ phase to form the orthogonal $1T_d$ phase as temperature decreases. While the structures of both the 1T′ and $1T_d$ phase are rather similar, their key differences lie with the dislocations between the stacking layers depicted in the layer distortion in Figure 1.1d. These seemingly trivial dislocations can lead to a significant symmetry change between the two structural phases [22, 23]. Nevertheless,

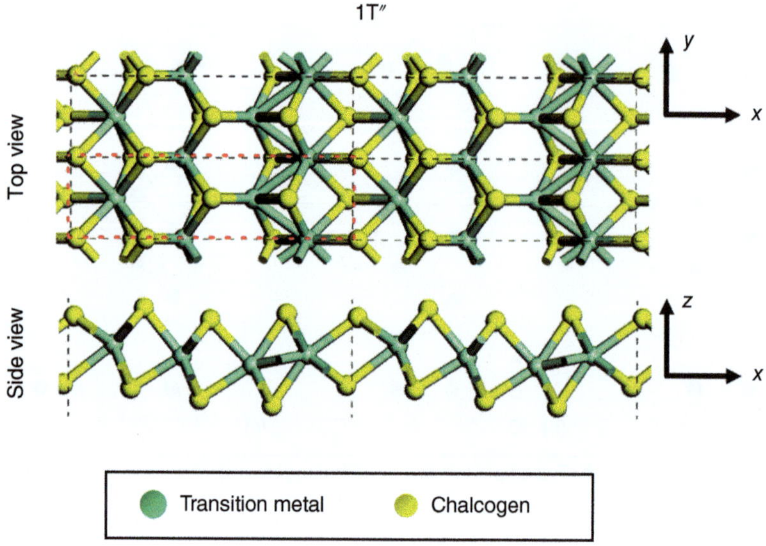

Figure 1.2 Atomic structures of the 1T″-phase 2D-TMD with the atomic planes indicated. Source: Reproduced with permission from Ma et al. [24]. Copyright 2016, The Royal Society of Chemistry.

with a similar structure present in the respective layer, $1T_d$ phase TMDs still possess similar quasi-metallic electronic properties as that of its 1T′ phase counterparts [9, 22, 23].

In the monolayer regime, apart from the formation of the 1T′ structural phase due to distortion from its 1T phase counterpart, other stable polymorphs have also been reported. This includes another octahedrally coordinated 1T″ phase (Figure 1.2) [25]. While there is an association between the transition metal atoms in the M–M configuration in the 1T′ and 1T″ distorted phases, the transition metal atoms are dimerized in the 1T′ phase while trimerization takes place in the 1T″ phase [26]. Besides, while the 1T′ phase possesses quasi-metallic electronic properties, computational studies have suggested that the 1T″ phase is a wide bandgap semiconductor where 1T″ phase monolayer-MoS_2 possesses an indirect bandgap of ~0.27 eV [24]. Nevertheless, no consensus has been reached in terms of the relative stability between the octahedral phases, particularly for monolayer-MoS_2, where different studies have reported different stability levels between the structural phases [24, 27, 28].

1.2.2 Phase Stability

Theoretical studies have shown that the 1H-1T′ energy differences vary between different Mo- and W-based 2D-TMD species, especially with the inclusion of spin-orbit coupling [29, 30]. Figure 1.3 displays the phase energetics of the respective monolayers in their respective phases as extracted from Ref. [30, 31], with $U = 0$ as the reference energy for the respective 2D-TMD in their 1H phase and 1T phase, which

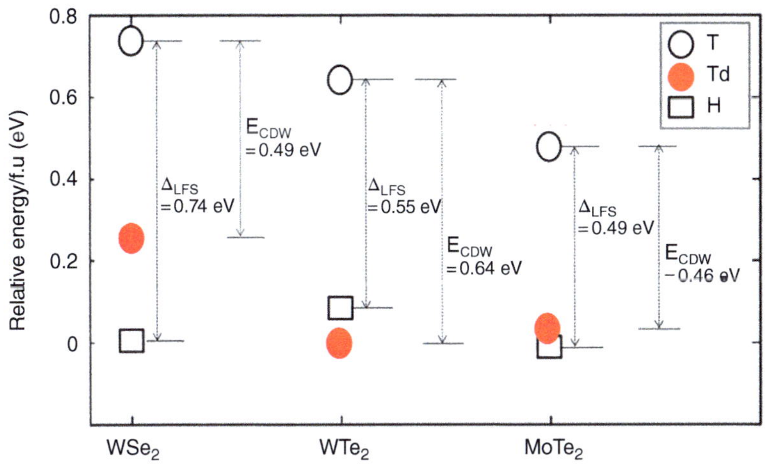

Figure 1.3 Relative stability of various phases of TMDs in the H, T, and T_d structural phases. The zero energy is the most stable phase for the TMDs. Source: Reproduced with permission from Santhosh et al. [31] © 2015 IOP Publishing Ltd.

are the most unstable of the polymorph. A smaller energy difference between the 1T′ and the 1H phase would mean a correspondingly smaller amount of external energy required to induce the 1H-1T′ phase transition. In the case of monolayer-WTe_2, the $U < 0$ registered for the 1T′ phase suggests that the compound is more stable in its 1T′ phase — i.e. 1H-1T′ phase transition takes place spontaneously without any external influence.

Even in the case of phase stability between the 1H and 1T phases, multiple studies have already been conducted specifically for $MoTe_2$ [32]. Under more general circumstances, simulations that involve the stability of different 2D-TMD species have also been extensively studied with a summary comprising 44 species displayed in Figure 1.4 [33]. When in their respective ground state polymorph, their different 2D-TMD species can possess semiconducting (T**/H**), metallic (T*/H*), or half-metallic (T^+) behaviors.

Based on the energy differences between the semiconducting 1H and quasi-metallic 1T′ phase of the respective Mo and W-based 2D-TMDs, the application of different electrostatic gating configurations can drive their 1H-1T′ phase transition. This is carried out by changing the carrier density or electron-chemical potential in the monolayer-TMD via a capacitor structure displayed in Figure 1.5 [29]. Specifically, in the case of undoped monolayer-$MoTe_2$ with a very small energy difference between the 1H and 1T′ phases, a mere surface charge density below $-0.04e$ or above $+0.09e$ per unit cell is required to drive the 1H-1T′ structural phase transition under pristine conditions. Whereas a significantly higher surface charge density of about $-0.29e$ or above $+0.35e$ per unit cell is required to drive the same structural transition for monolayer-MoS_2.

The control of pressure is a thermodynamic parameter that affects the phase stability of 2D-TMDs and induces a structural phase transition. As mentioned

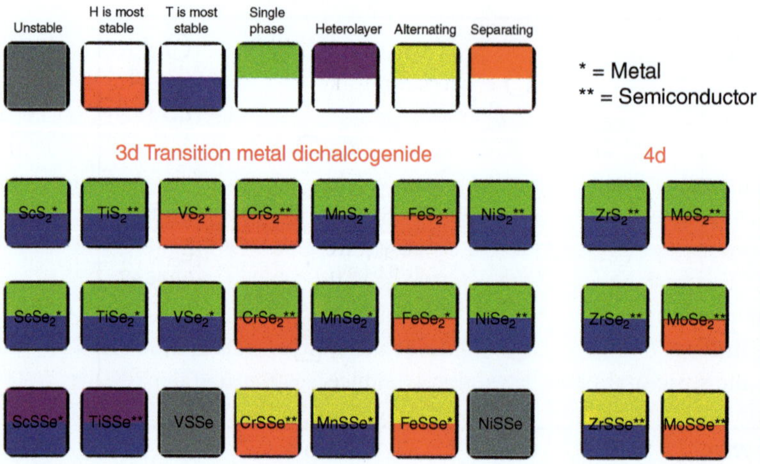

Figure 1.4 The structural stability and the electronic properties single phase and mixed phase of TMDs summarized in the table. The transition-metal atomic components are classified into the 3d, 4d, and 5d groups. With the gray colour boxes indicating phase separation. The resulting structural phases (T or H) could be metallic (*) or semiconducting (**). Source: Reproduced with permission.

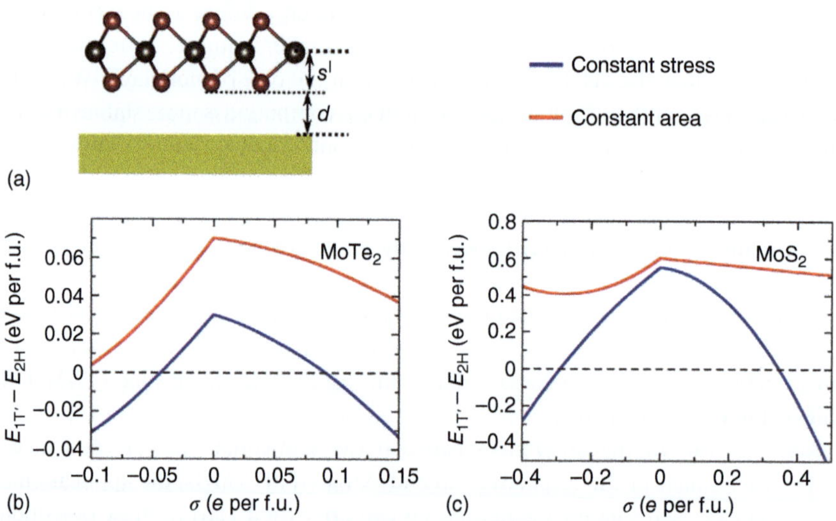

Figure 1.5 Phase boundary under constant charge in $MoTe_2$ and MoS_2 monolayers. (a) Schematic of a monolayer TMD lattice and an electron reservoir such as a metallic surface that is separated by a vacuum layer. Evolution in the energy difference of the respective 2D-TMDs between their 1H and 1T′ phases changes with respect to charge density, σ. The blue line represents situations where both the 1H and 1T′-phase lattices are in relaxed state while the red line represents the constant-area case where monolayer is kept to the area when under 1H-phase lattice constants. (b) 1H-phase $MoTe_2$ is a stable phase while 1T′-phase $MoTe_2$ is metastable when the monolayer is charge neutral. However, 1T′-phase $MoTe_2$ becomes increasingly stable when charged beyond the positive or negative threshold values. (c) 1H-phase monolayer-MoS_2 is stable when charge neutral. Nevertheless, the magnitude of charge required to induce 1H-1T′ transition is larger than its $MoTe_2$ counterpart. In both cases, transition at constant stress is more easily induced than the transition under constant area. Source: Li et al. [29]/Springer Nature/Licensed under CC BY 4.0.

earlier, the highly symmetric 1T phase structure for Mo and W-based 2D-TMDs is unstable under free-standing conditions and ambient pressure. They will therefore buckle and convert readily to the octahedral-like 1T′ phase structure. Density functional theory (DFT)- based studies to determine the phase diagrams of 2D-TMD with respect to tensile strain have shown that to maintain the stability of the metallic 1T phase structure, an equi-biaxial tensile strain of 10–15% is required for most 2D-TMDs (e.g. 13% for monolayer-MoS_2) except for $MoTe_2$. For the latter, a considerably smaller tensile strain <1.5% is required under appropriate constraints [30]. Given the small energy differences between the 1H and 1T′ phases for $MoTe_2$, minimal temperature changes also have profound effects on the transition dynamics due to the introduction of vibration effects to the system. Hence, the strain required to induce the structural phase transition is reduced at elevated temperatures [30]. Even monolayer-WTe_2, which is energetically more stable in the 1T′ phase under pristine conditions, will transform into semiconducting 1H phase when lattice compression is applied.

It is further discovered that the coexistence of multiple phases within a system – easily achievable under laboratory settings – is thermodynamically stable under reasonable thermodynamic conditions. Experimentally, such studies are achievable with a certain degree of restrictions. While tensile strain can easily be applied to 2D-TMDs by attaching it to a large elastic substrate where its deformations can be regulated [34, 35], the spontaneous formation of ripple patterns complicates the strain process and characterization [35–37]. As for bulk or multi-layer materials, deformations that are sufficiently significant can only be achieved via compression processes.

1.3 Electronic Band Structures of 2D-TMDs

Given the number of members in the 2D-TMD family, they exhibit a diverse range of macro properties related to their mechanical and electronic features. Their properties also differ between the monolayer, multilayer, and bulk counterparts. For example, in their bulk form, niobium, tantalum-based sulfides, and selenides as well as β-$MoTe_2$, and $1T_d$ phase WTe_2 possess metallic properties [38–41]. Conversely, TMD species such as the commonly studied MoS_2, WS_2, $MoSe_2$, WSe_2, α-$MoTe_2$, ReS_2, and ZrS_2 are semiconducting in nature [38, 39, 42–44]. At the molecular level, the interlayer van der Waals forces of interaction play a pivotal role in dictating their properties at both the macro and micro-electronic level. Even at the intralayer level, the intrinsic interactions taking place between the constituent atoms play a critical role in dictating the properties of the TMD species. Therefore, even though the lattice structures of the respective phases may be similar for different 2D-TMDs, the extensive interplay of the spin, charge, and orbital degrees of freedom result in very distinct electronic and optoelectronic properties.

In addition, apart from the chemical compositions of different 2D-TMDs, the diversity of structural phases and their respective stabilities brings about unique optoelectronic properties because of their unique electronic band structure, and the emergence of novel correlated and topological phases. These polymorphic electronic features are intimately related to the coupling between their charge, spin, orbital, and lattice degrees of freedom. A distinct example to be further elaborated on would be the presence of strong spin-orbit coupling that leads to the opening of the fundamental gap in 1T′ phase 2D-TMDs [8, 9].

1.3.1 Electronic Band Structures of the 1H, 1T, and 1T′ Phase

The electronic band structures of 2D-TMDs are dictated by multiple effects including strong Coulomb interaction between the constituent atoms and the effects of crystal symmetry. Meanwhile, spin-orbital coupling also has their distinct roles in determining the unique orbital characters, spin-valley properties, and unique optical selection rules. As discussed in the following section, the electronic band structures of multilayer TMD systems in the 2H-phase are a result of the weak interlayer van der Waals interactions. Structurally, 2H phase layered TMDs are characterized by the D_{6h} point symmetry group [45, 46]. Unlike their monolayer (1H) counterparts (Figure 1.6a), 2H phase multilayer and bulk TMDs are indirect bandgap semiconductors where the valence band maximum (VBM) is situated at the Γ-point of the hexagonal Brillouin zone, while the conduction band minimum (CBM) is approximately halfway between the Γ- and K-points (Figure 1.6b) [48, 49].

When examined in greater detail, by taking the example of monolayer-WSe_2 with spin-orbit coupling accounted for, the electronic bands at the Γ-point consist of contributions from both the Se (chalcogen) p_z-orbitals and the (transition mtal) d_{z^2}-orbitals in both the valence and conduction bands (Figures 1.7a,b, respectively). The energy bands of both the conduction and valence bands at the K/K'-points have a majority in-plane contribution from the localized transition metal orbitals. They include the $d_{x^2-y^2}$, and id_{xy}-states (VB) and a mixture of both the mtald$_{z^2}$ and chalcogen $p_x \mp ip_y$ orbitals at the conduction band [50–52]. The spatial overlap between adjacent layers and their corresponding effects on the band structure has

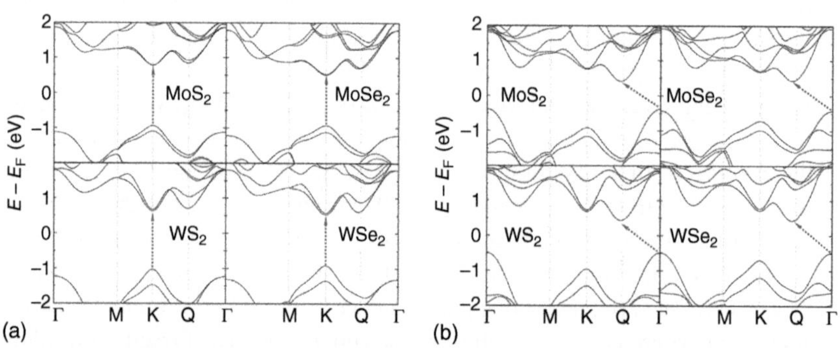

Figure 1.6 Comparing the band structures of the respective TMD materials in their monolayer and bulk form with SOC accounted for. (a) Band structure of MoS_2, $MoSe_2$, WS_2, and WSe_2 monolayers obtained from DFT calculations. Dashed arrows suggest that these are direct bandgap semiconductors. (b) Band structure of bulk MoS_2, $MoSe_2$, WS_2, and WSe_2 obtained from DFT calculations. Dashed arrows suggest that these are indirect bandgap semiconductors in their bulk form. Source: Roldán et al. [47]/With permission of John Wiley & Sons.

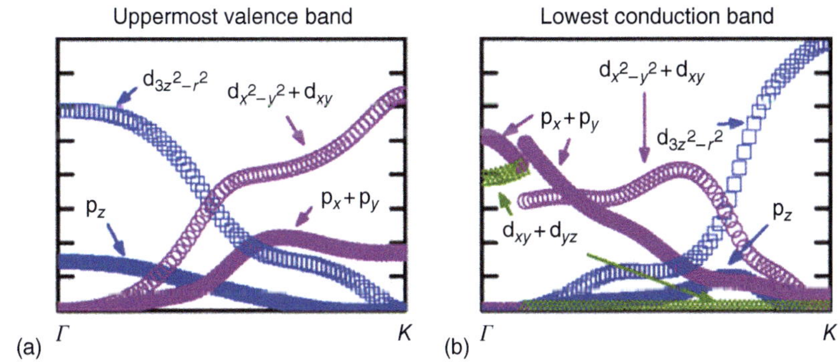

Figure 1.7 Spin-splitting in the valence and conduction bands as functions of $k_{//}$ along the $\Gamma-K$ points in the Brillouin zone with the contributions of the W (*d*) and Se (*p*) orbitals. (a) Uppermost valence band, and (b) lowermost conduction band. Source: Zhu et al. [50]/With permission of American Physical Society.

long-standing effects with variable layer thickness down to the monolayer limit [53, 54].

As for the electronic band structures of the commonly observed 1H, 1T, and 1T′ polymorphs, examples of the respective polymorphs of monolayer-MoS_2 are displayed in Figures. 1.8a,b,c, respectively [8, 54, 55]. The 1H-phase is a direct *K*–*K* bandgap semiconductor. The direct *K*–*K* transition has strong contributions from the transition metal d-orbitals (i.e. the Mo-atoms). With two contributing electrons from the transition metal d-orbitals, they will be paired and fully occupied in its lowest d_{z^2}-orbital state due to the hexagonal symmetry of the 1H phase structure. Hence, the 1H phase semiconducting structure is stabilized by the electron count of the transition metal d-orbitals. As discussed later, high-resolution scanning tunneling spectroscopy (STS) serves as an effective characterization technique that reveals a ~2.4 eV bandgap for chemical vapor deposition (CVD) synthesized monolayer-MoS_2 where it has been demonstrated that this bandgap decreases significantly with increasing thickness [57]. The thickness-dependent band structures have been further characterized using high-resolution angle-resolved photoemission spectroscopy (ARPES) as displayed in Figure 1.8d,g [56]. Note how the electronic structures of MoS_2 near the Fermi level evolves between its mono-, bi-, trilayer and its bulk form [56].

In terms of the quasi-stable metallic 1T-phase, there is a degeneracy in the electronic structure (lowest $d_{xy,yz,zx}$ degenerate orbitals) that is attributed to the structure's tetragonal symmetry. This results in the partially occupied transition metal d-orbitals that brings about the metallic character. In the regions near the Γ-point, the Fermi-level crosses both the Mo d- and S p-orbitals (Figure 1.8b). While the 1H-phase structure is stabilized by the electron count of the transition metal d-orbitals, charge-doping processes such as intercalation, alloying, gating, and defect engineering processes are expected to modify the transition metal d-orbital electron population. Consequently, this de-stabilizes the 1H-phase structure and leads to a structural transition into the 1T or 1T′-phase. Apart from modifying the electron count, external effects in the form of atomic distortion can also destabilize

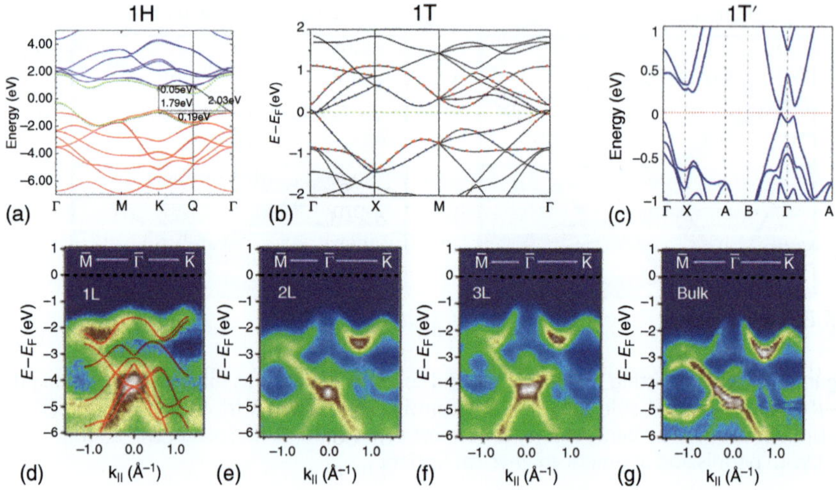

Figure 1.8 Theoretically and experimentally determined band structures of 2D-TMDs in their respective phases. (a) 1H, (b) 1T, and (c) 1T′-phase monolayer-MoS_2. The green lines in (a) show the top (bottom) valence (conduction) bands in bulk MoS_2. The lines marked by red and blue in (b) represent the bands, contributed mainly by the Mo 4d-orbitals and S 3p-orbitals, respectively. Band structure of 1T′-phase monolayer-MoS_2 in (c) with the fundamental, E_g, and inverted gap, 2δ. Inset: comparison of band structures with (red dashed line) and without (black solid line) spin-orbit coupling. ARPES band map of exfoliated (d) monolayer, (e) bilayer, (f) trilayer, and (g) bulk MoS_2 along the $M-\Gamma-K$ high symmetry lines. Source: (a) Kadantsev et al. [16]/Reproduced with permission from Elsevier. (b) Hu et al. [55]/Reproduced with permission from AIP Publishing. (c) Zhuang et al. [28]/Reproduced with permission of American Physical Society. (d)–(g) Reprinted with permission from Jin et al. [56], 2017 American Physical Society.

the electronic structure of the semiconducting 1H-phase which in turn leads to the formation of the 1T/1T′-phase.

As for the 1T′-phase 2D-TMDs, the zigzag chain structure lowers the transition metal d-orbitals below the chalcogen p-orbitals. This leads to a band inversion which in turn, leads to the formation of an inverted gap, 2δ, at the Γ-point in the Brillouin zone, along with two Dirac cones near the Γ-point in the 2D Brillouin zone. Under the influence of spin-orbit coupling, there is an opening of a fundamental gap, E_g, at the Dirac points (Figure 1.8(c)). ARPES study of 1T′-phase monolayer-WSe_2 synthesized by molecular beam epitaxy (MBE)[[58]] has shown a ~0.12 eV fundamental gap that is consistent with their band structure calculations. In this case the high-symmetry $\Gamma-Y$ and $\Gamma-P$ directional contributions have been detected in a single ARPES measurement. The STS and ARPES results exhibit the direct observation of the ~56 meV fundamental gap for the epitaxially grown 1T′-phase monolayer-WTe_2 [59]. Optical techniques have also been utilized to probe the optoelectronic properties of TMD materials. For instance, Fourier-transform infrared spectroscopy (FTIR) investigations by Keum et al. revealed an absorption feature at ~0.06 eV, which can be attributed to the fundamental gap of mechanically exfoliated 1T′-phase monolayer-$MoTe_2$ samples [9]. As presented later in the subsequent sections, spectroscopic ellipsometry has been used to probe the 1T′-phase MoS_2 and WSe_2 monolayers, where the emergence of the fundamental and inverted gaps has been observed [60, 61].

TMD systems in both their monolayer and multilayer forms exhibit myriads of rich physical phenomena, and they hold immense promise both in the realm of fundamental science and for future device and optoelectronic applications in their respective polymorphs. To effectively investigate and exploit the unique properties of each polymorph, efficient phase engineering strategies must be implemented to access the respective phases with minimal damage and interference to the system. The subsequent section will discuss the development of multiple phase transition strategies alongside the challenges and advantages each of them hold.

1.3.2 Indirect-to-Direct Bandgap Transition

In the natural state, TMD materials in their bulk and multilayer forms are indirect bandgap materials. Interestingly, as the sample thickness decreases gradually from a bulk system to one that is monolayer, not only does its bandgap increase in magnitude, but it also evolves into a direct bandgap semiconductor [38, 47, 51, 53, 54, 62, 63].

In the cases of TMD systems with constituent atoms from Group VI transition metals (Mo and W) and chalcogens such as S and Se, they are usually thermodynamically stable in the semiconducting 1H-phase [64]. Both monolayer-WSe_2 [65] and MoS_2 [38, 51] were first computationally predicted to possess such direct bandgap electronic structures at the K/K'-points of their hexagonal Brillouin zone. Figure 1.9 displays the calculated band structure of MoS_2 as it reduces from a bulk and eventually into the monolayer structure where it shows the gradual transition

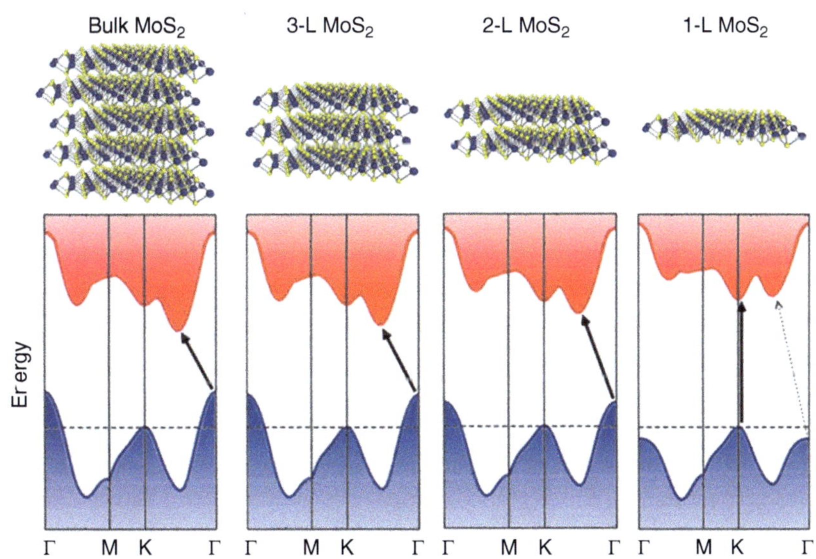

Figure 1.9 Thickness-dependent bandgap properties of semiconducting phase MoS_2 in its bulk, trilayer, bilayer, and monolayer form. With the arrows indicating the lowest energy transitions from the valence to the conduction band, one can see a gradual transition from an indirect (1.29 eV) to a direct bandgap (1.8 eV) with decreasing layer number. This accounts for the significant increase in photoluminescence efficiency of monolayer-MoS_2. Source: Reproduced with permission © 2016 WILEY-VCH Verlag GmbH & Co. KGaA, Weinheim Bonaccorso et al. [66].

from an indirect to a direct electronic bandgap [53, 54]. Besides, the computationally derived electronic bandgap for semiconducting phase MoS_2 shows an increase in magnitude from 0.88 eV in its bulk form to 1.71 eV for monolayer-MoS_2 [67]. The experimentally measured optical bandgap using photoluminescence measurements for monolayer-MoS_2 has been registered to be 2.16 eV where there is a distinct increase in the luminescence efficiency in the monolayer as compared to its bulk and multi-layer forms [53, 54]. Nevertheless, the disparity between the experimentally measured and theoretical value in the monolayer direct bandgap can be attributed to the presence of actual charge interactions in the real monolayer system.

The gradual transition of TMD systems from an indirect bandgap material in its multi-layer form to a direct bandgap system in its monolayer form can be attributed to the presence of interlayer hopping in their multilayer form. Figure 1.10 compares

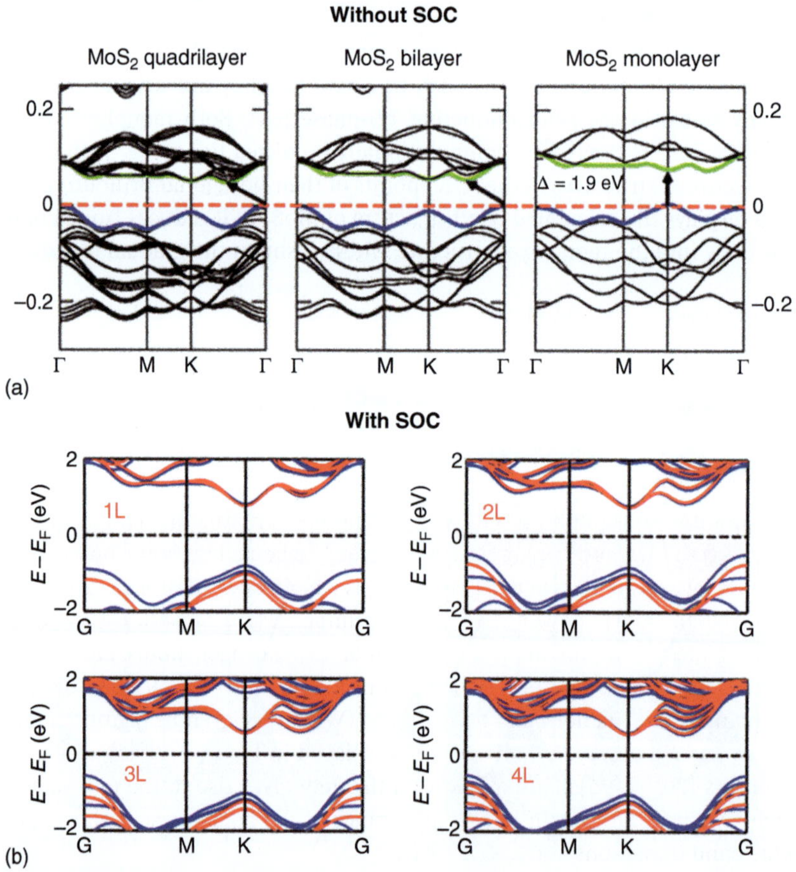

Figure 1.10 Band structures of (a) quadrilayer, bilayer and monolayer MoS_2 without the SOC effects, and (b) that with SOC in consideration. Source: (a) Reproduced with permission from Kuc et al. [62]. Copyright (2011) by the American Physical Society. (b) Reproduced with permission from Wickramaratne et al. [68]. AIP Publishing.

the band structures of WS_2 of different layer numbers with and without the effects of spin-orbit coupling (spin-orbit coupling effects to be discussed in the next section), the band edges at the respective high symmetry points K_V, Γ_V and Q_c (where subscripts C: conduction band; V: valence band) are shown to split with increasing layer number – these are direct evidence of interlayer hopping and the important role that Van der Waals interaction plays in such multilayer systems in dictating the system's electronic structures. The quantity of band splitting at each symmetry point also corresponds to the interlayer hopping strength. One should also note that contributions by the respective transition metal d- or chalcogen p-orbitals also vary at different regions of the electronic band structure. They result in different splitting magnitudes at different points in momentum space.

Notably, the Γ_V and Q_c-points display considerably more pronounced band splitting than K_V. This is because orbital contributions at the K_V-point are attributed mainly to the metal d-orbitals. Conversely, the chalcogen p_z-orbitals dominate the Γ_V and Q_c-points (c.f., Figure 1.7). With this in consideration, it is noted that interlayer charge hopping between the chalcogen p_z-orbitals plays a more significant role in comparison with other orbital interactions. Conversely, such interlayer hopping processes at the K_V-point is comparatively much weaker due to the large spatial separation between the transition metal planes. With increasing layer number, the energy level of the Γ_V-point registers a significant increase while that of Q_c is reduced. Meanwhile, the positions of both the K_C and K_V-points remain largely unchanged in the Brillouin zone. These relative evolution of the energy bands with contributions from both the metal and chalcogen orbitals results in the transformation of band structure from one that comprises an indirect band gap in the multilayer form to a direct bandgap system in the monolayer form.

1.3.3 Spin-Orbit Coupling and Its Effects and Optical Selection Rules

Readers who would like a very comprehensive treatment on this subject can also refer to the review [69]. Without the effects of interlayer van der Waals interaction, the effects of spin-orbit coupling (SOC) will become very prominent on the electronic properties of 2D-TMDs as compared to single-layer graphene. This is attributed to the heavy elemental components in the TMD systems where the SOC strength is directly proportional to the fourth power of the atomic number, Z [70]. Specifically, effects of the transition metal component outweighs that of the chalcogen atoms, particularly with the d-orbitals of the former [70]. For instance, there is a ~0.4 eV band splitting at the K-point in the valence band for W-based (WSe_2 displayed in Figure 1.11a,b) and ~0.2 eV for Mo-based 2D-TMDs (MoS_2 displayed in Figure 1.11c,d) as a result of SOC effects [50, 71–75]. This spin-split ultimately gives rise to the two valence sub-bands which in turn result in the appearance of the notable excitons A and B due to the band transitions at the K/K'-point.

Likewise, a significantly smaller spin-orbit split is also present at the CBM due contributions from the p- and d-orbitals [52, 76–78]. The polarity of the conduction band spin split may also differ depending on the transition metal atomic component. This leads to the lifting of the spin degeneracy of both the conduction and valence

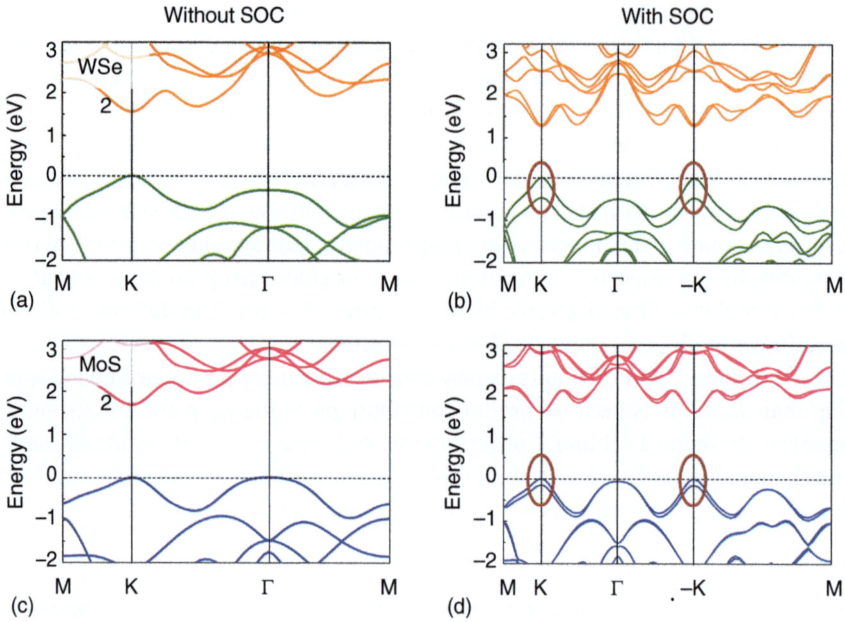

Figure 1.11 Band structure calculations of WSe_2 and MoS_2 monolayers with and without the effects of spin–orbit coupling. (a) Electronic band structure of monolayer-WSe_2 without, and (b) with the effects of spin–orbit coupling. (c) Electronic band structure of monolayer-MoS_2 without, and (d) with the effects of spin–orbit coupling. The valence band splits at the K(-K)-points are indicated by the red circles. Source: Tang et al. [71]/Reproduced with permission from American Chemical Society.

bands at the K/K'-point. Moreover, spin-splitting at the conduction band results in an energy separation between the spin-allowed and optically active (bright) transitions and the spin-forbidden and optically inactive transitions (dark).

Contributions from the electron-hole Coulomb exchange energy also affect the amplitude of the excitonic state splittings [79–81]. In the case of Mo-based 2D-TMDs, the lowest-energy transition will be the bright exciton [76, 77] as experimentally confirmed [82, 83]. Conversely, dark excitons in W-based 2D-TMDs are located at lower energy positions than their bright counterparts [82, 84–86].

With the spin-splitting effects, chiral optical selection rules apply for interband transitions at the K/K'-points. While the orbital Bloch functions at the conduction band transform with the angular momentum components of ± 1, those at the valence band remain invariant. This renders the optical selection rules for the interband transitions at the K/K' valley to be chiral. In other words, circularly polarized light of orientation $\sigma^{\pm}$ can only couple with interband transitions at the $K^{\pm}$ band, respectively [73, 87–90]. This unique optical selection behavior of 2D-TMDs makes them ideal for the investigation of valleytronic properties and applications [91–93].

While it is possible for the electrons to switch their valley, a spin-flip or a very energetically unfavorable transition is required. Hence, optically-generated electron-hole pairs are generally spin-valley polarized (locked). Hence, by conducting a spin-polarized excitation, exciton emission of 2D-TMDs can be co-polarized with the laser if the valley polarization lifetime is of similar order or longer than the exciton recombination time.

1.4 Excitons (Coulomb-Bound Electron-Hole Pairs)

An exciton is a quasiparticle bound state formed by an excited electron and hole due to their Coulomb interaction [94]. With the onset of photoexcitation, an electron will be excited from the filled valence band into the empty conduction band, and it will leave a hole in the valence band. With the electron and the hole possessing opposite polarity, they can attract each other via Coulomb interaction and such a many-body system involving a negatively charged conduction electron and a positively charged hole in the valence band can now be simplified. The formation of a bound state form between the electron-hole pair corresponds to a neutral *Exciton* which possesses a strongly-correlated relative spatial displacement between the constituents.

Based on the exciton radius and the Coulomb interactions within the electron-hole pair, two main classes of excitons can arise as a result of *long-range* and *short-range* coupling. With the large dielectric function within the crystal lattice, electric field screening reduces the Coulomb interaction between the electron-hole pairs. Hence, the exciton radii are significantly larger (long-range) in the scale of interparticle distances than the interatomic bond lengths. These are generally known as the *Wannier-Mott* or *large-radius-type* excitons commonly found in semiconducting systems such as GaAs [95] and Cu_2O [96]. With the smaller masses and screened Coulomb interaction, the binding energies of Wannier excitons are generally in the order of ~0.01 eV [97, 98].

Conversely, the short-range counterpart arises due to the overlapping wave functions of the constituent electron-hole pair at the scales in the order of the crystal lattice constant (e.g. $a_0 \sim 3$ Å for 2D-TMDs) and typically within the range of up to a few unit cells. This class of excitons correspond to the charge-transfer between nearest lattice sites. Besides, typically found in ionic crystals and organic molecular crystals, such electron-hole bound states are known as *Frenkel* Excitons which have binding energies in the order of ~0.1—1 eV [99–102].

Another important consideration regarding exciton formation due to Coulomb interaction is the self-energy of the electron-hole quasiparticle. This concept of self-energy is related to the repulsion between identical charges which results in an overall increase in the exciton bandgap of the system. It is the energy required to break the interaction between the electron-hole pair in the continuum similar to that of the Rydberg series of the hydrogen atom to create a free particle. Therefore, it is considered an excitonic Rydberg series (excited state of the electron-hole pairs) where the excitonic properties can be experimentally derived with the schematic of the optical absorption in the ideal 2D semiconductor displayed in Figure 1.12.

The presence of many-body interactions in the form of Coulomb interactions leads to the formation of the exciton resonances below the bandgap. Known as the *optical bandgap*, this energy gap is taken with reference to the lowest-energy excitonic

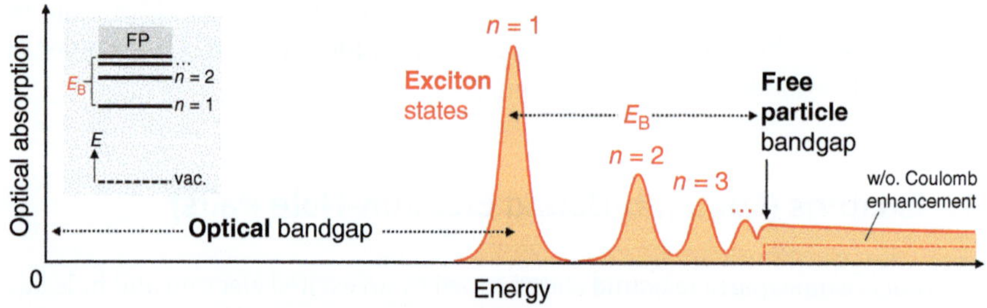

Figure 1.12 Schematic of the optical absorption of an ideal 2D semiconductor. It comprises a series of bright exciton transitions below the renormalized quasiparticle bandgap. Strong Coulomb interaction also enhances the continuum absorption in the energy range beyond the exciton binding energy, E_B. The inset shows the hydrogen atom-like energy level scheme of the exciton states, designated by the principal quantum number n, with E_B of the exciton ground state ($n = 1$) below the free-particle (FP) bandgap. Source: Wang et al. [69]/Reproduced with permission from American Physical Society.

feature in absorption at the exciton ground state ($n = 1$). As seen in Figure 1.12, the optical bandgap is distinct from the so-called *free-particle* band gap – corresponding to the continuum of free electrons and holes (where $n = \infty$). As a result of varying dielectric environments of different 2D systems, it plays a crucial role in tuning both the exciton binding energy and the free-particle bandgap of the system. Interestingly, this results in different systems possessing relatively similar optical bandgaps [103–105].

1.4.1 Exciton Binding Energy

Apart from the dielectric properties of the material in consideration, another key factor dictating the exciton binding energy would be the material's dimensionality. The exciton binding energy can thus be expressed based on the equation [106]:

$$E_n = -\frac{E_0}{\left(n + \frac{\alpha-3}{2}\right)^2} \tag{1.1}$$

where α denotes the material dimensionality, n the principal quantum number, and E_0 is the Rydberg constant.

As a result of this relation, the excitonic properties of 2D systems differ significantly from their 3D counterparts even of the same material. Strong electron-hole interaction results in a significantly different optical spectrum as compared to the 3D counterpart with the formation of a series of new excitonic levels below the quasiparticle bandgap as seen in Figure 1.12. A distinct feature of excitons in 2D systems is that they are strongly confined within the in-plane system where the electronic excitations are restricted within the 2D structure. Unlike excitons found in typical bulk three-dimensional systems, the large spatial overlap between the electron-hole orbitals can be attributed to the effects of quantum confinement. In addition, the electric field between the electron-hole pair can penetrate the

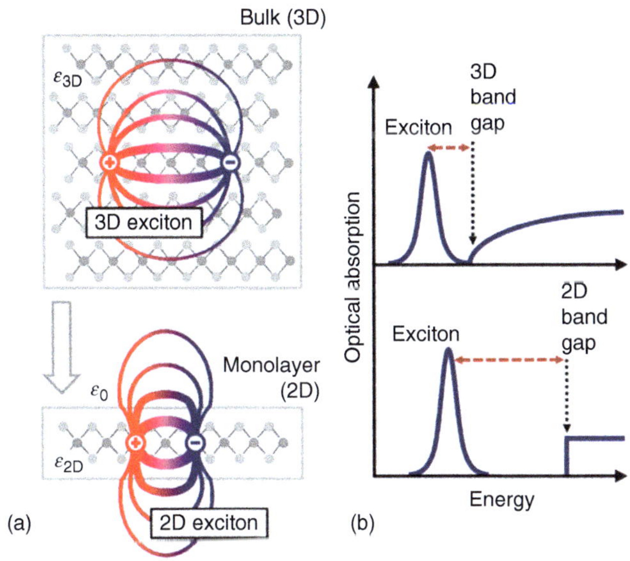

Figure 1.13 (a) Real-space representation of excitons in three-dimensional and two-dimensional systems. The changes in the dielectric environment are schematically represented by the respective dielectric constants ε_{3D} and ε_{2D} and the permittivity of free space ε_0. (b) The effects of dimensionality on the electronic and excitonic properties schematically represented by the optical absorption spectra. Change in dimensionality from 3D to 2D leads to an increase in the band gap and exciton binding energy represented by red dashed line. Source: Reproduced with permission Chernikov et al. [107]/Reproduced with permission from American Physical Society.

2D-plane where charge screening is absent. These features unique to 2D systems lead to the significant enhancement in the Coulomb attraction which in turn results in a larger excitonic binding energy as compared to the 3D counterpart (Figure 1.13a). For instance, while excitons in bulk WS_2 have binding energy of ~ 50 meV, excitons in monolayer-WS_2 have binding energies of ~ 320 meV [107].

As a result of the dimensionality, 2D excitons have unique properties with a combination of both Wannier-Mott and Frenkel features – they have both large Bohr radii and are strongly bound, respectively. In general, the exciton Bohr radii are in the order of ~1 nm [108–110]. Based on quasiparticle GW band structure calculations, the effective Bohr radius for MoS_2 is 9.3 and 13.0 Å in its monolayer and bilayer form, respectively, at the low-dimension limit [72]. While these may not appear to a significant magnitude, they are sufficiently large relative to the in-plane lattice constant of 2D-TMDs in general (~3 Å). Hence, the Wannier-Mott exciton theory is still applicable in such 2D systems. Correspondingly, the effective exciton binding energies stood at 0.224 and 0.106 eV in its monolayer and bilayer form, respectively. Interestingly, an alternative study has even reported an exciton spread over 65 Å [111]. Further unique excitonic effects attributed to the two-dimensional features come in the form of a step-function for gapped systems with dipole-allowed interband transitions (Figure 1.13b) [107].

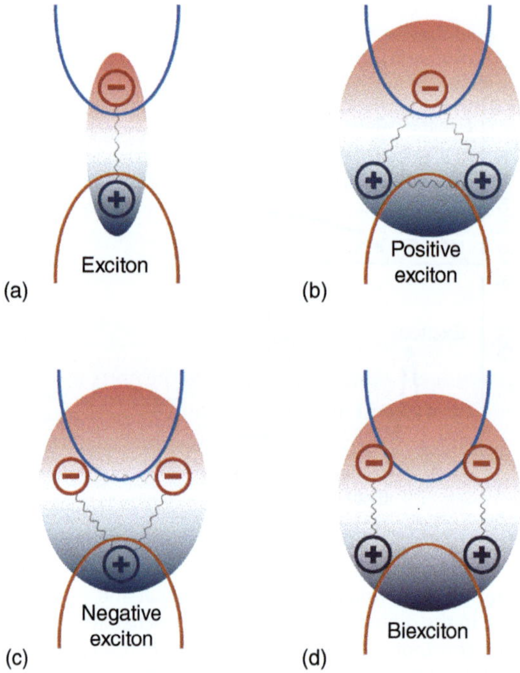

Figure 1.14 Schematic representation of quasiparticles. (a) Neutral exciton, (b) positive trion, (c) negative trion, and (d) neutral biexciton.

1.4.2 Excitons and Other Complex Quasiparticles

With increasing exciton (Figure 1.14a) and other charged-particle concentrations, complexes including the three-particle charged excitons known *trions* (Figure 1.14b,c) [71, 112, 113], and four-particle *bi-excitons* (Figure 1.14d) can also be formed [114, 115]. In the case of bi-excitons, their binding energies (~52 meV for monolayer-WSe_2) is more than an order of magnitude greater than those in conventional quantum-well systems [116]. This is attributed to the effects of both strong carrier confinement and reduced and non-local dielectric screening [108, 117].

While excitons and biexcitons are of neutral polarity due to their two-particle or four-particle composition, charged excitons known as a *trion* have odd-numbered (3) particles. Depending on its polarity, it can either comprise two holes and an electron (positive trion – Figure 1.14b) or two electrons and a hole (negative trion – Figure 1.14c) [112]. These charged quasiparticles are only formed in systems where excess charges are present and implies that the trion emission intensity serves to quantify the excess charge concentration.

While excitons in 2D-TMDs have binding energies in the order of ~102 meV, the binding energy of trions are usually an order of magnitude smaller but are still considerably larger than excitons in conventional semiconducting systems [116]. In many cases, both the exciton and trion peaks can be observed in photoluminescence experiments particularly at low temperatures. There are also instances where trions are observed in gated samples at an additional line 20–35 meV below the main exciton emission line and these can be attributed to the emission from charged excitons of both polarity [112, 118].

While exciton energy can vary as a result of varying Fermi energy induced by electronic gating process, the trion energy usually remained largely unchanged [112]. Besides, the energy split between the exciton and trion energy is expected to be linearly-dependent on the Fermi energy [119]. Hence, the trion binding energy can be estimated via the differences between the respective trion position and the associated exciton peak position based on the following expression [71, 112]:

$$E_X - E_{X'} = E_{\text{bin}} + E_F \tag{1.2}$$

where E_X and $E_{X'}$ denote the peak positions of the exciton and the associated trion, respectively; E_{bin} represents the trion binding energy while E_F is the shift in Fermi level relative to the conduction band maximum at the K/K'-point in the Brillouin zone [112]. In typical experimental studies, to obtain a reasonably good estimate of the trion binding energy, E_{bin}, is estimated for data sets at low temperature where E_F in Eq. (1.2) can be assumed to be negligible (i.e. $E_F \approx 0$) [71]. A series of analyses conducted show that the trion binding energy is in the range of ~18 meV [112] and some show marginally higher magnitudes at ~40 meV [71, 113] in systems such as MoS_2 and WSe_2 monolayers. The considerable magnitude of the trion binding energy suggests the importance of trion activities in 2D-TMD systems even at elevated temperatures of the order ~10^2 K.

1.4.3 Resonant Excitons in 2D-TMDs

While the aforementioned excitons typically occur within the direct bandgap of the relevant semiconducting materials and 2D systems, the presence of strong many-body interactions can also lead to the formation of unique high-energy regime quasiparticles known as high-energy *resonant excitons* – excitons that occur at energy levels well above the bandgap especially in the UV-Vis regime where higher energy bands are involved in their formation [120, 121]. In 2D systems, these unique resonant excitons are theoretically predicted [122] and have already been experimentally observed in monolayer-graphene [123]. Besides, such excitons have also been observed in 2D-TMDs including monolayer-WSe_2 [71]. The presence of resonant excitons in monolayer-WSe_2 and its absence from 2D-TMDs such as monolayer-MoS_2 where the atomic number of the constituent elemental components suggests the significant influence that SOC plays in the formation of this high-energy quasiparticle. This is attributed to the heavier atomic constituents of WSe_2 [70], where the SOC strength is directly proportional to the fourth power of the atomic number, Z. Therefore, it is specifically the transition metal atomic mass that plays a more significant role than the chalcogen atoms [70].

Specifically in the case of monolayer-WSe_2 as compared to systems such as monolayer-MoS_2, even though they share similar crystal structure in the semiconducting 1H-phase (Figure 1.15a), the strong SOC effects considerably transform the electronic and optical structures of monolayer-WSe_2. By combining this with the effects of strong electronic correlations, SOC facilitates the formation of

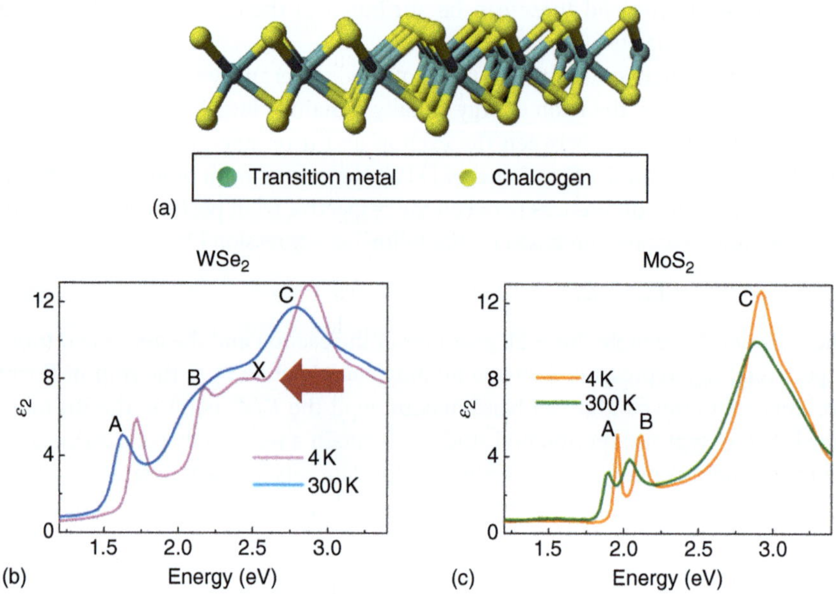

Figure 1.15 (a) The typical lattice structure of 1H-phase 2D-TMDs. Despite sharing similar lattice structures, the optical properties of different 2D-TMD species are significantly different. Source: Adapted from Tang et al. [71]. Examples shown here include the ε_2 spectra of (b) WSe_2 and, (c) MoS_2 monolayers. Source: Tang et al. [71]/Reproduced with permission from American Chemical Society.

three-dimensional resonant excitons in the high-energy visible regime (~2.4 eV) that can be directly observed via the optical response spectra (denoted by arrow in Figure 1.15b – c.f., Figure 1.20d). In contrast, this optical artifact is absent from monolayer-MoS_2 (Figure 1.15c – c.f., Figure 1.20a). As such, the fundamental understanding of high-energy resonant exciton is crucial in unraveling the intricate electronic structures of 2D materials in general, thus facilitating novel optoelectronic applications especially in the UV-Vis regime [71, 124–127].

1.5 Experimental Studies and Characterization of 2D-TMDs

Given the diverse range of potential applications that 2D-TMDs have promised, extensive characterization studies have been performed to uncover their intrinsic properties. This ranges from their optical bandgap to the electronic structures in the respective structural phases. With the unique low-temperature properties that several 2D-TMD systems possess, their morphologies have also been extensively reported especially in relation to their charge density wave structures. This section will delve slightly deeper into the respective techniques employed to uncover the unique properties of 2D-TMD systems. A brief overview of the synthesis techniques will also be provided.

1.5.1 Synthesis of 2D-TMDs

While MoS_2 and WS_2 are the main naturally-occurring TMD materials in the form layered crystals, single crystalline TMDs are good sources of monolayer TMDs for experimental characterizations and device fabrications [3, 64, 128]. Hence, 2D-TMDs are derived from bulk natural or synthetic crystals using mechanical exfoliation similar to how graphene is derived from graphite single crystals. While this is a convenient technique to obtain low-dimensional TMD samples for rapid experimental characterization and development of micro-device prototypes, the mono- or bilayer materials are generally non-scalable due to the random feature of the exfoliation technique and that the samples obtained are generally in the $\sim$100 μm^2 scale at the most.

An alternative to the mechanical exfoliation process would be the use of liquid-phase exfoliation with organic solvents [129, 130] and the Li-intercalation [42, 131] techniques. They allow for the synthesis of solutions containing TMD flakes at controllable thicknesses. Nevertheless, it may be difficult to regulate the phases of the synthesized 2D-TMDs particularly via the Li-intercalation technique for different species due to their relative stability. A mixture of different phases may be found over a large area of the 2D-TMDs due to the phase transition processes taking place within the sample [131–133] and the random stacking of the flakes in certain regions that results in inconsistency of the in-plane resistance may hinder the utilization of the films [131].

1.5.1.1 Chemical Vapour Deposition

Chemical vapour deposition (CVD) is a very efficient and cost-effective technique in the synthesis of high-quality and large-area 2D-TMDs (see schematic in Figure 1.16a). A very comprehensive treatment on this synthesis technique

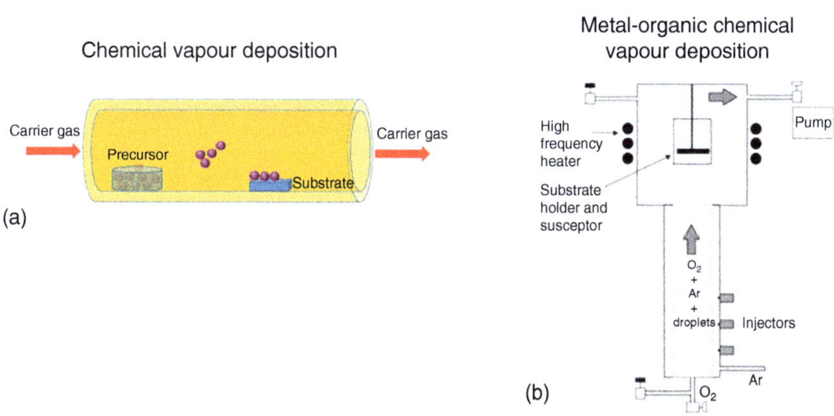

Figure 1.16 Schematics of CVD and MOCVD synthesis methods of TMD thin-films. (a) CVD is based on the direct chemical reaction between the precursor materials. (b) MOCVD works on the decomposition of metal–organic precursors and thereafter, their chemical reaction. Source: (a) Reproduced from Mittal et al. [134]/with permission of ELSEVIER. (b) Reproduced from Niu et al. [135]/with permission of ELSEVIER.

has been presented in Ref. [136] . For samples such as WSe_2 and MoS_2, WO_3 (MoO_3) and Se (S) powders react at elevated temperatures that lead to the eventual growth of the TMD monolayers on substrates such as SiO_2 and sapphire (Al_2O_3) [60, 137–139]. While CVD-grown 2D-TMDs on SiO_2 have greater instances of random lattice orientation which in turn would result in the formation of different grain boundaries [139, 140], those synthesized on Al_2O_3 have better control of domain orientations, thereby promoting their epitaxial growth [141]. The effective control and epitaxial growth of 2D-TMDs are crucial factors for the better control of the electrical and optical properties of large-area films.

Metal–organic CVD (MOCVD) is a further extension to the CVD technique based on their gas-phase precursors [142–145] where they are supplied to the main chamber with controlled flux (Figure 1.16b). Beside producing 2D-TMD samples with good electron mobility such as monolayer-MoS_2 of ~30 $cm^2\ V^{-1}\ s^{-1}$[143], a diverse variety of 2D-TMDs including $MoSe_2$ [146, 147], WS_2 [145, 148–152], WSe_2 [144, 153–155], ReS_2 [156–158], $ReSe_2$ [159], $MoTe_2$ [160, 161] and WTe_2 [161] have been grown.

1.5.1.2 Molecular Beam Epitaxy

Another notable method to synthesize high-quality monolayer TMDs at a scalable level would be the Molecular Beam Epitaxy (MBE) [162]. Figure 1.17 presents the schematic of this synthesis technique. In this case, several atomic sources are present in an ultra-high vacuum chamber (~10^{-10} mbar) to create the molecular

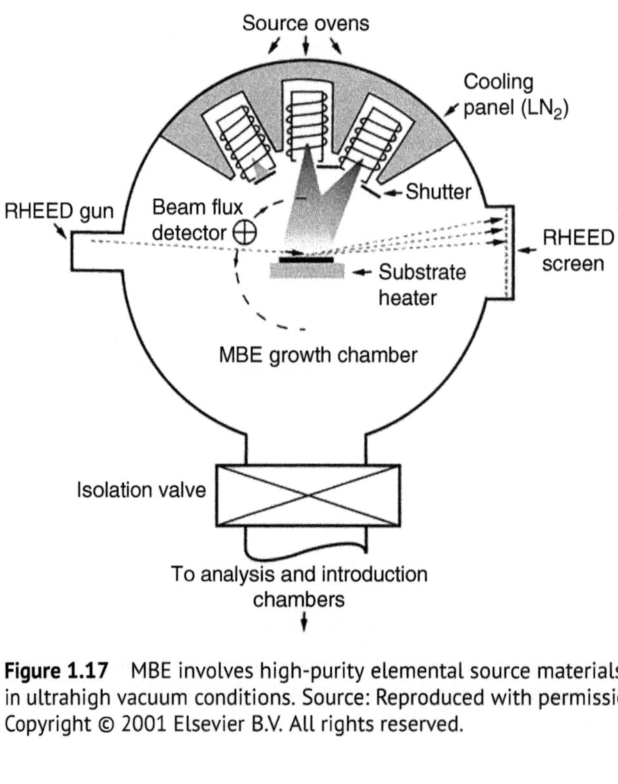

Figure 1.17 MBE involves high-purity elemental source materials that are co-evaporated in ultrahigh vacuum conditions. Source: Reproduced with permission Arthur [163].

beams. Molecular deposition is then conducted on a preheated substrate such as gold, Highly Ordered Pyrolytic Graphite (HOPG) while growth parameters including the film thickness and substrate crystallinity are monitored *in-situ* by reflection high-energy electron diffraction (RHEED) and low-energy electron diffraction.

With the availability of different substrates as mentioned earlier, 2D-TMDs can be synthesized on other layered materials and passivated surfaces without concern about lattice matching conditions due to their lack of dangling bonds. Besides, the van der Waals forces of interaction can be exploited for the epitaxial growth process where the substrate lattice structure is utilized to manipulate the monolayer crystal orientation [141]. Samples that are MBE-synthesized are generally of high-quality and crystallinity. Hence, they are generally characterized *in-situ* using high-precision systems such as angle-resolved photoelectron spectroscopy and scanning tunneling microscopy for fundamental studies and investigations [164, 165]. The effectiveness of this technique also allows for the synthesis of highly-ordered and crystalline vertical heterostructures such as $MoSe_2$/graphite [166], $SnSe_2/WSe_2$ [167] and $MoSe_2/Bi_2Se_3$ [168]. These exceptional features render MBE a highly versatile technique to synthesize 2D-TMDs and their heterostructures to study them at the fundamental level.

1.5.2 Optical Characterization

A diverse array of optical techniques is available to determine the optical properties of 2D-TMDs. Techniques such as photoluminescence and absorption edge spectroscopy serve as ideal methods to characterize optical band gaps [169]. Meanwhile, the absorption spectra of 2D-TMDs contain more information pertaining to their optoelectronic properties depending on their spectral regions. For instance, while the low-energy regions are generally dominated by the excitonic features, the high-energy regime contains information pertaining to interband transition states and band-nesting features related to the contours of the electronic band structures.

1.5.2.1 Photoluminescence

It has first been demonstrated that monolayer-MoS_2 possesses a direct bandgap and that results in a strong photoluminescence (PL) [53, 54]. Besides, PL can be utilized to demonstrate the indirect-to-direct bandgap transition of TMD materials when they are gradually reduced from a multi layer to a monolayer crystal. While techniques such as the scanning tunnelling spectroscopy the material's free-particle bandgap, E_0, PL characterizes the exciton transition energy, E_X. Of course, there are different types of excitons attributed to different forms of interband transitions in both atomically thin nanomaterials and heterostructures based on the schematic in Figure 1.18a [170]. By combining these two sets of measurements, one can arrive at the exciton binding energy which has been determined to be ~0.55 eV for monolayer-$MoSe_2$ [164] and ~0.37 eV for monolayer WSe_2 [172].

As mentioned in the earlier sections, spin-orbit coupling has a significant impact on the optical dichroic properties of 2D-TMDs in the form spin- and valley-selective excitonic states. Hence, PL excitation in the form of circularly polarized light can

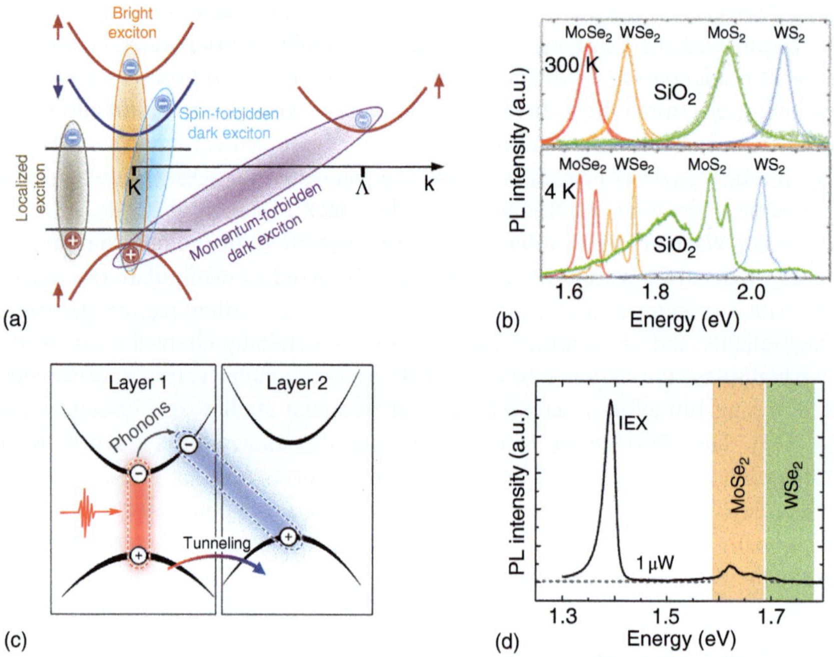

Figure 1.18 Different excitons in low-dimensional layers and heterostructures. (a) Schematic showing different types of excitons and how they take place attributed to different transitions. (b) (a) Comparing the PL spectra of the respective 2D-TMDs MLs at $T = 300$ K and $T = 4$ K when deposited directly onto SiO_2 and when capped with hBN. The appearances of additional PL peaks at low temperature indicate the presence of localized excitons due to lattice defect especially for the samples deposited directly on SiO_2. (c) Schematic showing how interlayer excitons appear where the electrons-holes are located in different constituent layers. (d) PL spectrum of a $MoSe_2$/WSe_2 heterostructure which shows the strongly PL response of interlayer excitons. Source: (a) and (c) Mueller and Malic [170]/Springer Nature/Licensed under CC BY 4.0. (b) [104]/American Physical Society/Licensed under CC BY 4.0. (d) Nagler et al. [171]/Reproduced with permission from IOP Publishing.

selectively induce excitonic transitions in either the K or K'-valley. Besides theoretical predictions [73], this unique effect has been well-verified experimentally [88–90, 173–178].

With the inevitable onset of surface impurities, defects, and lattice strain in 2D-TMD materials, PL serves as an effective technique to analyze electron-hole pairs (i.e. localized excitons – Figure 1.18a) that are bound in such defect-induced potential wells. During low-temperature PL measurements, apart from the usual bright exciton emission, a few other narrow optical resonances can also be observed (Figure 1.18b) and these are attributed to excitons trapped in local potential wells [104]. Nevertheless, such features are not visible at high temperature because the localized excitons are easily displaced by thermal excitations. These localized exciton states can be commonly found in monolayer TMDs [179–183] and crystal edges [179] where defects and discontinuities are commonly found.

The presence of Van der Waals interactions between stacked 2D-TMD layers [184] also yield interesting phenomena such as the formation of interlayer excitons where the components of the electron-hole pairs are situated in different TMD layers (Figure 1.18c) [171, 185–191]. The presence of interlayer excitons has been confirmed by multiple PL studies [171, 185, 186, 188]. Optical responses of interlayer excitons tend to be more prominent than their intralayer counterparts under weak excitation. In this study involving the time-resolved PL of time-resolved PL of the $MoSe_2/WSe_2$ heterostructure (Figure 1.18d), the long-lived excitonic response appears at the energy below the intralayer excitons. This can be attributed to the highly-efficient interlayer charge-tunneling processes that separate electron-hole pairs generated by optical absorption in the respective monolayers [171].

1.5.2.2 Spectroscopic Ellipsometry

Spectroscopic ellipsometry is a non-destructive and precise optical characterization technique that measures the thickness, optical properties, and surface morphology of samples. This is done by elucidating the complex reflection using optical modeling methods. Unlike other experimental techniques such as transmission and reflectivity measurements, both the real and imaginary components of the optical parameters can be derived simultaneously without the need for Kramers-Kronig transformations – an analytical technique that connects both the real and imaginary components of the complex optical function.

Having this as another major advantage, the collective features of spectroscopic ellipsometry make it a powerful and ideal system over a wide range of applications in both industrial and research disciplines.

Spectroscopic ellipsometry measures the change of light polarization upon its reflection from the sample. A schematic of the experimental setup is displayed in Figure 1.19a. The detector of spectroscopic ellipsometry measures the quantities, Ψ and Δ at each corresponding wavelength/photon energy. Parameter Ψ denotes the ratio between the amplitude of p- and s-polarized reflected light while Δ their phase difference. Specifically, p-polarized light has the electric field vector parallel to the plane of incidence while s-polarized light consists of electric field vector perpendicular to the incident plane (Figure 1.19b,c).

Typically, the energy range that is commonly used for spectroscopic ellipsometry measurement is the ultraviolet–visible (UV-Vis) regime (~0.5–6 eV). In this range, sample properties such as the optical band structures and bandgaps can be investigated. Nevertheless, other regions of the electromagnetic spectrum have also been used in spectroscopic ellipsometry measurements.

Spectroscopic ellipsometry serves as an effective technique in the characterization of 2D-TMDs. In this example, temperature-dependent high-resolution Spectroscopic Ellipsometry is conducted at temperatures between 4 and 300 K to systematically investigate the optical and excitonic properties of both 1H-phase WSe_2 and MoS_2 monolayers. Figure 1.20a,b display the real, ε_1, and imaginary, ε_2, components of dielectric function of monolayer-MoS_2 at 4 and 300 K. Meanwhile, the respective ε_1 and ε_2 spectra of monolayer-WSe_2 are displayed in Figure 1.20c,d. The optical parameters, ε_1 and ε_2 functions, of both the WSe_2 and MoS_2 monolayers are consistent with

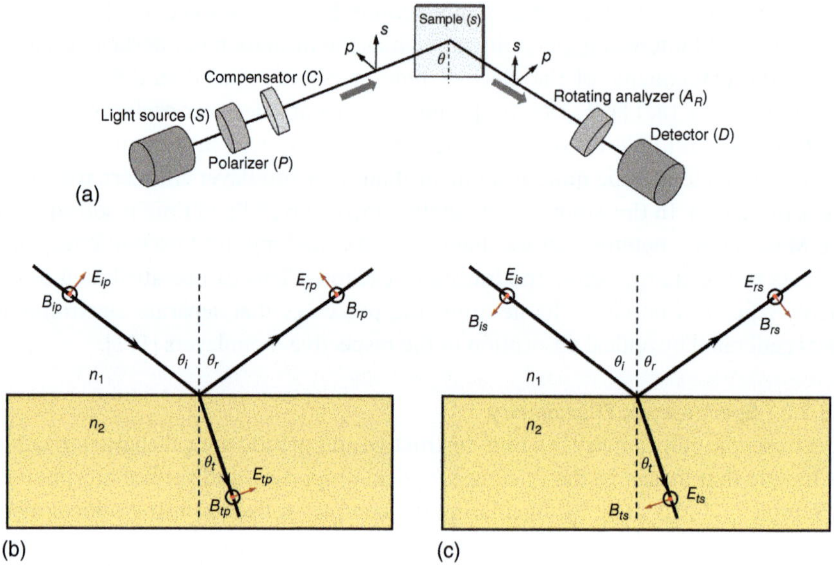

Figure 1.19 (a) Schematic of spectroscopic ellipsometry with the rotating-analyzer configuration. Electric and magnetic field for (b) *p*-polarized and (c) *s*-polarized waves during before and after reflection off the sample surface during the spectroscopic ellipsometry characterization process. Source: (a) Fujiwara [192]/Reproduced with permission from John Wiley & Sons.

other optical characterization reports of the same materials using either ellipsometric or reflectance techniques [53, 193].

Particularly, the ε_2-spectra of both MoS_2 and WSe_2 monolayers indicate that their optical and electronic structures differ significantly. As mentioned earlier, these distinctions are made more prominent in view of the similarities in their 1H-phase lattice structures. The differences in the optical properties must be attributed to other properties beyond the lattice structure for this class of materials. As such, a detailed investigation is needed to unravel the factors affecting the optoelectronic properties of 2D-TMDs.

To analyze the optical structures of both MoS_2 and WSe_2 monolayers, one can observe distinct peaks labeled A and B in the ε_2 spectrum (~1.96 and ~2.11 eV, respectively at 4 K) of monolayer-MoS_2 (Figure 1.20b – c.f., Figure 1.15c) have been ascribed to the excitonic transitions at the K/K'-points [194]. The strong inter-band transition (broad feature labelled as C) is attributed to the transitions in the region around the Γ- and M-points of the Brillouin zone where the conduction and valence bands are nested. This band-nesting feature arises due to the presence of van Hove singularities in the electronic band structures of MoS_2 and other 2D-TMDs (see also calculation results previously presented in Figure 1.11b) [195].

The ε_2 spectra of monolayer-WSe_2 also contain the distinctive peaks A, B, and C whether at low or ambient temperature (Figure 1.20d – c.f., Figure 1.15b). However, there is a distinct peak labeled as X located between peaks B and C. This feature is absent from monolayer-MoS_2 (Figure 1.20a). Interestingly, at the

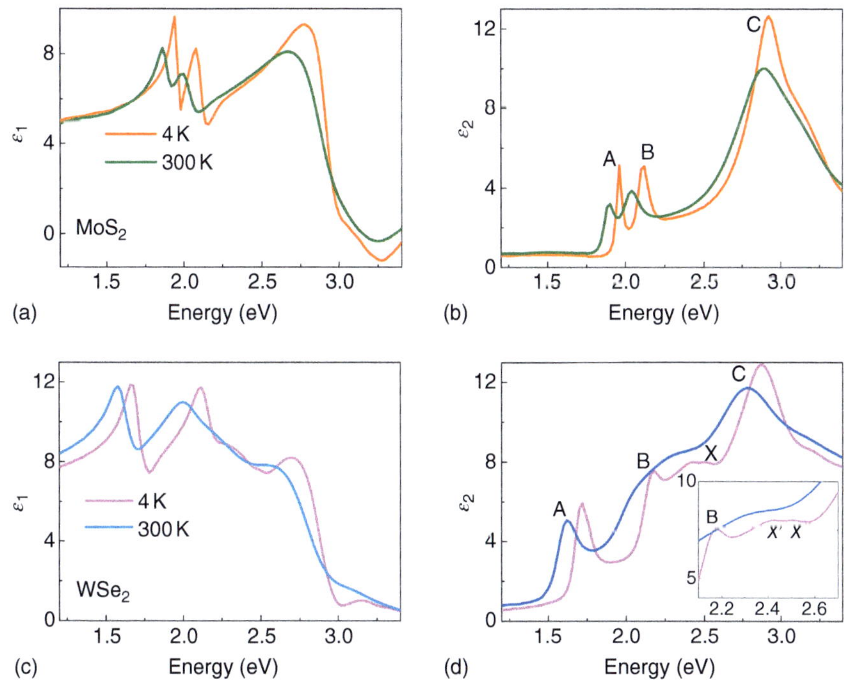

Figure 1.20 (a) ε_1 and (b) ε_2 dielectric functions of monolayer-MoS_2 at 4 and 300 K. (c) ε_1 and (d) ε_2 dielectric functions of monolayer WSe_2 at 4 and 300 K. Inset: Enlarged region of ε_2 indicating the presence of resonant exciton peak X and its associated trion X' at low temperature. Source: Tang et al. [71]/Reproduced with permission from American Chemical Society.

low-temperature below 100 K, peak X splits into two separate peaks – X and X' (~2.36 and ~2.49 eV, respectively – see inset Figure 1.20d). With reference to the ε_2 spectra of monolayer-MoS_2, peak A at ~1.72 eV and B at ~2.18 eV of monolayer-WSe_2 can be ascribed to the excitonic transitions at the K/K'-points in the Brillouin zone of monolayer-WSe_2 [50]. Meanwhile, the strong optical feature C at ~2.88 eV, corresponds to band-nesting around the Γ and M-points in the Brillouin zone of monolayer-WSe_2 [194].

While the origins of peaks A, B and C are well-accounted and explained in both the optical spectra of monolayer-MoS_2 and WSe_2, features X and X' (located at ~2.36 and ~2.49 eV, respectively) are only present in monolayer-WSe_2. While previous optical studies involving WSe_2 have attributed the appearance of X and X' to the effects of the Se p-orbitals or the intralayer effects that arose due to overlapping Se p- and W d-orbitals [195, 196], the explanation to properly account for the distinct features of X and X' remains unclear. This is because prior fundamental studies generally involved bulk or multilayer systems where interlayer interactions are considered. By combining temperature-dependent high-resolution spectroscopic ellipsometry down and high-energy Photoluminescence spectroscopy, feature X has been identified and attributed as a visible-range 3D resonant exciton in monolayer-WSe_2 [71].

The experimental study is also integrated with an extensive first-principles study to unravel the significant influence that spin-orbit coupling plays in the formation of this resonant exciton through the transformation of the electronic and optical structures of monolayer-WSe_2. With the presence of strong electronic correlations in monolayer-WSe_2, SOC facilitates the formation of three-dimensional resonant excitons in the visible regime that appear as peak X as confirmed in the collective analysis of both the spectroscopic ellipsometry and PL spectra.

Having witnessed the effectiveness of spectroscopic ellipsometry to characterize and identify important optical features of 1H-phase 2D-TMDs, it can also be employed to identify important optical structures in other structural phases as discussed in the next study.

A case in point would be the studies to monitor the 1H-1T′ phase transition process of monolayer-MoS_2 and WSe_2 through a high-temperature annealing process. After being transferred onto gold or other metallic substrates, optical characterization is conducted on monolayer-MoS_2 (MoS_2/Au) – see Figure 1.21 – and WSe_2 (WSe_2/Au) – see Figure 1.22 – after annealing at the respective temperatures. As presented in Figure 1.21b, at the optimal annealing temperature of MoS_2/Au, note the appearance of a broad mid-infrared pre-peak at the low-energy region (~0.6–1.5 eV)

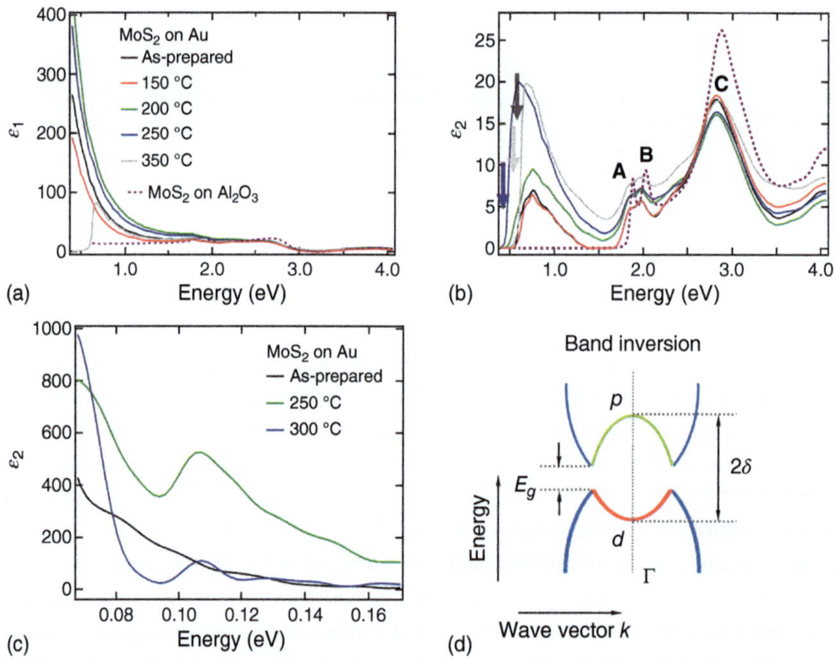

Figure 1.21 Dielectric functions (a) ε_1 and, (b) ε_2 of monolayer MoS_2/Au annealed at the respective temperatures and on Al_2O_3 substrate. (c) ε_2 spectra of monolayer-MoS_2/Au in the far-infrared regime measured using IR-VASE spectroscopic ellipsometry after annealing the sample at the stipulated temperatures. (d) Schematic band structure diagram of 1T′-phase 2D-TMD. Band inversion leads to the presence of the fundamental gap, E_g, and the inverted gap, 2δ. Source: Yin et al. [60]/Springer Nature / Licensed under CC BY 4.0.

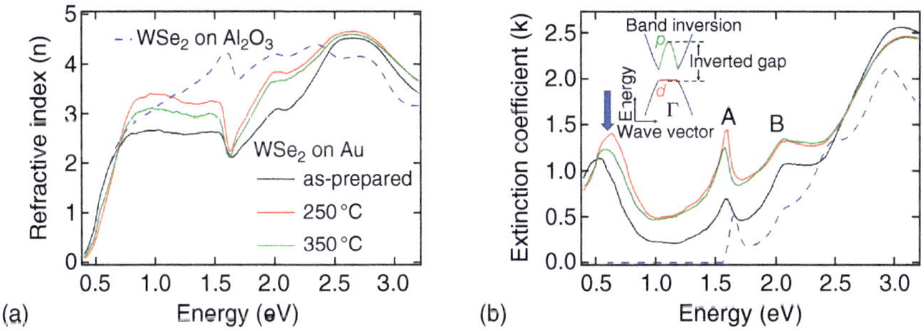

Figure 1.22 (a) Refractive index, $n(\omega)$, and (b) extinction coefficient, $\kappa(\omega)$, of WSe_2/Al_2O_3 and WSe_2/Au annealed at the stipulated temperatures measured by high-resolution spectroscopic ellipsometry. Inset shows the schematic band structure of 1T′-phase monolayer-WSe_2. Similar to monolayer-MoS_2, the lattice distortion together with the strong electron-electron correlations in 1T′-phase WSe_2 induces a band inversion in the vicinity of the Γ-point. This lowers the W d-orbital below the Se p-orbital and opens an inverted gap, 2δ. Source: Yin et al. [197]/John Wiley & Sons/Licensed under CC BY 4.0.

in the ε_2 spectra. Furthermore, another peak feature is observed in the far-infrared regime which can been attributed to the fundamental gap, E_g, of 1T′-phase MoS_2 at ~0.10 eV upon sample annealing at 250 °C (Figure 1.21c).

The appearances of the mid-infrared and far-infrared features correspond to the inverted gap, 2δ, and fundamental gap, E_g, of 1T′-phase MoS_2 (see schematic of the 1T′-phase band structure in Figure 1.21d). The presence of the pre-peaks along with peaks A, B, and C belonging to 1H-phase MoS_2 can be attributed to the coexistence of the monolayer in both 1H and 1T′-phase.

A similar result is registered for monolayer-WSe_2/Au undergoing the same annealing-induced phase transition process. Unlike the MoS_2/Au sample, the annealing temperature-dependent complex refractive indices ($N(\omega) = n(\omega) + i\kappa(\omega)$) is displayed (Figure 1.22). With the 1H 1T′ phase transition process taking place at higher annealing temperatures, the refractive index, $n(\omega)$, increases significantly in the spectral region between 0.8 and 2.5 eV when the sample is annealed at 250 °C (Figure 1.22a), and then reduces upon further annealing at 350 °C. Similar annealing temperature-dependent trend can be observed in the κ spectra as the ε_2 spectral where the mid-infrared feature attributed to the inverted gap can be observed (Figure 1.22b).

1.5.2.3 Raman Characterization

Similar to PL and Spectroscopic Ellipsometry, Raman spectroscopy is a fast, non-destructive and efficient optical characterization technique that provides structural and electronic information of materials under high spectral and spatial resolution where a comprehensive treatment of its characterization technique can be found in the review [198] and the previous book publication [199]. Based on the fundamental lattice vibrations of TMD materials in the monolayer, multilayer and heterostructure(al) form, one is able to derive essential and detailed information pertaining to properties such as interlayer coupling, spin–orbit splitting and external perturbations.

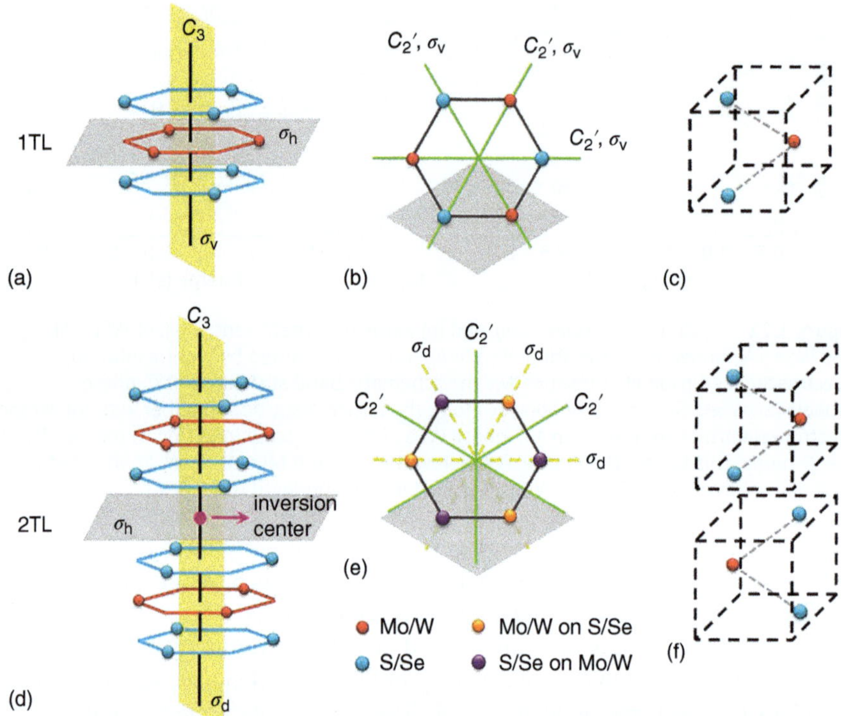

Figure 1.23 Comparing the symmetry operations between a monolayer (1TL) and bilayer (2TL) TMD systems. (a) Side view of a monolayer (odd number) where the axis of the C_3 operations (clockwise and anticlockwise) are denoted by the black vertical line. The horizontal (σ_h) and vertical (σ_v) reflection operations are depicted by the gray and yellow planes, respectively. (b) The monolayer plane view from the top where the axes of the C_2' operations are denoted by the green lines lying in the σ_h plane. The top view of the σ_v planes is also represented by the green lines. The gray diamond displays the unit cell of the crystal view from the top. (c) Side view of the unit cell monolayer, comprising an M and two X atoms. (d) Side view of a bilayer (even number) with the /AbA BaB/ stacking configuration where the C_3 operations axis is denoted by the black line. The inversion centre has been denoted by the pink circle. The horizontal (σ_h) and dihedral (σ_d) reflection operations are represented by the gray and yellow planes, respectively. (e) The bilayer viewed from the top where the purple spheres represent the X-atom above the M-atom and vice versa for the orange spheres. The axes of the three C_2' operations are denoted by the green lines lying in the σ_h plane. The σ_d planes viewed from the top are represented by the yellow dashed lines with the gray diamond demarcating the unit cell. (f) The bilayer unit cell viewed from the side containing four X atoms and two M atoms. The two layers are represented by two dashed boxes. Source: Zhao et al. [201]/Reproduced with permission from American Chemical Society.

Differences in the TMDs structural phases and symmetries in turn result in their distinct electronic structures. Given the unique symmetry of 2D-TMDs in their 1H/3R-phase (D_{3h}) and 1T-phase (D_{3d}), they can be easily distinguished based on Raman analysis [200].

While bulk TMDs have 24 symmetry operations, this number is significantly reduced in the mono- and few-layered systems of 12 in cases of either even or odd layer numbers. The symmetry operations of odd and even layer numbered systems are illustrated using a mono and bi-layer structure as shown in Figure 23 [201].

Experimentally, Raman scattering was first used to characterize monolayer-MoS_2 [201] where the evolution of the specific Raman modes is shown to vary with layer thickness. Notably, the Raman active vibrations E^1_{2g} and the A_{1g} modes are observed near 400 cm^{-1} at ~384 cm^{-1} and ~415 cm^{-1}, respectively. While the E^1_{2g} is an in-plane vibration mode that corresponds to the opposing vibrations of the in-plane Mo and S atoms, A_{1g} is an out-of-plane mode corresponding to the opposing vibrations of the two out-of-plane S atoms [201–203]

It is observed in Figure 1.24 that with decreasing layer number, the in-plane E^1_{2g}-mode registers a blue shift while the A_{1g}-mode registers an opposite trend with a red shift. These layer-dependent wave number evolutions are consistent with the trend as the MoS_2 transforms from a bulk material to a few-layered system [205]. Meanwhile, the opposite layer-dependent trends between the respective vibration modes suggest the effects of inter-layer interactions and stacking-induced structural changes and the long-range Coulomb interlayer interactions which has an effect even towards the intralayer bonding and lattice dynamics [206].

This similar technique is also utilized to analyze the thicknesses of different TMD species such as $MoSe_2$ [207, 208], WS_2 [209, 210] and WSe_2 [201, 210, 211] where both the E^1_{2g} and A_{1g} vibration modes are examined concurrently and similar thickness-dependent trends have been yielded with the example of WSe_2 in Figure 1.25.

Apart from determining the layer-dependent properties of TMD materials, Raman spectroscopy is also an effective means to distinguish their respective structural phases. Factors such as lattice strain and charge-doping have direct impacts on the phase transition dynamics in 2D-TMDs and Raman spectroscopy can directly analyze the evolving Raman modes as a result of these effects [212].

When considering the 1H-1T′ phase transition of monolayer-MoS_2, it has been demonstrated that upon electron-doping, there is a red-shift in the A_{1g}-mode while the E^1_{2g}-mode remains unchanged [213]. This indicates that A_{1g} mode is strongly sensitive to electron-doping due to the stronger electron-phonon coupling of the A_{1g}-mode [213]. Conversely, with increasing tensile (compressive) strain, the E^1_{2g}-mode is red-shifted (blue-shifted) while the A_{1g}-mode remains unchanged [214, 215]. In the same study involving the 1H-1T′ phase transition of MoS_2 on Au substrate induced by high-temperature annealing [60], both the E^1_{2g} and A_{1g}-modes are red-shifted – an indication of the concurrent increase in both the tensile strain and electron-doping to the monolayer-MoS_2 sample during the phase transition process. Figure 1.26a displays the evolution of both the Raman modes of MoS_2/Au with respect to the annealing temperature and Figure 1.26b,c display the position of the E^1_{2g} and A_{1g}-modes (with MoS_2 on SiO_2 as reference), respectively. The red-shift of the E-mode for MoS_2/Au from ~395 (as-prepared) to ~389 cm^{-1} (250 °C), indicates an increase in tensile strain by ~2.7% after annealing at 250 °C [214, 216]. Meanwhile, there is a redshift of the A_{1g}-mode from ~415 to ~409 cm^{-1} (Figure 1.26c), indicating an increase in electron-doping concentration by more than 1.5×10^{13} cm^{-2} from the Au substrate [57, 213]. Above the optimal annealing

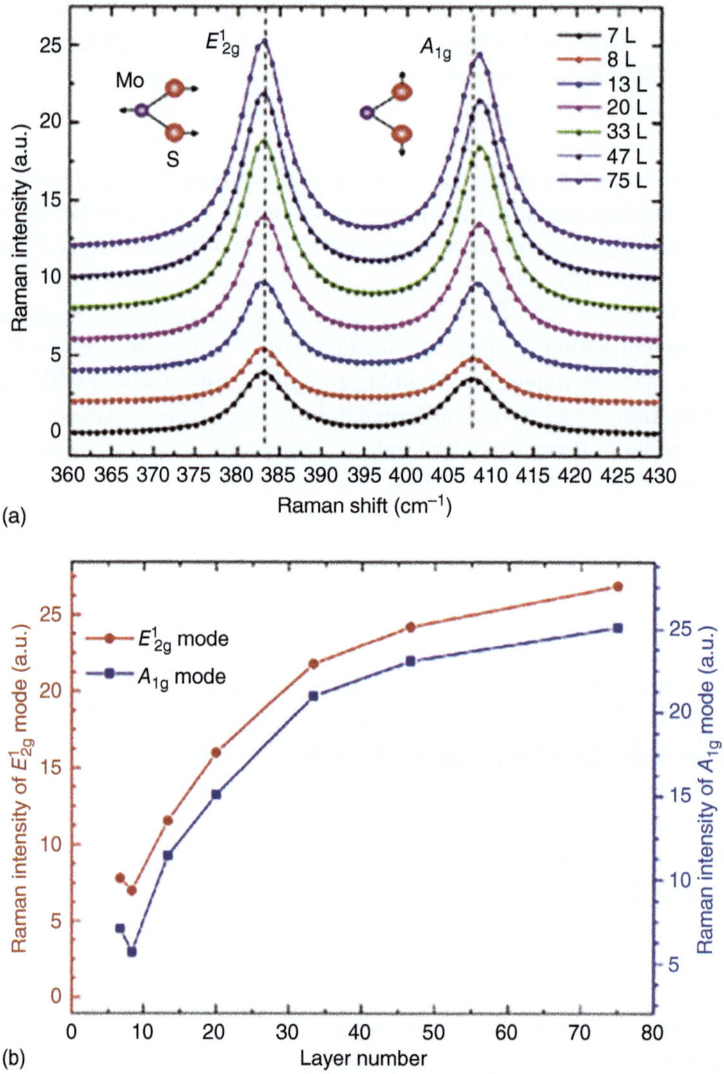

Figure 1.24 Raman profiles of MoS_2 of different thicknesses. (a) Raman spectra of MoS_2 films of different layer number and that of the bulk crystal. (b) Thickness-dependent frequency of E^1_{2g} (in-plane) and A_{1g} modes (out-of-plane) and their difference (right axis). Source: Reproduced with permission from Yuan et al. [204].

temperature of 250 °C, both the E^1_{2g} and A_{1g}-modes show gradual blue shift as the MoS_2 monolayer begins to undergo thermal decomposition.

These observations are consistent with a previous DFT study which suggested an increase in electron affinity of monolayer-MoS_2 with increasing lattice tensile strain [57]. This has allowed greater charge transfer from the electron-rich Au substrate to the MoS_2 monolayer. Besides, the frequency of both A_{1g} and E-modes can be

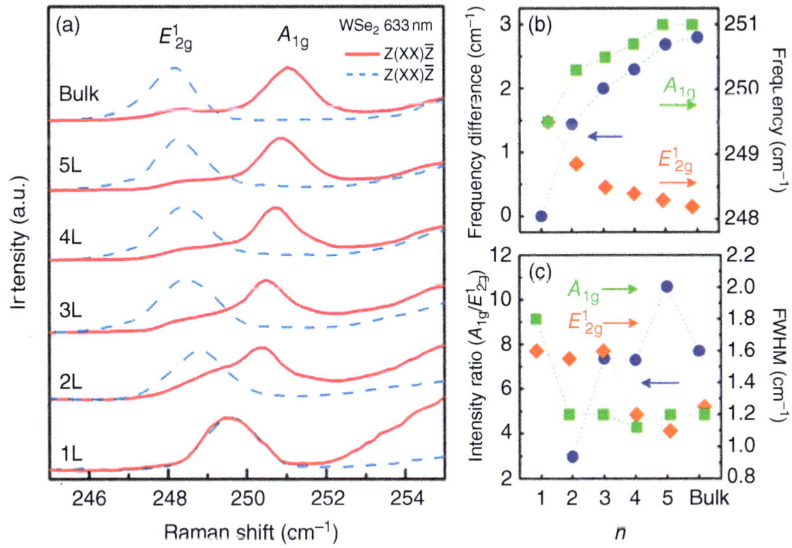

Figure 1.25 Raman profiles of WSe_2 of different thicknesses. (a) Raman spectra of WSe_2 between 1 to 5 layers and in its bulk form. These Raman spectra are obtained in the parallel ($Z(XX)\underline{Z}$) and cross ($Z(XY)\underline{Z}$) polarization conditions. (b) comparing the position of the A_{1g} and E^1_{2g} modes (right vertical axis) and their energy differences (left vertical axis) with respect to layer number. (c) the intensity ratio (left axis) and FWHM (right axis) of the A_{1g} and E^1_{2g} modes with respect to the layer number. Source: Zhao et al. [210]/With permission of Royal Society of Chemistry.

influenced by the presence of structural defects in the MoS_2 monolayer [217]. The frequency of A_{1g}-mode increases alongside a decrease in the frequency of E-mode with increasing lattice defects. However, this does not take place in the experimental observation as presented with increasing annealing temperature. Besides, the FWHM of the A_{1g}-mode ($\sim 5.7\,cm^{-1}$) and E-mode ($\sim 4.6\,cm^{-1}$) of the as-prepared monolayer- MoS_2 are close to those of pristine MoS_2 sample reported previously [217]. Hence, the effects of lattice defects can be ruled out as the main mechanism involved in this annealing-induced phase transition study.

Furthermore, doping-dependent energetics for large-strained (12%) and unstrained MoS_2 (Figure 1.26d) also show that with tensile strain in play, 1T′ phase MoS_2 is energetically favored at smaller electron doping concentrations as compared to one without any lattice strain. This shows that tensile strain and charge-transfer work concurrently to enable the 1H-1T′ phase transition of the 2D-TMD material. The effect of tensile strain facilitating a semiconductor-to-metal phase transition demonstrated recently in thin-film $MoTe_2$ further substantiate this claim [218]. As suggested in previous theoretical studies [219, 220], a system involving monolayer-MoS_2, and the electron-rich Au substrate alone is not a sufficient condition to induce the 1H–1T′ phase transition. This is due to the inefficiency in electron injection attributed to the formation of an interfacial tunnel barrier [219, 220].

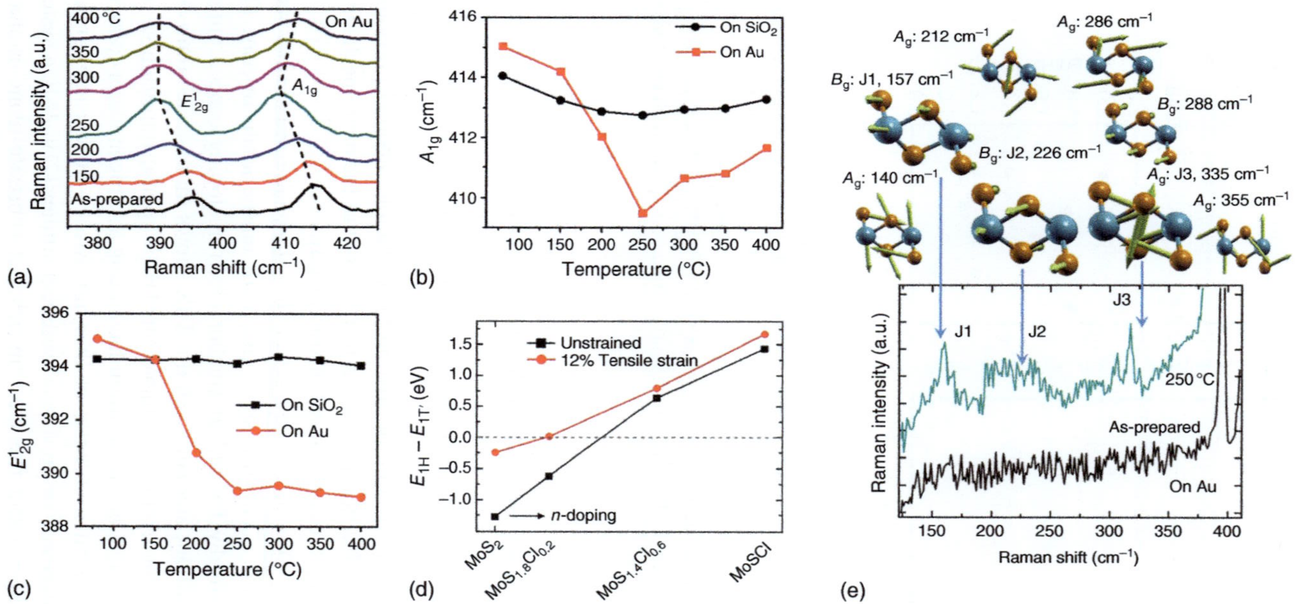

Figure 1.26 (a) Raman spectra of MoS_2/Au at the respective annealing temperatures. Extracted frequencies of (b) A_{1g} and (c) E^1_{2g}-modes as functions of annealing temperature with the respective modes for MoS_2/SiO_2 displayed as reference. The errors of the extracted energy positions are 0.5 cm^{-1}. (d) Calculated energy difference between 1H and 1T′ phases as a function of electron doping concentration for unstrained and strained (12% tensile) monolayer-MoS_2. (e) Top: Calculated Raman active eigenmodes and frequencies, with the corresponding irreducible representation assigned from group theory. Bottom: Comparing Raman spectra of MoS_2/Au before and after annealing at 250 °C. Source: Yin et al. [60]/Springer Nature/Licensed under CC BY 4.0.

Apart from the changes to the E^1_{2g} and the A_{1g} modes during the 1H-1T′ phase transition, there are three main features at ~157 (J1), ~224 (J2) and ~320 cm^{-1} (J3) which can be observed in the Raman spectra of the MoS_2/Au before and after annealing at 250 °C (Figure 1.26e). Distinct to the 1T′-phase [221, 222], these characteristic modes are absent from the semiconducting 1H-phase MoS_2. Besides, they are consistent with published theoretical and experimental studies that show this is the 1T′- rather than 1T-phase MoS_2 [55, 222, 223]. The top of Figure 1.27 displays the calculated Raman active modes and frequencies with their corresponding irreducible representation assigned from group theory of 1T′-phase MoS_2. Most of the experimental peaks correspond to the calculated Raman active eigenmodes.

1.5.3 Electronic Bandgap

Unlike the optical bandgap, the electronic bandgap, and the electronic structures of TMD systems can be characterized using techniques including angle-resolved photoemission spectroscopy (ARPES) and scanning tunneling spectroscopy (STS).

1.5.3.1 Angle-Resolved Photoemission Spectroscopy

As presented in the previous sections, factors including the layer number and the constituent atoms of the TMD materials have long-standing effects on their electronic structures. The effects of the diverse spatial distributions and orbital characters in both the valence and conduction bands previously restricted to mere theoretical calculations can be experimentally verified [225, 226].

ARPES characterization has been employed to investigate the thickness-dependence of the TMD band structure from an indirect to a direct bandgap semiconductor [56, 227, 228]. Owing to the effects of spin-orbit coupling, spin-resolved ARPES has also been involved in unraveling the spin textures of TMD materials [229–231].

Beyond investigating the band structures of TMD systems in their basic forms, ARPES has been utilized to investigate the band structures of TMD heterostructures, charge density waves and superconductivity with a very comprehensive treatment presented in the review [225].

Stacking of TMD monolayers can lead to effects beyond Van der Waals interfacial attraction. A case in point would be the onset of interfacial hybridization in a MoS_2/WSe_2 heterostructure (Figure 1.27) where unique band dispersions can be observed near the Γ-point [224]. Apart from TMD heterostructures, hybrid heterostructures such as MoS_2/graphene have also been investigated using ARPES where Van der Waals interactions result in the modification of the graphene band structure through the formation of several miniband-gaps. This can be attributed to the formation of a super-periodic potential formed by the MoS_2/graphene [232] or the graphene/WS_2 heterostructure [233]. With the immense interest generated by the discovery of superconductivity in "magic-angle" double-layer graphene [234], there is also considerable interest to investigate the effects of periodic Moiré superlattices in TMD heterostructures [235–237]. A case in point would be the evolution of the electronic structures into a highly localized system in the WS_2/WSe_2

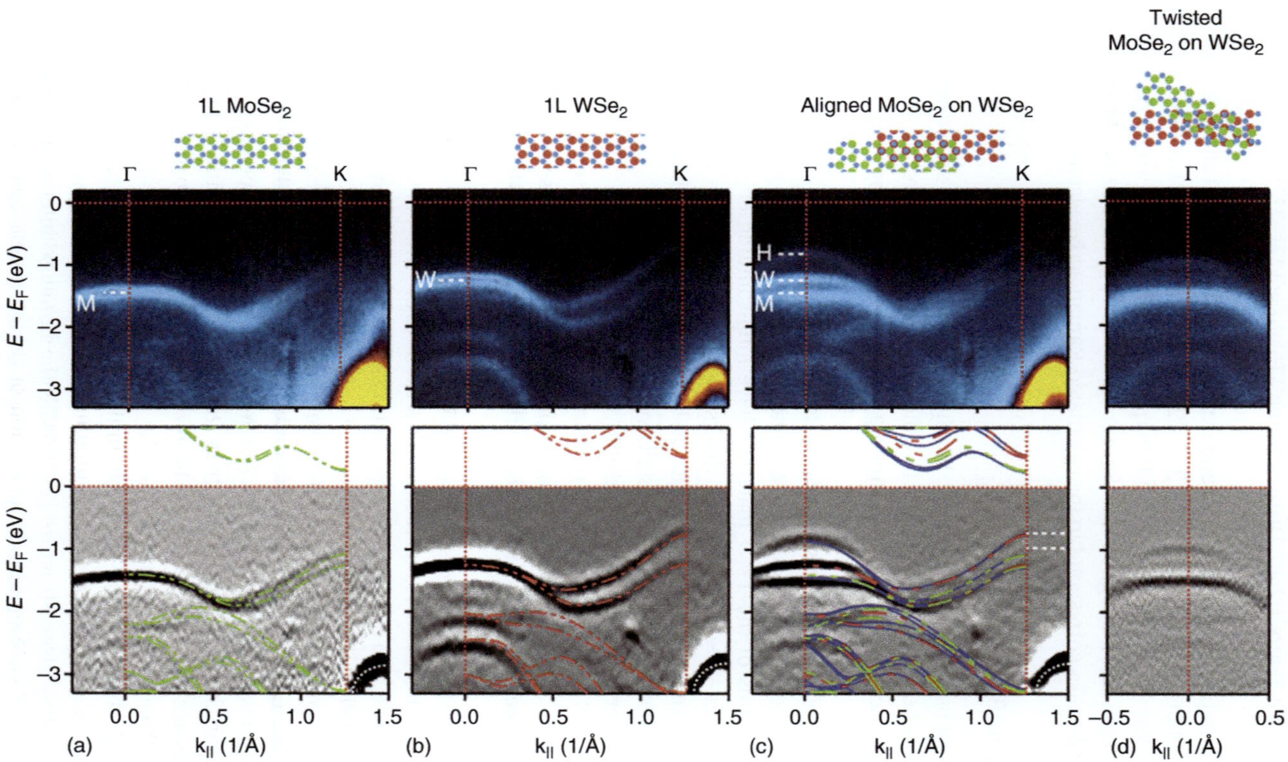

Figure 1.27 Comparing the ARPES-characterized band structures of monolayer and bilayer TMD heterostructures along the Γ–K momentum slices. (a) Band structure of monolayer-$MoSe_2$, (b) monolayer-WSe_2, (c) aligned $MoSe_2/WSe_2$, and (d) $MoSe_2/WSe_2$ heterostructure misaligned at ~30°. While bands near the K-point are very similar between the heterostructures and monolayers – weak interlayer hybridization near K – those at the Γ-point are substantially different, an onset of significant interlayer hybridization for the aligned system (c). Nevertheless, with a misalignment of ~30° between the $MoSe_2$ and WSe_2 layer (d), only two bands are now observed. This suggests that the third band near the Γ-point in the aligned heterostructure arises from commensurate domains. Source: Reproduced with permission from Wilson et al. [224]/American Association for the Advancement of Science/Licensed under CC BY 4.0.

Moiré superlattices with increasingly flat bands and renormalized effective mass. These emerging quantum phases have been monitored using APRES where a large gap opening and splitting exciton peaks have also been observed.

The complex interplay between exotic CDW and superconductive states in various TMD systems are equally intriguing where in some cases, they are found to be co-existing quantum orders [238–241]. ARPES serves as an ideal instrument for the direct observations of both CDW and superconductivity. For the former, Fermi surface reconstruction and folded band structures can be observed due to the enlargement of the lattice unit cells in systems such as $TaSe_2$, $TiSe_2$, VSe_2 and $NbSe_2$ [165, 242–246].

The CDW and superconductive states of 2H-phase $NbSe_2$ are particularly interesting due to their spatial anisotropy and orbital-selective properties [245, 246]. In addition to the anisotropic properties, low-temperature ARPES study has further suggested the role of electron-phonon coupling in dictating the spectral gap and band renormalization. Specifically, the spectral gap consists of two distinct anti-correlated components related to its superconductive and CDW properties [244]. Likewise, layer-dependent effects have also been observed in both the CDW and superconductive states where interlayer interaction and vertical quantum confinement are taken into consideration. For instance, few-layered TMDs have elevated T_{CDW} and weaker superconductivity than their bulk counterparts [232] where multiple theoretical models have been proposed to account for this phenomenon [247, 248].

1.5.3.2 Scanning Tunneling Spectroscopy (STS)

ST measurements have been conducted on a series of mono- and few-layer TMDs on substrates such as highly ordered pyrolytic graphite (HOPG) [164, 249–253] and gold substrates [254]. Comparing similar 2D-TMDs, the measured electronic gap for materials on HOPG tend to be higher than those grown on Au due to the presence of donor states in the latter. An example would be monolayer-MoS_2/HOPG (~2.15 eV) [249] compared to MoS_2/Au (~1.8 eV) [254].

Similar to techniques such as ARPES and PL, STS characterization can be employed to investigate the thickness dependence of TMD materials with the results of MoS_2 [57] (Figure 1.8de), $MoSe_2$ (Figure 1.28) [255] and WSe_2 (Figure 1.29) [253]. The spin texture of the WSe_2 system can be further derived by tracking the series of allowed and forbidden scattering channels with respect to energy. In this case, a large spin-splitting of the valence band can be observed, and this is attributed to the strong spin-orbit coupling. Besides, spin-valley-layer coupling can be observed in the conduction band of bilayer-WSe_2.

As 2D-TMD based semiconductor heterojunctions have also been put under scrutiny using both STS and scanning tunneling microscopy (STM) due to their immense potential for the fabrication of important vertical semiconductor structures and devices. By the detailed examination of the electronic band structures as a result of interfacial lattice strain, band offsets, formation of interfacial states and charge regions, the mechanisms underlying the carrier transport in the heterostructures can be unravelled [256].

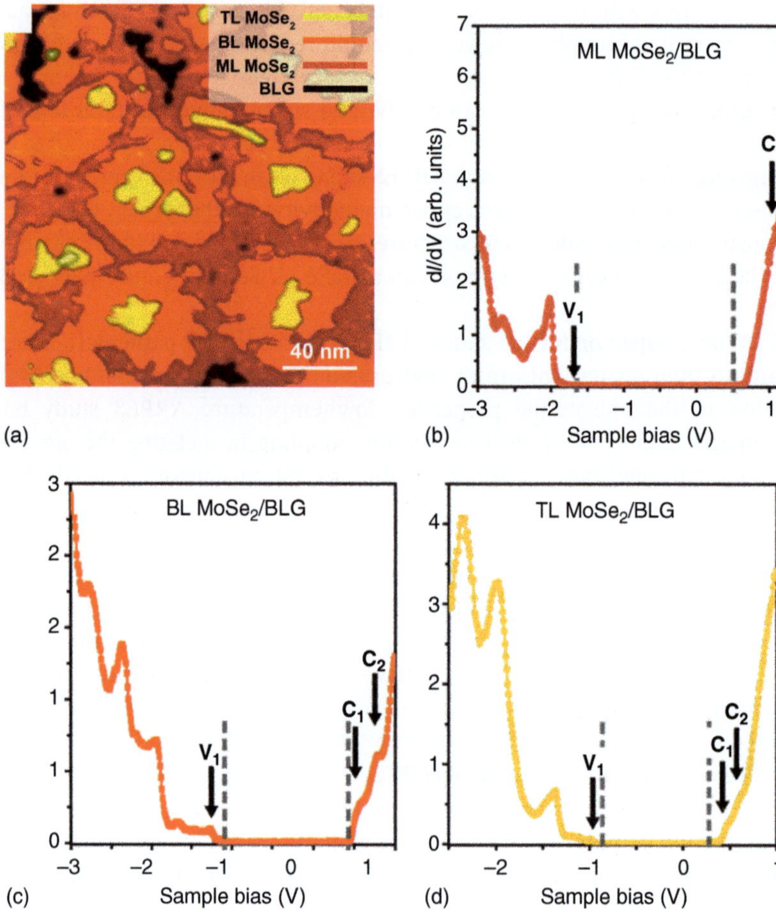

Figure 1.28 STM and STS characterization of mono-, bi- and trilayer $MoSe_2$ synthesized on bilayer graphene (BLG) using MBE. (a) STM image of monolayer-$MoSe_2$/BLG STS spectra for (b) monolayer-$MoSe_2$/BLG (b) bilayer $MoSe_2$/BLG, and (c) trilayer $MoSe_2$/BLG. Source: Reproduced with permission from Bradley et al. [255]/American Chemical Society/Licensed under CC BY 4.0.

The notable example of STM and STS study involving the WSe_2/MoS_2 heterostructure shows that the electronic properties of the bilayer heterostructure change correspondingly with the angle between the layers. Interfacial lattice strain has a direct impact on the system's electronic structures [257]. The band alignment of the vertical WSe_2/MoS_2 heterojunction can be converted from a type II to type I band alignment through lattice strain effects. This strategy involving the epitaxial growth of MoS_2 along the edge of monolayer-WSe_2 induced the strain field only in the MoS_2 region, thereby solely affecting the band structure of monolayer-MoS_2. A similar study can be conducted in the opposite order to examine the impact of lattice strain on monolayer-WSe_2 [256, 257].

The above finding is further substantiated by another study involving the same WSe_2/MoS_2 heterojunction which shows that it is an ideal system for creating a

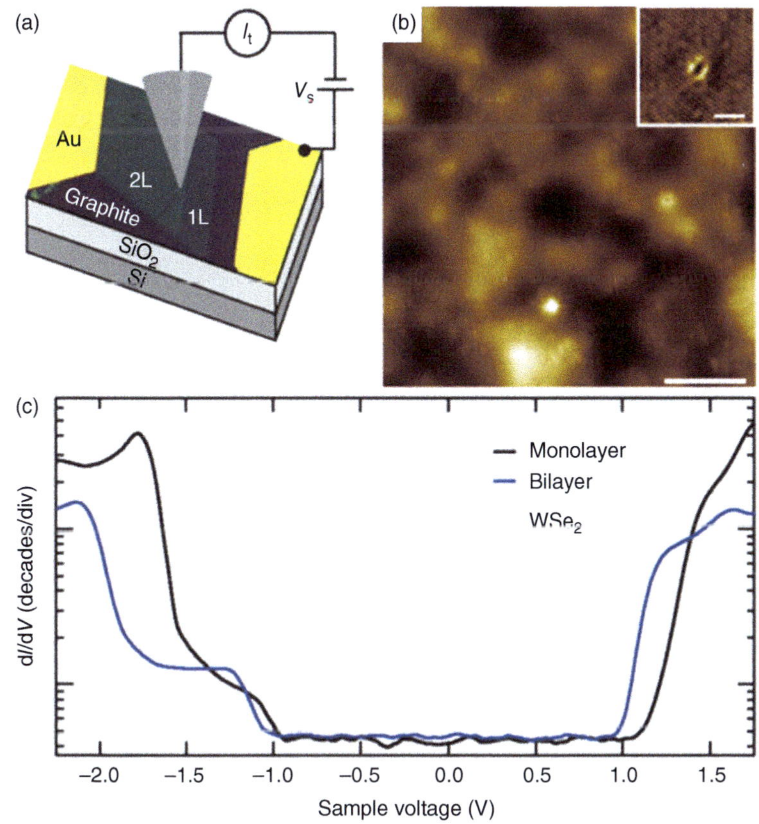

Figure 1.29 STM and STS characterization of WSe_2. (a) Schematic of the system setup to characterize the WSe_2/HOPG sample. (b) STM topography of bilayer-WSe_2. Inset: Atomic resolution of a small lattice defect. (c) dI/dV spectroscopy of both mono- and bilayer WSe_2. Source: Reproduced with permission from Yankowitz et al. [253]/Reproduced with permission from American Physical Society.

tunable interfacial electrical potential [258]. Such heterostructures can in turn can be utilized in piezoelectric devices [259, 260]. Even in cases where twist angle and lattice mismatch that results in the periodic moiré potentials and structures, the combination of STM and STS measurements can directly probe the magnitude and wavelength of the moiré potential and its electronic structure. A STM study on twisted bilayer-WSe_2 has revealed different localized flat bands between 0° and 60° stacking configureuration [261]. Meanwhile, a separate study on WSe_2/WS_2 has revealed the formation of flat bands near the valence band edge at low temperature [262]. Extensive characterization based on a combination of STM, AFM and TEM measurements has further revealed $MoSe_2/WSe_2$ with a 60° stacking to have a moiré wavelength ranging between 6 – 17 nm at highly symmetrical stacking locations at room temperature [263]. To better understand how the band structure evolves with respect to the moiré pattern and twisted angle, another study involving both STM and STS characterization of $MoSe_2/WSe_2$ hetero-bilayer has revealed a bandgap modulation between 0° and 60° stacking configureuration [264]. Characterization of local density of states further reveals different localization of electrons between the valence band and conduction band edges [264]. Beyond mere strain effects and mechanical lattice interactions, both STM and STS are effective probing tools to study the impacts of doping [265–268], temperature, structural phase [269–272], edge structure [271–275] on the electronic band structures of TMD heterostructures in both vertical and lateral setups.

1.5.4 Conclusions

This chapter offers a comprehensive overview of 2D-TMDs, their electronic and optical properties that arise due to the diverse polymorphic structures. Multiple experimental techniques have been developed over the years to characterize these unique properties that came about not only due to the structural features, but also intrinsic properties such as spin-orbit coupling. Having established these fundamentals related to 2D-TMDs, the subsequent chapters will further present to readers other important facets of this class of materials pertaining to the theoretical studies, experimental investigations and the potential applications this class of materials promises.

References

1 Novoselov, K.S., Geim, A.K., Morozov, S.V. et al. (2004). Electric field effect in atomically thin carbon films. *Science* 306: 666–669. https://doi.org/10.1126/science.1102896.

2 Frindt, R.F. (1966). Single crystals of MoS_2 several molecular layers thick. *J. Appl. Phys.* 37: 1928–1929. https://doi.org/10.1063/1.1708627.

3 Novoselov, K.S., Jiang, D., Schedin, F. et al. (2005). Two-dimensional atomic crystals. *Proc. Natl. Acad. Sci.* 102: 10451–10453. https://doi.org/10.1073/pnas.0502848102.

4 Wang, Q.H., Kalantar-Zadeh, K., Kis, A. et al. (2012). Electronics and optoelectronics of two-dimensional transition metal dichalcogenides. *Nat. Nanotechnol.* 7: 699–712. https://doi.org/10.1038/nnano.2012.193.

5 Xu, X., Yao, W., Xiao, D., and Heinz, T.F. (2014). Spin and pseudospins in layered transition metal dichalcogenides. *Nat. Phys.* 10: 343–350. https://doi.org/10.1038/nphys2942.

6 Xia, F., Wang, H., Xiao, D. et al. (2014). Two-dimensional material nanophotonics. *Nat. Photonics* 8: 899–907. https://doi.org/10.1038/nphoton.2014.271.

7 Eda, G., Fujita, T., Yamaguchi, H. et al. (2012). Coherent atomic and electronic heterostructures of single-layer MoS_2. *ACS Nano* 6: 7311–7317. https://doi.org/10.1021/nn302422x.

8 Qian, X., Liu, J., Fu, L., and Li, J. (2014). Quantum spin hall effect in two-dimensional transition metal dichalcogenides. *Science* 346: 1344–1347. https://doi.org/10.1126/science.1256815.

9 Keum, D.H., Cho, S., Kim, J.H. et al. (2015). Bandgap opening in few-layered monoclinic $MoTe_2$. *Nat. Phys.* 11: 482–486. https://doi.org/10.1038/nphys3314.

10 Calandra, M. (2013). Chemically exfoliated single-layer Mos_2: stability, lattice dynamics, and catalytic adsorption from first principles. *Phys. Rev., B* 88: 245428. https://doi.org/10.1103/PhysRevB.88.245428.

11 Tang, C.S., Yin, X., and Wee, A.T.S. (2021). 1D chain structure in 1T′-phase 2D transition metal dichalcogenides and their anisotropic electronic structures. *Appl. Phys. Rev.* 8: 011313. https://doi.org/10.1063/5.0023799.

12 Porer, M., Leierseder, U., Ménard, J.M. et al. (2014). Non-thermal separation of electronic and structural orders in a persisting charge density wave. *Nat. Mater.* 13: 857–861. https://doi.org/10.1038/nmat4042.

13 Liu, G., Debnath, B., Pope, T.R. et al. (2016). A charge-density-wave oscillator based on an integrated tantalum disulfide–boron nitride–graphene device operating at room temperature. *Nat. Nanotechnol.* 11: 845–850. https://doi.org/10.1038/nnano.2016.108.

14 Geremew, A.K., Rumyantsev, S., Kargar, F. et al. (2019). Bias-voltage driven switching of the charge-density-wave and normal metallic phases in 1T-TaS_2 thin-film devices. *ACS Nano* 13: 7231–7240. https://doi.org/10.1021/acsnano.9b02870.

15 Geremew, A.K., Kargar, F., Zhang, E.X. et al. (2019). Proton-irradiation-immune electronics implemented with two-dimensional charge-density-wave devices. *Nanoscale* 11: 8380–8386. https://doi.org/10.1039/C9NR01614G.

16 Kadantsev, E.S. and Hawrylak, P. (2012). Electronic structure of a single MoS_2 monolayer. *Solid State Commun.* 152: 909–913. https://doi.org/10.1016/j.ssc.2012.02.005.

17 Wilson, J.A., Di Salvo, F.J., and Mahajan, S. (1975). Charge-density waves and superlattices in the metallic layered transition metal dichalcogenides. *Adv. Phys.* 24: 117–201. https://doi.org/10.1080/00018737500101391.

18 Moncton, D.E., Axe, J.D., and DiSalvo, F.J. (1975). Study of superlattice formation in 2H-$NbSe_2$ and 2H-$TaSe_2$ by neutron scattering. *Phys. Rev. Lett.* 34: 734–737. https://doi.org/10.1103/PhysRevLett.34.734.

19 Fazekas, P. and Tosatti, E. (1979). Electrical, structural and magnetic properties of pure and doped 1T-TaS_2. *Philos. Mag. B* 39: 229–244. https://doi.org/10.1080/13642817908245359.

20 Sipos, B., Kusmartseva, A.F., Akrap, A. et al. (2008). From Mott state to superconductivity in 1T-TaS_2. *Nat. Mater.* 7: 960–965. https://doi.org/10.1038/nmat2318.

21 Tsen, A.W., Hovden, R., Wang, D. et al. (2015). Structure and control of charge density waves in two-dimensional 1T-TaS_2. *Proc. Natl. Acad. Sci.* 112: 15054–15059. https://doi.org/10.1073/pnas.1512092112.

22 Sun, Y., Wu, S.-C., Ali, M.N. et al. (2015). Prediction of Weyl semimetal in orthorhombic $MoTe_2$. *Phys. Rev., B* 92: 161107. https://doi.org/10.1103/PhysRevB.92.161107.

23 Ali, M.N., Xiong, J., Flynn, S. et al. (2014). Large, non-saturating magnetoresistance in WTe_2. *Nature* 514: 205–208. https://doi.org/10.1038/nature13763.

24 Ma, F., Gao, G., Jiao, Y. et al. (2016). Predicting a new phase (T″) of two-dimensional transition metal di-chalcogenides and strain-controlled topological phase transition. *Nanoscale* 8: 4969–4975. https://doi.org/10.1039/C5NR07715J.

25 Heising, J. and Kanatzidis, M.G. (1999). Structure of restacked MoS_2 and WS_2 elucidated by electron crystallography. *J. Am. Chem. Soc.* 121: 638–643. https://doi.org/10.1021/ja983043c.

26 Heising, J. and Kanatzidis, M.G. (1999). Exfoliated and restacked MoS_2 and WS_2: ionic or neutral species? Encapsulation and ordering of hard electropositive cations. *J. Am. Chem. Soc.* 121: 11720–11732. https://doi.org/10.1021/ja991644d.

27 Bruyer, E., Di Sante, D., Barone, P. et al. (2016). Possibility of combining ferroelectricity and Rashba-like spin splitting in monolayers of the 1T-type transition-metal dichalcogenides MX_2(M=Mo,W;X=S,Se,Te). *Phys. Rev., B* 94: 195402. https://doi.org/10.1103/PhysRevB.94.195402.

28 Zhuang, H.L., Johannes, M.D., Singh, A.K., and Hennig, R.G. (2017). Doping-controlled phase transitions in single-layer MoS_2. *Phys. Rev., B* 96: 165305. https://doi.org/10.1103/PhysRevB.96.165305.

29 Li, Y., Duerloo, K.-A.N., Wauson, K., and Reed, E.J. (2016). Structural semiconductor-to-semimetal phase transition in two-dimensional materials induced by electrostatic gating. *Nat. Commun.* 7: 10671. https://doi.org/10.1038/ncomms10671.

30 Duerloo, K.-A.N., Li, Y., and Reed, E.J. (2014). Structural phase transitions in two-dimensional Mo- and W-dichalcogenide monolayers. *Nat. Commun.* 5: 4214. https://doi.org/10.1038/ncomms5214.

31 Santhosh, K.C., Zhang, C., Hong, S. et al. (2015). Phase stability of transition metal dichalcogenide by competing ligand field stabilization and charge density wave. *2D Materials* 2: 035019. https://doi.org/10.1088/2053-1583/2/3/035019.

32 Kan, M., Nam, H.G., Lee, Y.H., and Sun, Q. (2015). Phase stability and Raman vibration of the molybdenum ditelluride ($MoTe_2$) monolayer. *Phys. Chem. Chem. Phys.* 17: 14866–14871. https://doi.org/10.1039/C5CP01649E.

33 Ataca, C., Şahin, H., and Ciraci, S. (2012). Stable, single-layer MX_2 transition-metal oxides and dichalcogenides in a honeycomb-like structure. *J. Phys. Chem., C* 116: 8983–8999. https://doi.org/10.1021/jp212558p.

34 Bertolazzi, S., Brivio, J., and Kis, A. (2011). Stretching and breaking of ultrathin MoS_2. *ACS Nano* 5: 9703–9709. https://doi.org/10.1021/nn203879f.

35 Cooper, R.C., Lee, C., Marianetti, C.A. et al. (2013). Nonlinear elastic behavior of two-dimensional molybdenum disulfide. *Phys. Rev., B* 87: 035423. https://doi.org/10.1103/PhysRevB.87.035423.

36 Miró, P., Ghorbani-Asl, M., and Heine, T. (2013). Spontaneous ripple formation in MoS_2 monolayers: electronic structure and transport effects. *Adv. Mater.* 25: 5473–5475. https://doi.org/10.1002/adma.201301492.

37 Singh, S.K., Neek-Amal, M., Costamagna, S., and Peeters, F.M. (2015). Rippling, buckling, and melting of single- and multilayer MoS_2. *Phys. Rev., B* 91: 014101. https://doi.org/10.1103/PhysRevB.91.014101.

38 Lebègue, S. and Eriksson, O. (2009). Electronic structure of two-dimensional crystals from ab initio theory. *Phys. Rev., B* 79: 115409. https://doi.org/10.1103/PhysRevB.79.115409.

39 Ding, Y., Wang, Y., Ni, J. et al. (2011). First principles study of structural, vibrational and electronic properties of graphene-like MX_2 (M=Mo, Nb, W, Ta; X=S, Se, Te) monolayers. *Physica B* 406: 2254–2260. https://doi.org/10.1016/j.physb.2011.03.044.

40 Dawson, W.G. and Bullett, D.W. (1987). Electronic structure and crystallography of $MoTe_2$ and WTe_2. *J. Phys. C: Solid State Phys.* 20: 6159–6174. https://doi.org/10.1088/0022-3719/20/36/017.

41 Augustin, J., Eyert, V., Böker, T. et al. (2000). Electronic band structure of the layered compound Td−WTe_2. *Phys. Rev., B* 62: 10812–10823. https://doi.org/10.1103/PhysRevB.62.10812.

42 Zeng, Z., Yin, Z., Huang, X. et al. (2011). Single-layer semiconducting nanosheets: high-yield preparation and device fabrication. *Angew. Chem. Int. Ed.* 50: 11093–11097. https://doi.org/10.1002/anie.201106004.

43 Tongay, S., Sahin, H., Ko, C. et al. (2014). Monolayer behaviour in bulk ReS_2 due to electronic and vibrational decoupling. *Nat. Commun.* 5: 3252. https://doi.org/10.1038/ncomms4252.

44 Böker, T., Severin, R., Müller, A. et al. (2001). Band structure of MoS_2 $MoSe_2$ and alpha-$MoTe_2$ angle-resolved photoelectron spectroscopy and ab initio calculations. *Phys. Rev., B* 64: 235305. https://doi.org/10.1103/PhysRevB.64.235305.

45 Wilson, J.A. and Yoffe, A.D. (1969). The transition metal dichalcogenides discussion and interpretation of the observed optical, electrical and structural properties. *Adv. Phys.* 18: 193–335. https://doi.org/10.1080/00018736900101307.

46 Ribeiro-Soares, J., Almeida, R.M., Barros, E.B. et al. (2014). Group theory analysis of phonons in two-dimensional transition metal dichalcogenides. *Phys. Rev., B* 90: 115438. https://doi.org/10.1103/PhysRevB.90.115438.

47 Roldán, R., Silva-Guillén, J.A., López-Sancho, M.P. et al. (2014). Electronic properties of single-layer and multilayer transition metal dichalcogenides MX_2 (M = Mo, W and X = S, Se). *Ann. Phys.* 526: 347–357. https://doi.org/10.1002/andp.201400128.

48 Yun, W.S., Han, S.W., Hong, S.C. et al. (2012). Thickness and strain effects on electronic structures of transition metal dichalcogenides: 2H-MX_2 semiconductors (M = Mo, W; X = S, Se, Te). *Phys. Rev., B* 85: 033305. https://doi.org/10.1103/PhysRevB.85.033305.

49 Zhao, W., Ribeiro, R.M., Toh, M. et al. (2013). Origin of indirect optical transitions in few-layer MoS_2, WS_2, and WSe_2. *Nano Lett.* 13: 5627–5634. https://doi.org/10.1021/nl403270k.

50 Zhu, Z.Y., Cheng, Y.C., and Schwingenschlögl, U. (2011). Giant spin-orbit-induced spin splitting in two-dimensional transition-metal dichalcogenide semiconductors. *Phys. Rev., B* 84: 153402. https://doi.org/10.1103/PhysRevB.84.153402.

51 Li, T. and Galli, G. (2007). Electronic properties of MoS_2 nanoparticles. *J. Phys. Chem., C* 111: 16192–16196. https://doi.org/10.1021/jp075424v.

52 Kormányos, A., Burkard, G., Gmitra, M. et al. (2015). k · p theory for two-dimensional transition metal dichalcogenide semiconductors. *2D Materials* 2: 022001. https://doi.org/10.1088/2053-1583/2/2/022001.

53 Mak, K.F., Lee, C., Hone, J. et al. (2010). Atomically thin MoS_2: a new direct-gap semiconductor. *Phys. Rev. Lett.* 105: 136805. https://doi.org/10.1103/PhysRevLett.105.136805.

54 Splendiani, A., Sun, L., Zhang, Y. et al. (2010). Emerging photoluminescence in monolayer MoS_2. *Nano Lett.* 10: 1271–1275. https://doi.org/10.1021/nl903868w.

55 Hu, T., Li, R., and Dong, J. (2013). A new (2 × 1) dimerized structure of monolayer 1T-molybdenum disulfide, studied from first principles calculations. *J. Chem. Phys.* 139: 174702. https://doi.org/10.1063/1.4827082.

56 Jin, W., Yeh, P.-C., Zaki, N. et al. (2013). Direct measurement of the thickness-dependent electronic band structure of MoS_2 using angle-resolved photoemission spectroscopy. *Phys. Rev. Lett.* 111: 106801. https://doi.org/10.1103/PhysRevLett.111.106801.

57 Huang, Y.L., Chen, Y., Zhang, W. et al. (2015). Bandgap tunability at single-layer molybdenum disulphide grain boundaries. *Nat. Commun.* 6: 6298. https://doi.org/10.1038/ncomms7298.

58 Ugeda, M.M., Pulkin, A., Tang, S. et al. (2018). Observation of topologically protected states at crystalline phase boundaries in single-layer WSe_2. *Nat. Commun.* 9: 3401. https://doi.org/10.1038/s41467-018-05672-w.

59 Tang, S., Zhang, C., Wong, D. et al. (2017). Quantum spin hall state in monolayer 1T'-WTe_2. *Nat. Phys.* 13: 683–687. https://doi.org/10.1038/nphys4174.

60 Yin, X., Wang, Q., Cao, L. et al. (2017). Tunable inverted gap in monolayer quasi-metallic MoS_2 induced by strong charge-lattice coupling. *Nat. Commun.* 8: 486. https://doi.org/10.1038/s41467-017-00640-2.

61 Tang, C.S., Yin, X., Yang, M. et al. (2020). Anisotropic collective charge excitations in quasimetallic 2D transition-metal dichalcogenides. *Adv. Sci.* 7: 1902726. https://doi.org/10.1002/advs.201902726.

62 Kuc, A., Zibouche, N., and Heine, T. (2011). Influence of quantum confinement on the electronic structure of the transition metal sulfide T_2. *Phys. Rev., B* 83: 245213. https://doi.org/10.1103/PhysRevB.83.245213.

63 Ramakrishna Matte, H.S.S., Gomathi, A., Manna, A.K. et al. (2010). MoS_2 and WS_2 analogues of graphene. *Angew. Chem. Int. Ed.* 49: 4059–4062. https://doi.org/10.1002/anie.201000009.

64 Radisavljevic, B., Radenovic, A., Brivio, J. et al. (2011). Single-layer MoS_2 transistors. *Nat. Nanotechnol.* 6: 147–150. https://doi.org/10.1038/nnano.2010.279.

65 Voß, D., Krüger, P., Mazur, A., and Pollmann, J. (1999). Atomic and electronic structure of WSe_2 from ab initio theory: bulk crystal and thin film systems. *Phys. Rev., B* 60: 14311–14317. https://doi.org/10.1103/PhysRevB.60.14311.

66 Bonaccorso, F., Bartolotta, A., Coleman, J.N., and Backes, C. (2016). 2D-crystal-based functional inks. *Adv. Mater.* 28: 6136–6166. https://doi.org/10.1002/adma.201506410.

67 Hill, H.M., Rigosi, A.F., Rim, K.T. et al. (2016). Band alignment in MoS_2/WS_2 transition metal dichalcogenide heterostructures probed by scanning tunneling microscopy and spectroscopy. *Nano Lett.* 16: 4831–4837. https://doi.org/10.1021/acs.nanolett.6b01007.

68 Wickramaratne, D., Zahid, F., and Lake, R.K. (2014). Electronic and thermoelectric properties of few-layer transition metal dichalcogenides. *J. Chem. Phys.* 140: 124710. https://doi.org/10.1063/1.4869142.

69 Wang, G., Chernikov, A., Glazov, M.M. et al. (2018). Colloquium: excitons in atomically thin transition metal dichalcogenides. *Rev. Mod. Phys.* 90: 021001. https://doi.org/10.1103/RevModPhys.90.021001.

70 Herman, F., Kuglin, C.D., Cuff, K.F., and Kortum, R.L. (1963). Relativistic corrections to the band structure of tetrahedrally bonded semiconductors. *Phys. Rev. Lett.* 11: 541–545. https://doi.org/10.1103/PhysRevLett.11.541.

71 Tang, C.S., Yin, X., Yang, M. et al. (2019). Three-dimensional resonant exciton in monolayer tungsten diselenide actuated by spin–orbit coupling. *ACS Nano* 13: 14529–14539. https://doi.org/10.1021/acsnano.9b08385.

72 Cheiwchanchamnangij, T. and Lambrecht, W.R.L. (2012). Quasiparticle band structure calculation of monolayer, bilayer, and bulk MoS_2. *Phys. Rev., B* 85: 205302. https://doi.org/10.1103/PhysRevB.85.205302.

73 Xiao, D., Liu, G.-B., Feng, W. et al. (2012). Coupled spin and valley physics in monolayers of MoS_2 and other group-VI dichalcogenides. *Phys. Rev. Lett.* 108: 196802. https://doi.org/10.1103/PhysRevLett.108.196802.

74 Zhang, Y., Chang, T.-R., Zhou, B. et al. (2014). Direct observation of the transition from indirect to direct bandgap in atomically thin epitaxial $MoSe_2$. *Nat. Nanotechnol.* 9: 111–115. https://doi.org/10.1038/nnano.2013.277.

75 Miwa, J.A., Ulstrup, S., Sørensen, S.G. et al. (2015). Electronic structure of epitaxial single-layer MoS_2. *Phys. Rev. Lett.* 114: 046802. https://doi.org/10.1103/PhysRevLett.114.046802.

76 Kośmider, K., González, J.W., and Fernández-Rossier, J. (2013). Large spin splitting in the conduction band of transition metal dichalcogenide monolayers. *Phys. Rev., B* 88: 245436. https://doi.org/10.1103/PhysRevB.88.245436.

77 Liu, G.-B., Shan, W.-Y., Yao, Y. et al. (2013). Three-band tight-binding model for monolayers of group-VIB transition metal dichalcogenides. *Phys. Rev., B* 88: 085433. https://doi.org/10.1103/PhysRevB.88.085433.

78 Kormányos, A., Zólyomi, V., Drummond, N.D., and Burkard, G. (2014). Spin-orbit coupling, quantum dots, and qubits in monolayer transition metal dichalcogenides. *Phys. Rev. X* 4: 011034. https://doi.org/10.1103/PhysRevX.4.011034.

79 Dery, H. and Song, Y. (2015). Polarization analysis of excitons in monolayer and bilayer transition-metal dichalcogenides. *Phys. Rev., B* 92: 125431. https://doi.org/10.1103/PhysRevB.92.125431.

80 Qiu, D.Y., Cao, T., and Louie, S.G. (2015). Nonanalyticity, valley quantum phases, and lightlike exciton dispersion in monolayer transition metal dichalcogenides: theory and first-principles calculations. *Phys. Rev. Lett.* 115: 176801. https://doi.org/10.1103/PhysRevLett.115.176801.

81 Echeverry, J.P., Urbaszek, B., Amand, T. et al. (2016). Splitting between bright and dark excitons in transition metal dichalcogenide monolayers. *Phys. Rev., B* 93: 121107. https://doi.org/10.1103/PhysRevB.93.121107.

82 Wang, G., Robert, C., Suslu, A. et al. (2015). Spin-orbit engineering in transition metal dichalcogenide alloy monolayers. *Nat. Commun.* 6: 10110. https://doi.org/10.1038/ncomms10110.

83 Zhang, X.-X., You, Y., Zhao, S.Y.F., and Heinz, T.F. (2015). Experimental evidence for dark excitons in monolayer WSe_2. *Phys. Rev. Lett.* 115: 257403. https://doi.org/10.1103/PhysRevLett.115.257403.

84 Arora, A., Koperski, M., Nogajewski, K. et al. (2015). Excitonic resonances in thin films of WSe_2: from monolayer to bulk material. *Nanoscale* 7: 10421–10429. https://doi.org/10.1039/C5NR01536G.

85 Withers, F., Del Pozo-Zamudio, O., Schwarz, S. et al. (2015). WSe_2 light-emitting tunneling transistors with enhanced brightness at room temperature. *Nano Lett.* 15: 8223–8228. https://doi.org/10.1021/acs.nanolett.5b03740.

86 Zhang, X.-X., Cao, T., Lu, Z. et al. (2017). Magnetic brightening and control of dark excitons in monolayer WSe_2. *Nat. Nanotechnol.* 12: 883–888. https://doi.org/10.1038/nnano.2017.105.

87 Yao, W., Xiao, D., and Niu, Q. (2008). Valley-dependent optoelectronics from inversion symmetry breaking. *Phys. Rev., B* 77: 235406. https://doi.org/10.1103/PhysRevB.77.235406.

88 Zeng, H., Dai, J., Yao, W. et al. (2012). Valley polarization in MoS_2 monolayers by optical pumping. *Nat. Nanotechnol.* 7: 490–493. https://doi.org/10.1038/nnano.2012.95.

89 Sallen, G., Bouet, L., Marie, X. et al. (2012). Robust optical emission polarization in MoS_2 monolayers through selective valley excitation. *Phys. Rev., B* 86: 081301. https://doi.org/10.1103/PhysRevB.86.081301.

90 Cao, T., Wang, G., Han, W. et al. (2012). Valley-selective circular dichroism of monolayer molybdenum disulphide. *Nat. Commun.* 3: 887. https://doi.org/10.1038/ncomms1882.

91 Pacchioni, G. (2020). Valleytronics with a twist. *Nat. Rev. Mater.* 5: 480–480. https://doi.org/10.1038/s41578-020-0220-2.

92 Mak, K.F., Xiao, D., and Shan, J. (2018). Light–valley interactions in 2D semiconductors. *Nat. Photonics* 12: 451–460. https://doi.org/10.1038/s41566-018-0204-6.

93 Schaibley, J.R., Yu, H., Clark, G. et al. (2016). Valleytronics in 2D materials. *Nat. Rev. Mater.* 1: 16055. https://doi.org/10.1038/natrevmats.2016.55.

94 Knox, R.S. (1983). Introduction to exciton physics. In: *Collective Excitations in Solids* (ed. B. Di Bartolo), 183–245. US: Springer.

95 Nam, S.B., Reynolds, D.C., Litton, C.W. et al. (1976). Free-exciton energy spectrum in GaAs. *Phys. Rev., B* 13: 761–767. https://doi.org/10.1103/PhysRevB.13.761.

96 Kazimierczuk, T., Fröhlich, D., Scheel, S. et al. (2014). Giant Rydberg excitons in the copper oxide Cu_2O. *Nature* 514: 343–347. https://doi.org/10.1038/nature13832.

97 Wannier, G.H. (1937). The structure of electronic excitation levels in insulating crystals. *Phys. Rev.* 52: 191–197. https://doi.org/10.1103/PhysRev.52.191.

98 Aldrich, C. and Bajaj, K.K. (1977). Binding energy of a Mott-Wannier exciton in a polarizable medium. *Solid State Commun.* 22: 157–160. https://doi.org/10.1016/0038-1098(77)90423-9.

99 Hjalmarson, H.P., Büttner, H., and Dow, J.D. (1981). Theory of core excitons. *Phys. Rev., B* 24: 6010–6019. https://doi.org/10.1103/PhysRevB.24.6010.

100 Stein, T., Kronik, L., and Baer, R. (2009). Reliable prediction of charge transfer excitations in molecular complexes using time-dependent density functional theory. *J. Am. Chem. Soc.* 131: 2818–2820. https://doi.org/10.1021/ja8087482.

101 Cho, S.W., Piper, L.F.J., DeMasi, A. et al. (2010). Electronic structure of C60/phthalocyanine/ITO interfaces studied using soft X-ray spectroscopies. *J. Phys. Chem., C* 114: 1928–1933. https://doi.org/10.1021/jp910504a.

102 Arkhipov, V.I. and Bässler, H. (2004). Exciton dissociation and charge photogeneration in pristine and doped conjugated polymers. *Phys. Status Solidi A* 201: 1152–1187. https://doi.org/10.1002/pssa.200404339.

103 Stier, A.V., Wilson, N.P., Clark, G. et al. (2016). Probing the influence of dielectric environment on excitons in monolayer WSe_2: insight from high magnetic fields. *Nano Lett.* 16: 7054–7060. https://doi.org/10.1021/acs.nanolett.6b03276.

104 Cadiz, F., Courtade, E., Robert, C. et al. (2017). Excitonic linewidth approaching the homogeneous limit in MoS_2-based van der Waals heterostructures. *Phys. Rev. X* 7: 021026. https://doi.org/10.1103/PhysRevX.7.021026.

105 Raja, A., Chaves, A., Yu, J. et al. (2017). Coulomb engineering of the bandgap and excitons in two-dimensional materials. *Nat. Commun.* 8: 15251. https://doi.org/10.1038/ncomms15251.

106 He, X.-F. (1991). Excitons in anisotropic solids: the model of fractional-dimensional space. *Phys. Rev., B* 43: 2063–2069. https://doi.org/10.1103/PhysRevB.43.2063.

107 Chernikov, A., Berkelbach, T.C., Hill, H.M. et al. (2014). Exciton binding energy and nonhydrogenic rydberg series in monolayer WS_2. *Phys. Rev. Lett.* 113: 076802. https://doi.org/10.1103/PhysRevLett.113.076802.

108 Berkelbach, T.C., Hybertsen, M.S., and Reichman, D.R. (2013). Theory of neutral and charged excitons in monolayer transition metal dichalcogenides. *Phys. Rev., B* 88: 045318. https://doi.org/10.1103/PhysRevB.88.045318.

109 Feng, J., Qian, X., Huang, C.-W., and Li, J. (2012). Strain-engineered artificial atom as a broad-spectrum solar energy funnel. *Nat. Photonics* 6: 866–872. https://doi.org/10.1038/nphoton.2012.285.

110 Qiu, D.Y., da Jornada, F.H., and Louie, S.G. (2013). Optical spectrum of MoS_2: many-body effects and diversity of exciton states. *Phys. Rev. Lett.* 111: 216805. https://doi.org/10.1103/PhysRevLett.111.216805.

111 Molina-Sánchez, A., Sangalli, D., Hummer, K. et al. (2013). Effect of spin-orbit interaction on the optical spectra of single-layer, double-layer, and bulk MoS_2. *Phys. Rev., B* 88: 045412. https://doi.org/10.1103/PhysRevB.88.045412.

112 Mak, K.F., He, K., Lee, C. et al. (2013). Tightly bound trions in monolayer MoS_2. *Nat. Mater.* 12: 207–211. https://doi.org/10.1038/nmat3505.

113 Gogoi, P.K., Hu, Z., Wang, Q. et al. (2017). Oxygen passivation mediated tunability of trion and excitons in MoS_2. *Phys. Rev. Lett.* 119: 077402. https://doi.org/10.1103/PhysRevLett.119.077402.

114 Thilagam, A. (2014). Exciton complexes in low dimensional transition metal dichalcogenides. *J. Appl. Phys.* 116: 053523. https://doi.org/10.1063/1.4892488.

115 You, Y., Zhang, X.-X., Berkelbach, T.C. et al. (2015). Observation of biexcitons in monolayer WSe_2. *Nat. Phys.* 11: 477–481. https://doi.org/10.1038/nphys3324.

116 Klingshirn, C.F. (2007). *Semiconductor Optics*, vol. 3. Springer.

117 Cudazzo, P., Tokatly, I.V., and Rubio, A. (2011). Dielectric screening in two-dimensional insulators: implications for excitonic and impurity states in graphane. *Phys. Rev., B* 84: 085406. https://doi.org/10.1103/PhysRevB.84.085406.

118 Ross, J.S., Wu, S., Yu, H. et al. (2013). Electrical control of neutral and charged excitons in a monolayer semiconductor. *Nat. Commun.* 4: 1474. https://doi.org/10.1038/ncomms2498.

119 Huard, V., Cox, R.T., Saminadayar, K. et al. (2000). Bound states in optical absorption of semiconductor quantum wells containing a two-dimensional electron gas. *Phys. Rev. Lett.* 84: 187–190. https://doi.org/10.1103/PhysRevLett.84.187.

120 Albrecht, S., Reining, L., Del Sole, R., and Onida, G. (1998). Ab initio calculation of excitonic effects in the optical spectra of semiconductors. *Phys. Rev. Lett.* 80: 4510–4513. https://doi.org/10.1103/PhysRevLett.80.4510.

121 Rohlfing, M. and Louie, S.G. (1998). Electron-hole excitations in semiconductors and insulators. *Phys. Rev. Lett.* 81: 2312–2315. https://doi.org/10.1103/PhysRevLett.81.2312.

122 Yang, L., Deslippe, J., Park, C.-H. et al. (2009). Excitonic effects on the optical response of graphene and bilayer graphene. *Phys. Rev. Lett.* 103: 186802. https://doi.org/10.1103/PhysRevLett.103.186802.

123 Mak, K.F., Shan, J., and Heinz, T.F. (2011). Seeing many-body effects in single- and few-layer graphene: observation of two-dimensional saddle-point excitons. *Phys. Rev. Lett.* 106: 046401. https://doi.org/10.1103/PhysRevLett.106.046401.

124 Zhang, Q., Jie, J., Diao, S. et al. (2015). Solution-processed graphene quantum dot deep-UV photodetectors. *ACS Nano* 9: 1561–1570. https://doi.org/10.1021/acsnano.5b00437.

125 Zeng, L., Tao, L., Tang, C. et al. (2016). High-responsivity UV-vis photodetector based on transferable WS_2 film deposited by magnetron sputtering. *Sci. Rep.* 6: 20343. https://doi.org/10.1038/srep20343.

126 Singh, R.S., Li, D., Xiong, Q. et al. (2016). Anomalous photoresponse in the deep-ultraviolet due to resonant excitonic effects in oxygen plasma treated few-layer graphene. *Carbon* 106: 330–335. https://doi.org/10.1016/j.carbon.2016.05.026.

127 Yang, L. (2011). Excitons in intrinsic and bilayer graphene. *Phys. Rev., B* 83: 085405. https://doi.org/10.1103/PhysRevB.83.085405.

128 Benameur, M.M., Radisavljevic, B., Héron, J.S. et al. (2011). Visibility of dichalcogenide nanolayers. *Nanotechnology* 22: 125706. https://doi.org/10.1088/0957-4484/22/12/125706.

129 Coleman, J.N., Lotya, M., O'Neill, A. et al. (2011). Two-dimensional nanosheets produced by liquid exfoliation of layered materials. *Science* 331: 568–571. https://doi.org/10.1126/science.1194975.

130 Nicolosi, V., Chhowalla, M., Kanatzidis, M.G. et al. (2013). Liquid exfoliation of layered materials. *Science* 340: 1226419. https://doi.org/10.1126/science.1226419.

131 Eda, G., Yamaguchi, H., Voiry, D. et al. (2011). Photoluminescence from chemically exfoliated MoS_2. *Nano Lett.* 11: 5111–5116. https://doi.org/10.1021/nl201874w.

132 Gordon, R.A., Yang, D., Crozier, E.D. et al. (2002). Structures of exfoliated single layers of WS_2, MoS_2, and $MoSe_2$ in aqueous suspension. *Phys. Rev., B* 65: 125407. https://doi.org/10.1103/PhysRevB.65.125407.

133 Kappera, R., Voiry, D., Yalcin, S.E. et al. (2014). Phase-engineered low-resistance contacts for ultrathin MoS_2 transistors. *Nat. Mater.* 13: 1128–1134. https://doi.org/10.1038/nmat4080.

134 Mittal, M., Sardar, S., and Jana, A. (2021). Chapter 7 - Nanofabrication techniques for semiconductor chemical sensors. In: *Handbook of Nanomaterials for Sensing Applications* (ed. C.M. Hussain and S.K. Kailasa), 119–137. Elsevier.

135 Niu, G., Saint-Girons, G., and Vilquin, B. (2013). Chapter 17 - Epitaxial systems combining oxides and semiconductors. In: *In Molecular Beam Epitaxy* (ed. M. Henini), 451–475. Elsevier.

136 Sun, L., Yuan, G., Gao, L. et al. (2021). Chemical vapour deposition. *Nat. Rev. Methods Primers* 1: 5. https://doi.org/10.1038/s43586-020-00005-y.

137 Huang, J.-K., Pu, J., Hsu, C.-L. et al. (2014). Large-area synthesis of highly crystalline WSe_2 monolayers and device applications. *ACS Nano* 8: 923–930. https://doi.org/10.1021/nn405719x.

138 Najmaei, S., Liu, Z., Zhou, W. et al. (2013). Vapour phase growth and grain boundary structure of molybdenum disulphide atomic layers. *Nat. Mater.* 12: 754–759. https://doi.org/10.1038/nmat3673.

139 van der Zande, A.M., Huang, P.Y., Chenet, D.A. et al. (2013). Grains and grain boundaries in highly crystalline monolayer molybdenum disulphide. *Nat. Mater.* 12: 554–561. https://doi.org/10.1038/nmat3633.

140 Zou, X., Liu, Y., and Yakobson, B.I. (2013). Predicting dislocations and grain boundaries in two-dimensional metal-disulfides from the first principles. *Nano Lett.* 13: 253–258. https://doi.org/10.1021/nl3040042.

141 Dumcenco, D., Ovchinnikov, D., Marinov, K. et al. (2015). Large-area epitaxial monolayer MoS_2. *ACS Nano* 9: 4611–4620. https://doi.org/10.1021/acsnano.5b01281.

142 Chung, J.W., Dai, Z.R., and Ohuchi, F.S. (1998). WS_2 thin films by metal organic chemical vapor deposition. *J. Cryst. Growth* 186: 137–150. https://doi.org/10.1016/S0022-0248(97)00479-X.

143 Dumcenco, D., Ovchinnikov, D., Sanchez, O.L. et al. (2015). Large-area MoS_2 grown using H_2 S as the sulphur source. *2D Materials* 2: 044005. https://doi.org/10.1088/2053-1583/2/4/044005.

144 Eichfeld, S.M., Hossain, L., Lin, Y.-C. et al. (2015). Highly scalable, atomically thin WSe_2 grown via metal–organic chemical vapor deposition. *ACS Nano* 9: 2080–2087. https://doi.org/10.1021/nn5073286.

145 Kang, K., Xie, S., Huang, L. et al. (2015). High-mobility three-atom-thick semiconducting films with wafer-scale homogeneity. *Nature* 520: 656–660. https://doi.org/10.1038/nature14417.

146 Xia, J., Huang, X., Liu, L.-Z. et al. (2014). CVD synthesis of large-area, highly crystalline $MoSe_2$ atomic layers on diverse substrates and application to photodetectors. *Nanoscale* 6: 8949–8955. https://doi.org/10.1039/C4NR02311K.

147 Chang, Y.-H., Zhang, W., Zhu, Y. et al. (2014). Monolayer $MoSe_2$ grown by chemical vapor deposition for fast photodetection. *ACS Nano* 8: 8582–8590. https://doi.org/10.1021/nn503287m.

148 Gao, Y., Liu, Z., Sun, D.-M. et al. (2015). Large-area synthesis of high-quality and uniform monolayer WS_2 on reusable au foils. *Nat. Commun.* 6: 8569. https://doi.org/10.1038/ncomms9569.

149 Kobayashi, Y., Sasaki, S., Mori, S. et al. (2015). Growth and optical properties of high-quality monolayer WS_2 on graphite. *ACS Nano* 9: 4056–4063. https://doi.org/10.1021/acsnano.5b00103.

150 Okada, M., Sawazaki, T., Watanabe, K. et al. (2014). Direct chemical vapor deposition growth of WS_2 atomic layers on hexagonal boron nitride. *ACS Nano* 8: 8273–8277. https://doi.org/10.1021/nn503093k.

151 Tan, H., Fan, Y., Zhou, Y. et al. (2016). Ultrathin 2D photodetectors utilizing chemical vapor deposition grown WS_2 with graphene electrodes. *ACS Nano* 10: 7866–7873. https://doi.org/10.1021/acsnano.6b03722.

152 Zhang, Y., Zhang, Y., Ji, Q. et al. (2013). Controlled growth of high-quality monolayer WS_2 layers on sapphire and imaging its grain boundary. *ACS Nano* 7: 8963–8971. https://doi.org/10.1021/nn403454e.

153 Chen, J., Liu, B., Liu, Y. et al. (2015). Chemical vapor deposition of large-sized hexagonal WSe_2 crystals on dielectric substrates. *Adv. Mater.* 27: 6722–6727. https://doi.org/10.1002/adma.201503446.

154 Liu, B., Fathi, M., Chen, L. et al. (2015). Chemical vapor deposition growth of monolayer WSe_2 with tunable device characteristics and growth mechanism study. *ACS Nano* 9: 6119–6127. https://doi.org/10.1021/acsnano.5b01301.

155 Yoshida, M., Iizuka, T., Saito, Y. et al. (2016). Gate-optimized thermoelectric power factor in ultrathin WSe_2 single crystals. *Nano Lett.* 16: 2061–2065. https://doi.org/10.1021/acs.nanolett.6b00075.

156 Keyshar, K., Gong, Y., Ye, G. et al. (2015). Chemical vapor deposition of monolayer rhenium disulfide (ReS_2). *Adv. Mater.* 27: 4640–4648. https://doi.org/10.1002/adma.201501795.

157 He, X., Liu, F., Hu, P. et al. (2015). Chemical vapor deposition of high-quality and atomically layered ReS_2. *Small* 11: 5423–5429. https://doi.org/10.1002/smll.201501488.

158 Hafeez, M., Gan, L., Li, H. et al. (2016). Large-area bilayer ReS_2 film/multilayer ReS_2 flakes synthesized by chemical vapor deposition for high performance photodetectors. *Adv. Funct. Mater.* 26: 4551–4560. https://doi.org/10.1002/adfm.201601019.

159 Hafeez, M., Gan, L., Li, H. et al. (2016). Chemical vapor deposition synthesis of ultrathin hexagonal $ReSe_2$ flakes for anisotropic Raman property and optoelectronic application. *Adv. Mater.* 28: 8296–8301. https://doi.org/10.1002/adma.201601977.

160 Naylor, C.H., Parkin, W.M., Ping, J. et al. (2016). Monolayer single-crystal 1T′-$MoTe_2$ grown by chemical vapor deposition exhibits weak antilocalization effect. *Nano Lett.* 16: 4297–4304. https://doi.org/10.1021/acs.nanolett.6b01342.

161 Zhou, J., Liu, F., Lin, J. et al. (2017). Large-area and high-quality 2D transition metal telluride. *Adv. Mater.* 29: 1603471. https://doi.org/10.1002/adma.201603471.

162 Joyce, B.A. (1985). Molecular beam epitaxy. *Rep. Prog. Phys.* 48: 1637–1697. https://doi.org/10.1088/0034-4885/48/12/002.

163 Arthur, J.R. (2002). Molecular beam epitaxy. *Surf. Sci.* 500: 189–217. ihttps://doi.org/https:/doi.org/10.1016/S0039-6028(01)01525-4.

164 Ugeda, M.M., Bradley, A.J., Shi, S.-F. et al. (2014). Giant bandgap renormalization and excitonic effects in a monolayer transition metal dichalcogenide semiconductor. *Nat. Mater.* 13: 1091–1095. https://doi.org/10.1038/nmat4061.

165 Barja, S., Wickenburg, S., Liu, Z.-F. et al. (2016). Charge density wave order in 1D mirror twin boundaries of single-layer $MoSe_2$. *Nat. Phys.* 12: 751–756. https://doi.org/10.1038/nphys3730.

166 Vishwanath, S., Liu, X., Rouvimov, S. et al. (2015). Comprehensive structural and optical characterization of MBE grown $MoSe_2$ on graphite, CaF_2 and graphene. *2D Materials* 2: 024007. https://doi.org/10.1088/2053-1583/2/2/024007.

167 Aretouli, K.E., Tsoutsou, D., Tsipas, P. et al. (2016). Epitaxial 2D $SnSe_2$/2D WSe_2 van der Waals heterostructures. *ACS Appl. Mater. Interfaces* 8: 23222–23229. https://doi.org/10.1021/acsami.6b02933.

168 Xenogiannopoulou, E., Tsipas, P., Aretouli, K.E. et al. (2015). High-quality, large-area $MoSe_2$ and $MoSe_2/Bi_2Se_3$ heterostructures on AlN(0001)/Si(111) substrates by molecular beam epitaxy. *Nanoscale* 7: 7896–7905. https://doi.org/10.1039/C4NR06874B.

169 Zhao, W., Ribeiro, R.M., and Eda, G. (2015). Electronic structure and optical signatures of semiconducting transition metal dichalcogenide nanosheets. *Acc. Chem. Res.* 48: 91–99. https://doi.org/10.1021/ar500303m.

170 Mueller, T. and Malic, E. (2018). Exciton physics and device application of two-dimensional transition metal dichalcogenide semiconductors. *npj 2D Mater. Appl.* 2: 29. https://doi.org/10.1038/s41699-018-0074-2.

171 Nagler, P., Plechinger, G., Ballottin, M.V. et al. (2017). Interlayer exciton dynamics in a dichalcogenide monolayer heterostructure. *2D Materials* 4: 025112. https://doi.org/10.1088/2053-1583/aa7352.

172 He, K., Kumar, N., Zhao, L. et al. (2014). Tightly bound excitons in monolayer WSe_2. *Phys. Rev. Lett.* 113: 026803. https://doi.org/10.1103/PhysRevLett.113.026803.

173 Mak, K.F., He, K., Shan, J., and Heinz, T.F. (2012). Control of valley polarization in monolayer MoS_2 by optical helicity. *Nat. Nanotechnol.* 7: 494–498. https://doi.org/10.1038/nnano.2012.96.

174 Jones, A.M., Yu, H., Ghimire, N.J. et al. (2013). Optical generation of excitonic valley coherence in monolayer WSe_2. *Nat. Nanotechnol.* 8: 634–638. https://doi.org/10.1038/nnano.2013.151.

175 Kim, J., Hong, X., Jin, C. et al. (2014). Ultrafast generation of pseudo-magnetic field for valley excitons in WSe_2 monolayers. *Science* 346: 1205–1208. https://doi.org/10.1126/science.1258122.

176 Sie, E.J., McIver, J.W., Lee, Y.-H. et al. (2015). Valley-selective optical stark effect in monolayer WS_2. *Nat. Mater.* 14: 290–294. https://doi.org/10.1038/nmat4156.

177 Mai, C., Semenov, Y.G., Barrette, A. et al. (2014). Exciton valley relaxation in a single layer of WS_2 measured by ultrafast spectroscopy. *Phys. Rev., B* 90: 041414. https://doi.org/10.1103/PhysRevB.90.041414.

178 Wang, G., Bouet, L., Lagarde, D. et al. (2014). Valley dynamics probed through charged and neutral exciton emission in monolayer WSe_2. *Phys. Rev., B* 90: 075413. https://doi.org/10.1103/PhysRevB.90.075413.

179 Tonndorf, P., Schmidt, R., Schneider, R. et al. (2015). Single-photon emission from localized excitons in an atomically thin semiconductor. *Optica* 2: 347–352. https://doi.org/10.1364/OPTICA.2.000347.

180 Srivastava, A., Sidler, M., Allain, A.V. et al. (2015). Optically active quantum dots in monolayer WSe_2. *Nat. Nanotechnol.* 10: 491–496. https://doi.org/10.1038/nnano.2015.60.

181 He, Y.-M., Clark, G., Schaibley, J.R. et al. (2015). Single quantum emitters in monolayer semiconductors. *Nat. Nanotechnol.* 10: 497–502. https://doi.org/10.1038/nnano.2015.75.

182 Koperski, M., Nogajewski, K., Arora, A. et al. (2015). Single photon emitters in exfoliated WSe_2 structures. *Nat. Nanotechnol.* 10: 503–506. https://doi.org/10.1038/nnano.2015.67.

183 Chakraborty, C., Kinnischtzke, L., Goodfellow, K.M. et al. (2015). Voltage-controlled quantum light from an atomically thin semiconductor. *Nat. Nanotechnol.* 10: 507–511. https://doi.org/10.1038/nnano.2015.79.

184 Geim, A.K. and Grigorieva, I.V. (2013). Van der Waals heterostructures. *Nature* 499: 419–425. https://doi.org/10.1038/nature12385.

185 Fang, H., Battaglia, C., Carraro, C. et al. (2014). Strong interlayer coupling in van der Waals heterostructures built from single-layer chalcogenides. *Proc. Natl. Acad. Sci.* 111: 6198–6202. https://doi.org/10.1073/pnas.1405435111.

186 Rivera, P., Schaibley, J.R., Jones, A.M. et al. (2015). Observation of long-lived interlayer excitons in monolayer $MoSe_2$–WSe_2 heterostructures. *Nat. Commun.* 6: 6242. https://doi.org/10.1038/ncomms7242.

187 Baranowski, M., Surrente, A., Klopotowski, L. et al. (2017). Probing the interlayer exciton physics in a $MoS_2/MoSe_2/MoS_2$ van der Waals heterostructure. *Nano Lett.* 17: 6360–6365. https://doi.org/10.1021/acs.nanolett.7b03184.

188 Fogler, M.M., Butov, L.V., and Novoselov, K.S. (2014). High-temperature superfluidity with indirect excitons in van der Waals heterostructures. *Nat. Commun.* 5: 4555. https://doi.org/10.1038/ncomms5555.

189 Yu, Y., Hu, S., Su, L. et al. (2015). Equally efficient interlayer exciton relaxation and improved absorption in epitaxial and nonepitaxial MoS_2/WS_2 heterostructures. *Nano Lett.* 15: 486–491. https://doi.org/10.1021/nl5038177.

190 Gao, S., Yang, L., and Spataru, C.D. (2017). Interlayer coupling and gate-tunable excitons in transition metal dichalcogenide heterostructures. *Nano Lett.* 17: 7809–7813. https://doi.org/10.1021/acs.nanolett.7b04021.

191 Hanbicki, A.T., Chuang, H.-J., Rosenberger, M.R. et al. (2018). Double indirect interlayer exciton in a $MoSe_2/WSe_2$ van der Waals heterostructure. *ACS Nano* 12: 4719–4726. https://doi.org/10.1021/acsnano.8b01369.

192 Fujiwara, H. (2007). *Spectroscopic Ellipsometry: Principles and Applications.* Wiley.

193 Li, D., Song, X., Xu, J. et al. (2017). Optical properties of thickness-controlled MoS_2 thin films studied by spectroscopic ellipsometry. *Appl. Surf. Sci.* 421: 884–890. https://doi.org/10.1016/j.apsusc.2016.09.069.

194 Carvalho, A., Ribeiro, R.M., and Castro Neto, A.H. (2013). Band nesting and the optical response of two-dimensional semiconducting transition metal dichalcogenides. *Phys. Rev., B* 88: 115205. https://doi.org/10.1103/PhysRevB.88.115205.

195 Kumar, R., Carvalho, A., Amara, K.K. et al. (2014). Photocarrier relaxation pathway in two-dimensional semiconducting transition metal dichalcogenides. *Nat. Commun.* 5: 4543. https://doi.org/10.1038/ncomms5543.

196 Zhao, W., Ghorannevis, Z., Chu, L. et al. (2013). Evolution of electronic structure in atomically thin sheets of WS_2 and WSe_2. *ACS Nano* 7: 791–797. https://doi.org/10.1021/nn305275h.

197 Yin, X., Tang, C.S., Wu, D. et al. (2019). Unraveling high-yield phase-transition dynamics in transition metal dichalcogenides on metallic substrates. *Adv. Sci.* 6: 1802093. https://doi.org/10.1002/advs.201802093.

198 Zhang, X., Qiao, X.-F., Shi, W. et al. (2015). Phonon and Raman scattering of two-dimensional transition metal dichalcogenides from monolayer, multilayer to bulk material. *Chem. Soc. Rev.* 44: 2757–2785. https://doi.org/10.1039/C4CS00282B.

199 Kolobov, A.V. and Tominaga, J. (2016). *Two-Dimensional Transition-Metal Dichalcogenides*, 1e. Springer International Publishing: Imprint: Springer.

200 Sandoval, S.J., Yang, D., Frindt, R.F., and Irwin, J.C. (1991). Raman study and lattice dynamics of single molecular layers of MoS_2. *Phys. Rev., B* 44: 3955–3962. https://doi.org/10.1103/PhysRevB.44.3955.

201 Zhao, Y., Luo, X., Li, H. et al. (2013). Interlayer breathing and shear modes in few-trilayer MoS_2 and WSe_2. *Nano Lett.* 13: 1007–1015. https://doi.org/10.1021/nl304169w.

202 Verble, J.L. and Wieting, T.J. (1970). Lattice mode degeneracy in $Mo\$_2$ and other layer compounds. *Phys. Rev. Lett.* 25: 362–365. https://doi.org/10.1103/PhysRevLett.25.362.

203 Bromley, R.A. (1971). The lattice vibrations of the MoS_2 structure. *Philos. Mag. A* 23: 1417–1427. https://doi.org/10.1080/14786437108217011.

204 Yuan, P., Li, C., Xu, S. et al. (2017). Interfacial thermal conductance between few to tens of layered-MoS_2 and c-Si: effect of MoS_2 thickness. *Acta Mater.* 122: 152–165. https://doi.org/10.1016/j.actamat.2016.09.045.

205 Bertrand, P.A. (1991). Surface-phonon dispersion of MoS_2. *Phys. Rev., B* 44: 5745–5749. https://doi.org/10.1103/PhysRevB.44.5745.

206 Wieting, T.J. and Verble, J.L. (1972). Interlayer bonding and the lattice vibrations of beta-GaSe. *Phys. Rev., B* 5: 1473–1479. https://doi.org/10.1103/PhysRevB.5.1473.

207 Tongay, S., Zhou, J., Ataca, C. et al. (2012). Thermally driven crossover from indirect toward direct bandgap in 2D semiconductors: $MoSe_2$ versus MoS_2. *Nano Lett.* 12: 5576–5580. https://doi.org/10.1021/nl302584w.

208 Tonndorf, P., Schmidt, R., Böttger, P. et al. (2013). Photoluminescence emission and Raman response of monolayer MoS_2, $MoSe_2$, and WSe_2. *Opt. Express* 21: 4908–4916. https://doi.org/10.1364/OE.21.004908.

209 Gutiérrez, H.R., Perea-López, N., Elías, A.L. et al. (2013). Extraordinary room-temperature photoluminescence in triangular WS_2 monolayers. *Nano Lett.* 13: 3447–3454. https://doi.org/10.1021/nl3026357.

210 Zhao, W., Ghorannevis, Z., Amara, K.K. et al. (2013). Lattice dynamics in mono- and few-layer sheets of WS_2 and WSe_2. *Nanoscale* 5: 9677–9683. https://doi.org/10.1039/C3NR03052K.

211 Li, H., Lu, G., Wang, Y. et al. (2013). Mechanical exfoliation and characterization of single- and few-layer nanosheets of WSe_2, TaS_2, and $TaSe_2$. *Small* 9: 1974–1981. https://doi.org/10.1002/smll.201202919.

212 Lee, J.E., Ahn, G., Shim, J. et al. (2012). Optical separation of mechanical strain from charge doping in graphene. *Nat. Commun.* 3: 1024. https://doi.org/10.1038/ncomms2022.

213 Chakraborty, B., Bera, A., Muthu, D.V.S. et al. (2012). Symmetry-dependent phonon renormalization in monolayer MoS_2 transistor. *Phys. Rev., B* 85: 161403. https://doi.org/10.1103/PhysRevB.85.161403.

214 Rice, C., Young, R.J., Zan, R. et al. (2013). Raman-scattering measurements and first-principles calculations of strain-induced phonon shifts in monolayer MoS_2. *Phys. Rev., B* 87: 081307. https://doi.org/10.1103/PhysRevB.87.081307.

215 Hui, Y.Y., Liu, X., Jie, W. et al. (2013). Exceptional tunability of band energy in a compressively strained trilayer MoS_2 sheet. *ACS Nano* 7: 7126–7131. https://doi.org/10.1021/nn4024834.

216 Liu, Z., Amani, M., Najmaei, S. et al. (2014). Strain and structure heterogeneity in MoS_2 atomic layers grown by chemical vapour deposition. *Nat. Commun.* 5: 5246. https://doi.org/10.1038/ncomms6246.

217 Mignuzzi, S., Pollard, A.J., Bonini, N. et al. (2015). Effect of disorder on Raman scattering of single-layer Mo_2. *Phys. Rev., B* 91: 195411. https://doi.org/10.1103/PhysRevB.91.195411.

218 Song, S., Keum, D.H., Cho, S. et al. (2016). Room temperature semiconductor–metal transition of $MoTe_2$ thin films engineered by strain. *Nano Lett.* 16: 188–193. https://doi.org/10.1021/acs.nanolett.5b03481.

219 Kang, J., Liu, W., Sarkar, D. et al. (2014). Computational study of metal contacts to monolayer transition-metal dichalcogenide semiconductors. *Phys. Rev. X* 4: 031005. https://doi.org/10.1103/PhysRevX.4.031005.

220 Popov, I., Seifert, G., and Tománek, D. (2012). Designing electrical contacts to MoS_2 monolayers: a computational study. *Phys. Rev. Lett.* 108: 156802. https://doi.org/10.1103/PhysRevLett.108.156802.

221 Kang, Y., Najmaei, S., Liu, Z. et al. (2014). Plasmonic hot electron induced structural phase transition in a MoS_2 monolayer. *Adv. Mater.* 26: 6467–6471. https://doi.org/10.1002/adma.201401802.

222 Nayak, A.P., Pandey, T., Voiry, D. et al. (2015). Pressure-dependent optical and vibrational properties of monolayer molybdenum disulfide. *Nano Lett.* 15: 346–353. https://doi.org/10.1021/nl5036397.

223 Tan, S.J.R., Abdelwahab, I., Ding, Z. et al. (2017). Chemical stabilization of 1T′ phase transition metal dichalcogenides with giant optical Kerr nonlinearity. *J. Am. Chem. Soc.* 139: 2504–2511. https://doi.org/10.1021/jacs.6b13238.

224 Wilson, N.R., Nguyen, P.V., Seyler, K. et al. Determination of band offsets, hybridization, and exciton binding in 2D semiconductor heterostructures. *Sci. Adv.* 3: e1601832. https://doi.org/10.1126/sciadv.1601832.

225 Yang, H., Liang, A., Chen, C. et al. (2018). Visualizing electronic structures of quantum materials by angle-resolved photoemission spectroscopy. *Nat. Rev. Mater.* 3: 341–353. https://doi.org/10.1038/s41578-018-0047-2.

226 Yuan, H., Liu, Z., Xu, G. et al. (2016). Evolution of the valley position in bulk transition-metal chalcogenides and their monolayer limit. *Nano Lett.* 16: 4738–4745. https://doi.org/10.1021/acs.nanolett.5b05107.

227 Mo, S.-K. (2017). Angle-resolved photoemission spectroscopy for the study of two-dimensional materials. *Nano Convergence* 4: 6. https://doi.org/10.1186/s40580-017-0100-7.

228 Alidoust, N., Bian, G., Xu, S.Y. et al. (2014). Observation of monolayer valence band spin-orbit effect and induced quantum well states in MoX_2. *Nat. Commun.* 5: 4673. https://doi.org/10.1038/ncomms5673.

229 Riley, J.M., Mazzola, F., Dendzik, M. et al. (2014). Direct observation of spin-polarized bulk bands in an inversion-symmetric semiconductor. *Nat. Phys.* 10: 835–839. https://doi.org/10.1038/nphys3105.

230 Suzuki, R., Sakano, M., Zhang, Y.J. et al. (2014). Valley-dependent spin polarization in bulk MoS_2 with broken inversion symmetry. *Nat. Nanotechnol.* 9: 611–617. https://doi.org/10.1038/nnano.2014.148.

231 Razzoli, E., Jaouen, T., Mottas, M.L. et al. (2017). Selective probing of hidden spin-polarized states in inversion-symmetric bulk MoS_2. *Phys. Rev. Lett.* 118: 086402. https://doi.org/10.1103/PhysRevLett.118.086402.

232 Pierucci, D., Henck, H., Avila, J. et al. (2016). Band alignment and minigaps in monolayer MoS_2-graphene van der Waals heterostructures. *Nano Lett.* 16: 4054–4061. https://doi.org/10.1021/acs.nanolett.6b00609.

233 Ulstrup, S., Koch, R.J., Singh, S. et al. Direct observation of minibands in a twisted graphene/WS_2 bilayer. *Sci. Adv.* 6: eaay6104. https://doi.org/10.1126/sciadv.aay6104.

234 Cao, Y., Fatemi, V., Fang, S. et al. (2018). Unconventional superconductivity in magic-angle graphene superlattices. *Nature* 556: 43–50. https://doi.org/10.1038/nature26160.

235 Wang, L., Shih, E.-M., Ghiotto, A. et al. (2020). Correlated electronic phases in twisted bilayer transition metal dichalcogenides. *Nat. Mater.* 19: 861–866. https://doi.org/10.1038/s41563-020-0708-6.

236 Ghiotto, A., Shih, E.-M., Pereira, G.S.S.G. et al. (2021). Quantum criticality in twisted transition metal dichalcogenides. *Nature* 597: 345–349. https://doi.org/10.1038/s41586-021-03815-6.

237 Pei, D., Wang, B., Zhou, Z. et al. (2022). Observation of gamma-valley moiré bands and emergent hexagonal lattice in twisted transition metal dichalcogenides. *Phys. Rev. X* 12: 021065. https://doi.org/10.1103/PhysRevX.12.021065.

238 Ganguli, S.C., Vaňo, V., Kezilebieke, S. et al. (2022). Confinement-engineered superconductor to correlated-insulator transition in a van der Waals monolayer. *Nano Lett.* 22: 1845–1850. https://doi.org/10.1021/acs.nanolett.1c03491.

239 Kusmartseva, A.F., Sipos, B., Berger, H. et al. (2009). Pressure induced superconductivity in pristine $1T-TiSe_2$. *Phys. Rev. Lett.* 103: 236401. https://doi.org/10.1103/PhysRevLett.103.236401.

240 Wei, M.J., Lu, W.J., Xiao, R.C. et al. (2017). Manipulating charge density wave order in monolayer $1T-TiSe_2$ by strain and charge doping: a first-principles investigation. *Phys. Rev., B* 96: 165404. https://doi.org/10.1103/PhysRevB.96.165404.

241 Yokota, K.-i, Kurata, G., Matsui, T., and Fukuyama, H. (2000). Superconductivity in the quasi-two-dimensional conductor 2H-$TaSe_2$. *Physica B* 284–288: 551–552. https://doi.org/10.1016/S0921-4526(99)02166-3.

242 Terashima, K., Sato, T., Komatsu, H. et al. (2003). Charge-density wave transition of $1T-VSe_2$ studied by angle-resolved photoemission spectroscopy. *Phys. Rev., B* 68: 155108. https://doi.org/10.1103/PhysRevB.68.155108.

243 Li, Y.W., Jiang, J., Yang, H.F. et al. (2018). Folded superstructure and degeneracy-enhanced band gap in the weak-coupling charge density wave system $2H-TaSe_2$. *Phys. Rev., B* 97: 115118. https://doi.org/10.1103/PhysRevB.97.115118.

244 Rahn, D.J., Hellmann, S., Kalläne, M. et al. (2012). Gaps and kinks in the electronic structure of the superconductor 2H-$NbSe_2$ from angle-resolved photoemission at 1 K. *Phys. Rev., B* 85: 224532. https://doi.org/10.1103/PhysRevB.85.224532.

245 Ugeda, M.M., Bradley, A.J., Zhang, Y. et al. (2016). Characterization of collective ground states in single-layer $NbSe_2$. *Nat. Phys.* 12: 92–97. https://doi.org/10.1038/nphys3527.

246 Nakata, Y., Sugawara, K., Ichinokura, S. et al. (2018). Anisotropic band splitting in monolayer $NbSe_2$: implications for superconductivity and charge density wave. *Npj 2D Mater. Appl.* 2: 12. https://doi.org/10.1038/s41699-018-0057-3.

247 Calandra, M., Mazin, I.I., and Mauri, F. (2009). Effect of dimensionality on the charge-density wave in few-layer 2H-$NbSe_2$. *Phys. Rev., B* 80: 241108. https://doi.org/10.1103/PhysRevB.80.241108.

248 Chen, P., Chan, Y.H., Fang, X.Y. et al. (2015). Charge density wave transition in single-layer titanium diselenide. *Nat. Commun.* 6: 8943. https://doi.org/10.1038/ncomms9943.

249 Zhang, C., Johnson, A., Hsu, C.-L. et al. (2014). Direct imaging of band profile in single layer MoS_2 on graphite: quasiparticle energy gap, metallic edge states, and edge band bending. *Nano Lett.* 14: 2443–2447. https://doi.org/10.1021/nl501133c.

250 Chiu, M.-H., Zhang, C., Shiu, H.-W. et al. (2015). Determination of band alignment in the single-layer MoS_2/WSe_2 heterojunction. *Nat. Commun.* 6: 7666. https://doi.org/10.1038/ncomms8666.

251 Liu, H., Jiao, L., Yang, F. et al. (2014). Dense network of one-dimensional midgap metallic modes in monolayer $MoSe_2$ and their spatial undulations. *Phys. Rev. Lett.* 113: 066105. https://doi.org/10.1103/PhysRevLett.113.066105.

252 Lezama, I.G., Ubaldini, A., Longobardi, M. et al. (2014). Surface transport and band gap structure of exfoliated 2H-$MoTe_2$ crystals. *2D Materials* 1: 021002. https://doi.org/10.1088/2053-1583/1/2/021002.

253 Yankowitz, M., McKenzie, D., and LeRoy, B.J. (2015). Local spectroscopic characterization of spin and layer polarization in WSe_2. *Phys. Rev. Lett.* 115: 136803. https://doi.org/10.1103/PhysRevLett.115.136803.

254 Sørensen, S.G., Füchtbauer, H.G., Tuxen, A.K. et al. (2014). Structure and electronic properties of in situ synthesized single-layer MoS_2 on a gold surface. *ACS Nano* 8: 6788–6796. https://doi.org/10.1021/nn502812n.

255 Bradley, A.J., Ugeda, M.M., da Jornada, F.H. et al. (2015). Probing the role of interlayer coupling and coulomb interactions on electronic structure in few-layer $MoSe_2$ nanostructures. *Nano Lett.* 15: 2594–2599. https://doi.org/10.1021/acs.nanolett.5b00160.

256 Peng, W., Wang, H., Lu, H. et al. (2021). Recent progress on the scanning tunneling microscopy and spectroscopy study of semiconductor heterojunctions. *Small* 17: 2100655. https://doi.org/10.1002/smll.202100655.

257 Zhang, C., Li, M.-Y., Tersoff, J. et al. (2018). Strain distributions and their influence on electronic structures of WSe_2–MoS_2 laterally strained heterojunctions. *Nat. Nanotechnol.* 13: 152–158. https://doi.org/10.1038/s41565-017-0022-x.

258 Kobayashi, Y., Yoshida, S., Sakurada, R. et al. (2016). Modulation of electrical potential and conductivity in an atomic-layer semiconductor heterojunction. *Sci. Rep.* 6: 31223. https://doi.org/10.1038/srep31223.

259 Duerloo, K.-A.N., Ong, M.T., and Reed, E.J. (2012). Intrinsic piezoelectricity in two-dimensional materials. *J. Phys. Chem. Lett.* 3: 2871–2876. https://doi.org/10.1021/jz3012436.

260 Wu, W., Wang, L., Li, Y. et al. (2014). Piezoelectricity of single-atomic-layer MoS_2 for energy conversion and piezotronics. *Nature* 514: 470–474. https://doi.org/10.1038/nature13792.

261 Zhang, Z., Wang, Y., Watanabe, K. et al. (2020). Flat bands in twisted bilayer transition metal dichalcogenides. *Nat. Phys.* 16: 1093–1096. https://doi.org/10.1038/s41567-020-0958-x.

262 Li, H., Li, S., Naik, M.H. et al. (2021). Imaging moiré flat bands in three-dimensional reconstructed WSe_2/WS_2 superlattices. *Nat. Mater.* 20: 945–950. https://doi.org/10.1038/s41563-021-00923-6.

263 Shabani, S., Halbertal, D., Wu, W. et al. (2021). Deep moiré potentials in twisted transition metal dichalcogenide bilayers. *Nat. Phys.* 17: 720–725. https://doi.org/10.1038/s41567-021-01174-7.

264 Nieken, R., Roche, A., Mahdikhanysarvejahany, F. et al. (2022). Direct STM measurements of R-type and H-type twisted $MoSe_2/WSe_2$. *APL Mater.* 10: 031107. https://doi.org/10.1063/5.0084358.

265 Liu, Y., Qiu, Z., Carvalho, A. et al. (2017). Gate-tunable giant stark effect in few-layer black phosphorus. *Nano Lett.* 17: 1970–1977. https://doi.org/10.1021/acs.nanolett.6b05381.

266 Song, Z., Schultz, T., Ding, Z. et al. (2017). Electronic properties of a 1D intrinsic/p-doped heterojunction in a 2D transition metal dichalcogenide semiconductor. *ACS Nano* 11: 9128–9135. https://doi.org/10.1021/acsnano.7b03953.

267 Blades, W.H., Frady, N.J., Litwin, P.M. et al. (2020). Thermally induced defects on WSe_2. *J. Phys. Chem., C* 124: 15337–15346. https://doi.org/10.1021/acs.jpcc.0c04440.

268 Biadala, L., Peng, W., Lambert, Y. et al. (2019). Trap-free heterostructure of PbS nanoplatelets on InP(001) by chemical epitaxy. *ACS Nano* 13: 1961–1967. https://doi.org/10.1021/acsnano.8b08413.

269 Cheng, F., Hu, Z., Xu, H. et al. (2019). Interface engineering of Au(111) for the growth of 1T′-$MoSe_2$. *ACS Nano* 13: 2316–2323. https://doi.org/10.1021/acsnano.8b09054.

270 Lu, J., Bao, D.-L., Qian, K. et al. (2017). Identifying and visualizing the edge terminations of single-layer $MoSe_2$ island epitaxially grown on Au(111). *ACS Nano* 11: 1689–1695. https://doi.org/10.1021/acsnano.6b07512.

271 Lauritsen, J.V., Nyberg, M., Vang, R.T. et al. (2003). Chemistry of one-dimensional metallic edge states in MoS_2 nanoclusters. *Nanotechnology* 14: 385–389. https://doi.org/10.1088/0957-4484/14/3/306.

272 Bollinger, M.V., Lauritsen, J.V., Jacobsen, K.W. et al. (2001). One-dimensional metallic edge states in MoS_2. *Phys. Rev. Lett.* 87: 196803. https://doi.org/10.1103/PhysRevLett.87.196803.

273 Kang, S., Lee, D., Kim, J. et al. (2020). 2D semiconducting materials for electronic and optoelectronic applications: potential and challenge. *2D Materials* 7: 022003. https://doi.org/10.1088/2053-1583/ab6267.

274 Chen, Y.-C., Cao, T., Chen, C. et al. (2015). Molecular bandgap engineering of bottom-up synthesized graphene nanoribbon heterojunctions. *Nat. Nanotechnol.* 10: 156–160. https://doi.org/10.1038/nnano.2014.307.

275 Rizzo, D.J., Wu, M., Tsai, H.-Z. et al. (2019). Length-dependent evolution of type II heterojunctions in bottom-up-synthesized graphene nanoribbons. *Nano Lett.* 19: 3221–3228. https://doi.org/10.1021/acs.nanolett.9b00758.

2

Synthesis and Phase Engineering of Low-Dimensional TMDs and Related Material Structures

Bijun Tang[1], Jiefu Yang[1], and Zheng Liu[1,2]

[1] *Nanyang Technological University, School of Materials Science and Engineering, 50 Nanyang Avenue, Singapore 639798, Singapore*
[2] *CNRS-International-NTU-Thales Research Alliance (CINTRA), UMI 3288, Research Techno Plaza, 50 Nanyang Drive, Singapore 637553, Singapore*

2.1 Introduction

Since the discovery of monolayer graphene in 2004, two-dimensional (2D) materials have attracted substantial attention owing to their unprecedented properties and potential applications in various fields such as electronics, optoelectronics, spintronics, and catalysis [1–8]. Numerous materials have been developed in the past decade, such as black phosphorus, hexagonal boron nitride (h-BN), and transition metal dichalcogenides (TMDs), just to name a few [1, 9–11]. Particularly, 2D TMDs have emerged as a class of outstanding 2D materials by virtue of their broad range of fascinating properties along with their diverse polymorph crystal structures [12–15].

Although the prevailing phase is primarily determined by the coordination environment and thermodynamics, many TMDs are found to possess multiple phases, including the thermodynamically less stable ones [16, 17]. More importantly, the presence of polymorphism in TMDs leads to very different physicochemical properties. For example, 2H MoS_2 is a typical n-type semiconductor, whereas 1T′ MoS_2 exhibits metallic behavior [6, 18]. In addition to the distinct electronic properties, the engineered 1T′ MoS_2 also demonstrates improved catalytic performance compared to the conventional 2H phase [18, 19]. Recent studies have also revealed many other interesting properties in 1T′ MoS_2, such as magnetism, superconductivity, and ferromagnetism, thus demonstrating the great potential of phase engineering [20–24]. Therefore, engineering 2D TMDs into designated phases is of great importance not only for fundamental research but also for meeting the requirement for different applications.

In the following sections, we introduce the structures (Section 2.2) as well as techniques to synthesize (Sections 2.3) and engineer the phase (Sections 2.4) of 2D TMDs. The chapter is structured as follows: We first describe the typical crystal structure

Two-Dimensional Transition-Metal Dichalcogenides: Phase Engineering and Applications in Electronics and Optoelectronics, First Edition. Edited by Chi Sin Tang, Xinmao Yin, and Andrew T. S. Wee.

of 2D TMDs and summarize their electronic properties. Then, we provide a brief introduction to the commonly used synthesis methods for preparing TMDs' atomic layers. Subsequently, we introduce and discuss several representative phase engineering strategies for 2D TMDs, which are classified into two types – direct synthesis and phase transformation. We highlight the key factors governing the successful synthesis of specific phases, including precursor selection, catalyst, temperature control, and alloying. For phase transformation strategies, emphasis is given to ion intercalation and thermal treatment.

2.2 Structure of 2D TMDs

TMDs have a general formula of MX_2, where M refers to a transition metal spanning from group 4 to group 10, and X refers to a chalcogen (S, Se, or Te). Most TMDs crystallize in graphite-like layered structures. The constituent layers in TMDs covalently bond in the 2D plane, but only weakly couple with neighboring layers through van der Waals (vdWs) interaction, rendering them perfect for obtaining monolayer 2D materials. The metal atoms contribute four electrons to fill the bonding states of TMDs, resulting in metal and chalcogen atoms with oxidation states of +4 and −2, respectively. The surface of the layers is terminated by the lone-pair electrons of chalcogen atoms, and the absence of dangling bonds renders the layers resistant to interactions with ambient species [12]. Given the large number of combinations of transition metal and chalcogen elements, TMDs exhibit a variety of electronic structures and phases.

A single layer of TMD consists of three layers of atoms, including two layers of chalcogen atoms (X), and one layer of metal atoms (M) sandwiched in between, resulting in an X–M–X layer configuration [17, 25]. There are two general phases reported for monolayer TMDs – the trigonal prismatic phase and the octahedral phase (Figure 2.1). The trigonal prismatic phase with hexagonal symmetry (the D_{3h} group) corresponds to the trigonal prismatic coordination of the metal atoms. It is also known as 1H phase – for monolayer TMDs – or 2H phase – for bulk crystals. The octahedral phase with tetragonal symmetry (the D_{3d} group) corresponds to the octahedral coordination of metal atoms, which is also termed the 1T phase in the context of monolayer TMDs. Unlike the 1H phase, where the X atoms are vertically aligned along the z-axis, i.e. AbA stacking, the letters in capital and lowercase represent M and X atoms, the X atoms in the 1T phase are shifted, leading to an AbC stacking sequence. For some TMDs, such as WTe_2, the 1T phase is not thermodynamically stable, and the distorted octahedral structure is demonstrated, which is conventionally known as the 1T′ phase. It is worth mentioning that apart from the single-layer arrangement of the M and X atoms, different stacking sequences of layers can also result in different phases. In addition to the 2H phase with a stacking sequence of AbA BaB, stacking the 1H layer in AbA CaC BcB sequence can yield a rhombohedral phase denoted as 3R [26]. With recent advances in characterization techniques such as annular dark-field scanning transmission electron microscopy (ADF-STEM), these crystal phases can be clearly resolved and differentiated, providing a solid foundation for phase engineering of 2D TMDs at the atomic level [27, 28].

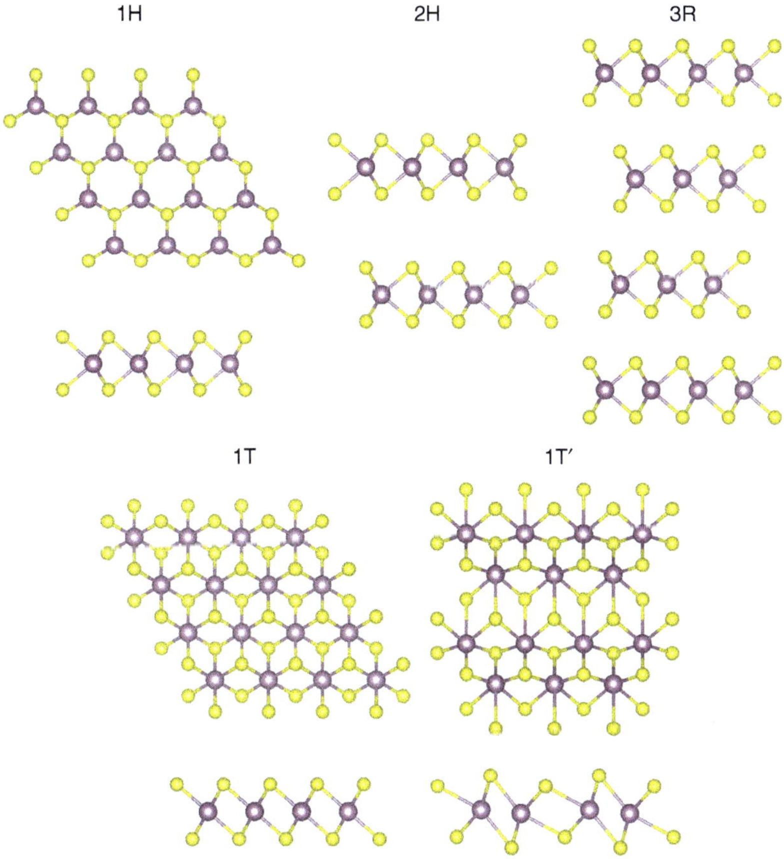

Figure 2.1 Five reported structure prototype of TMDs. Yellow, chalcogen atoms; purple, transition metal atoms.

The electronic structure of TMDs depend sensitively on the coordination environment, i.e. filling of the d-orbitals of the transition metal as the p-orbitals of chalcogen atoms are located at a much lower energy level than the Fermi level. For the 1H phase, the d-orbital of the transition metal splits into three different energy levels: d_{z^2}, $d_{x^2-y^2,\ xy}$, and $d_{xy,\, yz}$, as shown in Figure 2.2a. The first two levels possess a sizable energy gap of ~1 eV. On the other hand, the d-orbital of the octahedrally coordinated transition metal, i.e. 1T phase, splits into two degenerate states: $d_{xy,\, xz,\, yz}$ – collectively referred to as t_{2g} – and $d_{x^2-y^2, z^2}$ – collectively referred to as e_g, as demonstrated in Figure 2.2b. A maximum of six electrons can fill the e_g orbital.

The variation of the d-electron count in the transition metal, from group 4 to group 10, together with its coordination environment, has given rise to diverse electronic and magnetic properties of TMDs, as summarized in Table 2.1 [1, 12, 30]. Metallic behavior is generally demonstrated for TMDs with partially filled orbitals such as 1T-ReS_2. When orbitals are fully occupied, the materials are semiconducting, such as 1H MoS_2 and 1H WS_2. Even though chalcogen atoms have a minor effect on the electronic structure of TMDs, compared to the metal atoms, a trend can still be identified – as the atomic number of chalcogen increases, the bandgap

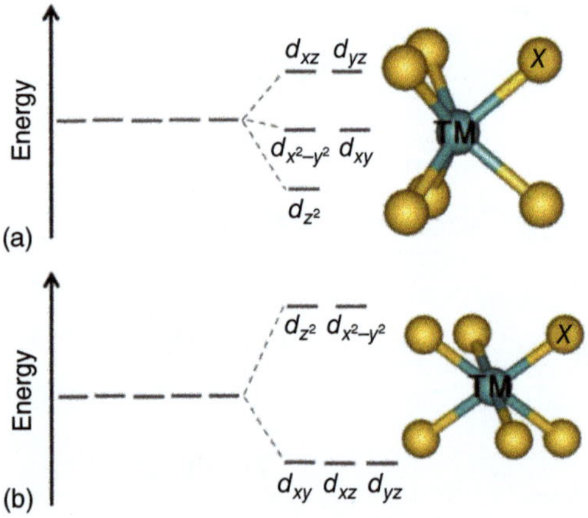

Figure 2.2 The atomic arrangement and corresponding d-orbital splitting of the transition metal in (a) trigonal prismatic phase (1H) and (b) octahedral phase (1T). Source: Reproduced with permission from Gong et al. [29]. Copyright 2013, AIP Publishing.

Table 2.1 Electronic properties of different TMDs.

Group	M	X	Properties
4	Ti, Hf, Zr	S, Se, Te	Semiconducting (E_g = 0.2–2 ev). Diamagnetic
5	V, Nb, Ta	S, Se, Te	Narrow band metals ($\rho = 10^{-4}\ \Omega$ cm) or semimetals. Superconducting. Charge density wave (CDW). Paramagnetic, antiferromagnetic, or diamagnetic
6	Mo, W	S, Se, Te	Sulfides and selenides are semiconducting (E_g = 1 eV). Tellurides are semimetallic ($\rho = 10^{-3}\ \Omega$ cm). Diamagnetic
7	Tc, Re	S, Se, Te	Small-gap semiconductors. Diamagnetic
10	Pd, Pt	S, Se, Te	Sulfides and selenides are semiconducting (E_g = 0.4 eV) and diamagnetic. Tellurides are metallic and paramagnetic. $PdTe_2$ is superconducting

Note: ρ, in-plane electrical resistivity.
Source: Reproduced with permission from [12]. Copyright 2013, Springer Nature.

of the resulting TMDs generally decreases due to the broadening of the d-bands. For example, the bandgap of 1H WS_2, 1H WSe_2, and 1H WTe_2 decreases gradually from 2 to 1.03 eV [31, 32].

2.3 Synthesis of 2D TMDs

Nowadays, the reliable production of high-quality, large-scale, atomically thin 2D layer forms one of the major research fields of TMD. To date, two predominant strategies to obtain TMD monolayers have been employed – one is the top-down method, such as liquid or mechanical exfoliation from bulk crystals, and the other is the bottom-up method, such as physical vapor deposition (PVD) and chemical vapor deposition (CVD) [2, 33, 34].

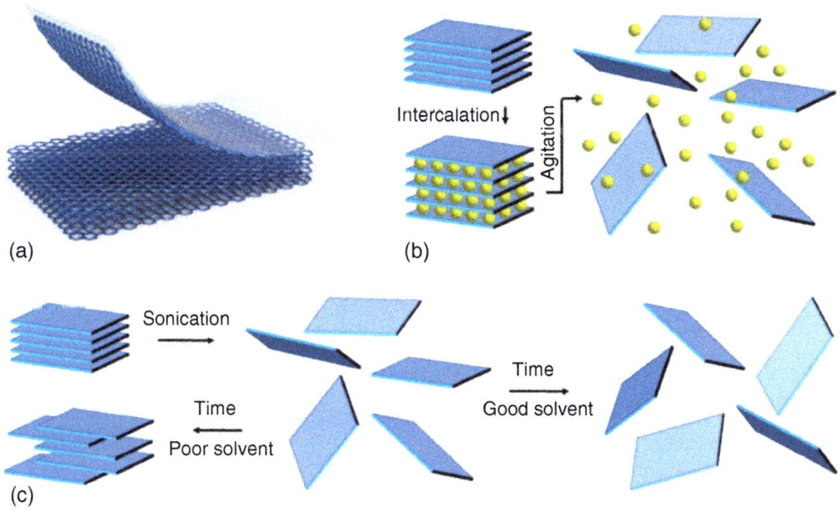

Figure 2.3 Commonly used top-down methods for preparing 2D TMDs. (a) Mechanical exfoliation. Source: Adapted from Hu et al. [35]. (b) Intercalation-based exfoliation. (c) Ultrasonic exfoliation. Source: Reproduced with permission from Nicolosi et al. [36]. Copyright 2013, American Association for the Advancement of Science.

2.3.1 Top-Down Method

Since the first exfoliation of monolayer graphene from graphite in 2004, the mechanical exfoliation method has been recognized as one of the most efficient and universal methods of obtaining fresh atomic layers (Figure 2.3a) [37, 38]. Thereafter, a large number of 2D TMDs were prepared, such as MoS_2 and WS_2 [39–41]. Exfoliated monolayers are highly appreciated by laboratories as they generally possess a high quality, which is ideal for the exploration of their intrinsic properties [42–44]. However, mechanical exfoliation suffers from problems like low yield, poor reproducibility, arduous procedures, and limited size, hindering the application of exfoliated 2D layers.

Liquid-phase exfoliation, mainly including ultrasonic exfoliation and intercalation-based exfoliation, is considered a viable alternative to mechanical exfoliation for producing large quantities of solution-dispersed 2D layers (Figure 2.3b,c) [36, 42, 45]. With different solvents, a variety of 2D TMDs have been obtained by direct sonication of layered crystals, such as MoS_2, $MoSe_2$, $MoTe_2$, WS_2, $TaSe_2$, and $NbSe_2$ [36, 42]. Notably, whether the solvent can overcome the cohesive energy between the neighboring layers is the most important consideration when applying direct sonication techniques. Solvents must be chosen to have surface energies comparable to those of the material. However, the solvent choices to date are still limited. Moreover, the low yield of monolayers and the relatively small size of samples prepared by ultrasonic exfoliation limit their further application. Intercalation-based exfoliation, in which foreign species are intercalated into the interlayers of bulk crystals to increase the interlayer spacing, is recognized as a mature technique for producing 2D materials with high-yield monolayers [12, 36].

So far, a host of intercalation species have been successfully applied, including cations, anions, and molecules [46–50].

It should be noted that all of the abovementioned exfoliation techniques are only applicable to those bulk crystals that have layered structures and are subject to the availability of crystals. Chemical vapor transport (CVT) is one of the most commonly used methods to obtain TMD crystals, in which a transport agent is used to transport the metal and chalcogen elements across a thermal gradient under vacuum [51, 52]. This process generally requires days or even weeks to complete.

2.3.2 Bottom-Up Method

Unlike the top-down method, which requires both crystal formation and exfoliation steps, bottom-up methods, based on vapor deposition, provide great opportunities to directly synthesize large-scale atomically thin 2D TMD layers, which is crucial for meeting the requirement of industrial applications, especially in electronics and optoelectronics. Vapor deposition normally takes place in the tube furnace. Precursors containing the element of target 2D materials are evaporated and reacted at high temperatures, and then transported onto the substrate where they nucleate, grow, and are ultimately formed into targeted 2D flakes. Common vapor deposition techniques include PVD and CVD.

Although both PVD and CVD have been successfully applied to produce high-quality and large-scale 2D materials, the PVD method is limited by the choice of precursors. Nowadays, the CVD technique has gradually become the dominant method for preparing various kinds of 2D materials. Numerous TMD layers such as MoS_2, WS_2, $MoSe_2$, WSe_2, $MoTe_2$, and WTe_2 as well as their alloys and heterostructures have been successfully grown on the silicon substrate and explored on other substrates with CVD, as displayed in Figure 2.4a [53–58].

A typical CVD setup for the synthesis of 2D TMD is illustrated in Figure 2.4b. CVD is a chemical process involving multiple parameters, such as reaction temperature, reaction time, types of precursors, and carrier gas flow rate [3, 59]. This multi-variable nature has given rise to great tunability when synthesizing 2D TMDs. In a typical CVD process, by adjusting the related growth parameters, the thickness, size, and shape of 2D flakes can be easily controlled [3, 14, 60]. Recently, the CVD method has also demonstrated great potential in phase engineering, which will be detailed in the later part.

2.4 Phase Engineering of 2D TMDs

As discussed in Section 2.2, the crystal phase plays a vital role in determining the fascinating physicochemical properties of 2D TMDs. Even with the same composition, different phases might lead to distinct characteristics. For example, MoS_2 and $MoSe_2$ are well-known semiconducting materials that crystallize in the 2H phase; whereas recent studies have revealed that MoS_2 and $MoSe_2$ can also crystallize into metallic 1T′ phase and demonstrate superior performance in

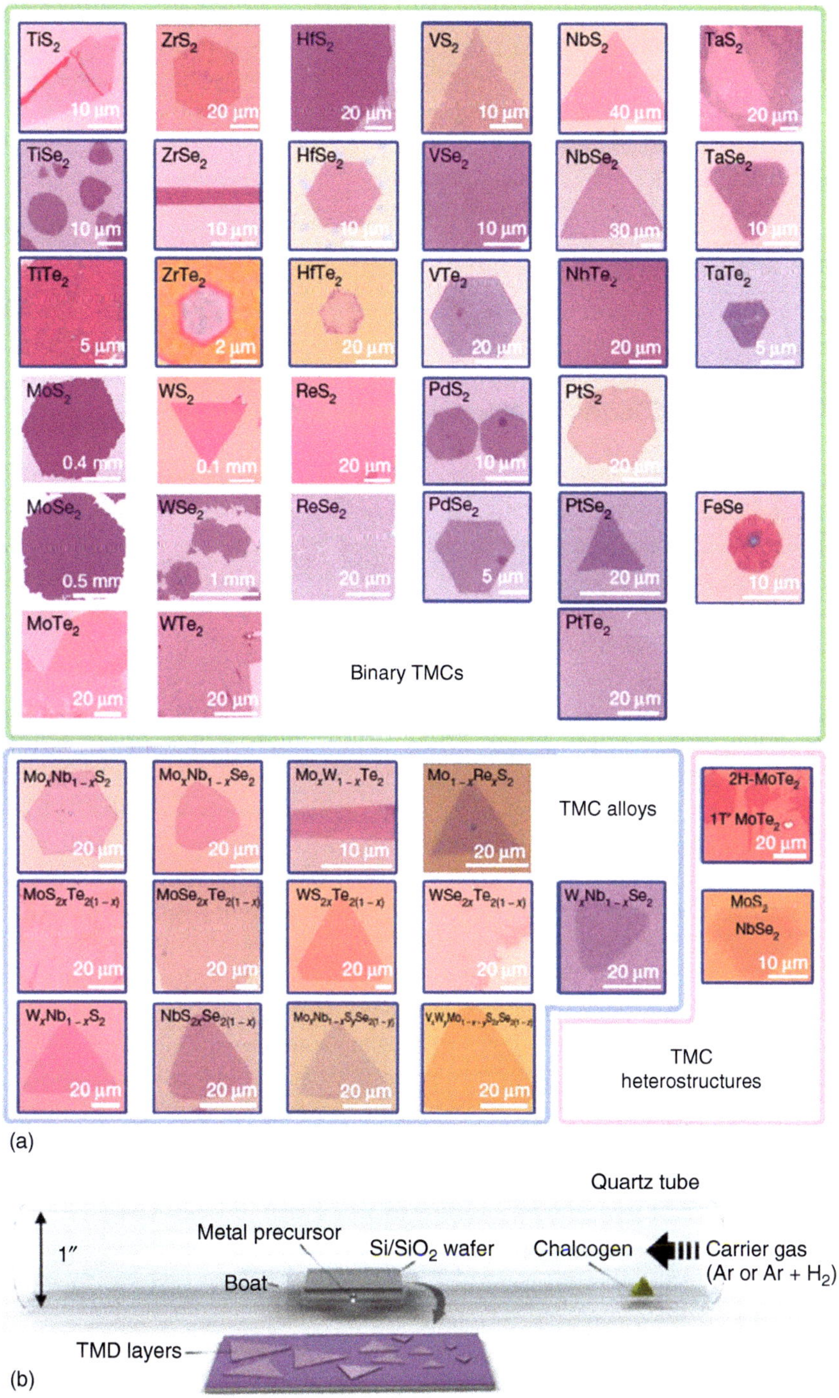

Figure 2.4 The CVD synthesis of 2D TMDs. (a) A collection of 2D TMDs produced by CVD. (b) Schematic of the typical CVD setup for the growth of 2D TMDs. Source: Reproduced with permission from Jiadong Zhou et al. [14], © 2018, Springer Nature.

electrocatalysis [18]. Considering the profound influence of crystal structure on TMDs' properties, the rational experimental design to realize the phase engineering of TMD is thus of great importance in the further exploration and application of 2D TMDs. In this section, we expound on phase engineering strategies, which are classified into two categories: direct synthesis of TMDs with targeted phases and external factor-induced phase transformation.

2.4.1 Direct Synthesis of TMDs with Targeted Phases

The key to realizing direct phase-selective synthesis of TMDs is to control the experimental conditions and establish a favorable environment for the formation of the targeted phase. Here, the key factors governing the successful phase-selective synthesis of 2D TMDs are introduced, and some representative works are discussed in detail.

2.4.1.1 Precursor Selection

Precursors, also known as reactants, form the basis for material synthesis. The proper selection of the type of precursor and the way to supply it into the reaction not only ensures the successful growth of ultrathin TMD layers, but also provides opportunities for phase engineering. To illustrate, by using conventional precursors such as MoO_3, controllable and large-scale synthesis of 2H MoS_2 has been achieved [53, 61–63]. However, the preparation of metastable 1T (1T′) MoS_2 is challenging due to the higher formation of energy and stability issue. Liu et al. reported the usage of K_2MoS_4 as the precursor to selectively prepare 1T′ MoS_2 monolayers [19]. It was found that the stability of 1T′- and 2H-K_xMoS_2 is inversely proportional to the increase in K concentration – when K concentration exceeds 44%, the 1T′ phase becomes more stable than the 2H phase, as illustrated in Figure 2.5a. K_2MoS_4 was thus selected to realize the K-enriched CVD growth of 1T′ MoS_2. Specifically, under a reducing/inert atmosphere, K_2MoS_4 undergoes two different reactions, leading to the inclusion/exclusion of K in the reaction product of MoS_2, thereby providing the desirable phase tunability. By tuning the K concentration by adjusting the reaction gas atmosphere (e.g. H_2 concentration and reaction temperature), monolayer 1T′ MoS_2 was selectively obtained, which also demonstrated enhanced hydrogen evolution reaction (HER) performance (Figure 2.5b,c). Apart from MoS_2, this approach has also been applied to grow 1T′ WS_2 monolayers with K_2WS_4 used as the precursor.

The importance of incorporating K atoms into the fabrication of metastable 1T′ TMD crystals is also reported by Yu et al. and Lai et al., respectively [18, 64]. Large-scale synthesis of metastable 1T′-phase group VIB TMDs, including WS_2, WSe_2, MoS_2, $MoSe_2$, $WS_{2x}Se_{2(1-x)}$, and $MoS_{2x}Se_{2(1-x)}$, has been achieved by using K_2WO_4 and K_2MoO_4 as the metal precursors. Some intriguing physical properties were discovered in these metastable TMDs. For example, thickness-dependent intrinsic superconductivity was revealed in 1T′ WS_2 owing to the high intrinsic carrier concentration and the semi-metallic nature of 1T′ WS_2, with critical transition temperatures of 8.6 and 5.7 K for a thickness of 90.1 nm and the monolayer, respectively.

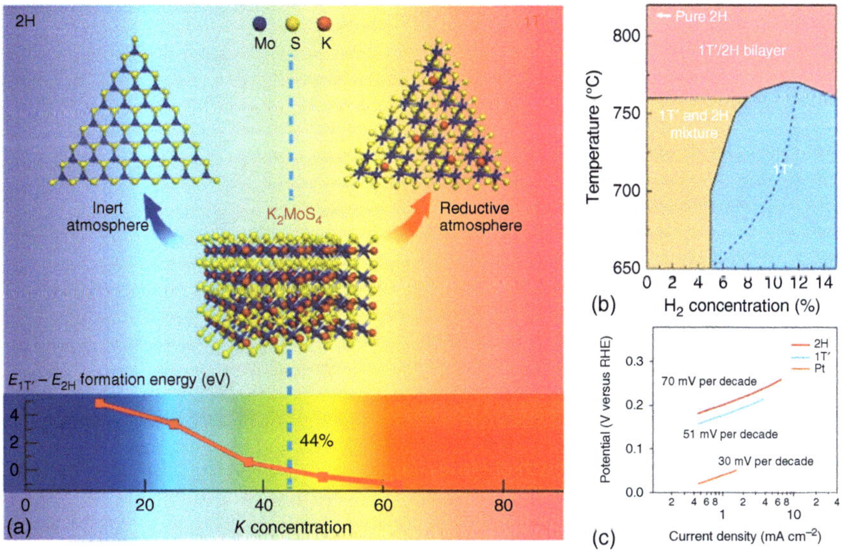

Figure 2.5 Phase-selective growth of MoS_2. (a) Schematic for the phase-selective growth of 2H and 1T′ MoS_2 using K_2MoS_4 as the precursor. (b) Phase diagram of MoS_2 with respect to the growth temperature and H_2 concentration. (c) Tafel plots of 1T′ and 2H MoS_2, suggesting enhancing HER performance of the 1T′ phase. Source: Reproduced with permission from Liu et al. [19]. Copyright 2018, Springer Nature.

In addition to transition metal sulfides and selenides, researchers have also devoted great effort to finding suitable precursors in order to engineer the phase of transition metal tellurides and explore novel phenomena. A typical example is the work reported by Zhang et al., in which the profound influence of precursor stoichiometry on the sample phase is demonstrated [65]. The phase of the CVD-grown $MoTe_2$ thin film can be simply adjusted by changing the metal precursor between MoO_3 and $MoO_{2-2.5}$, as summarized in the phase diagram shown in Figure 2.6a. For example, 2H and 1T′ $MoTe_2$ were obtained via the direct tellurization of $MoO_{2-2.5}$ and MoO_3 at 650 °C, respectively, which manifested distinct Raman fingerprints (Figure 2.6b). By tuning the stoichiometry of the precursor, the defect densities in the obtained $MoTe_2$ vary, thereby resulting in different stable phases. It is noteworthy that the precursor-controlled, phase-selective growth provides an excellent opportunity for the phase-patterned growth of $MoTe_2$ atomic layers (Figure 2.6c), which is conducive to constructing high-performance, field-effect transistors (FETs) and arrays of logic devices.

On top of the phase transition between the 2H and 1T′ phase, it is also possible to obtain 3R $MoTe_2$ with proper precursor selection. In a recent study by Yang et al., $MoCl_5$ was adopted to synthesize 3R-$MoTe_2$ (Figure 2.7a) [66]. Compared to other Mo sources such as metal Mo and MoO_3, $MoCl_5$ has a much lower melting point; moreover, the binding energy of Mo–Cl is also lower than that of Mo–O, which is favorable for the formation of Mo–Te at a relatively low temperature, thereby enhancing the controllability of growing 3R $MoTe_2$.

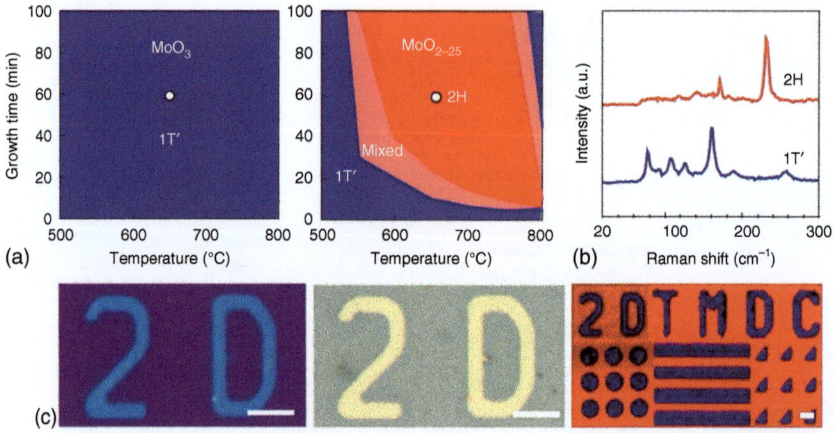

Figure 2.6 Phase-selective growth of $MoTe_2$. (a) Phase diagram of the CVD-grown $MoTe_2$ thin film as a function of tellurization temperature and growth time with MoO_3 and $MoO_{2-2.5}$ as precursors. (b) Typical Raman spectra of 2H and 1T′ $MoTe_2$. (c) Phase-patterned growth of $MoTe_2$ atomic layers: the optical image of the MoO_x pattern before (left) and after (middle) tellurization, in which the precursor used for the "2D" pattern and the background are MoO_3 and $MoO_{2-2.5}$, respectively. Raman mapping images of the hetero-phase pattern are shown in the right panel. Scale bars: 5 μm. Source: Reproduced with permission from Zhang et al. [65], © 2019, Springer Nature.

It should be noted that apart from the metal precursor, the chalcogen precursor plays an essential role in the phase engineering of 2D TMDs as well. For example, Te flux was found to have an extraordinary impact on the CVD-grown $MoTe_2$ phases, as reported by Zhou et al. [67]. With sufficient Te supply, 2H $MoTe_2$ will be obtained. In contrast, insufficient Te vapor will lead to the formation of 1T′ $MoTe_2$ (Figure 2.7b).

2.4.1.2 Catalyst

Catalysts, such as alkali metal halides, have recently demonstrated great potential in the preparation of a wide range of 2D TMDs, by reducing the melting points of precursors and lowering the reaction barrier via the formation of intermediate products [14]. On top of the stable phase, introducing the catalyst into the synthesis reaction can also help to realize the phase engineering of 2D TMDs. For example, by introducing iodine into the reaction, a few layers of metastable 1T′-$MoTe_2$ with high crystallinity and uniformity can be obtained, as reported by Zhang et al. [68]. Iodine atoms mainly contribute to three aspects of the growth process – first, they change the length of the Mo—Te bond, making 1T′ $MoTe_2$ more stable than the 2H phase; second, they induce the formation of volatile intermediates like MoO_2I_2, which is important for the controllable supply of Mo and Te, as well as the subsequent nucleation of 1T′ $MoTe_2$ crystal flakes; third, they can be easily removed thanks to their high volatility at elevated temperatures, minimizing residue in the $MoTe_2$ flakes. In another report, KCl was incorporated into the CVD process to tune the phase of the grown TaS_2 via the intercalation of K^+ between TaS_2 layers, as illustrated in Figure 2.8a [69]. By controlling the TaS_2:KCl ratio, the phase of the deposited 2D layers can be effectively changed from the 1T phase to the 2H phase (Figure 2.8b).

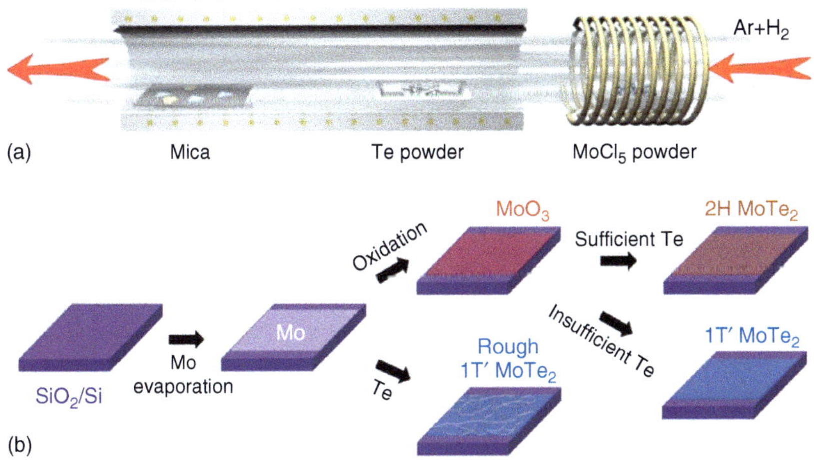

Figure 2.7 (a) Schematic illustration of the CVD setup for synthesizing 3R $MoTe_2$ using $MoCl_5$ as the precursor. Source: Yang et al. [66]/Reproduced with permission from John Wiley & Sons. (b) Schematic illustration of the Te flux effect on the growth of 1T′ and 2H $MoTe_2$. Source: Zhou et al. [67]/Reproduced with permission from John Wiley & Sons.

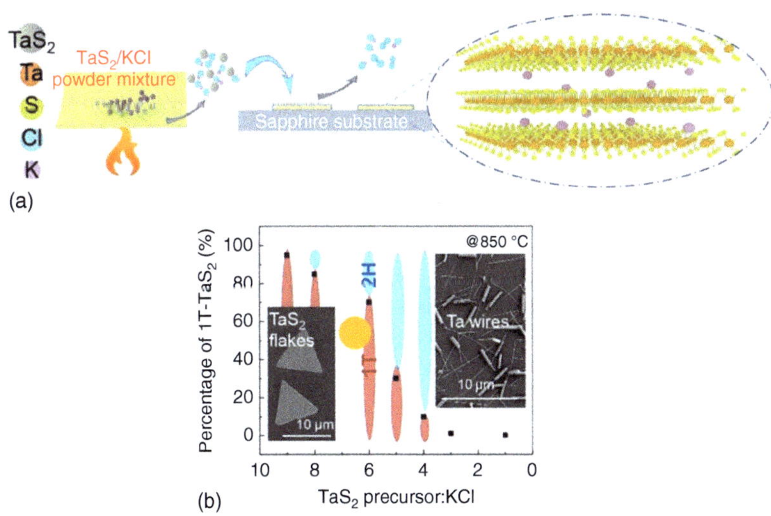

Figure 2.8 (a) A diagram demonstrating the proposed KCl-enhanced deposition process. (b) Phase evolution of the TaS_2 by tuning the precursor: catalyst ratio. Source: Reproduced with permission from Zhao et al. [69], © 2019, Royal Society of Chemistry.

In addition to the single catalyst, synergistic catalysts have recently demonstrated great potential in phase engineering. A representative example is the work by Lin et al., in which a metal oxide (i.e. Fe_3O_4), together with an alkali metal halide (i.e. NaCl), was introduced as a synergistic catalyst in the CVD process to grow 1T WS_2 (Figure 2.9a) [70]. With the help of the synergistic catalyst, 1T WS_2 can nucleate in the early stage of the growth to form a unique butterfly-shaped single crystal with

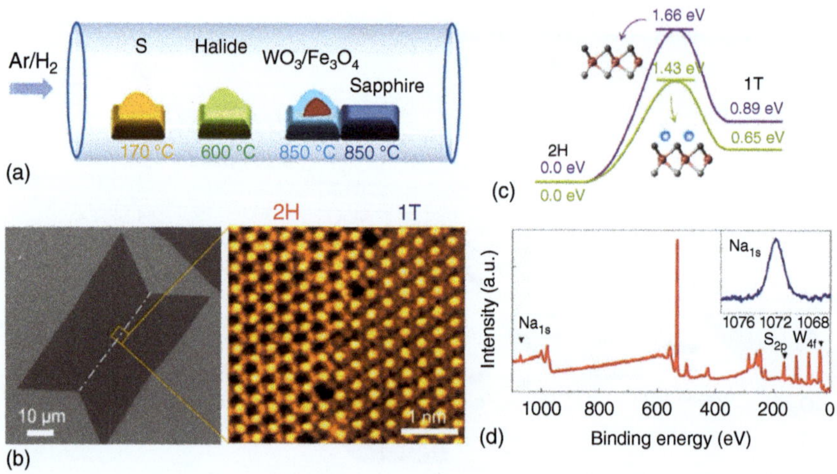

Figure 2.9 (a) Schematic illustration of the CVD growth of 1T WS_2 with Fe_3O_4 and NaCl being used as the synergistic catalysts. (b) SEM image of the butterfly-shaped WS_2 atomic layer with 1T phase on one side and the 2H phase on the other. Scale bar: 10 μm. (c) Calculated formation energies of 2H and 1T WS_2 with (green) and without (purple) the adsorption of alkali metal atoms. (d) XPS survey spectrum of WS_2 nuclei formed in the early stage of nucleation, indicating Na signal on the WS_2 nuclei. Source: Lin et al. [70]/from American Chemical Society.

the 1T phase on one side and the 2H phase on the other (Figure 2.9b). Density Functional Theory (DFT) calculations revealed that the presence of alkali metal atoms on the WS_2 surface could considerably reduce the formation energy of the 1T phase by 0.24 eV (Figure 2.9c). The X-ray photoelectron spectroscopy (XPS) measurements on the nuclei indicate the formation of the volatile intermediate product Na_xWO_y (Figure 2.9d). The inhomogeneous charge doping induced by the Na atoms should be responsible for the formation of nuclei with two distinct phases. Although it is clear that Fe_3O_4 is conducive to the formation of the intriguing 1T/2H butterfly-shaped WS_2 lattice, the underlying mechanism remains elusive.

2.4.1.3 Temperature Control

In addition to the proper selection of precursors and the addition of catalysts, temperature control is also of great importance in synthesizing 2D crystals with designated phases. Temperature control is a straightforward approach for 2D TMD phase engineering, as it is a fundamental parameter governing the thermodynamics and kinetics of phase transitions. Generally speaking, phase selectivity achieved by temperature control exploits the differences in the thermodynamic stabilities of the phases, by hindering or promoting certain kinetic pathways for phase transition.

By controlling the growth temperature, Sung et al. achieved the phase-selective CVD growth of 1T′ and 2H $MoTe_2$ with MoO_3 and Te being used as precursors [71]. $MoTe_2$ possesses a relatively small ground-state energy difference between 1T′ and 2H phases, which is ideal for phase transformation. It was found that a higher growth temperature of 710 °C tends to give rise to 1T′ $MoTe_2$ exhibiting rectangular

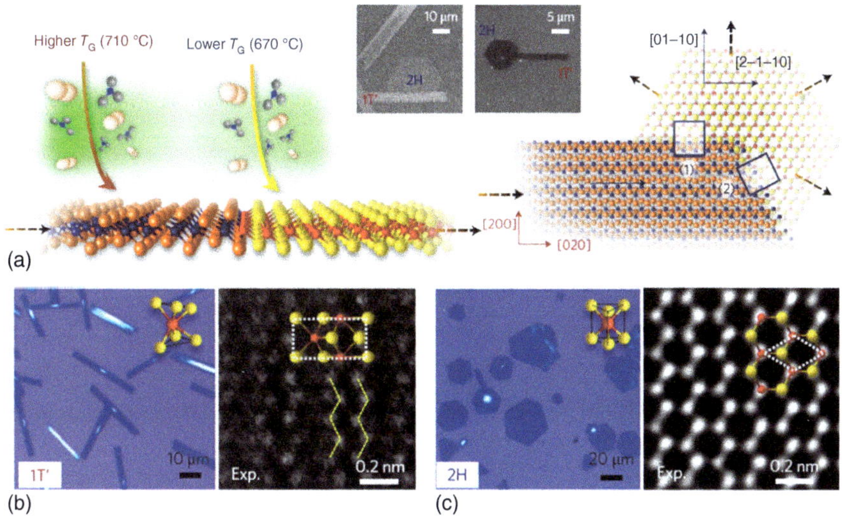

Figure 2.10 (a) Scheme of the sequential growth for coplanar heteroepitaxy of 1T′/2H $MoTe_2$ polymorphs. (b) Optical image of 1T′ $MoTe_2$ crystals in rectangular (monoclinic) facets and the corresponding high-magnification HAADF-STEM image. (c) Optical image of 2H $MoTe_2$ crystals in hexagonal shapes and the corresponding high-magnification HAADF-STEM image. Source: Reproduced with permission from Sung et al. [71], ©2017, Springer Nature.

or trapezoidal shapes, while a lower growth temperature of 670 °C, generally leads to the formation of 2H $MoTe_2$ with hexagonal or triangular shapes (Figure 2.10).

Such temperature-governed phase tunability of $MoTe_2$ crystals is also reported by Keum et al. with the flux method [72]. When the temperature exceeds 500 °C, $MoTe_2$ begins to transform from the stable 2H phase into the 1T′ phase. It is worth mentioning that the phase transition between the two phases is reversible, as slowing the cooling of 1T′ phase can lead to the recovery of the 2H phase. Empante and colleagues have particularly investigated the effect of cooling rate in the CVD process on phase formation [73]. It was revealed that naturally cooling, quenching at 350 °C, and quenching at 450 °C will lead to the formation of 2H, 1T′, and 1T $MoTe_2$, respectively (Figure 2.11a). Theoretical studies further supported the experimental findings. Figure 2.11b demonstrates the temperature-dependent free energies of the three phases, and the one with the lowest free energy at each temperature is thermodynamically preferred.

Besides growth temperature and cooling rate, substrate temperature is another important temperature factor to consider in phase engineering. A typical example is the work by Chen et al., in which the 1T′ and 1H WSe_2 were selectively obtained by controlling the substrate temperature during the molecular beam epitaxy (MBE) process [74]. It is illustrated that the optimal substrate temperature for monolayer 1T′ WSe_2 is 280 °C, whereas that for 1H phase is above 400 °C. The prominent effect of substrate temperature is also reported by Nakata et al. in the study of $NbSe_2$ [75]. They observed that optimal ranges of substrate temperature for obtaining

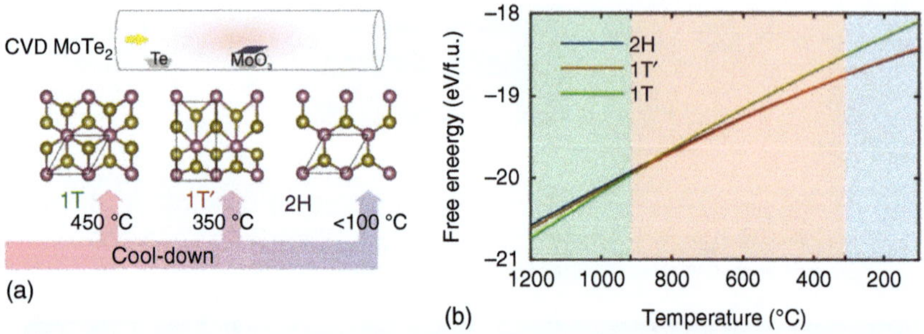

Figure 2.11 (a) Effect of cooling rate on the phase formation of CVD-grown $MoTe_2$. (b) Calculated free energies of 2H, 1T′, and 1T $MoTe_2$ with respect to different temperatures. Source: Reproduced with permission from Empante et al. [73]. Copyright 2017, American Chemical Society.

metallic 2H and insulating 1T phases are 500–515 and 580–600 °C, respectively (Figure 2.12a).

In fact, not only the intrinsic structure of a layer, but temperature control has also recently proven essential in determining the stacking sequences of 2D TMDs. For instance, Zeng et al. reported the PVD of few-layer WS_2 and WSe_2 on SiO_2/Si substrate [76]. They found that the high-temperature deposition region is more favorable for the growth of 2H flakes exhibiting a hexagonal shape, whereas the 3R phase manifesting a triangle shape tends to be obtained in the lower temperature region (Figure 2.12b). The different layered shape was believed to originate from the varied concentration ratios of W and S/Se atoms in the vapor precursor, thereby giving rise to different crystal edge terminations, growth speeds, and morphology transitions.

Recent studies have also shown that in addition to TMDs, temperature has a prominent effect on the phase structure of numerous transition metal chalcogenides. For example, phase-selective growth of ferromagnetic Cr_5Te_8 was reported by Tang et al. very recently. The thermodynamically more favorable trigonal Cr_5Te_8 was found to dominate at a higher temperature, whereas the structurally more distorted monoclinic phase was favorable at a lower temperature [8]. Moreover, the crystal phase of CVD-grown FeTe was also found to be temperature-dependent [77, 78]. The reaction of $FeCl_2$ and Te at higher temperatures led to the formation of ferromagnetic hexagonal FeTe, whereas the same reaction performed at lower temperatures led to the crystallization of antiferromagnetic tetragonal FeTe.

2.4.1.4 Alloying

Alloying is another effective method to engineer the structure and phase of 2D TMDs as well as the associated properties. Recently, alloying 2D TMDs has attracted increasing research attention owing to their more interesting properties via stoichiometric variation [79–83]. Compared to their well-studied binary counterparts, the additionally introduced degree of freedom in 2D TMD alloys may give rise to more appealing phenomena like bandgap engineering and phase transformation. As discussed above, the phase and associated electronic properties of 2D TMDs depend strongly on the d-orbital electrons and coordination environment. Alloying different

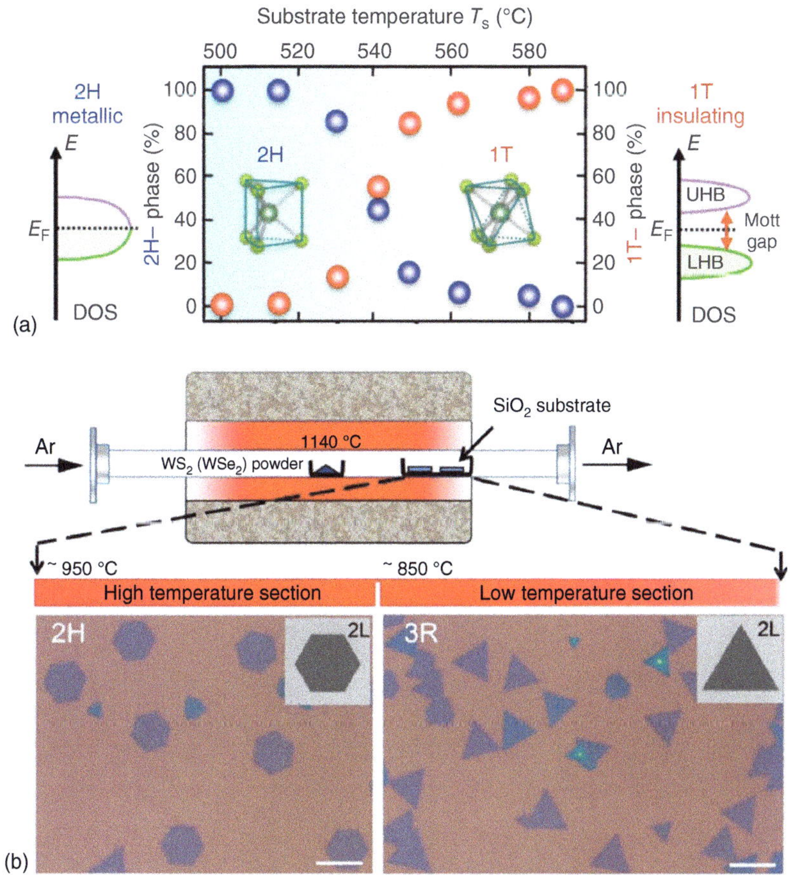

Figure 2.12 (a) Surface area ratio between the 2H and 1T domains, with respect to the substrate temperature. The left and right panels display the schematic density of states (DOS) of monolayer 2H and 1T $NbSe_2$, respectively. Source: Nakata et al. [75]/Springer Nature/Licensed under CC BY 4.0. (b) Schematic illustration of the PVD setup for the controlled growth of 3R and 2H TMDs layers. 3R and 2H WS_2 with different shapes are obtained in different deposition regions. Source: Reproduced with permission from Zeng et al. [76], © 2019, John Wiley & Sons.

TMDs could therefore provide more possibilities and unlock more potential of 2D TMDs. For instance, effective bandgap engineering has been attained in various alloying systems such as $MoS_{2(1-x)}Se_{2x}$, $WS_{2x}Se_{2(1-x)}$, $Mo_{1-x}W_xS_2$, $ReS_{2x}Se_{2(1-x)}$, etc., by manipulating the chemical composition of each component [84–89], as shown in Figure 2.13. In addition, n-type to p-type carrier-type modulation has been realized in $Mo_{1-x}W_xSe_2$ and $WS_{2x}Se_{2-2x}$ systems, which also exhibits greatly enhanced photoluminescence intensity [86, 90].

In fact, by alloying two TMDs with distinct structures, the phase of the TMD alloy can also be modulated by adjusting the chemical composition. In 2017, Yu et al. reported the realization of semiconductor (2H)-metal (1Td) phase transition in the $WSe_{2(1-x)}Te_{2x}$ single crystal prepared by CVT (Figure 2.14a) [92].

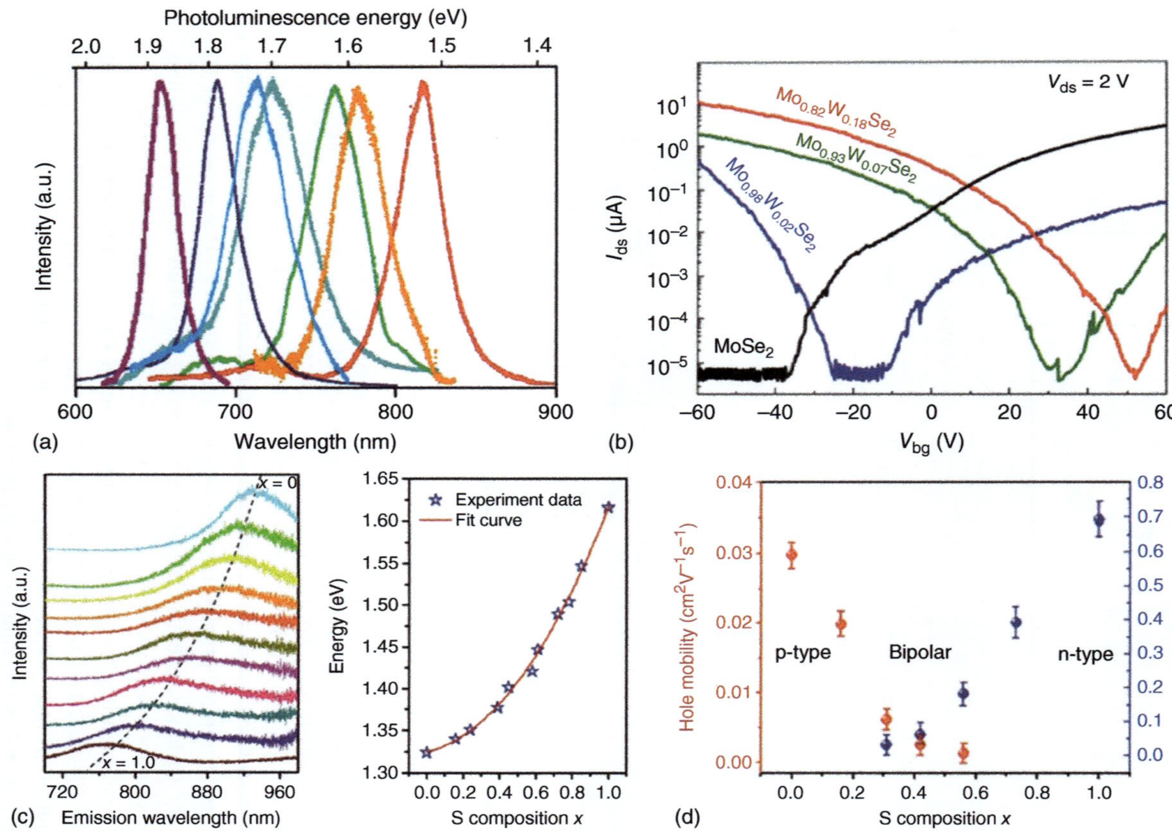

Figure 2.13 Bandgap engineering and carrier type modulation achieved in 2D TMDs alloys. (a) Normalized photoluminescence (PL) spectra of a series of $MoS_{2x}Se_{2(1-x)}$ with different compositions. Source: Reproduced with permission from Mann et al. [85]. Copyright 2013, Wiley-VCH. (b) Transfer characteristics of $Mo_{1-x}W_xSe_2$ with different W concentrations, indicating a n-type to p-type transition as W content increases. Source: Reproduced with permission from Li et al. [90]. Copyright 2016, Wiley-VCH. (c) PL spectra and corresponding optical bandgap of a series of monolayer $ReS_{2x}Se_{2(1-x)}$ with different compositions. (d) Field-effect carrier mobility as a function of the composition of $ReS_{2x}Se_{2(1-x)}$, revealing a systematic carrier type modulation. Source: Reproduced with permission from Cui et al. [91]. Copyright 2017, Wiley-VCH.

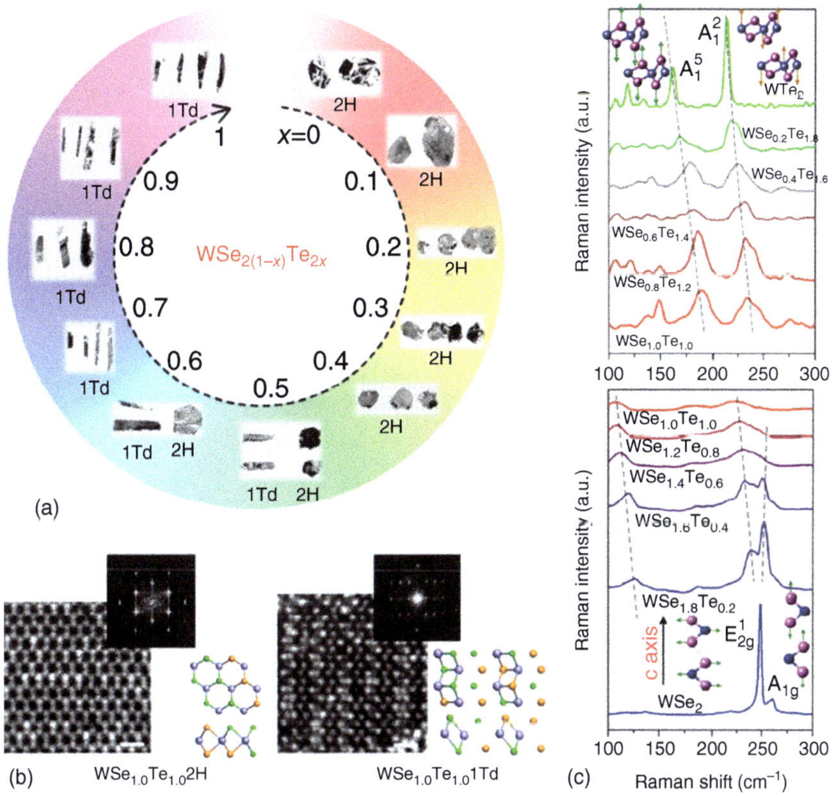

Figure 2.14 (a) Photographs of the CVT-grown $WSe_{2(1-x)}Te_{2x}$ single crystals. 2H and 1Td types TMD alloys exhibit circularly platy and bar shapes, corresponding to hexagonal and orthorhombic phases, respectively. (b) Z-contrast STEM images exhibiting the atomic structure of 2H and 1T $WSe_{1.0}Te_{1.0}$ monolayers. (c) Raman intensity of monolayer 1Td $WSe_{2(1-x)}Te_{2x}$ ($x = 0.6-1$) and 2H $WSe_{2(1-x)}Te_{2x}$ ($x = 0-0.6$) alloys with different Te compositions. Source: Reproduced with permission from Yu et al. [92], © 2017, John Wiley & Sons.

By controlling the Te concentration, the phase of $WSe_{2(1-x)}Te_{2x}$ can be effectively tuned from the 2H phase to the coexistence of 2H and 1Td phase, and eventually to the pristine 1Td phase, as confirmed by atomic-resolution STEM (Figure 2.14b). Raman spectroscopy analysis further confirmed the composition-dependent phase transformations in alloy (Figure 2.14c). A similar strategy has also been adopted by Ho et al. to engineer the phase of WSe_2. With a large amount of Cr dopants, a local phase transition from the 2H to the 1T phase was realized [93].

Considering the complex exfoliation process, followed by CVT and poor controllability over the sample size and thickness, Tang et al. adopted the facile CVD method and achieved the phase-controlled synthesis of monolayers 1H and 1T′ $WTe_{2x}S_{2(1-x)}$ (Figure 2.15a) [31]. Owing to different structure symmetry, distinct sample morphologies of 1H and 1T′ samples were obtained (Figure 2.15c). Like $WSe_{2(1-x)}Te_{2x}$, $WTe_{2x}S_{2(1-x)}$ alloy also manifests phase-dependent electronic

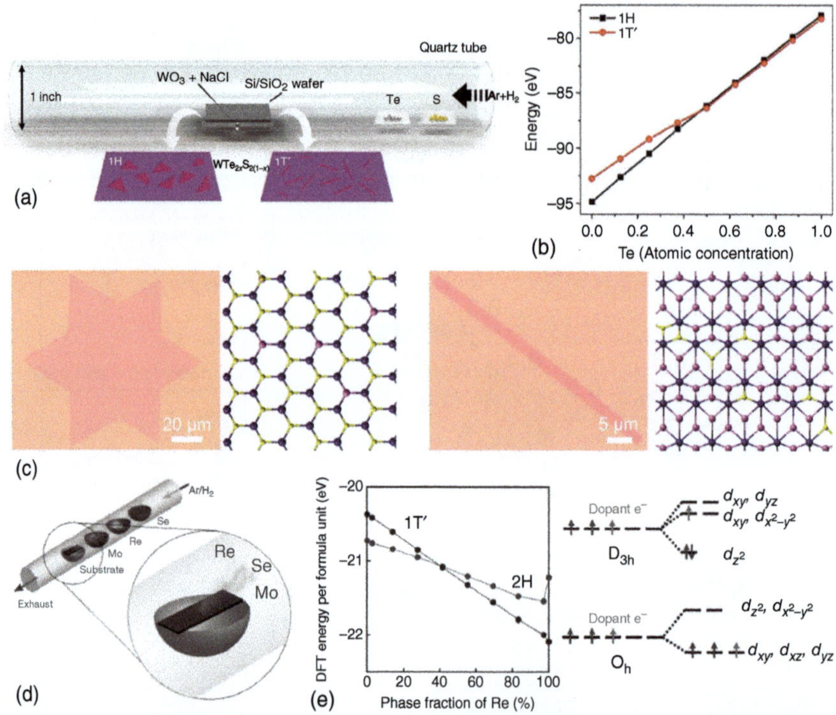

Figure 2.15 (a) Schematic representation of the experimental setup for synthesizing $WTe_{2x}S_{2(1-x)}$ alloy. (b) Energy shift of the $WTe_{2x}S_{2(1-x)}$ monolayer as a function of Te concentration x. (c) OM image and atomic structures of 1H and 1T′ $WTe_{2x}S_{2(1-x)}$ monolayers. Source: Reproduced with permission from Tang et al. [31], © 2019, John Wiley & Sons. (d) Schematic of the CVD setup for the growth of Re-doped $MoSe_2$. (e) Energy of the 2H and 1T′ $MoSe_2$ with different Re doping content. For 2H and 1T′ phases, the splitting of the degenerate 4d orbitals of Mo is demonstrated with the filling of an extra electrons from the Re dopant. Source: Reproduced with permission from Kochat et al. [94]. Copyright 2017, Wiley-VCH.

properties. The n-type semiconducting nature of the 1H phase and the metallic character of the 1T′ phase were verified by FET measurements. By comparing the energy of different phases obtained from first-principles calculations, it was found that 1H phase is more favorable at the small Te concentration and the 1T′ phase is more stable with large Te content, and the critical Te concentration inducing phase transition is around $x = 0.45$ (Figure 2.15b). Different from the conventionally known displacement of metal atoms in the 1T′ phase, an interesting phenomenon was reported in the $WTe_{2x}S_{2(1-x)}$ alloy, i.e. the local displacement of Te atoms from the original 1H lattice sites, attributed to structural distortion. This may provide new insight into the structural formation of 2D TMD alloys.

In another report by Kochat et al., CVD method was also used to engineer the phase of monolayer Re-doped $MoSe_2$ alloy (Figure 2.15d) [94]. DFT calculations revealed that the critical Re concentration is around 0.42, above which the $Mo_{1-x}Re_xSe_2$ alloy is stable in the 1T′ phase. Crystal field theory was used to interpret the underlying mechanism of phase transformation (Figure 2.15e). The replacement of Mo with Re will bring in an extra electron into the $MoSe_2$ lattice. In trigonal prismatic coordination geometry, this extra electron is promoted to a higher energy level; however, the resulting incomplete filling would destabilize the 2H phase. In contrast, this extra electron fills the lower energy state in the octahedral coordination geometry, stabilizing the 1T′ phase. In addition to the enriched electronic phases, novel phenomena such as magnetic ordering were also observed in the $Mo_{1-x}Re_xSe_2$ system in both the 2H and 1T′ phases.

2.4.2 External Factor-Induced Phase Transformation

Apart from the direct synthesis of TMDs with desired phases, phase engineering of 2D TMDs can also be achieved by transforming thermodynamically stable phases into unstable ones, or vice versa, with the aid of external factors such as ion intercalation and thermal treatment. Notably, phase transformation offers an efficient alternative to obtain TMDs in certain phases that may be difficult to attain by conventionally direct synthesis methods.

2.4.2.1 Ion Intercalation

Ion intercalation is one of the most commonly used approaches to induce phase transformation of 2D TMDs and attain their metastable phases. Early in 1983, Py and Haering reported the Li intercalation-induced structural phase transition of 2H MoS_2 [95]. The Mo coordination changed from prismatic to octahedral, attributed to the dramatic increase in the electronic energy of the semiconducting host by charge transfer from the lithium intercalant. Sun et al. employed DFT calculations to investigate the origin of phase transformation in MoS_2 and found that electron doping, accompanied by ion intercalation, can destabilize the crystal structure of the 2H MoS_2, and cause its structural transformation into 1T′ phase through the redistribution of the Mo 4d-orbitals (Figure 2.16a) [96]. Moreover, it is revealed that it is easier for phase transition to take place by increasing the layer thickness, which is further supported by the work of Sun et al. [97]. This dependency should originate from the layer-dependent density of states of 2H MoS_2, leading to the reduced phase stability of 2H MoS_2 along with an increase of layer numbers upon electron doping (Figure 2.16b). Moreover, compared to multilayer MoS_2, monolayer MoS_2 requires a longer reaction time due to its much longer energy barrier for phase transition (Figure 2.16c).

As one of the most widely used intercalation reagents, *n*-butyllithium has been successfully applied to obtain a series of metastable TMDs with enhanced properties. For instance, by locally converting the semiconducting 2H MoS_2 to metallic 1T MoS_2 with *n*-butyllithium, Kappera and colleagues achieved the fabrication of FETs with low-resistance contacts [6]. Specifically, PMMA was used as the mask, and EBL was applied to remove PMMA from the region where 1T conversion was desired. The sample was then immersed into *n*-butyllithium for phase transition to occur in the exposed region. Apart from MoS_2, *n*-butyllithium, together with ball milling,

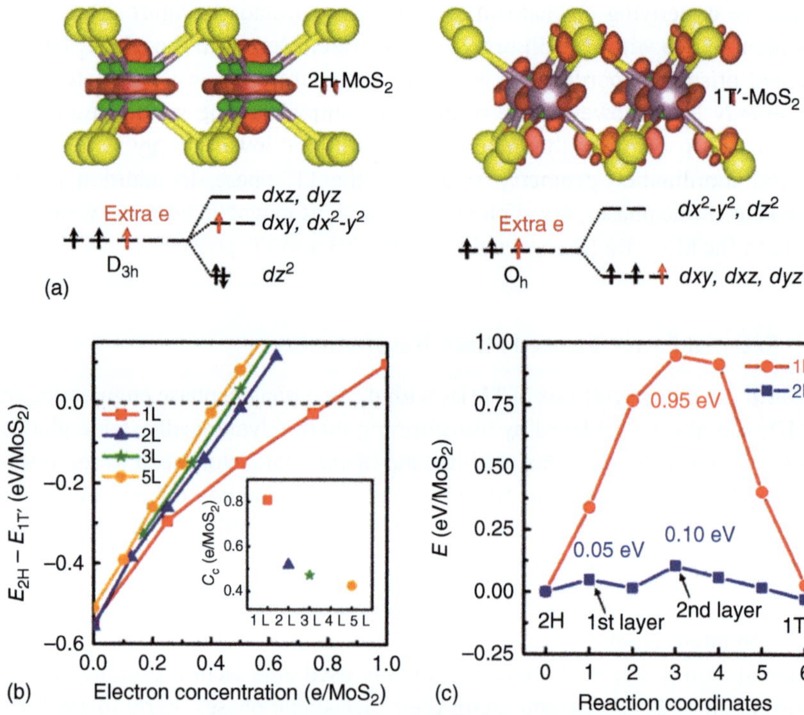

Figure 2.16 (a) Isosurface (0.003 e/Å3) of the charge distributions of 2H (left) and 1T′ (right) MoS_2 doped with 1.00 e/f.u and corresponding d-orbital splitting of Mo 4d orbitals. Source: Sun et al. [96] Springer Nature/Licensed under CC BY 4.0. (b) Layer-dependent energy difference between 2H and 1T′ MoS_2 as a function of injected electron concentration. (c) Phase transition barriers for negatively charged monolayer (red) and bilayer (blue) MoS_2. Source: Reproduced with permission from Sun et al. [97]. Copyright 2018, American Chemical Society.

was also applied to obtain other 1T TMDs, including WS_2, $MoSe_2$, $Mo_{0.5}W_{0.5}S$, and MoSSe, as reported by Tan et al. [98].

Some solvent-free methods have also been proposed in the past. A typical example is the work done by Voiry, in which lithium borohydride ($LiBH_4$) was used as a lithium intercalant to exfoliate bulk MoS_2 powder into 1T phase nanosheets by mixing and heating with MoS_2 powder at 300 °C in an Ar-filled glovebox [99]. The exfoliated 1T MoS_2 nanosheets manifested excellent catalytic performance for hydrogen evolution, with a notably low Tafel slope. By lithiation in an ultrahigh vacuum, Tan et al. also achieved the 2H-1T′ phase evolution in TMDs such as MoS_2 and WS_2 [100]. It was found that 1T′ phase TMDs with superb stability can be obtained by hydrogenating the intercalated Li to form lithium hydride (LiH). DFT calculations revealed that LiH is a good electron donor, which can stabilize the 1T′ phase against 2H conversion by forming significantly enhanced interlayer dipole–dipole interactions.

In addition to Li ions, other intercalants such as Na and K have been employed to engineer the TMD phase as well [24, 101]. For example, K-intercalation-induced 2H to 1T and 1T′ phase transitions were demonstrated by Zhang and colleagues [24].

Remarkably, K intercalation not only caused structural phase transformation but also gave rise to superconducting phase transitions in MoS_2. All three intercalated phases exhibited superconductivity with critical temperatures of 6.9, 2.8, and 4.6 K, respectively. To achieve better control of the intercalant amount, Wang et al. adopted an electrochemical approach to intercalate Na into MoS_2 [101]. It was revealed that an appropriate amount of intercalant could lead to phase transition, whereas excessive concentration would cause structural degradation.

2.4.2.2 Thermal Treatment

Thermal treatment is another effective and convenient post-treatment strategy to induce phase transition in 2D TMDs [102–104]. As one of the most common approaches, thermal annealing has been widely adopted for phase engineering, especially to transform the metastable phase into a thermodynamically more stable phase via atomic rearrangement at high temperatures. The activation energy provided by thermal annealing can overcome the energy barrier of phase transition. After annealing at above 250 °C in an ultrahigh vacuum, 1T to 2H polymorph transition was induced on the surface of bulk 1T-TaS_2, as reported by Wang et al. [104]. Particularly, phase transition was limited to the topmost monolayer, probably due to the easier structural relaxation of surface atoms.

As discussed previously, by using K_2MoO_4 as the metal precursors, Yu et al. have successfully synthesized metastable 1T′ MoS_2 and $MoSe_2$ crystals [18]. Not only that, but they also achieved the restoration of the 2H phase from the 1T′ phase via annealing the sample at 200 °C for two hours in Ar, as evidenced by the Raman and PL spectra of 1T′ MoS_2 before and after annealing (Figure 2.17a,b). For better controllability, laser irradiation was implemented for local phase patterning through a local heating effect. The irradiated flake manifested evident Raman and PL signal changes. The J1 mode signals representing 1T′ MoS_2 were extinct after irradiation, whereas the E_{2g} mode signals could be observed. PL mapping results further indicated the irradiation-induced metallic 1T′ to semiconducting 2H phase transitions.

As an effective local polymorph engineering method, laser-induced phase patterning was also adopted by Cho et al. to fabricate an ohmic heterophase homojunction between semiconducting 2H and metallic 1T′ MoTe2 [102]. The optical images and

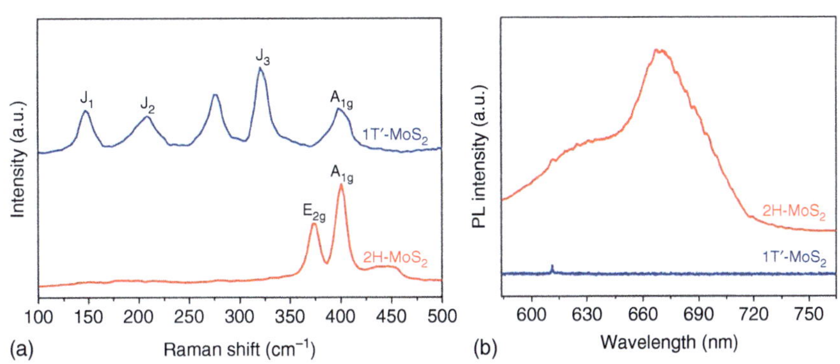

Figure 2.17 (a, b) Raman and PL spectra of 1T′ MoS_2 and 2H MoS_2 obtained by annealing the 1T′ flakes. Source: Reproduced with permission from Yu et al. [18]. Copyright 2018, Springer Nature.

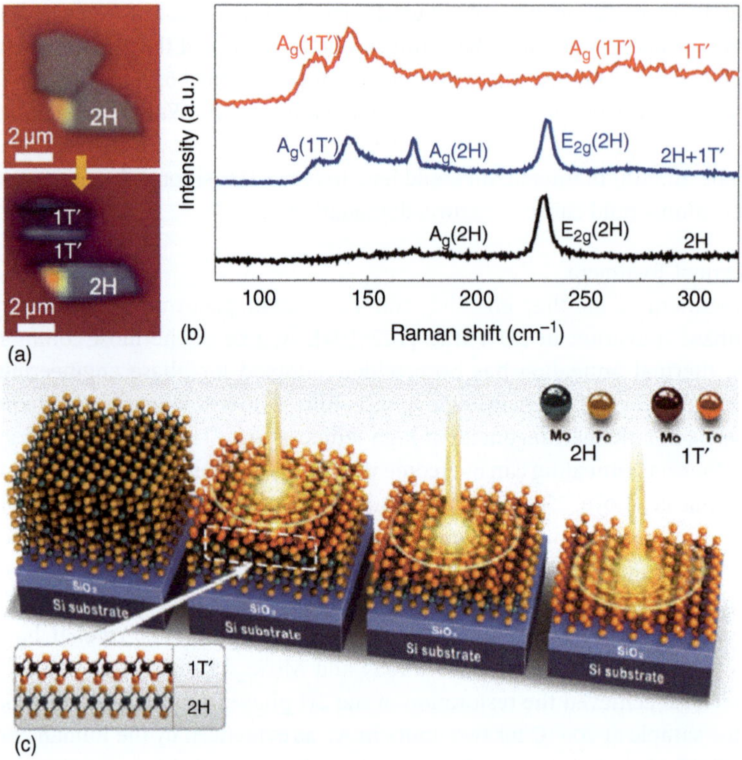

Figure 2.18 (a) Optical images of the mechanically exfoliated few-layer 2H $MoTe_2$ flake before (top) and after (bottom) laser irradiation. (b) The evolution of Raman spectra from 2H phase (bottom) to 1T′ phase (top). (c) Schematic illustration of the laser-irradiation process. Source: Reproduced with permission from Cho et al. [102], © 2015, American Association for the Advancement of Science (AAAS).

the Raman spectra reveal the reduced flake thickness and structural evolution in the irradiated regions (Figure 2.18a,b). The thinning effect occurs along with the phase transition in the flake, which initially takes place at the top layer, as illustrated in Figure 2.18c. During the laser irradiation, the irradiated region in the 2H MoTe2 flake continued to thin until the flake reached a few-layer thickness with the transformed 1T′ phase. The remaining few layers are probably due to the heat-sink effect towards the bottom of the SiO_2 substrate during the irradiation process. The local phase transition in $MoTe_2$ from 2H to 1T′ is believed to originate from the formed Te vacancies at elevated temperatures under laser irradiation.

2.5 Conclusion

Phase engineering is of great research importance for the further exploration and application of 2D TMDs, on account of the profound influence of crystal structure on TMDs' properties. In this chapter, a comprehensive overview of the state-of-the-art research activities in the synthesis and phase engineering of 2D TMDs has been provided, including crystal structures, synthesis techniques, and phase engineering strategies. Different polymorphs of TMDs (i.e. 1H, 2H, 3R, 1T, and 1T′) have been introduced. Various synthesis methods, including top-down approaches (e.g. mechanical exfoliation and liquid-phase exfoliation) and bottom-up approaches (e.g. PVD and CVD), have been summarized. To achieve phase engineering, several strategies have been highlighted and introduced in detail – direct synthesis of TMDs with targeted phases (e.g. precursor selection, catalyst, temperature control, and alloying) as well as external factor-induced phase transformation (e.g. ion intercalation and thermal treatment). Although most studies have focused on Mo and W-based TMDs, extending phase engineering concepts and strategies to other 2D TMDs is expected to lead to more intriguing properties and enhanced performance for various promising applications.

References

1 Bhimanapati, G.R., Lin, Z., Meunier, V. et al. (2015). *ACS Nano* 9: 11509.
2 Chen, Y., Fan, Z., Zhang, Z. et al. (2018). *Chem. Rev.* 118: 6409.
3 Cai, Z., Liu, B., Zou, X., and Cheng, H.-M. (2018). *Chem. Rev.* 118: 6091.
4 Kim, S., Konar, A., Hwang, W.-S. et al. (2012). *Nat. Commun.* 3: 1.
5 Lin, X., Yang, W., Wang, K.L., and Zhao, W. (2019). *Nat. Electron.* 2: 274.
6 Kappera, R., Voiry, D., Yalcin, S.E. et al. (2014). *Nat. Mater.* 13: 1128.
7 Lopez-Sanchez, O., Lembke, D., Kayci, M. et al. (2013). *Nat. Nanotechnol.* 8: 497.
8 Tang, B., Wang, X., Han, M. et al. (2022). *Nat. Electron.* 5: 224.
9 Zhang, K., Feng, Y., Wang, F. et al. (2017). *J. Mater. Chem. C* 5: 11992.
10 Liu, H., Du, Y., Deng, Y., and Peide, D.Y. (2015). *Chem. Soc. Rev.* 44: 2732.
11 Chang, C., Chen, W., Chen, Y. et al. (2021). *Acta Phys. Chim. Sin.* 37: 2108017.
12 Chhowalla, M., Shin, H.S., Eda, G. et al. (2013). *Nat. Chem.* 5: 263.
13 Wang, Q.H., Kalantar-Zadeh, K., Kis, A. et al. (2012). *Nat. Nanotechnol.* 7: 699.
14 Zhou, J., Lin, J., Huang, X. et al. (2018). *Nature* 556: 355.
15 Tang, B., Che, B., Xu, M. et al. (2021). *Small Struct.* 2: 2000153.
16 Chen, Y., Lai, Z., Zhang, X. et al. (2020). *Nat. Rev. Chem.* 4: 243.
17 Voiry, D., Mohite, A., and Chhowalla, M. (2015). *Chem. Soc. Rev.* 44: 2702.
18 Yu, Y., Nam, G.-H., He, Q. et al. (2018). *Nat. Chem.* 10: 638.
19 Liu, L., Wu, J., Wu, L. et al. (2018). *Nat. Mater.* 17: 1108.
20 Acerce, M., Voiry, D., and Chhowalla, M. (2015). *Nat. Nanotechnol.* 10: 313.
21 Yan, S., Qiao, W., He, X. et al. (2015). *Appl. Phys. Lett.* 106: 012408.
22 Shirodkar, S.N. and Waghmare, U.V. (2014). *Phys. Rev. Lett.* 112: 157601.
23 Acerce, M., Akdoğan, E.K., and Chhowalla, M. (2017). *Nature* 549: 370.
24 Zhang, R., Tsai, I.-L., Chapman, J. et al. (2016). *Nano Lett.* 16: 629.
25 Lin, Z., Carvalho, B.R., Kahn, E. et al. (2016). *2D Mater.* 3: 022002.
26 Wilson, J.A. and Yoffe, A. (1969). *Adv. Phys.* 18: 193.
27 Krivanek, O.L., Chisholm, M.F., Nicolosi, V. et al. (2010). *Nature* 464: 571.

28 Zhao, X., Ning, S., Fu, W. et al. (2018). *Adv. Mater.* 30: 1802397.
29 Gong, C., Zhang, H., Wang, W. et al. (2013). *Appl. Phys. Lett.* 103: 053513.
30 Chhowalla, M., Jena, D., and Zhang, H. (2016). *Nat. Rev. Mater.* 1: 16052.
31 Tang, B., Zhou, J., Sun, P. et al. (2019). *Adv. Mater.* 31: 1900862.
32 Chen, J., Liu, B., Liu, Y. et al. (2015). *Adv. Mater.* 27: 6722.
33 Shi, Y., Li, H., and Li, L.-J. (2015). *Chem. Soc. Rev.* 44: 2744.
34 Wei, Z., Li, B., Xia, C. et al. (2018). *Small Methods* 2: 1800094.
35 Hu, Z., Liu, Z.B., and Tian, J.G. (2020). *Chin. J. Chem.* 38: 981.
36 Nicolosi, V., Chhowalla, M., Kanatzidis, M.G. et al. (2013). *Science* 340: 1226419.
37 Huang, J.Y., Ding, F., Yakobson, B.I. et al. (2009). *Proc. Natl. Acad. Sci.* 106: 10103.
38 Novoselov, K.S., Jiang, D., Schedin, F. et al. (2005). *Proc. Natl. Acad. Sci.* 102: 10451.
39 Mak, K.F., Lee, C., Hone, J. et al. (2010). *Phys. Rev. Lett.* 105: 136805.
40 Radisavljevic, B., Radenovic, A., Brivio, J. et al. (2011). *Nat. Nanotechnol.* 6: 147.
41 Zhao, W., Ghorannevis, Z., Chu, L. et al. (2013). *ACS Nano* 7: 791.
42 Coleman, J.N., Lotya, M., O'Neill, A. et al. (2011). *Science* 331: 568.
43 Yin, Z., Li, H., Li, H. et al. (2011). *ACS Nano* 6: 74.
44 Han, S.A., Bhatia, R., and Kim, S.-W. (2015). *Nano Converg.* 2: 17.
45 Bourlinos, A.B., Georgakilas, V., Zboril, R. et al. (2009). *Small* 5: –1841.
46 Jeong, S., Yoo, D., Ahn, M. et al. (2015). *Nat. Commun.* 6: 1.
47 Kovtyukhova, N.I., Wang, Y., Berkdemir, A. et al. (2014). *Nat. Chem.* 6: 957.
48 Sun, Y., Sun, Z., Gao, S. et al. (2012). *Nat. Commun.* 3: 1.
49 Ding, Y., Chen, Y.-P., Zhang, X. et al. (2017). *J. Am. Chem. Soc.* 139: 9136.
50 Li, F., Xue, M., Zhang, X. et al. (2018). *Adv. Energy Mater.* 8: 1702794.
51 Wang, D., Luo, F., Lu, M. et al. (2019). *Small* 15: 1804404.
52 Hu, D., Xu, G., Xing, L. et al. (2017). *Angew. Chem. Int. Ed.* 56: 3611.
53 Lee, Y.H., Zhang, X.Q., Zhang, W. et al. (2012). *Adv. Mater.* 24: 2320.
54 Gao, Y., Liu, Z., Sun, D.-M. et al. (2015). *Nat. Commun.* 6: 8569.
55 Zhou, J., Liu, F., Lin, J. et al. (2017). *Adv. Mater.* 29: 1603471.
56 Chen, K., Chen, Z., Wan, X. et al. (2017). *Adv. Mater.* 29: 1700704.
57 Zhou, J., Tang, B., Lin, J. et al. (2018). *Adv. Funct. Mater.* 28: 1801568.
58 Tang, B., Lu, Y., Zhou, J. et al. (2020). *Mater. Today* 41: 72.
59 Xu, M., Tang, B., Lu, Y. et al. (2021). *J. Am. Chem. Soc.* 143: 18103.
60 Zhang, J., Wang, F., Shenoy, V.B. et al. (2020). *Mater. Today* 40: 132.
61 Najmaei, S., Liu, Z., Zhou, W. et al. (2013). *Nat. Mater.* 12: 754.
62 Wang, S., Rong, Y., Fan, Y. et al. (2014). *Chem. Mater.* 26: 6371.
63 Chen, J., Tang, W., Tian, B. et al. (2016). *Adv. Sci.* 3: 1500033.
64 Lai, Z., He, Q., Tran, T.H. et al. (2021). *Nat. Mater.* 20: 1113.
65 Zhang, Q., Wang, X.-F., Shen, S.-H. et al. (2019). *Nat. Electron.* 2: 164.
66 Yang, D., Hu, X., Zhuang, M. et al. (2018). *Adv. Funct. Mater.* 28: 1800785.
67 Zhou, L., Zubair, A., Wang, Z. et al. (2016). *Adv.·Mater.* 28: 9526.
68 Zhang, Q., Xiao, Y., Zhang, T. et al. (2017). *Chem. Mater.* 29: 4641.
69 Zhao, R., Grisafe, B., Ghosh, R.K. et al. (2019). *Nanoscale* 11: 6016.

70 Lin, Y.-C., Yeh, C.-H., Lin, H.-C. et al. (2018). *ACS Nano* 12: 12080.
71 Sung, J.H., Heo, H., Si, S. et al. (2017). *Nat. Nanotechnol.* 12: 1064.
72 Keum, D.H., Cho, S., Kim, J.H. et al. (2015). *Nat. Phys.* 11: 482.
73 Empante, T.A., Zhou, Y., Klee, V. et al. (2017). *ACS Nano* 11: 900.
74 Chen, P., Pai, W.W., Chan, Y.-H. et al. (2018). *Nat. Commun.* 9: 1.
75 Nakata, Y., Sugawara, K., Shimizu, R. et al. (2016). *NPG Asia Mater.* 8: e321.
76 Zeng, Z., Sun, X., Zhang, D. et al. (2019). *Adv. Funct. Mater.* 29: 1806874.
77 Kang, L., Ye, C., Zhao, X. et al. (2020). *Nat. Commun.* 11: 1.
78 Wang, X., Bian, C., He, Y. et al. (2021). *Mater. Today* 45: 35.
79 Gao, T., Zhang, Q., Li, L. et al. (2018). *Adv. Opt. Mater.* 6: 1800058.
80 Wang, L.N., Hu, P., Long, Y. et al. (2017). *J. Mater. Chem. A* 5: 22855.
81 Xie, L.M. (2015). *Nanoscale* 7: 18392.
82 Pham, V.P. and Yeom, G.Y. (2016). *Adv. Mater.* 28: 9024.
83 Tedstone, A.A., Lewis, D.J., and O'Brien, P. (2016). *Chem. Mater.* 28: 1965.
84 Gong, Y.J., Liu, Z., Lupini, A.R. et al. (2014). *Nano Lett.* 14: 442.
85 Mann, J., Ma, Q., Odenthal, P.M. et al. (2014). *Adv. Mater.* 26: 1399.
86 Duan, X.D., Wang, C., Fan, Z. et al. (2016). *Nano Lett.* 16: 264.
87 Cui, F.F., Feng, Q.L., Hong, J.H. et al. (2017). *Adv. Mater.* 29: 1705015.
88 Chen, Y.F., Xi, J.Y., Dumcenco, D.O. et al. (2013). *Acs Nano* 7: 4610.
89 Park, J., Kim, M.S., Park, B. et al. (2018). *Acs Nano* 12: 6301.
90 Li, X., Lin, M.W., Basile, L. et al. (2016). *Adv. Mater.* 28: 8240.
91 Cui, F., Feng, Q., Hong, J. et al. (2017). *Adv. Mater.* 29: 1705015.
92 Yu, P., Lin, J., Sun, L. et al. (2017). *Adv. Mater.* 29: 1603991.
93 Ho, C.-H., Chen, W.-H., Tiong, K.K. et al. (2017). *ACS Nano* 11: 11162.
94 Kochat, V., Apte, A., Hachtel, J.A. et al. (2017). *Adv. Mater.* 29: 1703754.
95 Py, M. and Haering, R. (1983). *Can. J. Phys.* 61: 76.
96 Sun, X., Wang, Z., Li, Z., and Fu, Y.Q. (2016). *Sci. Rep.* 6: 1.
97 Sun, L., Yan, X., Zheng, J. et al. (2018). *Nano Lett.* 18: 3435.
98 Tan, C., Luo, Z., Chaturvedi, A. et al. (2018). *Adv. Mater.* 30: 1705509.
99 Voiry, D., Salehi, M., Silva, R. et al. (2013). *Nano Lett.* 13: 6222.
100 Tan, S.J., Abdelwahab, I., Ding, Z. et al. (2017). *J. Am. Chem. Soc.* 139: 2504.
101 Wang, X., Shen, X., Wang, Z. et al. (2014). *ACS Nano* 8: 11394.
102 Cho, S., Kim, S., Kim, J.H. et al. (2015). *Science* 349: 625.
103 Guo, Y., Sun, D., Ouyang, B. et al. (2015). *Nano Lett.* 15: 5081.
104 Wang, Z., Sun, Y. Y., Abdelwahab, I. et al. (2018). *ACS Nano* 12: 12619.

3

Thermoelectric Properties of Polymorphic 2D-TMDs

H. K. Ng[1], *Yunshan Zhao*[2], *Dongzhi Chi*[1], *and Jing Wu*[1]

[1] *Institute of Materials Research and Engineering (IMRE), Agency for Science, Technology and Research (A*STAR), 2 Fusionopolis Way, Innovis, #08-03, Singapore 138634, Republic of Singapore*
[2] *Nanjing Normal University, Phonon Engineering Research Center of Jiangsu Province, Center for Quantum Transport and Thermal Energy Science, Institute of Physics Frontiers and Interdisciplinary Sciences, School of Physics and Technology, Nanjing 210023, China*

3.1 Introduction to 2D Thermoelectrics

Even with the global innovation and revolution on technology to improve the quality of life, there has not been much improvement in our ability to reduce the amount of energy we waste. Statistically, the global energy conversion efficiency is estimated to be only about one-third of the total energy we produce. The rest of the two-thirds enters the climate system as residual heat or waste heat [1]. With growing concerns on climate change and global warming as a consequence of destruction and pollution from fossil fuels, energy production that is still mainly dependent on fossil fuels must shift towards renewable energy sources (such as solar, wind, geothermal, tidal, and more). To decrease dependency on fossil fuels, energy production from renewable energy sources must be ramped up in order to keep up with the surging global energy consumption that is projected to increase 50% by the year 2050. Thermoelectrics represent a renewable energy source capable of directly converting heat energy into electrical energy, and vice versa. Heat energy is ubiquitous and is obtainable from anywhere ranging from industrial to residential processes and even the human body. Contradictory, while thermal processes generate more than 90% of the energy we use, the energy we waste is primarily in the form of heat. To this end, thermoelectrics provide an ideal platform to harvest and recycle this colossal amount of residual/waste heat energy.

Thermoelectrics was discovered by Thomas J. Seebeck in 1821 via the Seebeck effect which describes the generation of a potential difference across a material under an applied temperature gradient. The magnitude of the potential difference generated (Seebeck voltage, V_{TEP}) under the applied temperature gradient (ΔT) is defined as the Seebeck coefficient (S) or thermopower, as in Eq. (3.1).

$$S = \frac{-V_{\mathrm{TEP}}}{\Delta T} \tag{3.1}$$

Two-Dimensional Transition-Metal Dichalcogenides: Phase Engineering and Applications in Electronics and Optoelectronics, First Edition. Edited by Chi Sin Tang, Xinmao Yin, and Andrew T. S. Wee.

The efficiency (η) of a thermoelectric material is defined in Eq. (3.2), where the pre-factor represents the Carnot efficiency and $T_{h,c}$ are the temperatures at the hot and cold side, respectively. The only other factor affecting η is the parameter ZT, termed as the figure of merit, as in Eq. (3.3).

$$\eta = \frac{T_h - T_c}{T_h} \frac{\sqrt{1+ZT}-1}{\sqrt{1+ZT}+\frac{T_h-T_c}{T_h}} \tag{3.2}$$

$$ZT = \frac{S^2\sigma}{\kappa}T = \frac{S^2\sigma}{(\kappa_e + \kappa_{ph})}T \tag{3.3}$$

where σ, κ, κ_e, and κ_{ph} are the electrical conductivity, thermal conductivity, electronic and phonon contributions to thermal conductivity at a thermodynamic temperature, T. The term $S^2\sigma$ represents the thermoelectric power factor (PF) which determines the material's ability to generate electrical energy under a temperature gradient. It can be observed that a higher ZT favors a larger η, hence ZT has been employed as the primary measure to identify and screen for thermoelectric materials with high conversion efficiency.

3.1.1 Why 2D over 3D?

2D materials offer a unique platform where relatively high electrical conductivity coexists with high Seebeck coefficient. One advantage of 2D materials is the quantum size effect [2, 3], where quantum confinement leads to the discretization of the electronic density of states (DOS), in contrast to the smoothly varying DOS in 3D materials as illustrated in Figure 3.1 [4–6]. Such rapidly varying DOS (with respect to electron energy E, i.e. high dDOS/dE) is advantageous for thermoelectricity because when the Fermi level (E_F) is tuned to these positions in the DOS, a large asymmetry exists between hot ($E > E_F$) and cold ($E < E_F$) electrons. This results in a more effective transport of entropy per charge and thus a higher Seebeck coefficient without large sacrifice in the electrical conductivity. Meanwhile, the tunability of the E_F in 2D materials is significantly more accessible than in bulk materials, for example, by the introduction of intercalated ions or an electric field, which assists in the optimization of the Seebeck coefficient and thus the power factor [7–11]. Another advantage of 2D materials is the possibility of separately engineering the transport of thermal phonons from that of free charge carriers. Furthermore, the scattering mechanism and hence scattering time of electrons and phonons can be judiciously tuned by the change in dimensionality. This can be done in the classical picture by directly affecting the mean free path (average distance between collisions of quasiparticles) or in

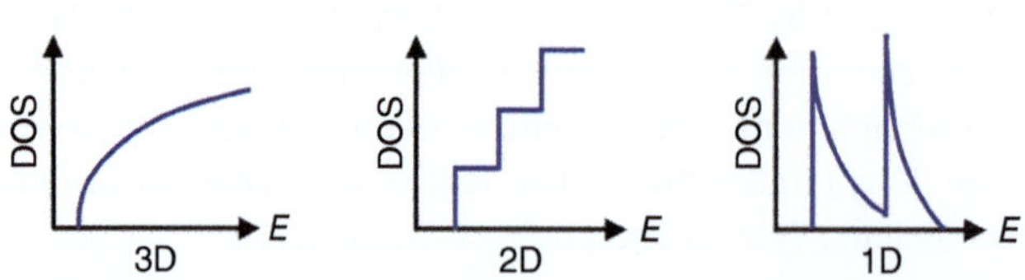

Figure 3.1 Discretization of density of states from 3D to 2D and 1D due to quantum confinement.

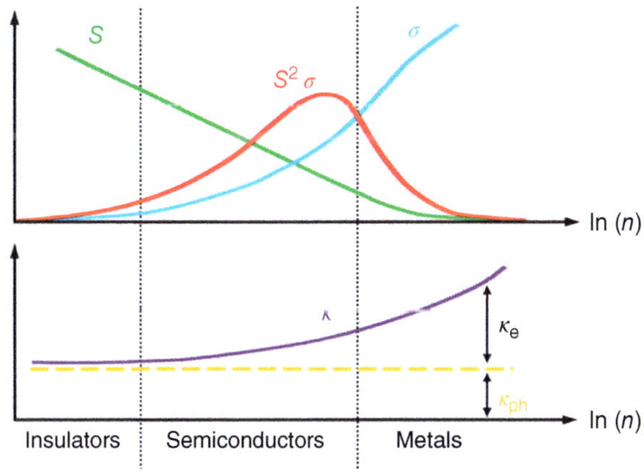

Figure 3.2 Interdependence between Seebeck coefficient (S), electrical conductivity (σ), and electronic (κ_e) and lattice (κ_{ph}) thermal conductivity as a function of carrier concentration (n).

coherent and/or hydrodynamic ways by controlling feature sizes close to the phonon and/or electron wavelengths.

3.1.2 Why 2D Semiconductors?

In this chapter, we focus on the thermoelectric performance of transition metal chalcogenides which represent one of the most commonly studied family of 2D semiconductors, and employ recent theoretical and experimental results to suggest novel ideas and approaches towards uncovering TE physics and technology. Maximizing TE efficiency (ZT) has always been a perplexing task owing to the interdependent relation between TE parameters (Figure 3.2). For instance, the increase in carrier concentration within a material results in the increase of σ but reduces S because of their inverse relationship to carrier concentration. Simultaneously, κ_e also increases with σ according to the Wiedemann–Franz law, ultimately deteriorating ZT. Due to the interplay between these properties, the best performing TE materials are typically narrow-bandgap semiconductors. This is because wide-bandgap semiconductors are relatively difficult to be highly doped [12], hence limiting their ability to optimize electrical conductivity for high TE power factor. They are also typically composed of light elements, which translates to relatively high thermal conductivity. In the other extreme, metals or semi-metals have too large a carrier density or sometimes bipolar carriers, and their DOS typically varies slowly at the Fermi level, which results in a low Seebeck coefficient. Therefore, narrow-bandgap semiconductors composed of heavy elements have been the focus in the search for high-performance thermoelectric materials, especially for low-to-intermediate temperature applications. The TE power factor is the product of $S^2\sigma$ and it represents the material's ability to generate electrical energy under a temperature gradient.

3.2 Thermoelectric Transport

Thermoelectric transport is commonly described from Boltzmann Transport Equations (BTEs) [13, 14] or Landauer [15, 16] perspective. In this section, we dive into the thermoelectric coefficients using BTEs and discuss different scattering mechanisms that can be present within a material.

3.2.1 Boltzmann Transport Equation

The transport properties relevant to a TE material can be modelled and studied using the linearized BTE under the relaxation-time approximation, where the deviation of the local distribution of charge carriers, $f(E)$, from the steady-state under the perturbation of a small electric field ($\vec{\varepsilon}$), temperature gradient (ΔT), and concentration gradient can be written as [13, 14]

$$f(E) - f_0(E) = -\tau(E)v(E)\frac{\mathrm{d}f_0}{\mathrm{d}E}\left[q \cdot \vec{\varepsilon} - \left(\frac{E - E_\mathrm{F}}{T}\right)\Delta T\right] \tag{3.4}$$

where q, $v(E)$, $\tau(E)$, E_F, T represent the unit charge, carrier velocity, carrier relaxation time, Fermi level and temperature, respectively. The reference system sets $E_\mathrm{C} = 0$. Here, f_0 is the equilibrium Fermi-Dirac distribution given as

$$f_0(E) = \frac{1}{e^{\frac{(E-E_\mathrm{F})}{k_\mathrm{B}T}} + 1} \tag{3.5}$$

where k_B is the Boltzmann constant. Using Eq. (3.4), the charge- (j) and heat- (j_Q) current densities can then be determined as

$$j \equiv nqv = \int_{E=0}^{\infty} g(E)\, v(E)\, q\, [f(E) - f_0(E)]\, \mathrm{d}E \tag{3.6}$$

$$j_Q \equiv n(E - E_\mathrm{F})v = \int_{E=0}^{\infty} g(E)\, (E - E_\mathrm{F})\, v(E)\, [f(E) - f_0(E)]\, \mathrm{d}E \tag{3.7}$$

where $g(E)$ is the DOS. Note that at steady-state, the average velocity of carriers is zero, that is, $\int_{-\infty}^{\infty} g(E)\, v(E)\, q\, f_0(E)\, \mathrm{d}E = 0$. By substituting Eq. (3.4) into Eqs. (3.6) and (3.7), the corresponding definitions for Seebeck coefficient (S), electrical conductivity (σ) and electronic thermal conductivity (κ_e) can be derived as

$$S = \left.\frac{\varepsilon}{\Delta T}\right|_{j=0} = \frac{1}{qT}\left[\frac{X_1}{X_0} - E_\mathrm{F}\right] \tag{3.8}$$

$$\sigma = \left.\frac{j}{\varepsilon}\right|_{\Delta T=0} = q^2 X_0 \tag{3.9}$$

$$\kappa_\mathrm{e} = \left.\frac{Q}{\Delta T}\right|_{j=0} = \frac{1}{T}\left[X_2 - \frac{(X_1)^2}{X_0}\right] \tag{3.10}$$

where $X_i = \int_{E=0}^{\infty} g(E)\, \tau(E)\, v^2(E)\, E^i \left(\frac{\mathrm{d}f_0}{\mathrm{d}E}\right) \mathrm{d}E$.

Taking assumptions of (i) single-band material with parabolic band structure, (ii) power law relaxation time, and (iii) materials under small electric field, the DOS $g(E)$ can be written as

$$g(E) = \frac{N_\mathrm{V}}{g_D a^{3-D}}\left(\frac{2m_d}{\hbar^2}\right)^{\frac{D}{2}} (k_\mathrm{B}T)^{\frac{r+D}{2+i}} \left[\frac{2(r+i)}{D} + 1\right] F_{r+\frac{D}{2}+i-1}(\eta) \tag{3.11}$$

$$\tau(E) = \tau_0 E^r \tag{3.12}$$

$$v^2(E) = \frac{2}{m_\sigma D} E \tag{3.13}$$

and the integral X_i can be then expressed as

$$X_{i,D} = \frac{N_V \tau_0}{g_D a^{3-D} m_\sigma} \left(\frac{2m_d}{\hbar^2}\right)^{\frac{D}{2}} (k_B T)^{\frac{r+D}{2+i}} \left[\frac{2(r+i)}{D} + 1\right] F_{r+\frac{D}{2}+i-1}(\eta) \tag{3.14}$$

where N_V, τ_0, D, a, r represents the number of degenerate conduction valleys, scattering time constant, dimensionality, confinement length scale and scattering parameter, respectively. $g_D = 2\pi^2(2\pi)$ for $D = 3$ (2), respectively. Here $F_j(\eta) = \int_0^\infty \left[\frac{x^j}{\exp(x-\eta)} + 1\right] dx$ is the jth-order Fermi integral and can be evaluated through numerical integration [17, 18], and η is the reduced Fermi potential defined as $\eta = \frac{(E_F - E_C)}{k_B T}$ where E_C is the energy at the conduction band minimum (CBM). m_d and m_σ are the DOS effective mass and conductivity effective mass defined as

$$m_d = \left\{ \sqrt[3]{m_x m_y m_z}\ , D = 3\ \sqrt[2]{m_x m_y}, D = 2\ m_x, D = 1 \right. \tag{3.15}$$

$$m_\sigma^{-1} = \sum_{i=1}^{N} m_{\sigma,i}^{-1} \tag{3.16}$$

The steady-state solutions to the linearized BTE can then be written as

$$S = -\frac{k_B}{q}\left[\eta - \frac{\left(r + \frac{D}{2} + 1\right) F_{r+\frac{D}{2}}(\eta)}{\left(r + \frac{D}{2}\right) F_{r+\frac{D}{2}-1}(\eta)}\right] \tag{3.17}$$

$$n = \frac{N_V}{g_D a^{3-D}} \left(\frac{2k_B T m_d}{\hbar^2}\right)^{\frac{D}{2}} F_{\frac{D}{2}-1}(\eta) \tag{3.18}$$

$$\sigma = \frac{N_V}{g_D a^{3-D} m_\sigma} \left(\frac{2m_d}{\hbar^2}\right)^{\frac{D}{2}} (k_B T)^{r+\frac{D}{2}} \left(\frac{2r}{D} + 1\right) F_{r+\frac{D}{2}-1}(\eta) \tag{3.19}$$

For degenerate semiconductors and metals ($\eta \gg 0$), the Seebeck coefficient ($q = -e$) can then be approximated using the Mott formula [19]

$$S_{\text{degenerate}} \cong -\frac{\pi^2 k_B}{3e\eta}\left[\frac{\partial \ln(g(E))}{\partial \ln(E)} + \frac{\partial \ln(\tau(E))}{\partial \ln(E)} + \frac{\partial \ln(v^2(E))}{\partial \ln(E)}\right] \cong -\frac{\pi^2 k_B}{3e\eta}\left(\frac{D}{2} + r\right) \tag{3.20}$$

For nondegenerate semiconductors, E_F is located within the bandgap with a distance from the conduction or valence band edges larger than 2 to 3 $k_B T$ such that the Fermi integral become

$$F_j(\eta) = \exp(\eta)\, \Gamma(n+1) \tag{3.21}$$

where $\Gamma(n+1)$ is the gamma function that has the property $\Gamma(n+1) = n\Gamma(n)$ and can be evaluated analytically. Substituting Eqs. (1.21) into (1.17), the Seebeck coefficient becomes

$$S_{\text{non-degenerate}} = -\frac{k_B}{q}\left[\eta - \left(r + \frac{D}{2} + 1\right)\right] \tag{3.22}$$

Plotting σ ($\propto n$) and S as a function of η, the counterbalanced interdependency between S and σ can be evidently seen in Figure 3.3. There are two avenues to

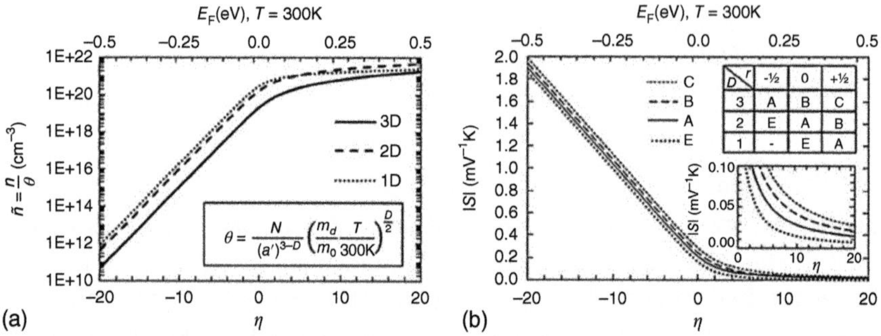

Figure 3.3 (a) Carrier concentration (n) and (b) Seebeck coefficient (S) as a function of η. Equivalent value of E_F at 300 K is labeled on the top horizontal axis [14]. Source: Reproduced with permission from Pichanusakorn and Bandaru [14]. Copyright © 2009, Elsevier B.V.

improve ZT: First, maximizing the TE power factor ($S^2\sigma$) of a material, which has been shown to occur in the degenerate semiconductor region [14]. Other factors contributing to an enhancement in $S^2\sigma$ include larger valley degeneracy (N_V), larger DOS effective mass (m_d), larger electron-hole asymmetry and smaller confinement length (a). Second, reducing κ by employing phononic crystals or nanostructuring [20].

3.2.2 Scattering Parameter for Different Mechanism

From Eqs. (3.17)–(3.19), the TE coefficients are strongly dependent on the scattering parameter (r), which is determined by the type of scattering processes dominant within a material. Within a material, carriers encounter perturbations arising from impurity atoms, lattice vibrations (phonons), crystal defects or with other carriers [13, 21]. Such a perturbation then scatters the carrier, causing collisions which can be beneficial or detrimental to TE transport. In this section, we described the effect of r and its scattering momentum relaxation time (τ_m) on TE coefficients for different types of scattering mechanisms. By solving the Hamiltonian under a perturbing potential, the key results for τ_m of common scattering mechanisms (impurity atoms, phonons, crystal defects, and with other carriers) are summarized, with detailed derivation for scattering rates available in literature [13, 14].

3.2.2.1 Ionized/Charged Impurity Scattering

When carriers encounter the electric fields of an ionized impurity, scattering due to Coulombic interactions occur. By solving the Poisson equation, the perturbing potential for such ionized impurity scattering is

$$U_{ii}(r) = \frac{q^2}{4\pi\varepsilon_r\varepsilon_0 r} e^{\frac{-r}{L_D}} \tag{3.23}$$

where r is the distance from the perturbation. q, ε_r, ε_0 represents electric charge, relative and vacuum permittivity, respectively. $L_D \left(\equiv \sqrt{\frac{\varepsilon_r\varepsilon_0 k_B T}{q^2 n_0}}\right)$ is the Debye length.

Using Fermi's Golden Rule, the corresponding τ_m for ionized impurity scattering is derived as in Eq. (3.24) with the scattering parameter, r determined to be 3/2.

$$\tau_m(p) \propto \frac{\varepsilon_r{}^2\varepsilon_0{}^2}{N_I q^4} E^{\frac{3}{2}}(p) \tag{3.24}$$

3.2.2.2 Phonons Scattering

Majority of the scattering occurring in semiconductors arises from phonons which are quantized modes of vibrations occurring within a rigid crystal lattice. There are two types of phonons, namely acoustic and optical phonons. For acoustic (optical) phonons, neighbouring atoms are displaced in the same (transverse) direction which translates to changes in lattice spacing due to acoustic (optical) deformation potential that then leads to the scattering of carriers. In the case of polar semiconductors, the polar nature of bonds creates strong interactions which results in an electric dipole that can be perturbed by acoustic or optical phonons that scatters carriers. Polar acoustic phonon scattering (piezoelectric scattering, PZ) becomes important at very low temperatures in very pure semiconductors. On the other hand, polar optical phonon (POP) scattering is substantial in polar materials and dominates charge transport near room temperature [22]. Another type of phonon scattering prominent in certain semiconductors is intervalley (IV) scattering where a carrier is moved from a given valley to another valley, thereby producing significant changes in momentum such that phonons with wave vectors near the zone boundary are required.

The resulting perturbing potential for the different phonon scattering can be expressed as

$$U_S = K_\beta u_\beta \tag{3.25}$$

where β is the phonon wave vector, u_β is the Fourier component of lattice vibration and K_β is defined as follows

$$\text{(ADP)} \qquad |K_\beta|^2 = \beta^2 D_A{}^2 \tag{3.26}$$

$$\text{(ODP)} \qquad |K_\beta|^2 = D_O{}^2$$

$$\text{(POP)} \qquad |K_\beta|^2 = \frac{\rho q^2 \omega_o{}^2}{\beta^2 \varepsilon_{r,f\to 0}\varepsilon_0}\left(\frac{\varepsilon_{r,f\to 0}}{\varepsilon_{r,f\to\infty}} - 1\right)$$

$$\text{(PZ)} \qquad |K_\beta|^2 = \left(\frac{q e_{PZ}}{\varepsilon_r \varepsilon_0}\right)^2$$

$$\text{(IV)} \qquad |K_\beta|^2 = D_{if}{}^2$$

where D_A, D_O, e_{PZ}, D_{if}, ω_o, ρ represents the acoustic deformation potential (ADP), optical deformation potential (ODP), piezoelectric (PZ) constant, intervalley (IV) deformation potential, optical phonon energy, and mass density respectively. D_{if} characterizes the strength of scattering from an initial valley "i" to a final valley "f." The corresponding $\tau_m(p)$ for each type of phonon scattering are then derived to be

$$\text{(ADP)} \qquad \frac{1}{\tau_m(p)} = \frac{\pi D_A{}^2 k_B T}{\hbar c_l} g_C(E) \tag{3.27}$$

$$\text{(ODP)} \qquad \frac{1}{\tau_m(p)} = \frac{\pi D_O{}^2}{2\rho\omega_0}\left(N_o + \frac{1}{2} \mp \frac{1}{2}\right) g_C(E \pm \hbar\omega_0)$$

$$\text{(POP)} \qquad \frac{1}{\tau_m(p)} = \frac{q^2\omega_0\left(\frac{\varepsilon_{r,f\to 0}}{\varepsilon_{r,f\to\infty}} - 1\right)}{2\pi\varepsilon_{r,f=0}\varepsilon_0\sqrt{\frac{2E(p)}{m^*}}}$$

$$\times\left[N_o \sinh^{-1}\left(\frac{E(p)}{\hbar\omega_0}\right)^{\frac{1}{2}} + (N_o + 1)\sinh^{-1}\left(\frac{E(p)}{\hbar\omega_0} - 1\right)^{\frac{1}{2}}\right]$$

$$\text{(IV)} \qquad \frac{1}{\tau_m(p)} = \frac{\pi D_{if}{}^2 Z_f}{2\rho\omega_{if}}\left(N_{if} + \frac{1}{2} \mp \frac{1}{2}\right) g_{Cf}(E \pm \hbar\omega_{if} - \Delta E_{fi})$$

where g_C, ω_{if}, Z_f, N_{if} represents the DOS, intervalley phonon energy, the number of final valleys available for scattering and the number of intervalley phonons, respectively. Here, it is important to note that the $\tau_m(p)$ for intervalley scattering is due to optical phonon scattering.

3.2.2.3 Carrier–Carrier Scattering

In materials with high levels of carrier density, collision between carriers can become prominent and adversely affect charge transport. There are two types of carrier–carrier scattering: (i) binary process in which one carrier collides with another and (ii) collective process in which a carrier interacts with the plasma comprised by the carriers. When viewed in the center-of-mass reference frame, the binary process is similar to an ionized impurity scattering event (Section 2.2.1). In the collective process, carriers interact with oscillations in the carrier density which are accompanied by opposing electric fields at a plasma frequency, $\omega_p = \sqrt{\frac{q^2 n}{\varepsilon_r\varepsilon_0 m^*}}$. These oscillations can be sustained if $\omega_p\tau \geq 1$ such that the carrier-carrier collisions do not dampen out the oscillations. The interaction potential and momentum relaxation time for carrier-plasmon scattering are

$$U_{PL} = \frac{q\rho_\beta}{\varepsilon_r\varepsilon_0\beta^2} \tag{3.28}$$

$$\frac{1}{\tau(p)} = \frac{q^2\omega_p\left(N_p + \frac{1}{2} \mp \frac{1}{2}\right)}{4\pi\varepsilon_r\varepsilon_0\hbar\sqrt{\frac{2E(p)}{m^*}}}\ln\left[\frac{\hbar\beta_{PL}}{p\left(\mp 1 \pm \sqrt{1 \pm \frac{\hbar\omega_p}{E(p)}}\right)}\right]$$

where ρ_β is the charge density oscillation of the plasma that produces the interacting electric field. At high carrier density, it is also important to note that while plasmon scattering rate may be high, its effect on transport may be subtle, since usually the plasma oscillations are heavily dampened by phonon scattering. Importantly, carrier–carrier scattering plays a relatively minor role in scattering as it is usually only dominant in crystals of exceptional low defects and impurities. In 2D semiconductors, scattering is instead dominated by phonons at high temperatures, and by charged-impurities or defects at low temperatures.

3.2.2.4 Surface Roughness Scattering

Surface roughness scattering is the elastic scattering of a charged particle with an interfacial disorder between an imperfect interface between two different materials. Surface roughness scattering can be categorized as long-range and short-range. Long-range surface roughness scattering mimics coulombic interactions of charged impurity scattering, while short-range surface roughness scattering is associated with surface disorders. Such surface roughness scattering becomes detrimental to carrier mobility at high fields or at high carrier concentrations [23, 24].

3.3 Experimental Characterization TE in 2D

Due to the nanoscale nature of layered 2D materials, electrical and thermoelectric measurement techniques are different from the bulk material community. So before starting the discussion about the thermoelectric properties, we would like to give a brief introduction for the experimental measurement techniques of 2D materials.

3.3.1 Electrical Measurements

3.3.1.1 FET Measurements

Electrical measurements are performed using two source-meters (Keithley 6430 and 2400), from which the sample's transfer characteristics and conductance can be obtained. Electric field modulation is applied via the back gate through the dielectric layer of the substrate to tune the sample's carrier density. Two- and four-probe electrical conductivity can be extracted at different backgate voltages, as illustrated in Figure 3.4a. By using the four-probe technique, contact resistances can be eliminated. Electrical conductivity is then calculated by considering the geometry of the measured sample using $\sigma = G\frac{L}{Wt}$ where G, L, W and t represents conductance, length, width and thickness, respectively. The FET mobility is extracted using the transconductance method via the transfer characteristics $\mu = \frac{dI_{sd}}{dV_g}\frac{1}{V_{sd}}\frac{1}{C_{ox}}\frac{L}{W}$ where C_{ox} is the capacitance of the dielectric employed ($C_{ox} = 11.5\ \mathrm{nF\,cm^{-2}}$ for 300 nm SiO_2). Other extraction methods of mobility include the Y-function method and transmission line method, but are not covered in this chapter and have been

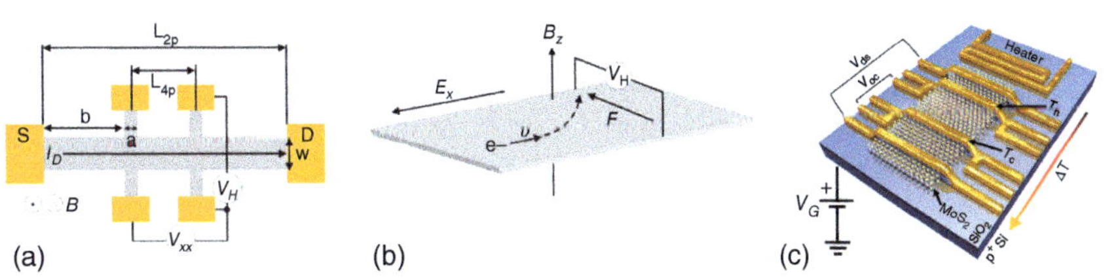

Figure 3.4 Schematic diagram of (a) four-probe technique for electrical transport and Hall effect measurements. (b) Illustration of the Hall effect of an electron. (c) Thermoelectric measurement device structure. Source: Mitta et al. [25]/IOP Publishing/Licensed under CC BY 4.0.

discussed in details elsewhere [25–29]. The field-effect carrier density (n_{FET}) of a 2D semiconductor can also be estimated via the transfer characteristics given by

$$n_{FET} = C_{ox}\frac{V_g - V_{th}}{q} \tag{3.29}$$

where V_{th} is the threshold voltage with the assumption that the device is channel-dominated ($V_g > V_{th}$) [25]. However, it is important to note that n_{FET} does not fully represent the intrinsic carrier density present within a 2D semiconductor, which is a commonly known limitation of the field-effect technique [30].

3.3.1.2 Hall Measurements

Hall effect measurements are commonly employed to extract the intrinsic material properties of a semiconductor which include carrier density, type, and mobility. Figure 3.4b illustrates how an electron moves within a conductive channel under applied longitudinal electric and transverse magnetic fields, which is based on the Lorentz force [31]. Under a transverse external magnetic field (B_z), electrons experience a Lorentz force which leads to a voltage difference (Hall voltage, V_H) transverse to the flow of the electrons. The sign of V_H depends on the carrier type and its magnitude depends on the carrier density, current and magnetic field. Hall effect measurements can be conducted with a sinusoidal AC current or a DC current flowing through the channel of the device. AC measurements is usually performed with lock-in amplifiers to achieve larger signal-to-noise ratio. For DC measurements to attain large signal-to-noise ratio, a higher current is required which can simultaneously results in undesirable effects due to threshold voltage shift, Joule heating-induced breakdown, and phase transition [32–34]. The 2D sheet carrier density (n_{2D}), sheet resistance (R_{SH}) and Hall mobility (μ_H) can be calculated using the following equations:

$$n_{2D} = \frac{I_{ds}}{q}\frac{\Delta B_z}{\Delta V_H} \tag{3.30}$$

$$R_{SH} = \frac{V_{xx}}{I_{ds}}\frac{W}{L_{4p}} \tag{3.31}$$

$$\mu_H = \frac{V_H}{V_{xx}}\frac{L_{4p}}{W}\frac{1}{B_z} \tag{3.32}$$

3.3.2 Seebeck Measurement

The main challenge in obtaining the TE powerfactor ($S^2\sigma$) of a material is the measurement of Seebeck coefficient ($S = -\frac{V_{TEP}}{\Delta T}$), which is defined as the ratio of open-circuit voltage (V_{TEP}) to the temperature difference (ΔT) along the 2D material. The direct measurement of Seebeck coefficient in 2D materials is made possible by fabricating local micro-resistance thermometers across the 2D material as shown in Figure 3.4c. This technique is commonly used for many 2D materials such as semiconductors (Bi_2O_2Se [22], MoS_2 [35–39], $PdSe_2$ [40], etc…), graphene [9, 11] as well as 1D nanowires [41–43] and nanotubes [44]. By applying a heating current (I_h) through the micro-heater, a temperature gradient is generated along the 2D material via Joule heating, which then induces a thermoelectric voltage (V_{TEP}) along the 2D material. The measurement of Seebeck coefficient is divided into two parts, the measurement for V_{TEP} and ΔT.

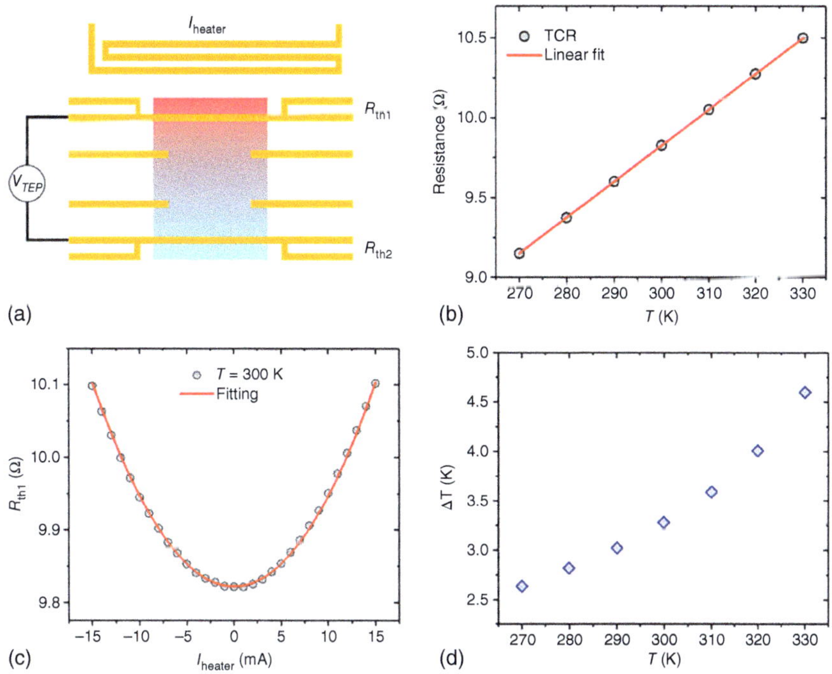

Figure 3.5 (a) Schematic diagram of temperature gradient (ΔT) measurement. (b) Temperature coefficient of resistance curve for thermometer. (c) Thermometer resistance (R_{th1}) as a function of heater current (I_{heater}) and (d) corresponding ΔT as a function of global temperature [39].

3.3.2.1 ΔT Calibration

The measurement of the temperature gradient across the 2D material channel (ΔT) is based on the four-probe resistances ($R_{th1,2}$) of the fabricated thermometers which are sensitive to changes in the local temperature as in Figure 3.5a. These thermometers are initially calibrated in a cryostat with a highly accurate temperature controller, where the temperature coefficient of resistance (TCR) of each thermometer is recorded as a function of global temperature as illustrated in Figure 3.5b. These TCR curves then work as temperature calibration data. During the ΔT measurement, a series of DC currents (I_{heater}) is applied to the heater where the corresponding R of each thermometer is recorded at different heating currents. Given the relation $R \propto \Delta T \propto {I_{heater}}^2$, a parabolic $R - I_{heater}$ curve is obtained as in Figure 3.5c. Based on the differences between the thermometer resistance and $R - T$ calibration data, the local temperature at each thermometer can be extracted to determine ΔT as shown in Figure 3.5d.

3.3.2.2 V_{TEP} Measurement

There are two different methods for V_{TEP} measurement depending on whether a DC or AC heating current is applied to the micro-heater. By sweeping a DC current in the micro-heater, a voltage difference V_{TEP} can be measured according to the heating current. A parabolic $V_{TEP} - I_{heater}$ curve is obtained owing to the relation

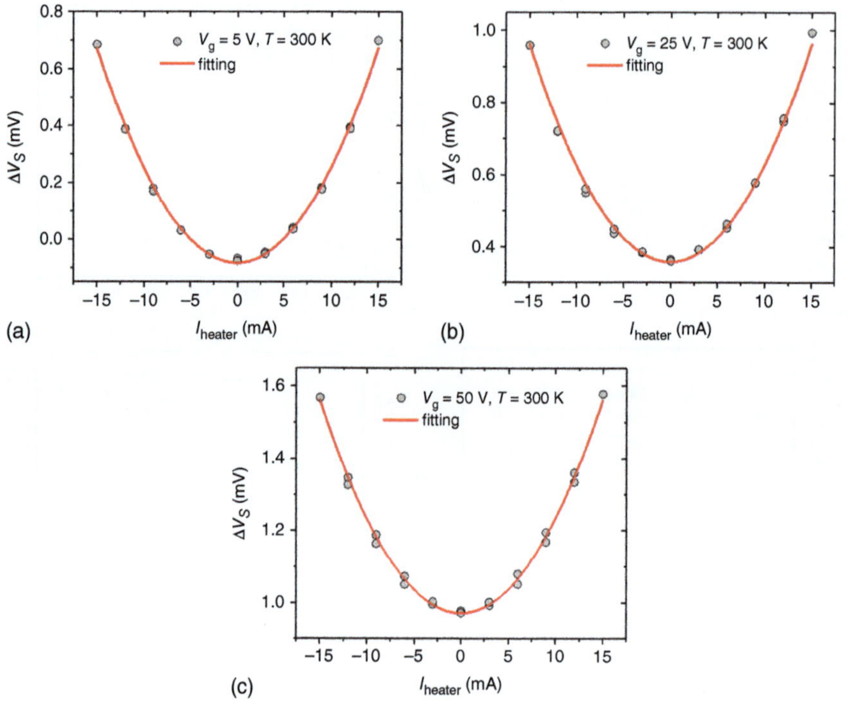

Figure 3.6 Open-circuit voltage (V_{TEP}) of a n-type sample as a function of heater current (I_{heater}) at different backgate voltages (a) 5 V, (b) 25 V, and (c) 50 V. All V_{TEP} exhibit parabolic behavior with fittings of R-square ≥0.994 [39].

$V_{TEP} \propto \Delta T \propto {I_{heater}}^2$ as illustrated in Figure 3.6. A very accurate V_{TEP} measurement is obtainable using this DC sweeping method although it requires a very slow sweep of I_{heater} to ensure thermal equilibrium and can be quite time consuming. The AC method which employs a lock-in amplifier to directly lock onto the V_{TEP} signal frequency induced by I_{heater} requires significantly less time. By applying an AC heating current $I_{heater} = I\sin(\omega t)$ with the frequency of ω, the temperature difference along the 2D material $\Delta T \propto {I_{heater}}^2 = I^2(\omega t)$, one can hence measure the V_{TEP} by sensing the 2ω signal, which is $\frac{\pi}{2}$ out-of-phase with the heater signal [37]. However, for high resistance 2D materials, this measurement is limited by the input impedance of the lock-in amplifier [7]. After the measurement of ΔT and V_{TEP}, the Seebeck coefficient of the measured 2D materials can be extracted as $S = -\frac{V_{TEP}}{\Delta T}$.

3.3.3 Thermal Conductivity

For thermal transport measurement in 2D materials, there are several popular techniques like Raman spectrometer, time-domain thermoreflectance (TDTR), thermal bridge method and e-beam heating technique that are appropriate for measuring the in-plane thermal conductivity of 2D materials, which is a key parameter for evaluating their figure of merit (ZT). Other methods such as three-omega method and H-type and T-type method are briefly discussed as well. Their measurement principles are discussed in Sections 3.3.3.1–3.3.3.4.

3.3.3.1 Raman Spectrometer

Raman spectrometer is a non-contact optical technique to explore both the intrinsic thermal conductance and interfacial thermal conductance of 2D and 2D heterostructures [45, 46]. Typically, a laser is used as a heater source and a temperature gradient along a radial direction of the measured 2D flakes is subsequently generated, as shown in Figure 3.7a [47] and Figure 3.7c [48]. By measuring the frequency shift of Raman sensitive peaks in Figure 3.7b [47] and Figure 3.7d [48], the temperature gradient could be obtained, and the thermal conductance is further extracted once the input heating is known, either from the laser heating itself or from the electrical Joule heating. In 2008, Balandin et al. reported the first Raman measurement on in-plane thermal conductivity of suspended monolayer graphene with a super-high value around 5300 W mK^{-1} [47], which ignited the research interest in thermal conductance of other 2D materials like TMDCs and h-BN [48–50], etc., by using Raman characterization. For TMDCs like MoS_2, both in-plane E_{2g}^{1} and out-of-plane A_{1g} modes are observed to have a linear evolution with the local heating temperature due to the sensitive anharmonic lattice vibrations under a temperature change [51, 52]. Comparing with other thermal measurement techniques, the samples for Raman

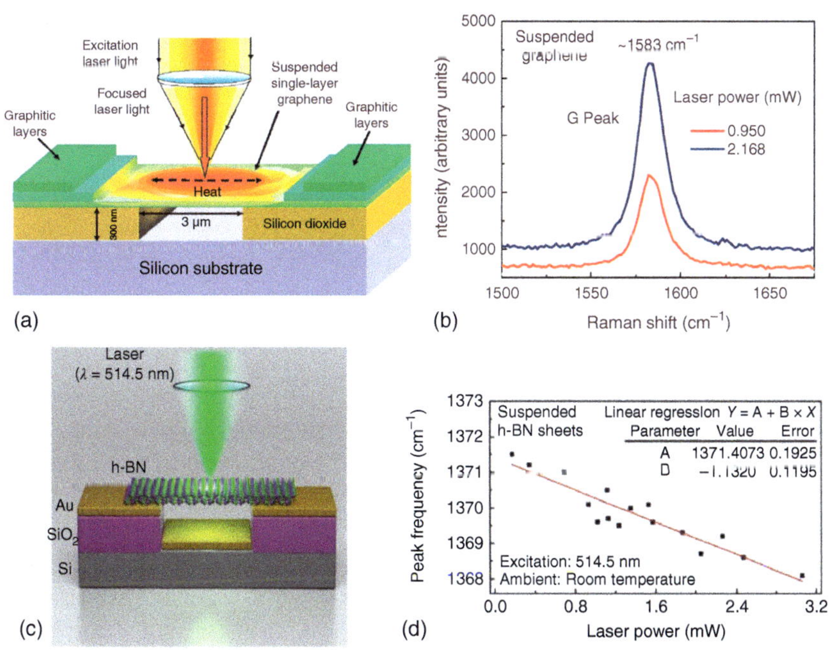

Figure 3.7 (a) Sketch of Raman laser heating a suspended monolayer graphene [47]. (b) G peak shift of monolayer graphene under different input laser power. Source: Balandin et al. [47]/Reproduced with permission from American Chemical Society. (c) Schematic of suspended h-BN under laser illumination [48]. (d) Raman sensitive peak of suspended h-BN under different laser power [48].

measurement are easy to fabricate and the measurement is simple for operation. Nevertheless, one issue related to Raman spectrometer is the accuracy in determining the optical absorption of the specimen. Due to the disparate calculated or measured power absorption, the variation between reported thermal conductivity values for even the same material can be quite large.

For a suspended 2D flake neglecting the radiation heat loss, the diffusive thermal dissipation equation can be expressed as [51, 53]

$$\kappa \frac{1}{r}\frac{\partial}{\partial r}\left(r\frac{\partial T(r)}{\partial r}\right) + q(r) = 0 \quad \text{for} \ r < R \tag{3.33}$$

where κ is the thermal conductivity of the measured specimen, r is the radial position from the center of the hole, R is the radius of the suspended area and $T(r)$ is the temperature distribution profile in the suspended region. The volumetric optical Gaussian beam heating, $q(r)$, is given as [54]

$$q(r) = \frac{P}{\pi r_0^2}\alpha\frac{1}{t}\exp\left(-\frac{r^2}{r_0^2}\right) \tag{3.34}$$

where P is the laser power, α is the absorption coefficient, t is the thickness of the sample and $r_0$is the radius of the Raman beam. By combining Eqs. (3.33) and (3.34), $T(r)$ is related the thermal conductivity, κ of the measured sample under different laser power. On the other hand, the average temperature inside the beam spot from the incident laser is written as [51, 54]

$$T_m \approx \frac{\int_0^R T(r)q(r)rdr}{\int_0^R q(r)rdr} \tag{3.35}$$

Therefore, by fitting Eq. (3.35) with the relation between T_m and P obtained experimentally, the thermal conductivity of the suspended flake could be extracted [51].

Moreover, the Raman characterization is useful to extract the interfacial thermal resistance of a heterostructure device [55–57]. Considering that for a 2D TE device, the channel is typically encapsulated by a dielectric material like h-BN, the heat dissipation across these two adjacent layers is a critical parameter for further applying an in-plane temperature bias and the interfacial thermal resistance of 2D flakes supported on different substrates should be thus known. For interfacial thermal conductance measurement, a temperature gradient is created by an electrical Joule heating or an electrical Joule heating, as shown in Figure 3.8a,d [55, 56]. The temperature gradient across the heterostructures can be accurately characterized due to their different power dependent Raman peaks of each layer, where the temperature drops across different interface are shown for graphene, MoS_2, WSe_2 and the silicon substrate. The thermal resistances across the different heterostructure interfaces by Raman measurement agree well with that of SThM [57]. For both laser power and temperature dependent Raman peak shift measurement, the process is much similar to that of in-plane thermal conductivity measurement discussed previously shown in Figure 3.7b,c [55].

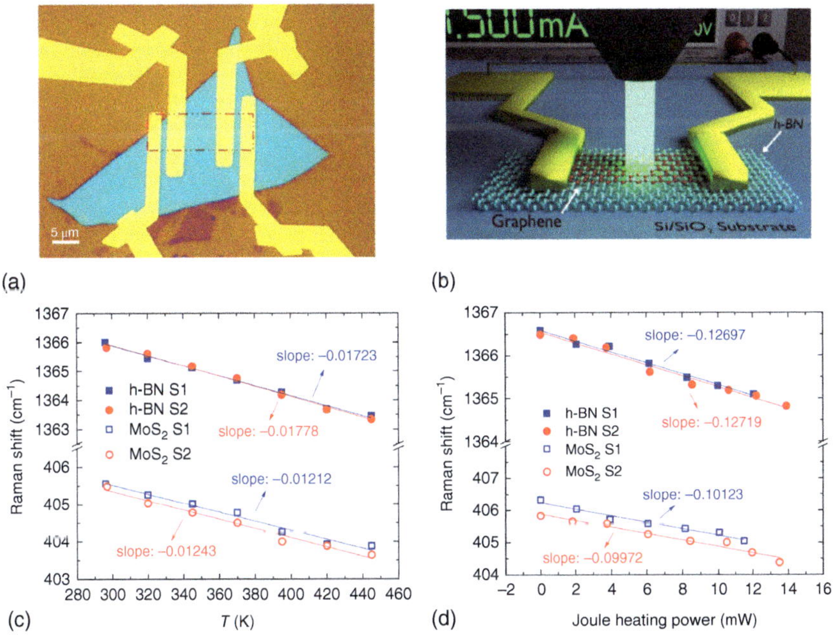

Figure 3.8 (a) Optical image of monolayer MoS_2 on h-BN. Source: Liu et al. [55]/Springer Nature/Licensed under CC BY 4.0. (b) Schematic of Raman technique for interfacial thermal resistance measurement. Source: Chen et al. [56]/Reproduced with permission from AIP Publishing. (c) and (d) is the temperature and laser power dependent Raman modes shift for MoS_2/h-BN heterostructures, respectively. Source: Liu et al. [55]/Springer Nature/CC BY 4.0.

3.3.3.2 TDTR (FDTR)

TDTR measures the thermal properties by using picosecond transient thermoreflectance, which can measure both thermal conductivity and heat capacity of bulk materials and thin films [58–60]. Due to its ability to measure the large range thermal conductivity and an advantage of simple sample preparation, TDTR becomes a popular technique for thermal properties measurement. Normally, a pulsed laser produces an optical pulse with a certain wavelength and frequency, which is split into a pump beam and a probe beam by using a polarizing beam splitter (PBS). The pump pulse is further modulated by an electro-optic modulator (EOM) before focusing on the sample. By using a delay stage, the path of probe beam is changed and the arrival time on the sample is delayed with respect to the pump beam to detect the temperature change. The signal of probe beam is then collected by a photodiode defector and further captured by a standard lock-in amplifier, as shown in Figure 3.9a,b. Rather than modulating the time delay of the two pulses, the frequency-domain thermoreflectance (FDTR) measures the thermoreflectance response as a function of modulation frequency of the pump pulse, where a constant continuous wave (CW) laser can be used instead of the complex delay state that in TDTR. For both TDTR and FDTR measurement, a metal film is deposited on the surface of the samples to well transduce the pulse, where the surface of the sample should be optically smooth with a roughness less than 15 nm.

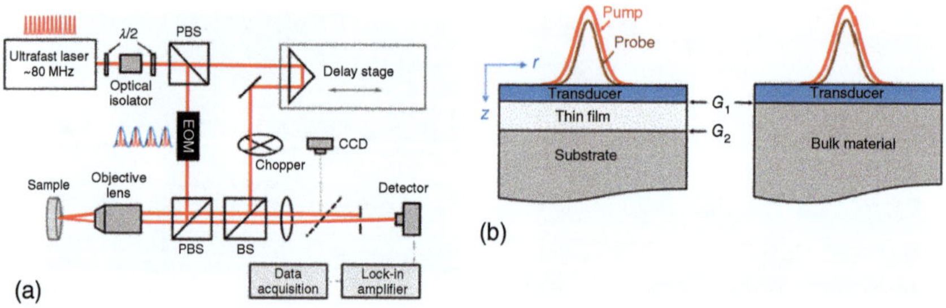

Figure 3.9 (a) Schematic of TDTR setup. (b)Thin films and bulk materials measurement by TDTR. The pump and probe beams are shown. Source: Reproduced with permission from Jiang et al. [58].

By using TDTR measurement, the intrinsic thermal conductivity of various 2D materials like graphene, black phosphorus and MoS_2, etc. could be obtained [61–63] and the interfacial thermal resistance between 2D flakes and different substrates is measured as well [58, 64]. Jiang et al. measured the thermal conductance of different groups in TMDs and found that the thermal conductivity is dominated by the chalcogen species [63]. Both WS_2 and MoS_2 show higher thermal conductivity than that of WSe_2 and $MoSe_2$, which agrees well with the theoretical calculations [65], as shown in Figure 3.10a. By conventional and beam offset TDTR, Rai et al. studied the thermal transport in layered InSe, and the in-plane thermal conductivity shows an unconspicuous dependence on the flake thickness [66] shown in Figure 3.10b. For black phosphorus with a puckered lattice structure, the anisotropic thermal transport is studied by TDTR measurement [67] in Figure 3.10c and the anisotropic ratio between zigzag (ZZ) and armchair (AC) is comparable with that from other thermal characterization techniques like Raman spectrometer [25] and electron beam heating technique [68] discussed later. For both TDTR and FDTR thermal conductance measurement, there is still no work about measuring the thermal conductivity of 2D materials with thickness down to few layers and attempts should be made to further extend the applications of transient thermoreflectance techniques.

3.3.3.3 Thermal Bridge Method (Electron Beam Heating Technique)

The thermal bridge method or suspended pads method was developed by Kim et al. in 2003 to measure the thermal conductance of one-dimensional carbon nanotubes [69], which was later extended to characterize the transport properties of various nanostructures like 2D materials [61–63], including their thermal conductivity, electrical conductivity and Seebeck coefficient. The method relies on a specially customized suspended setup, which involves a series of nanofabrication processes [64]. The suspended device, micro-electro-thermal system (MEMS), includes two suspended SiN_x membranes and the built-in platinum (Pt) coils are designed as thermometers atop as shown in Figure 3.11. When conducting the measurement, an AC current is applied to the two thermometers to measure the electrical resistance of both heater and sensor and a DC current is applied to the heater side to create a temperature bias. Under a thermal equilibrium state, the thermal conductance of

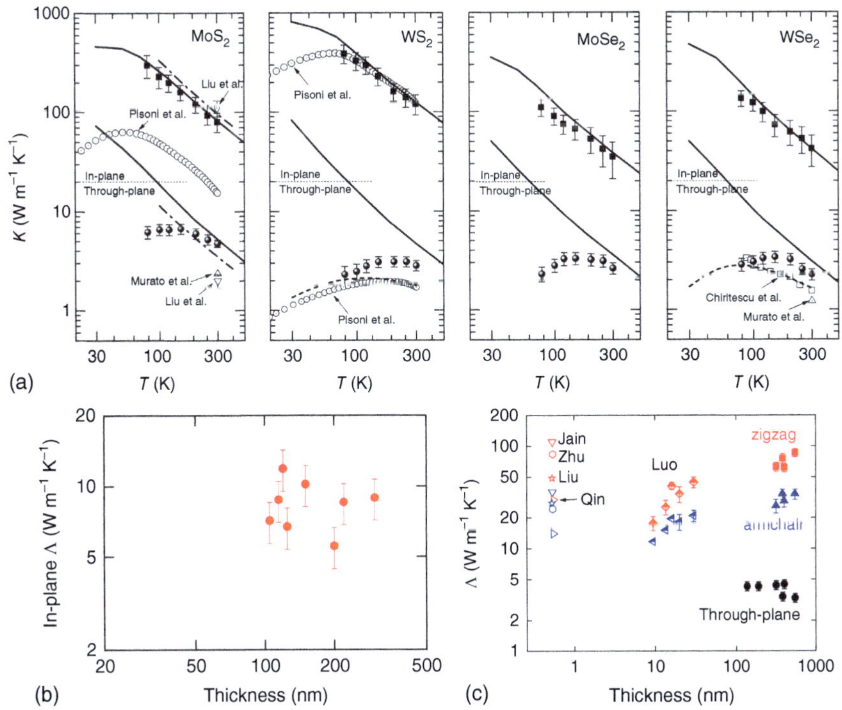

Figure 3.10 (a) Temperature dependent in-plane and out-of-plane thermal conductivities of MoS_2, WS_2, $MoSe_2$ and WSe_2. Source: Jiang et al. [63]/Reproduced with permission from John Wiley & Sons. (b) In-plane thermal conductivity of InSe as a function of thickness [66]. (c) Thickness dependent thermal conductivity of black phosphorus for zigzag, armchair directions and through-plane. Source: Jang et al. [67]/Reproduced with permission from John Wiley & Sons.

the measured nanostructures can be calculated as

$$G_s = G_b \frac{\Delta T_s}{\Delta T_h - \Delta T_s} \tag{3.36}$$

where G_s is the thermal conductance of the nanostructures, G_b is the thermal conductance of the suspended Pt beams, ΔT_h and ΔT_s is the temperature rise for heater and sensor, respectively [46, 52]. The thermal conductivity can thus be obtained once the dimension of the sample is known. Since the temperature gradient is established across the sample, its Seebeck coefficient can be subsequently obtained by measuring the open-circuit voltage together with the known temperature bias ($\Delta T_h - \Delta T_s$). While the measurement of Seebeck coefficient is limited by the fact that the carrier concentration of the flake cannot be tuned by a back- or top-gating like treatment in the FET device and the final power factor can never be modulated to an optimal value.

For thermal bridge measurement, it is challenging to get rid of the effect of the contact thermal resistance, which is typically obtained by either intercepting method or theoretical calculations [46]. To deal with this contact issue, the electron-beam

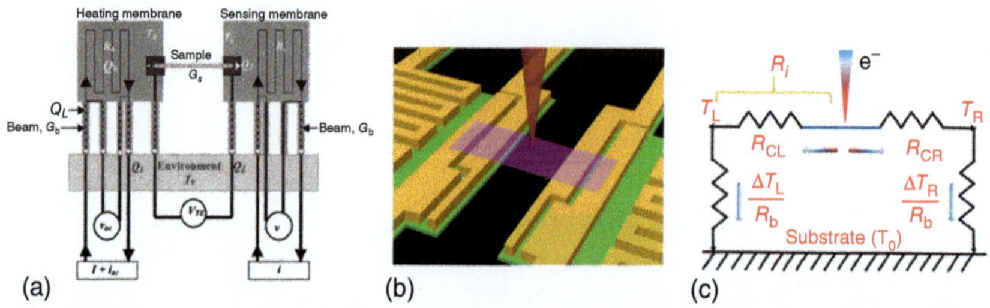

Figure 3.11 Schematic diagram of experimental setup for (a) thermal bridge method. Source: Shi et al. [70]/American Society of Mechanical Engineers. (b) Schematic diagram and (c) equivalent thermal resistance circuit for the e-beam heating technique [71]. Source:

heating technique was developed by Wang et al. in 2011 [72], which provides a spatial resolution measurement as well. Electron beam heating technique was initially developed for the thermal conductance measurement of one dimensional nanostructures [73, 74]. Recently, this technique has been applied for two dimensional materials like few layers MoS_2 and black phosphorus [68, 75]. Different from the thermal bridge method, an electron beam is employed to heat up the sample in the suspended region, as shown in Figure 3.11b. The heat is thus conducted to the two thermometers and further to the substrate with temperature T_{sub}. At the thermal equilibrium state, the cumulative thermal resistance is derived as [71]

$$R_i(x) = R_b \left\{ \frac{a_0 - a_i(x)}{1 + a_i(x)} \right\} \tag{3.37}$$

$$a_i = \frac{\Delta T_h}{\Delta T_s} \tag{3.38}$$

$$a_0 = \Delta T_{h0}/\Delta T_{s0} \tag{3.39}$$

where R_b is the thermal resistance of the suspended Pt beams same with the thermal bridge method, $\Delta T_h(\Delta T_{h0})$ and $\Delta T_s(\Delta T_{s0})$ are the temperature rises in the heater and sensor when the electron beam is on (off), respectively. As shown in Figure 3.12a,c, the cumulative thermal resistance of 2D flakes like few-layer MoS_2 [71] and black phosphorus [68] is obtained when an electron beam shines along the suspended sample. The thermal conductivity of MoS_2 could therefore be extracted once its cross-sectional area is characterized as shown in Figure 3.12b [71]. Another advantage of electron-beam heating technique is its ability to measure the inhomogeneity or the interfacial thermal resistance along the nanostructures [73, 76]. Additional measurement details can be found in literatures [72–74].

3.3.3.4 Other Thermal Property Measurement Methods

Other than the measurement techniques discussed above, there are other methods like H-&T-type method and 3-omega that are used for thermal property measurement of 2D. For H-type (T-type) method, a 2D flake is suspended between two metal nanofilms to form a H (T) shape structure. The metal nanofilm works as

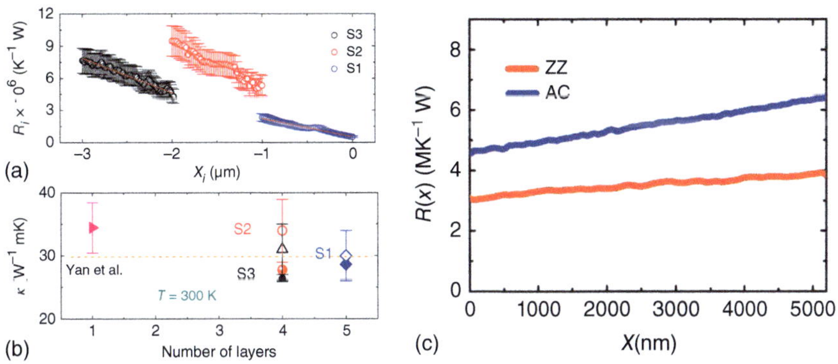

Figure 3.12 (a) Length dependent cumulative thermal resistance of few layer MoS_2 [71]. (b) Thermal conductivity of MoS_2 with different layers obtained by electron beam heating technique [71]. (c) Cumulative thermal resistance of black phosphorus as a function of its length. Source: Zhao et al. [68]/Reproduced with permission from John Wiley & Sons.

both a heater and temperature sensor, as shown in Figure 3.13a [77]. A larger electrical current is applied on nanofilm A and temperature gradient along the 2D flake is thus established. A relatively small current is introduced on the nanofilm B to measure its electrical resistance. By measuring the electrical resistance change of each electrode, the average temperature across the two ends of the sample is obtained. The work principle is much similar to that of thermal bridge, where the thermal conductivity, electrical conductivity and Seebeck coefficient of one 2D flake could be obtained. Readers can refer to literature for more measurement details [77, 81–83]. The H-&T-type method was initially employed to measure the thermal transport in one dimensional nanostructures [82]. Recently, Wang et al. has developed to characterize the thermal conductivity of suspended monolayer graphene [78] as shown in Figure 3.13b. By introducing well-controlled defects in one side of the graphene, a thermal rectification factor around 26% is achieved.

For 3-omega (3ω) method, Cahill and Pohl firstly developed to measure both the thermal conductivity and the specific heat of bulk amorphous materials [84, 85]. Basically, an AC current with a frequency of 1ω is applied the metal electrode, which services as both a heater and thermometers, as shown in Figure 3.13c [79]. Due to Joule heating, a temperature rise at a frequency of 2ω is created in the metal electrode, leading to a further AC voltage fluctuation with a frequency of 3ω. By measuring the 3ω voltage, the heat that diffuses into the substrate is obtained, which contains the information related to the thermal property of measured samples. By using the 3ω method, the interface thermal resistance across the 2D flakes with various substrates is obtained, as shown in Figure 3.13d [80]. One issue related to 3ω method measurement is that the simple one-dimensional heat transfer model and diffusive heat transport are assumed. Both the dimensions of metal electrode and the thickness of the specimen should be thus well-designed. Considering the fact that the metal electrode directly contacts with sample and a dielectric film is typically employed on top of the sample if it is electrically conductive, which unavoidable decreases the measurement resolutions [86, 87]. All of these factors

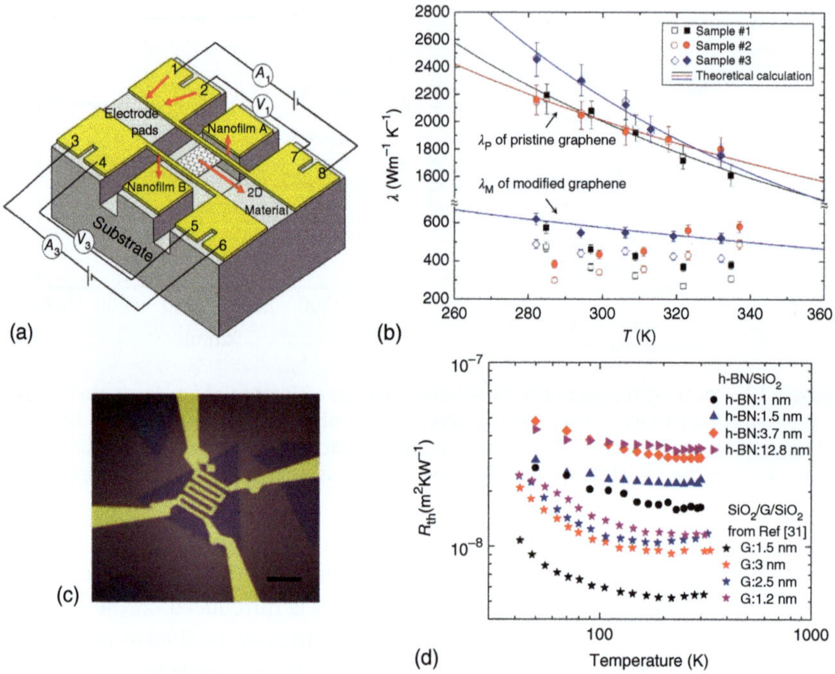

Figure 3.13 (a) Schematic of H-type measurement for a suspended 2D flake. Source: Zhao and Wang [77]/Reproduced with permission from Engineered Science Publisher. (b) Temperature thermal conductivity of monolayer graphene characterized by H-type method. Source: Wang et al. [78]/Springer Nature/Licensed under CC BY 4.0. (c) Setup of three omega measurement with Au electrodes [79]. (d) Interfacial thermal resistance of h-BN/SiO_2 and SiO_2/graphene/SiO_2 [80].

should be considered for further measuring the in-plane thermal conductivity of suspended 2D materials using 3ω method.

3.4 Manipulation of TE Properties in 2D

2D materials provide an attractive platform to study the effects of discretized DOS on TE performance owing to the atomically clean surfaces, lack of dangling bonds, and if prepared carefully, no roughness of the 2D surface and hence minimal quantum well thickness variation. Particularly, the family of TMDs is favorable for thermoelectrics because of their unique structural properties, chemical stability and attractive semiconducting characteristics [88]. The discretization of DOS is expected to enhance the power factor ($S^2\sigma$), and the relatively high electrical conductivity and low thermal conductivity in TMDs which makes them suitable for thermoelectric applications [3, 89]. In addition, the atomically clean and smooth surfaces and the large surface-to-thickness ratio of TMDs allows effective modulation or modification of their thermoelectric properties through various techniques such as carrier concentration modulation [7, 8, 10, 22, 36, 90], strain engineering [91–93], phase engineering [37, 90, 94] and moire superlattices [95–97]. The ease of modulating or modifying thermoelectric properties thus provides an effective means to decouple the interdependency between S and σ, which leads to new scattering physics and physical phenomena, thereby offering new pathways to tune and optimize thermoelectric properties and performance. In following sections, strategies that can modulate and modify thermoelectric properties are introduced and summarized with their key findings, with more complete discussions available in literature [2, 3, 89, 98].

3.4.1 Tuning of Carrier Concentration

The tuning of carrier concentration in TMDs can be easily achieved by the application of an external electric field (field-effect), providing an additional degree of freedom to modulate and optimize TE properties and performance, as illustrated in Figure 3.14. Modulation of carrier concentration by field-effect is also a reversible process that does not alter the intrinsic properties of the material. Consequently, the dependence of thermoelectric properties as a function of carrier concentration can be easily studied in 2D materials. Field-effect modulation of carrier concentration is commonly achieved by the back-gate through a dielectric layer (Figure 3.14a) or by the use of ionic gating (Figure 3.14b) as demonstrated in TMD layers where power factors as high as 8.5 and 4 mW m^{-1} K^{-2} were reported as shown in Figure 3.14c,d, respectively [8, 10, 90]. Using field-effect modulation, different levels of carrier concentrations can be achieved in a single material/device, typically ranging up to 10^{14} cm^{-2} for TMDs. Additionally, the field-effect modulation can also act to tune the scattering mechanism dominating charge transport through screening of Coulombic interactions and reducing effects of electron-phonon scattering on electronic and thermoelectric transport. For example, F. Yang et al. demonstrated a gate-tunable transition from POP scattering to piezoelectric scattering in Bi_2O_2Se [22]. At high doping levels, electrostatically injected electrons has also been demonstrated to carry enough energy to induce a structural phase transition [90], to attain a superconducting state [99], and even recently predicted to alter electron–phonon interactions and thus electron-phonon scattering to modulate electronic properties [100].

3.4.2 Strain Engineering

Strain engineering has been demonstrated as an effective strategy to tune electronic and thermal properties of TMDs. Owing to their extraordinary mechanical properties, TMD monolayers can withstand lattice deformations of more than 10% before breaking [101]. Therefore, strain offers a promising avenue for improving electronic, thermal and thermoelectric properties of TMDs. A comprehensive review on the effects of strain on thermoelectric properties is available in literature [93, 102, 103]. For example, strain has been demonstrated to reduce thermal conductivity by softening phonon modes [104, 105], modifying electronic band structure [105, 106], induce phase transitions [94, 107] and enhance carrier mobility [108]. Strain can be tensile or compressive, each can be introduced via many sources that can include lattice mismatch, surface roughness, differences in thermal expansion coefficient, or when deposited on flexible substrates (bending, stretching or twisting) [109]. Generally, by controlling the amount of deformation of the substrate, the amount of strain exerted on 2D materials can be controlled. Separate studies by Conley et al.

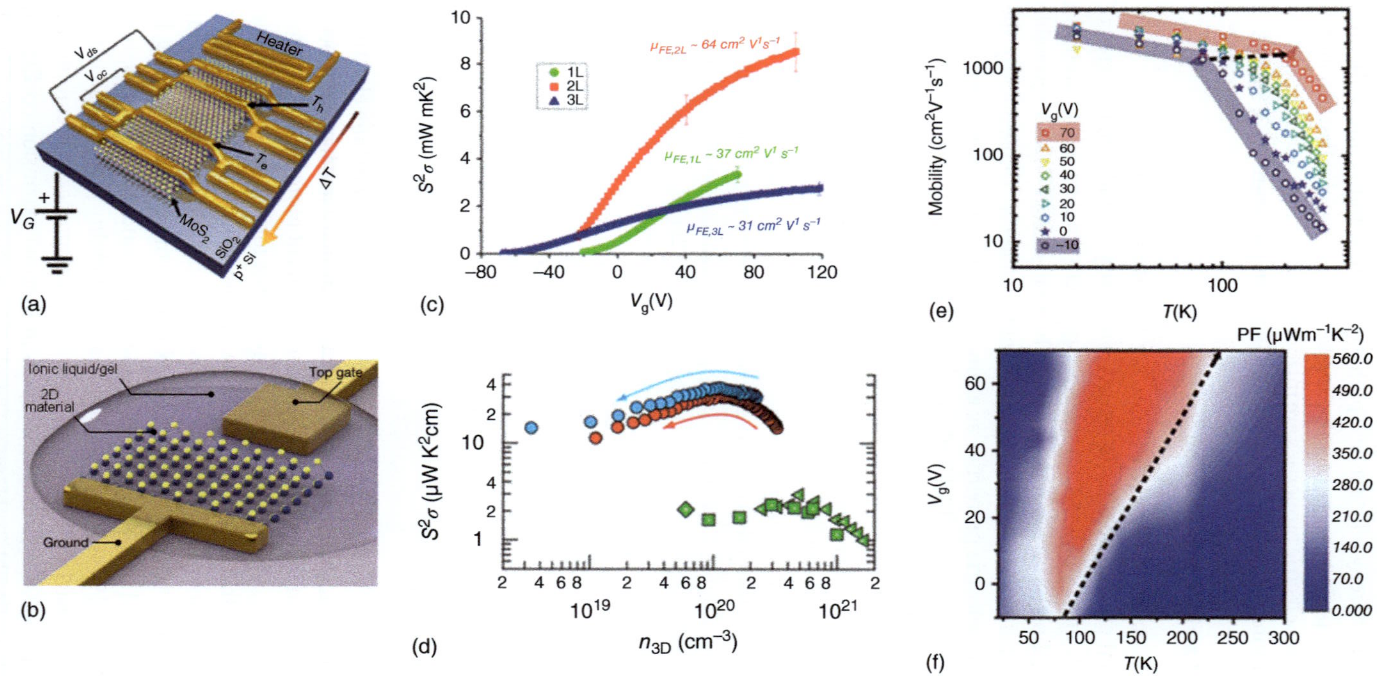

Figure 3.14 Typical thermoelectric devices employing the field-effect modulation technique via (a) SiO_2 dielectric layer and (b) ionic liquid. Source: Ng et al. and Wang et al. [36, 90]/Reproduced with permission from AIP Publishing (c) Power factor of monolayer, bilayer and trilayer MoS_2 as a function of the backgate voltage (V_g). The bilayer MoS_2 exhibits the highest power factor at 300 K owing to its larger valley degeneracy and higher effective mass [8]. Source: Reproduced with permission from Hippalgaonkar et al. [8]. Copyright © 2017, American Physical Society. (d) Power factor of n-type (red circles) and p-type (blue circles) trilayer WSe_2 as a function of gating realized by ionic gating [10]. Source: Reproduced with permission from Yoshida et al. [10]. Copyright © 2016, American Chemical Society. (e) Gate-tunable mobility and (f) thermoelectric power factor as a function of temperature revealing the transition of scattering mechanism in Bi_2O_2Se [22]. Source: Yang et al. [22]/Reproduced with permission from John Wiley & Sons.

and Christopher et al. have shown that, the application of uniaxial strain leads to phonon softening and the linear reduction of optical band gap and Raman modes (A′ and E′) by ~38 meV/% strain and ~0.75 cm^{-1}/% as well as ~2.1 cm^{-1}/% strain for monolayer MoS_2, respectively [105, 110].

Strain has also been predicted and demonstrated to not only influence the band gap but also enhance carrier mobility in TMDs by lowering effective mass, suppressing electron-phonon coupling and reducing lattice scattering [91, 111–113]. Hosseini et al. have shown in a study relating the mobility of TMDs as a function of uniform biaxial tensile strain and found that carrier mobility can be enhanced by up to 84% [108]. More recently, strain have been employed to induce lattice distortion that breaks lattice symmetry by using bulged substrates to create ripples in MoS_2 (Figure 3.15a). Such symmetry breaking was reported to simultaneously reduce electron-phonon scattering (Figure 3.15b,c) and enhance intrinsic dielectric screening (Figure 3.15d) to dramatically enhance carrier mobility (~900 $cm^2\ V^{-1}\ s^{-1}$) in bilayer rippled MoS_2 (Figure 3.15e). Such magnitudes of carrier mobility in rippled MoS_2 are more than two orders of magnitude higher than in flat MoS_2 at low carrier concentrations, which is also well above its phonon-limited mobility (~200–410 $cm^2\ V^{-1}\ s^{-1}$) [114–116]. The enhanced carrier mobility then dramatically improves electrical conductivity while maintaining high magnitudes of Seebeck coefficient, ultimately leading to improvements of up one order of magnitude in thermoelectric power factor at room temperature reaching 7.5 $mW\ m^{-1}\ K^{-2}$ at a carrier concentration of about $10^{19}\ cm^{-3}$ in bilayer rippled MoS_2 (Figure 3.15f,g) [39]. Such high values of thermoelectric power factor also represents the highest value ever reported in the measured carrier concentration range. In addition to this excellent electronic improvement, the rippled MoS_2 based thermoelectric devices are expected to reduce lattice thermal conductivity and further enhance the figure of merit values. Lattice strain have been reported to greatly reduce lattice thermal conductivity by substantially shortening the phonon lifetime [113, 117, 118]. Remarkably, the strategy of rippled MoS_2 effectively results in large lattice strain, which breaks the strong trade-off among the various parameters that jointly determine the thermoelectric performance, simultaneously realizing high electrical conductivity, large Seebeck coefficient, and suppressing lattice thermal conductivity. They synergistically contribute to the drastically improved performance in rippled MoS_2 based thermoelectric devices. Furthermore, such strain-induced symmetry breaking was also demonstrated to modulate spin dynamics in rippled bilayer MoS_2 where an enhanced spin-orbit splitting was observed [119]. The ability of strain to tune and modulate the electronic and thermoelectric properties of TMDs makes it a promising strategy to decouple interdependencies between thermoelectric coefficients for enhancing thermoelectric performance.

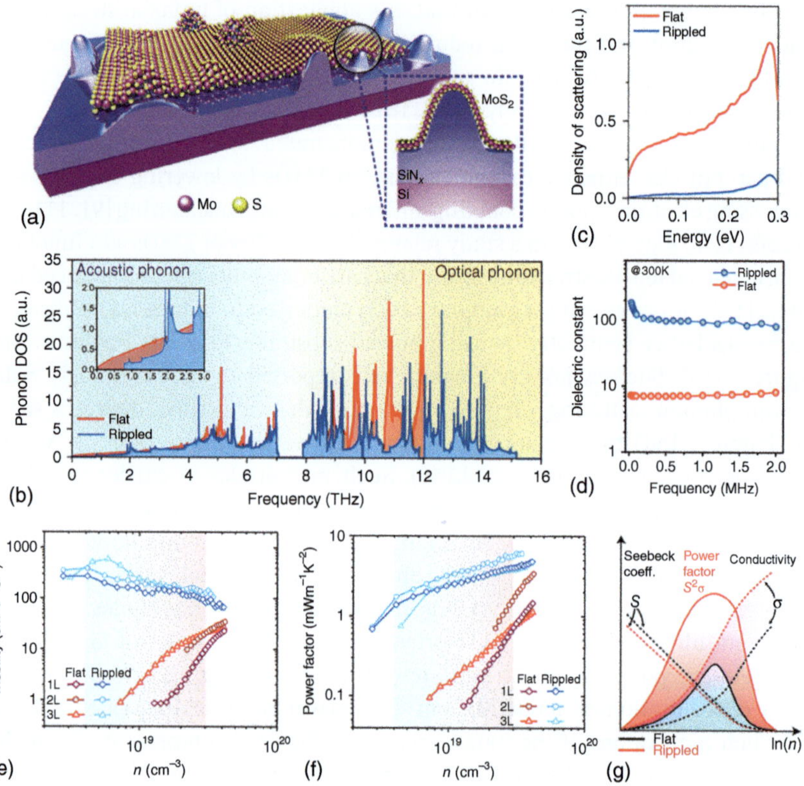

Figure 3.15 (a) Illustration of rippled-MoS_2 on a bulged substrate with a root-mean-square surface roughness of 2 nm. (b) Renormalization of the phonon density of states (DOS) in rippled bilayer MoS_2 (red) in comparison with flat MoS_2 (blue), as well as its corresponding (c) density of scattering and (d) intrinsic dielectric constant $\varepsilon_{\perp}$. (e) Field-effect mobility and (f) thermoelectric power factor of rippled (red) and flat (blue) MoS_2 as a function of carrier concentration at 300 K. (g) Illustration of enhanced thermoelectric power factor in rippled MoS_2 as a result of enhanced carrier mobility. All panels are taken from. Source: Ng et al. [39]/Springer Nature.

3.4.3 Band Engineering

The electronic band structure of 2D semiconductors is crucial because TE performance depends on the material's band gap, valley degeneracy and carrier effective mass, all of which can be modified with band engineering. The electronic band structure is typically engineered using two strategies namely layer thickness and band convergence.

3.4.3.1 Layer Thickness and Band Convergence

The layer thickness (number of layers) of 2D materials have also been demonstrated to tune valley degeneracy as well as the band structure. As an example, Figure 3.16 illustrates the evolution of the band structure of MoS_2 as its thickness decreases from a bulk to four layers, bilayer and then monolayer, where the widely known indirect-to-direct bandgap transition can be observed [120]. Importantly, as the

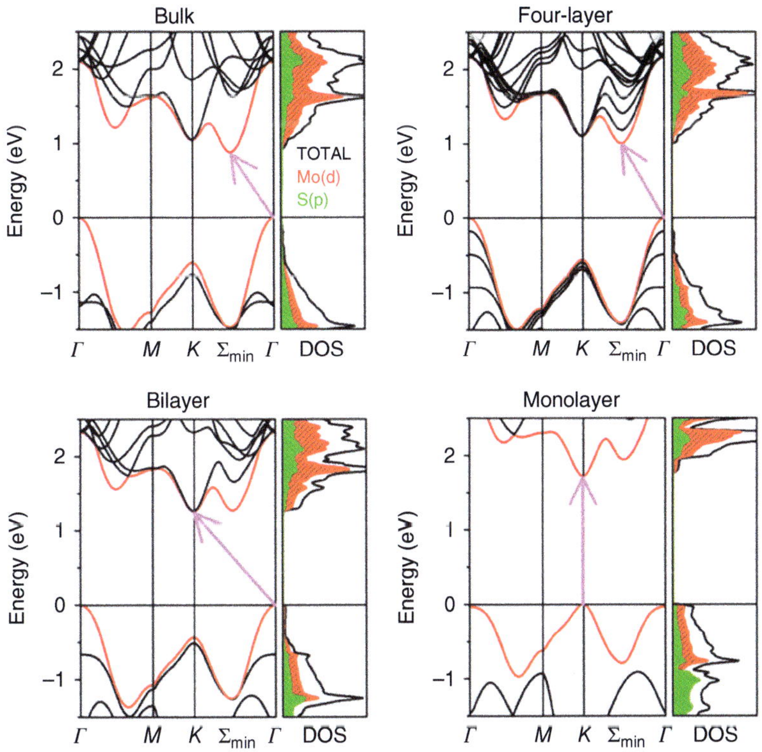

Figure 3.16 Band structures and density of states of bulk, four-layer, bilayer, and monolayer MoS_2. Direct/indirect band gaps are represented by the pink arrows [120]. Source: Reproduced with permission from Hong et al. [120]. Copyright © 2016, American Physical Society.

number of layers decreases, a shift in the CBM and valence band maximum (VBM) from the Σ_{min} and Γ to the κ-point can be observed. In the conduction band, the valley degeneracy of the κ-point is only 2 but the same for the Σ_{min}-point is 6, which results in a larger effective mass that can yield a larger value of Seebeck coefficient without deteriorating carrier mobility. The increase in valley degeneracy also corresponds to the convergence of conducting bands which enhances the electrical conductivity [120, 121]. From Figure 3.14c, the effect of layer thickness on thermoelectric power factor is demonstrated in bilayer MoS_2 where the κ-point and Σ_{min}- point can result in a net 10-fold degeneracy that can enhance Seebeck coefficient and electrical conductivity from a larger effective mass and the convergence of conducting bands, respectively [8]. Similar layer thickness dependence has also been observed in other TMD crystals [10, 122]. The layer thickness dependence of MoS_2 on thermoelectric performance was systematically studied where bilayer MoS_2 was demonstrated to exhibit a thermoelectric performance 200% better than its monolayer and trilayer counterparts as shown in Figure 3.17 [7, 8]. This enhancement in bilayer MoS_2 is attributed to larger valley degeneracy and higher accessible DOS [7, 8]. Another excellent example of band convergence is demonstrated in

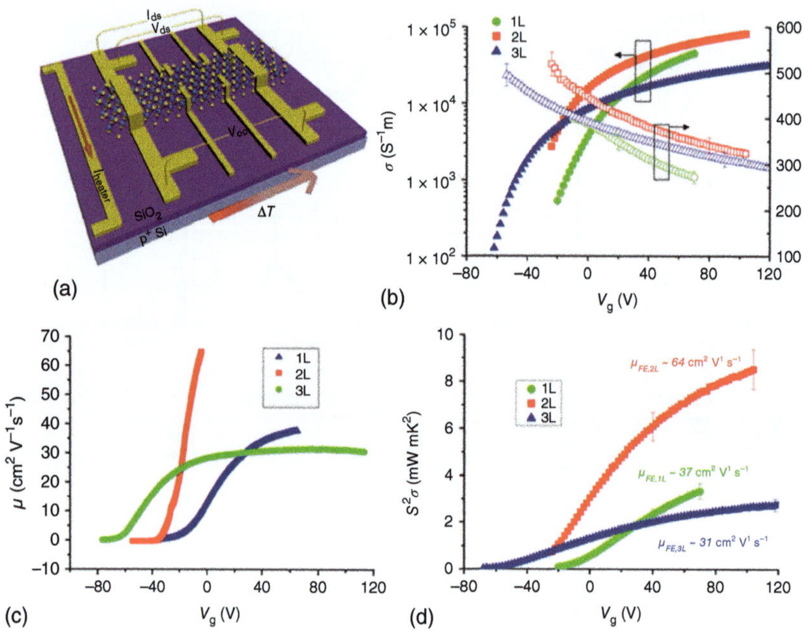

Figure 3.17 (a) Schematic of device structure for thermoelectric measurements [8]. Comparison of thermoelectric properties between monolayer, bilayer and trilayer MoS_2 on SiO_2/Si substrate: (b) Electrical conductivity (σ) and Seebeck coefficient (S), (c) Field-effect mobility (μ) and (d) power factor ($S^2\sigma$). Source: All panels are taken from [8].

$PdSe_2$ as shown in Figure 3.18a,b. Figure 3.18c illustrates the thinning down of $PdSe_2$ leads to the reduction of the energy difference between two conduction band minima (ΔCBM) and thus improved band convergence. The thermoelectric power factor in thin $PdSe_2$ (5 nm) is enhanced by up to 300% compared to thick $PdSe_2$ (9 nm) [40]. Therefore, band engineering in TMDs provides an effective pathway to manipulate their electronic band structure to decouple thermoelectric coefficients and enhance thermoelectric performance.

3.4.4 Phase Transition

The polymorphism of TMDs provides another pathway towards optimizing thermoelectric performance by phase engineering [123]. A TMD monolayer can either exhibit the 2H phase or 1T phase. The 2H phase has a hexagonal symmetry which corresponds to a trigonal prismatic coordination of the metal atoms, which is semiconducting in nature and thermodynamically stable. The 1T phase has a tetragonal symmetry which corresponds to an octahedral coordination of the metal atoms, which is metallic in nature but is thermodynamically unstable. The 1T phase is usually stabilized by a Mott transition accompanied by lattice distortion that results in a distorted-1T phase (1T′ phase) that is semi-metallic in nature and achievable by the intercalation of alkali ions [124, 125], direction solution synthesis [126, 127], electrostatic doping [90, 120, 128] or strain [94]. The different phases of TMDs are shown in Figure 3.19 [129].

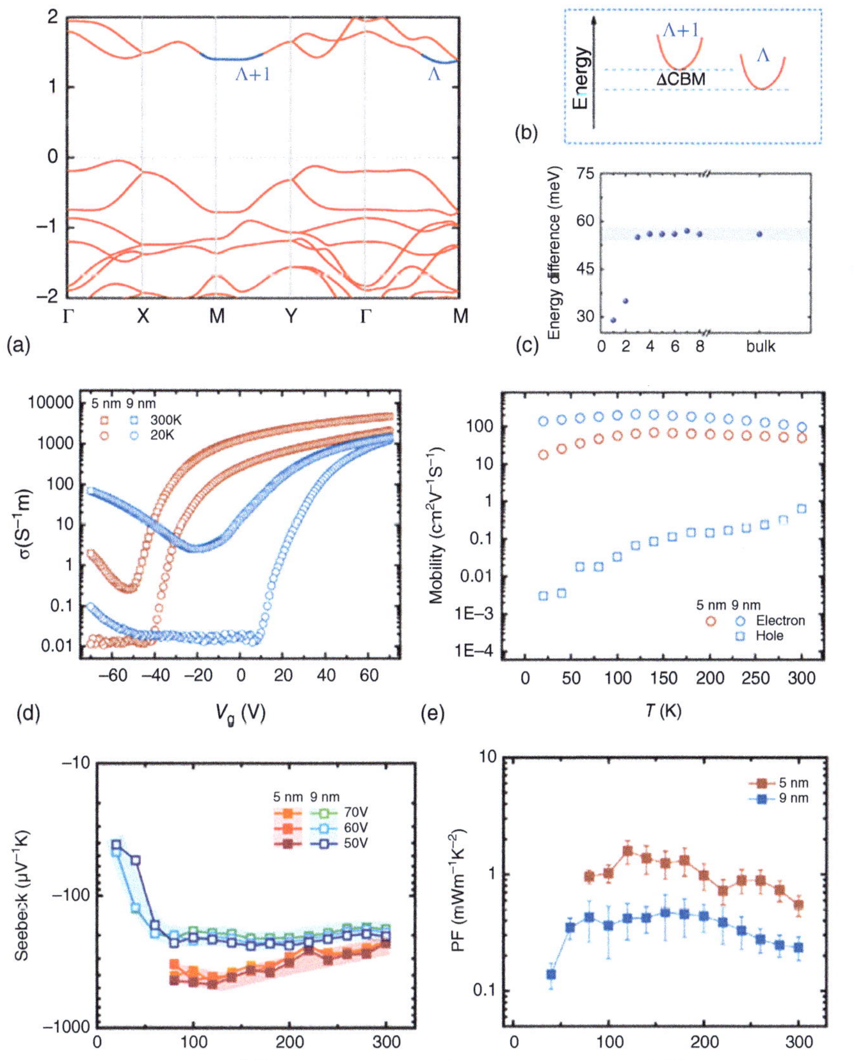

Figure 3.18 (a) Electronic band structure of monolayer $PdSe_2$. (b) Illustration of converging conduction band minimum (ΔCBM) marked in (a). (c) Layer-dependent energy difference of ΔCBM. Thermoelectric properties of $PdSe_2$ flakes with different thicknesses: (d) Electrical conductivity (*σ*), (e) Field-effect mobility, (f) Seebeck coefficient and (g) Power factor. Source: Zhao et al. [40]/Reproduced with permission from John Wiley & Sons.

The semi-metallic nature of 1T′-phase can be beneficial for electron transport owing to their small electronic band gap (usually tens of meV), therefore promoting electrical conductivity due to the small activation energy required. This semi-metallic property of 1T′-phase has been employed to achieve low-resistance contacts for transistors and electronic applications [130]. Moreover, the 1T′-phase has a lower lattice symmetry compared to the 2H phase which can potentially reduce phonon contribution to thermal conductivity [131]. The 1T′ phase also possesses

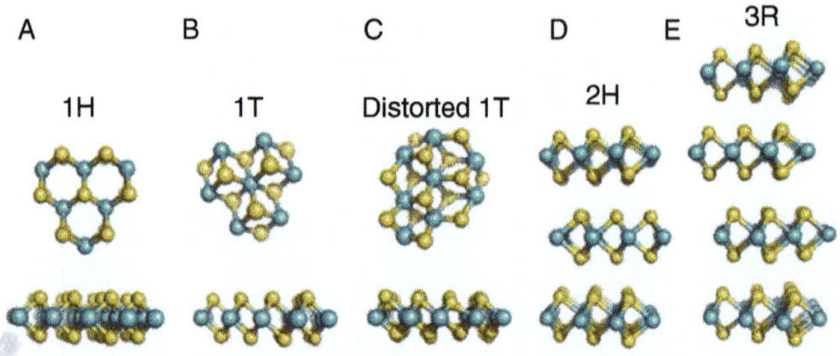

Figure 3.19 Different polymorphs or phases of single-layer and multi-layer TMDs ranging from (A) 1H phase, (B) 1T phase, (C) distorted 1T phase (1T′ phase), (D) 2H phase and (E) 3R phase [129]. Source: Reproduced with permission from Voiry et al. [129]. Copyright © 2015, Royal Society of Chemistry.

reduced effective mass and deformation potential, thus increasing the drift velocity of carriers and reducing carrier-phonon interaction and hence scattering to promote electrical conductivity [123]. However, the asymmetry of the DOS is limited in a semi-metallic system and thus the charges have a lower (almost metal-like) Seebeck coefficient. Theoretical calculations on 1T′-phase TMDs predicted power factors of up to $0.4\,W\,m^{-1}\,K^{-2}$ and ZT values as high as 0.6, attributing these enhancements mostly to the reduction in thermal conductivity and enhancements in electrical conductivity [123]. In addition, experimentally, thermoelectric performance is constrained by ambipolar conduction from the small electronic band gap which quenches Seebeck coefficient albeit excellent electrical conductivity [125].

Another exciting and promising aspect of phase engineering for enhanced thermoelectrics is its potential to induce carrier filtering effect, also termed the energy filtering effect. From Eq. (3.20), the Seebeck coefficient is dependent on the function, $\frac{df_0}{dE}$ where cold carriers ($E < E_F$) have negative contribution to Seebeck coefficient than hot carriers ($E > E_F$) as illustrated in Figure 3.20. The carrier filtering effect enhances Seebeck coefficient by preferentially filtering away cold carriers which alters the carrier scattering mechanism to enhance Seebeck coefficient without adversely compromising electrical conductivity. Carrier filtering effect can be induced by introducing dopants or grain boundaries that act as energy barriers to scatter cold carriers so that only hot carriers have sufficient energy to pass through these barriers (Figure 3.21) [132, 134, 135]. These energy barriers can also act as phonon scatterers to reduce the phonon mean free path and thus thermal conductivity. In this regard, by designing a system capable of tuning the relative phase compositions of 2H phase and 1T′ phase within a single TMD material can produce phase boundaries that act as energy barriers to enhance thermoelectric performance. For comprehensive reviews on phase engineering and carrier filtering effect on thermoelectric performance, please refer to references [123, 131, 132, 134, 136].

Phase engineering and its relative phase composition in TMD systems can be beneficial for TE properties. Ng et al. examined the effects of phase composition in mixed-phase Li_xMoS_2 and how it regulated TE properties as displayed in Figure 3.22a [37]. It is shown that with a decreasing proportion of the 1T/1T′ phase and a corresponding increase of the 2H-phase, the system registers a crossover from p- to n-type carrier conduction where the Seebeck coefficient converts from positive

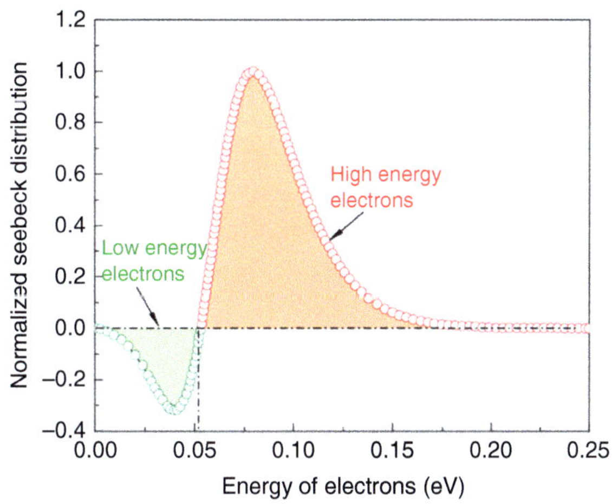

Figure 3.20 Relationship between the energy of carriers (electrons in this case) and the normalized Seebeck distribution for a degenerately doped n-type SiGe alloy [132, 133]. Source: Mao et al. and Minnich et al. [132, 133]/Reproduced with permission from Royal Society of Chemistry.

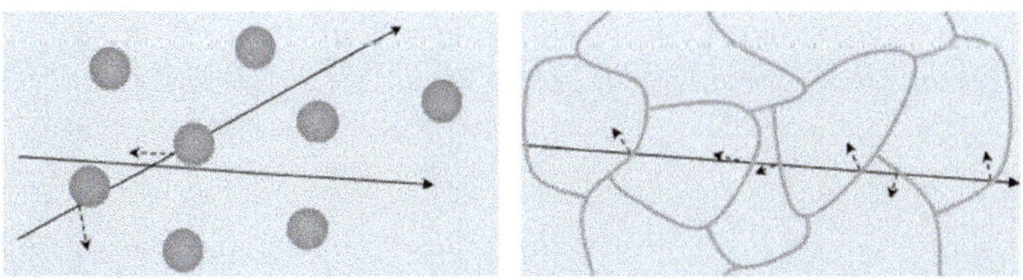

Figure 3.21 Carrier filtering by (left) randomly distributed dopants or nanoprecipitates within the material and (right) grain boundaries in polycrystalline and nanostructured materials [134]. Source: Narducci et al. [134]/Reproduced with permission from ELSEVIER.

to negative. From 20 mW m^{-1} K^{-2}, the thermoelectric power factor decreased nearly to zero with an initial increase in H-phase proportion. This is followed by an increase to 27 mW m^{-1} K^{-2} with further annealing (Figure 3.22b). Modulation of thermoelectric performance via phase engineering is also demonstrated in 300 nm thick Li_xMoS_2 nanosheets with dominant 1T′-phase, where 1T′-to-2H phase transformation is obtained through thermal annealing as shown in Figure 3.22c [125]. In thicker mixed-phase Li_xMoS_2 nanosheets, the relative composition of 1T′-2H phases becomes a challenge to control. Moreover, the higher concentration of structural disorder and defects due to phase transformation led to poorer thermoelectric performance as observed in Figure 3.22d. These findings suggest that the hybrid electronic properties in thin mixed-phase Li_xMoS_2 system allow for the control of thermoelectric performance by tuning between the 1T/1T′ and 2H phases.

3.5 Future Outlook and Perspective

In the past decade, significant progress has been achieved in the development of thermoelectric materials, where theoretical and experimental findings have inspired novel and rich physics in layered and 2D materials, not only in fundamental thermoelectric studies but also for its practical purposes. In this chapter, we have highlighted some key achievements in thermoelectric studies of TMD systems, mostly

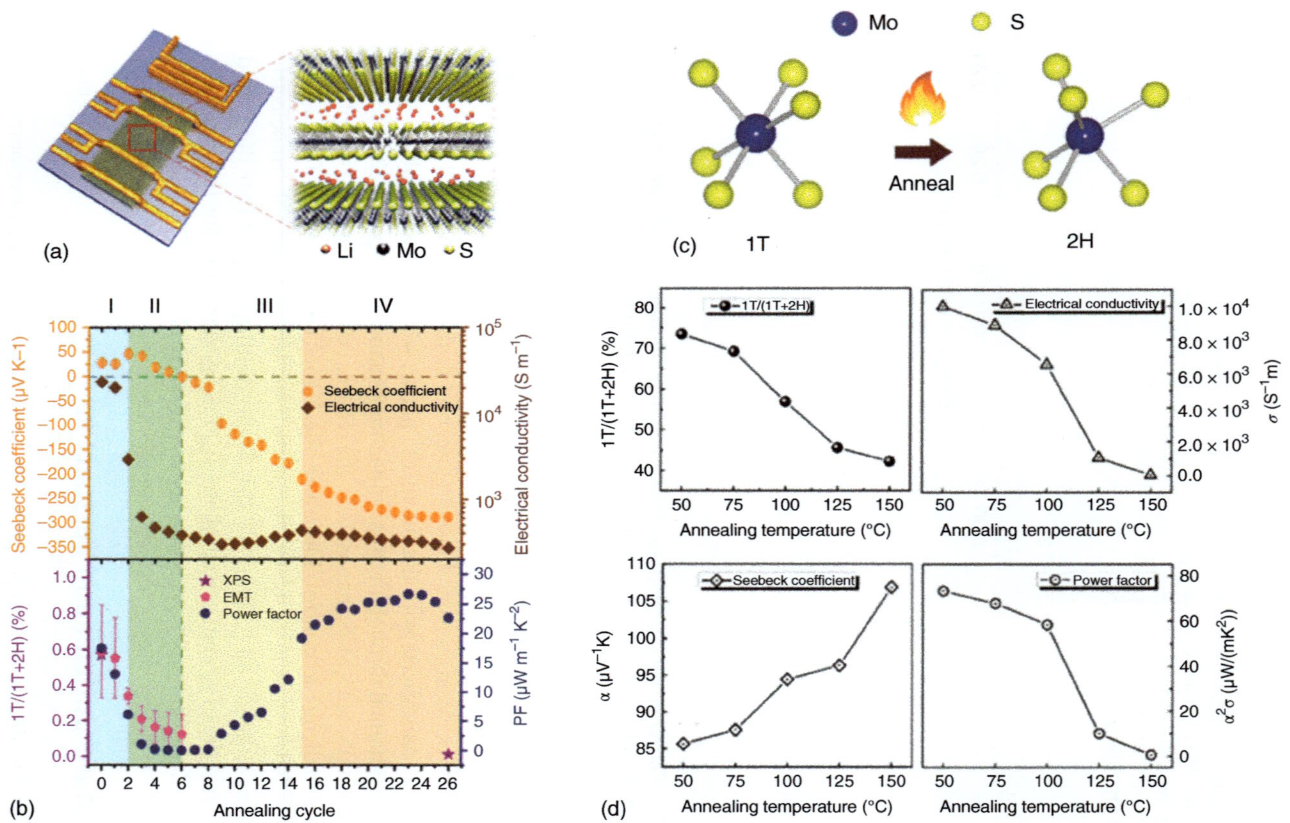

Figure 3.22 Experimental study on the thermoelectric properties of metallic phase TMD systems. (a) Schematic depicting the Li_xMoS_2 device with Li^+ ions intercalated within the interlayer van der Waals gaps. Regions near the intercalated Li^+ ions are the 1T′-phase, while the remaining are in the 2H-phase [37]. (b) Top: Seebeck coefficient and electrical conductivity; Bottom: thermoelectric power factor (PF) of the Li_xMoS_2 device across the annealing cycles. The PF maximizes to ~27 mW m^{-1} K^{-2} with further annealing cycles [37]. (c) Schematic diagram of 1T-to-2H phase transition via thermal annealing [125]. (d) Thermoelectric properties of Li_xMoS_2 nanosheet as a function of relative 1T-phase composition [125].

in the past decade. On top of the intrinsic advantages brought about in 2D materials which include confinement of DOS, electron correlations and energy-dependent scattering, formulated strategies ranging from carrier concentration tuning, band convergence and degeneracy, ripples-induced-strain to phase engineering create exciting avenues and exploration space for manipulating thermoelectric transport.

Understanding and then employing the new physics in charge transport in 2D materials will become beneficial for next-generation electronics such as FETs, thermoelectric devices and other novel functional devices. The development of such next-generation electronics will contribute largely towards the realization of a smart nation, ubiquitous and self-powered electronics. To this end, the 2D platform offers a possible solution towards wearable devices with maintenance-free batteries by employing thermoelectric energy conversion. In recent years, machine learning and computational methods has been developed to be one of the most exciting tools for materials science. These high-throughput computational methods have proven its potential of speeding up both fundamental and applied research by identifying potential 2D materials as candidates for specific applications in transistors, sensors, or thermoelectric devices [123, 137–143]. Despite the above, a systematic and combined database that allows for the comparison of bulk and monolayer thermoelectric properties of materials is still lacking. In addition, such a systematic database of thermoelectric properties is necessary to develop machine learning models for predicting the thermoelectric properties of new materials, which would circumvent the high computational cost of additional DFT calculations and potentially guide materials discovery. While there have been a few recent reports on the applications of machine learning for thermoelectric properties, the field is still developing and its translation from academia to industry remains to be seen [138, 144, 145].

While the vast number of possibilities to optimize and attain breakthrough in electronic, thermal and thermoelectric performance in 2D TMD systems remain promising, substantial further studies in the future are required to continue discovering new physics, especially with the advent of high-throughput machine learning tools available today. These new capabilities will surely spawn many exciting discoveries in the near future.

References

1 Jouhara, H., Khordehgah, N., Almahmoud, S. et al. (2018). Waste heat recovery technologies and applications. *Therm. Sci. Eng. Prog.* 6: 268–289.

2 Li, D., Gong, Y., Chen, Y. et al. (2020). *Recent progress of two-dimensional thermoelectric materials. Nano-Micro Lett.* 12: 36.

3 Wu, J., Chen, Y., Wu, J., and Hippalgaonkar, K. (2018). Perspectives on thermoelectricity in layered and 2D materials. *Adv. Electron. Mater.* 4: 1800248.

4 Hicks, L.D. and Dresselhaus, M.S. (1993). Thermoelectric figure of merit of a one-dimensional conductor. *Phys. Rev. B* 47: 16631–16634.

5 Hicks, L.D. and Dresselhaus, M.S. (1993). Effect of quantum-well structures on the thermomagnetic figure of merit. *Phys. Rev. B* 47: 727–731.

6 Rowe, D.M. (2006). *Thermoelectrics Handbook: Macro to Nano*. CRC Press.

7 Kayyalha, M., Maassen, J., Lundstrom, M. et al. (2016). Gate-tunable and thickness-dependent electronic and thermoelectric transport in few-layer MoS_2. *J. Appl. Phys.* 120: 134305.

8 Hippalgaonkar, K., Wang, Y., Ye, Y. et al. (2017). High thermoelectric power factor in two-dimensional crystals of MoS_2. *Phys. Rev. B* 95: 115407.

9 Wei, P., Bao, W., Pu, Y. et al. (2009). Anomalous thermoelectric transport of dirac particles in graphene. *Phys. Rev. Lett.* 102: 1–4.

10 Yoshida, M., Iizuka, T., Saito, Y. et al. (2016). Gate-optimized thermoelectric power factor in ultrathin WSe_2 single crystals. *Nano Lett.* 16: 2061–2065.

11 Zuev, Y.M., Chang, W., and Kim, P. (2009). Thermoelectric and magnetothermoelectric transport measurements of graphene. *Phys. Rev. Lett.* 102: 1–4.

12 Walsh, A., Buckeridge, J., Catlow, C.R.A. et al. (2013). Limits to doping of wide band gap semiconductors. *Chem. Mater.* 25: 2924–2926.

13 Lundstrom, M. (2000). *Fundamentals of Carrier Transport*. Cambridge University Press https://doi.org/10.1017/CBO9780511618611.

14 Pichanusakorn, P. and Bandaru, P. (2010). Nanostructured thermoelectrics. *Mater. Sci. Eng., R* 67: 19–63.

15 Lundstrom, M. and Jeong, C. (2011). Colloquium : A Landauer Perspective on Thermoelectrics. Purdue University.

16 Jeong, C., Kim, R., and Lundstrom, M.S. (2012). On the best bandstructure for thermoelectric performance: a Landauer perspective. *J. Appl. Phys.* 111: 113707.

17 Mohankumar, N. and Natarajan, A. (1995). The accurate numerical evaluation of half-order Fermi-Dirac integrals. *Phys. Status Solidi* 188: 635–644.

18 Kim, R., Wang, X., and Lundstrom, M. (2019). *Notes on Fermi-Dirac Integrals*, 4e. arXiv, Purdue University, 1–15. https://arxiv.org/abs/0811.0116.

19 Cutler, M. and Mott, N.F. (1969). Observation of Anderson localization in an electron gas. *Phys. Rev.* 181: 1336–1340.

20 Zhao, Y., Yang, L., Kong, L. et al. (2017). Ultralow thermal conductivity of single-crystalline porous silicon nanowires. *Adv. Funct. Mater.* 27: 1–8.

21 Shuai, J., Mao, J., Song, S. et al. (2017). Tuning the carrier scattering mechanism to effectively improve the thermoelectric properties. *Energy Environ. Sci.* 10: 799–807.

22 Yang, F., Wu, J., Suwardi, A. et al. (2021). Gate-tunable polar optical phonon to piezoelectric scattering in few-layer Bi_2O_2Se for high-performance thermoelectrics. *Adv. Mater.* 33: 2004786.

23 Das Sarma, S., Adam, S., Hwang, E.H., and Rossi, E. (2011). Electronic transport in two-dimensional graphene. *Rev. Mod. Phys.* 83: 407–470.

24 Ishigami, M., Chen, J.H., Cullen, W.G. et al. (2007). Atomic structure of graphene on SiO_2. *Nano Lett.* 7: 1643–1648.

25 Mitta, S.B., Choi, M.S., Nipane, A. et al. (2021). Electrical characterization of 2D materials-based field-effect transistors. *2D Mater* 8: 012002.

26 Ortiz-Conde, A., García Sánchez, F.J., Liou, J.J. et al. (2002). A review of recent MOSFET threshold voltage extraction methods. *Microelectron. Reliab.* 42: 583–596.

27 Cheng, Z., Pang, C.-S., Wang, P. et al. (2022). How to report and benchmark emerging field-effect transistors. *Nat. Electron.* 5: 416–423.

28 Chien, Y.C., Feng, X., Chen, L. et al. (2021). Charge carrier mobility and series resistance extraction in 2D field-effect transistors: toward the universal technique. *Adv. Funct. Mater.* 31: 2105003.

29 Nasr, J.R., Schulman, D.S., Sebastian, A. et al. (2019). Mobility deception in nanoscale transistors: an untold contact story. *Adv. Mater.* 31: 1–9.

30 Baugher, B.W.H., Churchill, H.O.H., Yang, Y., and Jarillo-herrero, P. (2013). Intrinsic electronic transport properties of high quality monolayer and bilayer MoS_2. *Nano Lett.* 13: 4212–4216.

31 Hall, E.H. (1879). On a new action of the magnet on electric currents. *Am. J. Math.* 2: 287–292.

32 Cho, K., Park, W., Park, J. et al. (2013). Electric stress-induced threshold voltage instability of multilayer MoS_2 field effect transistors. *ACS Nano* 7 (9): 7751, 7751–7758, 7758.

33 Kim, C., Issarapanacheewin, S., Moon, I. et al. (2020). High-electric-field-induced phase transition and electrical breakdown of $MoTe_2$. *Adv. Electron. Mater.* 6: 1–7.

34 Ahmed, F., Kim, Y.D., Choi, M.S. et al. (2017). High electric field carrier transport and power dissipation in multilayer black phosphorus field effect transistor with dielectric engineering. *Adv. Funct. Mater.* 27: 1–9.

35 Wu, J., Schmidt, H., Amara, K.K. et al. (2014). Large thermoelectricity via variable range hopping in CVD grown single-layer MoS_2. *Nano Lett.* 14: 2730–2734.

36 Ng, H.K., Chi, D., and Hippalgaonkar, K. (2017). Effect of dimensionality on thermoelectric powerfactor of molybdenum disulfide. *J. Appl. Phys.* 121: 204303.

37 Ng, H.K., Abutaha, A., Voiry, D. et al. (2019). Effects of structural phase transition on thermoelectric performance in lithium- intercalated molybdenum disulfide (Li_xMoS_2). *ACS Appl. Mater. Interfaces* 11: 12184–12189.

38 Wu, J., Liu, Y., Liu, Y. et al. (2020). Large enhancement of thermoelectric performance in MoS_2/h-BN heterostructure due to vacancy-induced band hybridization. *Proc. Natl. Acad. Sci.* 117: 13929–13936.

39 Ng, H.K., Xiang, D., Suwardi, A. et al. (2022). Improving carrier mobility in two-dimensional semiconductors with rippled materials. *Nat. Electron.* https://doi.org/10.1038/s41928-022-00777-z.

40 Zhao, Y., Yu, P., Zhang, G. et al. (2020). Low-symmetry $PdSe_2$ for high performance thermoelectric applications. *Adv. Funct. Mater.* 30: 2004896.

41 Zhang, Y., Li, H., Wang, L. et al. (2015). Photothermoelectric and photovoltaic effects both present in MoS_2. *Sci. Rep.* 5: 7938.

42 Roddaro, S., Ercolani, D., Safeen, M.A. et al. (2013). Giant thermovoltage in single InAs nanowire field-effect transistors. *Nano Lett.* 13: 3638–3642.

43 Tian, Y., Sakr, M.R., Kinder, J.M. et al. (2012). One-dimensional quantum confinement effect modulated thermoelectric properties in InAs nanowires. *Nano Lett.* 12: 6492–6497.

44 Yu, C., Shi, L., Yao, Z. et al. (2005). Thermal conductance and thermopower of an individual single-wall carbon nanotube. *Nano Lett.* 5: 1842–1846.

45 Malekpour, H. and Balandin, A.A. (2018). Raman-based technique for measuring thermal conductivity of graphene and related materials. *J. Raman Spectrosc.* 49: 106–120.

46 Zhao, Y., Zeng, X., Ren, L. et al. (2021). Heat conduction of electrons and phonons in thermal interface materials. *Mater. Chem. Front.* 5: 5617–5638.

47 Balandin, A.A., Ghosh, S., Bao, W. et al. (2008). Superior thermal conductivity of single-layer graphene. *Nano Lett.* 8: 902–907.

48 Zhou, H., Zhu, J., Liu, Z. et al. (2014). High thermal conductivity of suspended few-layer hexagonal boron nitride sheets. *Nano Res.* 7: 1232–1240.

49 Taube, A., Judek, J., Łapińska, A., and Zdrojek, M. (2015). Temperature-dependent thermal properties of supported MoS_2 monolayers. *ACS Appl. Mater. Interfaces* 7: 5061–5065.

50 Zhang, X., Sun, D., Li, Y. et al. (2015). Measurement of lateral and interfacial thermal conductivity of single- and bilayer MoS_2 and $MoSe_2$ using refined optothermal Raman technique. *ACS Appl. Mater. Interfaces* 7: 25923–25929.

51 Yan, R., Simpson, J.R., Bertolazzi, S. et al. (2014). Thermal conductivity of monolayer molybdenum disulfide obtained from temperature-dependent Raman spectroscopy. *ACS Nano* 8: 986–993.

52 Zhao, Y., Zheng, M., Wu, J. et al. (2021). Modification of thermal transport in few-layer MoS_2 by atomic-level defect engineering. *Nanoscale* 13: 11561–11567.

53 Yin, S., Zhang, W., Tan, C. et al. (2021). Thermal conductivity of few-layer PtS_2 and $PtSe_2$ obtained from optothermal Raman spectroscopy. *J. Phys. Chem. C* 125: 16129–16135.

54 Cai, W., Moore, A.L., Zhu, Y. et al. (2010). Thermal transport in suspended and supported monolayer graphene grown by chemical vapor deposition. *Nano Lett.* 10: 1645–1651.

55 Liu, Y., Ong, Z.-Y., Wu, J. et al. (2017). Thermal conductance of the 2D MoS_2/h-BN and graphene/h-BN interfaces. *Sci. Rep.* 7: 1–8.

56 Chen, C.C., Li, Z., Shi, L., and Cronin, S.B. (2014). Thermal interface conductance across a graphene/hexagonal boron nitride heterojunction. *Appl. Phys. Lett.* 104: 081908.

57 Vaziri, S., Yalon, E., Rojo, M.M. et al. (2019). Ultrahigh thermal isolation across heterogeneously layered two-dimensional materials. *Sci. Adv.* 5: 1–8.

58 Jiang, P., Qian, X., and Yang, R. (2018). Tutorial: Time-domain thermoreflectance (TDTR) for thermal property characterization of bulk and thin film materials. *J. Appl. Phys.* 124: 161103.

59 Wang, Y., Park, J.Y., Koh, Y.K., and Cahill, D.G. (2010). Thermoreflectance of metal transducers for time-domain thermoreflectance. *J. Appl. Phys.* 108: 043507.

60 Wilson, R.B. and Cahill, D.G. (2014). Anisotropic failure of Fourier theory in time-domain thermoreflectance experiments. *Nat. Commun.* 5: 1–11.

61 Schmidt, A.J., Collins, K.C., Minnich, A.J., and Chen, G. (2010). Thermal conductance and phonon transmissivity of metal-graphite interfaces. *J. Appl. Phys.* 107: 104907.

62 Smith, B., Vermeersch, B., Carrete, J. et al. (2017). Temperature and thickness dependences of the anisotropic in-plane thermal conductivity of black phosphorus. *Adv. Mater.* 29: 1603756.

63 Jiang, P., Qian, X., Gu, X., and Yang, R. (2017). Probing anisotropic thermal conductivity of transition metal dichalcogenides MX_2 (M = Mo, W and X = S, Se) using time-domain thermoreflectance. *Adv. Mater.* 29: 1–7.

64 Koh, Y.K., Bae, M.H., Cahill, D.G., and Pop, E. (2010). Heat conduction across monolayer and few-layer graphenes. *Nano Lett.* 10: 4363–4368.

65 Lindroth, D.O. and Erhart, P. (2016). Thermal transport in van der Waals solids from first-principles calculations. *Phys. Rev. B* 94: 1–11.

66 Rai, A., Sangwan, V.K., Gish, J.T. et al. (2021). Anisotropic thermal conductivity of layered indium selenide. *Appl. Phys. Lett.* 118: 073101.

67 Jang, H., Wood, J.D., Ryder, C.R. et al. (2015). Anisotropic thermal conductivity of exfoliated black phosphorus. *Adv. Mater.* 27: 8017–8022.

68 Zhao, Y., Zhang, G., Nai, M.H. et al. (2018). Probing the physical origin of anisotropic thermal transport in black phosphorus nanoribbons. *Adv. Mater.* 30: 1–7.

69 Kim, P., Shi, L., Majumdar, A., and McEuen, P.L. (2001). Thermal transport measurements of individual multiwalled nanotubes. *Phys. Rev. Lett.* 87: 215502-1-215502–4.

70 Shi, L., Li, D., Yu, C. et al. (2003). Measuring thermal and thermoelectric properties of using a microfabricated device. *J. Heat Transfer* 125: 881–888.

71 Aiyiti, A., Bai, X., Wu, J. et al. (2018). Measuring the thermal conductivity and interfacial thermal resistance of suspended MoS_2 using electron beam self-heating technique. *Sci. Bull.* 63: 452–458.

72 Wang, Z., Xie, R., Bui, C.T. et al. (2011). Thermal transport in suspended and supported few-layer graphene. *Nano Lett.* 11: 113–118.

73 Liu, D., Xie, R., Yang, N. et al. (2014). Profiling nanowire thermal resistance with a spatial resolution of nanometers. *Nano Lett.* 14: 806–812.

74 Zhao, Y. et al. (2017). Engineering the thermal conductivity along an individual silicon nanowire by selective helium ion irradiation. *Nat. Commun.* 8: 1–8.

75 Aiyiti, A., Hu, S., Wang, C. et al. (2018). Thermal conductivity of suspended few-layer MoS_2. *Nanoscale* 10: 2727.

76 Zhao, Y., Liu, X., Rath, A. et al. (2020). Probing thermal transport across amorphous region embedded in a single crystalline silicon nanowire. *Sci. Rep.* 1–8: https://doi.org/10.1038/s41598-020-57514-9.

77 Zhao, S. and Wang, H. (2020). An integrated h-type method to measure thermoelectric properties of two-dimensional materials. *ES Energy Environ.* 9: 59–66.

78 Wang, H., Hu, S., Takahashi, K. et al. (2017). Experimental study of thermal rectification in suspended monolayer graphene. *Nat. Commun.* 8: 1–8.

79 Guo, J., Yang, F., Xia, M. et al. (2019). Conformal interface of monolayer molybdenum diselenide/disulfide and dielectric substrate with improved thermal dissipation. *J. Phys. D. Appl. Phys.* 52: 385306.

80 Li, X., Yan, Y., Dong, L. et al. (2017). Thermal conduction across a boron nitride and SiO_2 interface. *J. Phys. D. Appl. Phys.* 50: 104002.

81 Wang, H., Kurata, K., Fukunaga, T., and Takamatsu, H. (2016). A simple method for fabricating free-standing large area fluorinated single-layer graphene with size-tunable nanopores. *Carbon N.Y.* 99: 564–570.

82 Ma, W., Miao, T., Zhang, X. et al. (2016). A T-type method for characterization of the thermoelectric performance of an individual free-standing single crystal Bi_2S_3 nanowire. *Nanoscale* 8: 2704–2710.

83 Fujii, M., Zhang, X., Xie, H. et al. (2005). Measuring the thermal conductivity of a single carbon nanotube. *Phys. Rev. Lett.* 065502: 8–11.

84 Cahill, D.G. and Pohl, R.O. (1987). Thermal conductivity of amorphous solids above the plateau. *Phys. Rev. B* 35: 4067–4073.

85 Cahill, D.G. and Pohl, R.O. (1988). Thermal properties of a tetrahedrally bonded amorphous solid: $CdGeAs_2$. *Phys. Rev. B* 37: 8773.

86 Dames, C. and Chen, G. (2005). 1ω, 2ω, and 3ω methods for measurements of thermal properties. *Rev. Sci. Instrum.* 76: 1–14.

87 Xian, Y., Zhang, P., Zhai, S. et al. (2018). Experimental characterization methods for thermal contact resistance : a review. *Appl. Therm. Eng.* 130: 1530–1548.

88 Manzeli, S., Ovchinnikov, D., Pasquier, D. et al. (2017). 2D transition metal dichalcogenides. *Nat. Rev. Mater.* 2: 17033.

89 Zhang, G. and Zhang, Y.-W. (2017). Thermoelectric properties of two-dimensional transition metal dichalcogenides. *J. Mater. Chem. C* 5: 7684–7698.

90 Wang, Y., Xiao, J., Zhu, H. et al. (2017). Structural phase transition in monolayer $MoTe_2$ driven by electrostatic doping. *Nature* 550: 487–491.

91 Liu, T., Liu, S., Tu, K.-H. et al. (2019). Crested two-dimensional transistors. *Nat. Nanotechnol.* 14: 223–226.

92 Bhattacharyya, S., Pandey, T., and Singh, A.K. (2014). Effect of strain on electronic and thermoelectric properties of few layers to bulk MoS_2. *Nanotechnology* 25: 465701.

93 Zhang, G. and Zhang, Y.W. (2015). Strain effects on thermoelectric properties of two-dimensional materials. *Mech. Mater.* 91: 382–398.

94 Duerloo, K.A.N., Li, Y., and Reed, E.J. (2014). Structural phase transitions in two-dimensional Mo- and W-dichalcogenide monolayers. *Nat. Commun.* 5: 1–9.

95 Li, Y., Dietrich, S., Forsythe, C. et al. (2021). Anisotropic band flattening in graphene with one-dimensional superlattices. *Nat. Nanotechnol.* https://doi.org/10.1038/s41565-021-00849-9.

96 Wu, K., Rademaker, L., and Zaanen, J. (2014). Bilayer excitons in two-dimensional nanostructures for greatly enhanced thermoelectric efficiency. *Phys. Rev. Appl.* 2: 1–8.

97 Kim, S.E., Mujid, F., Rai, A. et al. (2021). Extremely anisotropic van der Waals thermal conductors. *Nature* 597: 660–665.

98 Pallecchi, I., Manca, N., Patil, B. et al. (2020). Review on thermoelectric properties of transition metal dichalcogenides. *Nano Futur.* 4: 1–19.

99 Ye, J.T., Zhang, J., Akashi, R. et al. (2012). Superconducting dome in a gate-tuned band insulator. *Science* 338: 1193–1196.

100 Ma, J., Yang, R., and Chen, H. (2021). A large modulation of electron-phonon coupling and an emergent superconducting dome in doped strong ferroelectrics. *Nat. Commun.* 12: 1–9.

101 Cooper, R.C., Lee, C., Marianetti, C.A. et al. (2013). Nonlinear elastic behavior of two-dimensional molybdenum disulfide. *Phys. Rev. B* 87: 035423.

102 Peng, Z., Chen, X., Fan, Y. et al. (2020). Strain engineering of 2D semiconductors and graphene: from strain fields to band-structure tuning and photonic applications. *Light Sci. Appl.* 9: 190.

103 Yang, S., Chen, Y., and Jiang, C. (2021). Strain engineering of two-dimensional materials: methods, properties, and applications. *InfoMat* 3: 397–420.

104 Ding, Z., Pei, Q.X., Jiang, J.W., and Zhang, Y.W. (2015). Manipulating the thermal conductivity of monolayer MoS_2 via lattice defect and strain engineering. *J. Phys. Chem. C* 119: 16358–16365.

105 Conley, H.J., Wang, B., Ziegler, J.I. et al. (2013). Bandgap engineering of strained monolayer and bilayer MoS_2. *Nano Lett.* 13: 3626–3630.

106 Carrascoso, F., Frisenda, R., and Castellanos-Gomez, A. (2021). Biaxial versus uniaxial strain tuning of single-layer MoS_2. *Nano Mater. Sci.* https://doi.org/10.1016/j.nanoms.2021.03.001.

107 Ghorbani-Asl, M., Borini, S., Kuc, A., and Heine, T. (2013). Strain-dependent modulation of conductivity in single-layer transition-metal dichalcogenides. *Phys. Rev. B* 87: 1–6.

108 Hosseini, M., Elahi, M., Pourfath, M., and Esseni, D. (2015). Strain-induced modulation of electron mobility in single-layer transition metal dichalcogenides MX_2 (M = Mo, W; X = S, Se). *IEEE Trans. Electron Devices* 62: 3192–3198.

109 Martella, C., Mennucci, C., Lamperti, A. et al. (2018). Designer shape anisotropy on transition-metal-dichalcogenide nanosheets. *Adv. Mater.* 30: 1–9.

110 Christopher, J.W., Vutukuru, M., Lloyd, D. et al. (2019). Monolayer MoS_2 strained to 1.3% with a microelectromechanical system. *J. Microelectromech. Syst.* 28: 254–263.

111 Shi, H., Pan, H., Zhang, Y.W., and Yakobson, B.I. (2013). Quasiparticle band structures and optical properties of strained monolayer MoS_2 and WS_2. *Phys. Rev. B* 87: 1–8.

112 Deng, S., Che, S., Debbarma, R., and Berry, V. (2019). Strain in a single wrinkle on an MoS_2 flake for in-plane realignment of band structure for enhanced photo-response. *Nanoscale* 11: 504–511.

113 Wu, Y., Chen, Z., Nan, P. et al. (2019). Lattice strain advances thermoelectrics. *Joule* 3: 1276–1288.

114 Yu, Z., Ong, Z.-Y., Pan, Y. et al. (2016). Realization of room-temperature phonon-limited carrier transport in monolayer MoS_2 by dielectric and carrier screening. *Adv. Mater.* 28: 547–552.

115 Li, X., Mullen, J.T., Jin, Z. et al. (2013). Intrinsic electrical transport properties of monolayer silicene and MoS_2 from first principles. *Phys. Rev. B* 87: 115418.

116 Kaasbjerg, K., Thygesen, K.S., and Jacobsen, K.W. (2012). Phonon-limited mobility in n-type single-layer MoS_2 from first principles. *Phys. Rev. B* 85: 115317.

117 Jiang, B., Yu, Y., Cui, J. et al. (2021). High-entropy-stabilized chalcogenides with high thermoelectric performance. *Science* 371: 830–834.

118 Roychowdhury, S., Ghosh, T., Arora, R. et al. (2021). Enhanced atomic ordering leads to high thermoelectric performance in $AgSbTe_2$. *Science* 371: 722–727.

119 Liu, T., Xiang, D., Ng, H.K. et al. (2022). Modulation of spin dynamics in 2D transition-metal dichalcogenide via strain-driven symmetry breaking. *Adv. Sci.* 2200816: 1–8.

120 Hong, J., Lee, C., Park, J.-S., and Shim, J.H. (2016). Control of valley degeneracy in MoS_2 by layer thickness and electric field and its effect on thermoelectric properties. *Phys. Rev. B* 93: 6.

121 Pei, Y., Shi, X., LaLonde, A. et al. (2011). Convergence of electronic bands for high performance bulk thermoelectrics. *Nature* 473: 66–69.

122 Wickramaratne, D., Zahid, F., and Lake, R.K. (2014). Electronic and thermoelectric properties of few-layer transition metal dichalcogenides. *J. Chem. Phys.* 140: 124710.

123 Ouyang, B., Chen, S., Jing, Y. et al. (2018). Enhanced thermoelectric performance of two dimensional MS_2 (M = Mo, W) through phase engineering. *J. Mater.* 4: 329–337.

124 Goldsmid, H.J. (1964). *Thermoelectric Refrigeration*. Springer https://doi.org/10.1017/CBO9781107415324.004.

125 Huang, H., Cui, Y., Li, Q. et al. (2016). Metallic 1T phase MoS_2 nanosheets for high-performance thermoelectric energy harvesting. *Nano Energy* 26: 172–179.

126 Lai, Z., He, Q., Tran, T.H. et al. (2021). Metastable 1T′-phase group VIB transition metal dichalcogenide crystals. *Nat. Mater.* https://doi.org/10.1038/s41563-021-00971-y.

127 Yu, Y., Nam, G.-H., He, Q. et al. (2018). High phase-purity 1T′-MoS_2- and 1T′-$MoSe_2$-layered crystals. *Nat. Chem.* 10: 638–643.

128 Li, Y., Duerloo, K.A.N., Wauson, K., and Reed, E.J. (2016). Structural semiconductor-to-semimetal phase transition in two-dimensional materials induced by electrostatic gating. *Nat. Commun.* 7: 1–8.

129 Voiry, D., Mohite, A., and Chhowalla, M. (2015). Phase engineering of transition metal dichalcogenides. *Chem. Soc. Rev.* 44: 2702–2712.

130 Kappera, R., Voiry, D., Yalcin, S.E. et al. (2014). Phase-engineered low-resistance contacts for ultrathin MoS_2 transistors. *Nat. Mater.* 13: 1128–1134.

131 Ouyang, B., Lan, G., Guo, Y. et al. (2015). Phase engineering of monolayer transition-metal dichalcogenide through coupled electron doping and lattice deformation. *Appl. Phys. Lett.* 107: 191903.

132 Mao, J., Liu, Z., Zhou, J. et al. (2018). Advances in thermoelectrics. *Adv. Phys.* 67: 69–147.

133 Minnich, a.J., Dresselhaus, M.S., Ren, Z.F., and Chen, G. (2009). Bulk nanostructured thermoelectric materials: current research and future prospects. *Energy Environ. Sci.* 2: 466.

134 Narducci, D., Selezneva, E., Cerofolini, G. et al. (2012). Impact of energy filtering and carrier localization on the thermoelectric properties of granular semiconductors. *J. Solid State Chem.* 193: 19–25.

135 Gayner, C. and Amouyal, Y. (2019). Energy filtering of charge carriers: current trends, challenges, and prospects for thermoelectric materials. *Adv. Funct. Mater.* 18: 1901789.

136 Narducci, D. (2017). Energy filtering and thermoelectrics: artifact or artifice? *J. Nanosci. Nanotechnol.* 17: 1663–1667.

137 Chen, W., Pöhls, J.H., Hautier, G. et al. (2016). Understanding thermoelectric properties from high-throughput calculations: trends, insights, and comparisons with experiment. *J. Mater. Chem. C* 4: 4414–4426.

138 Choudhary, K., Garrity, K.F., and Tavazza, F. (2020). Data-driven discovery of 3D and 2D thermoelectric materials. *J. Phys. Condens. Matter* 32: 475501.

139 Toher, C., Plata, J.J., Levy, O. et al. (2014). High-throughput computational screening of thermal conductivity, Debye temperature, and Grüneisen parameter using a quasiharmonic Debye model. *Phys. Rev. B* 90: 1–14.

140 Zhu, H., Hautier, G., Aydemir, U. et al. (2015). Computational and experimental investigation of $TmAgTe_2$ and XYZ_2 compounds, a new group of thermoelectric materials identified by first-principles high-throughput screening. *J. Mater. Chem. C* 3: 10554–10565.

141 Ricci, F., Chen, W., Aydemir, U. et al. (2017). An ab initio electronic transport database for inorganic materials. *Sci. Data* 4: 170085.

142 Prashun, G., Vladan, S., and Eric, S.T. (2017). Computationally guided discovery of thermoelectric materials. *Nat. Rev. Mater.* 2: 1–16.

143 Antunes, L.M., Vikram, P., Jose, J. et al. (2022). Machine learning approaches for accelerating the discovery of thermoelectric materials. In: *Machine Learning in Materials Informatics: Methods and Applications, ACS Symposium Series*, vol. 1416 (ed. Y. An), 1–32. ACS Publications https://doi.org/10.1021/bk-2022-1416.ch001.

144 Deng, T., Recatala-Gomez, J., Ohnishi, M. et al. (2021). Electronic transport descriptors for the rapid screening of thermoelectric materials. *Mater. Horiz.* 8: 2463–2474

145 Laugier, L., Bash, D., Recatala-Gomez, J. (2018). Predicting thermoelectric properties from crystal graphs and material descriptors - first application for functional materials. https://arxiv.org/abs/1811.06219.

4

Emerging Electronic Properties of Polymorphic 2D-TMDs

Tong Yang[1], *Zishen Wang*[2], *Jiaren Yuan*[3], *Jun Zhou*[4], *and Ming Yang*[1]

[1] The Hong Kong Polytechnic University, Department of Applied Physics, Hung Hom, Kowloon, Hong Kong SAR, China
[2] National University of Singapore, Department of Physics, Faculty of Science, Singapore 117542, Singapore
[3] Nanchang University, School of Physics and Materials Science, Nanchang, Jiangxi, 330031, China
[4] Institute of Materials Research and Engineering, Agency for Science, Technology and Research (A*STAR), 2 Fusionopolis Way Singapore 138634, Singapore

4.1 Electronic Structure and Optical Properties of 2D-TMDs

TMDs feature sandwiched X–M–X structures with hexagonal lattices, where M represents transition metal elements such as Ti, Hf, Pt, Mo, and W, and X represents chalcogen elements such as S, Se, and Te [1–3]. For the bulk TMDs, strong covalent bonds are found within the MX_2 layer, while weak van der Waals interaction is found between MX_2 layers, which binds the MX_2 layers to form the TMD bulks. The varied arrangement of the MX_2 layers gives rise to different phases of the TMD bulks, which include 2H, 1T, and 3R phases [1, 2]. In the monolayer limit, due to the different arrangements among top, middle, and bottom atom planes, as well as distorted atom positions, various phases such as H, T and T$'$ have been found in monolayer TMDs, showing a rich variety of electronic and optical properties [4–6]. In this section, using two-dimensional group-VIB TMDs as the prototype of 2D-TMDs, we will give a brief introduction of the electronic band structure and optical properties of H-, T- and T$'$-phase TMDs.

4.1.1 Electronic and Optical Properties of 1H-Phase 2D-TMDs

The 1H-phase 2D group-VIB TMDs such as monolayer MoS_2 and WS_2 have the most stable structure, where the three atomic planes (chalcogen-metal-chalcogen) stack in A-B-A order as shown in Figure 4.1a. The triangular lattices in the atomic planes are in trigonal prismatic coordination, forming 2D hexagonal lattices where transition metal elements occupy A sublattice sites and the two chalcogen elements fill B sublattice sites. Consequently, the first Brillouin zone of 2D TMDs is quite similar to that of graphene, as illustrated in Figure 4.1b.

Two-Dimensional Transition-Metal Dichalcogenides: Phase Engineering and Applications in Electronics and Optoelectronics, First Edition. Edited by Chi Sin Tang, Xinmao Yin, and Andrew T. S. Wee.

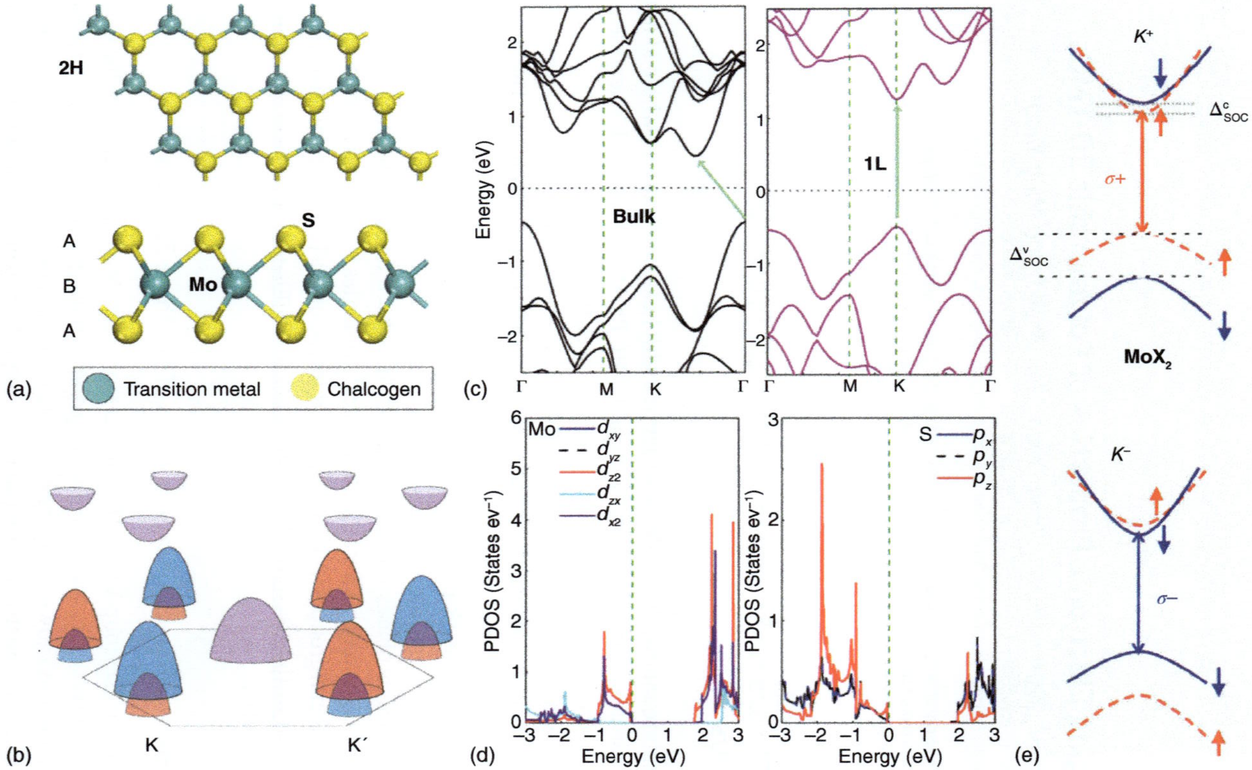

Figure 4.1 **Structure and electronic band structure of monolayer TMDs**. (a) Atomic structure and (b) the first Brillouin zone of monolayer TMDs, showing spin-splitting of the bands at the K and K′ points at the corner of the Brillouin zone. Orange and blue color denotes spin-up and spin-down polarization, respectively. (c) Calculated layer dependent band structure of MoS_2. (d) Projected density of states for monolayer MoS_2. (e) Schematic band edge splitting induced by SOC at the two inequivalent K and K′ (K^+ and K^-) points. Source: Adapted from Xiao et al. [7].

The thickness dependent band structure of 2D TMDs is shown in Figure 4.1c, from which we can see that both valence band maximum (VBM) and conduction band minimum (CBM) in monolayer H-phase VIB TMDs are located at the two inequivalent K and K′ points (the Brillouin zone corners), resulting in a direct band gap in the visible light range.

The orbital resolved density of states for monolayer MoS_2 is shown in Figure 4.1d. From Figure 4.1c and d, we can see that the valence band maximum at the K point is mainly contributed by d_{xy} and $d_{x^2-y^2}$ orbitals of Mo atoms [8]. These orbitals are strongly localized in the Mo atomic plane (in-plane) and hybridize with the weak contribution from p_x and p_y orbitals of S atoms. The valence band maximum at the Brillouin zone centre (the Γ point) is only slightly lower in energy (~10 meV) than that at the K points, which mainly derives from the d_{z^2} orbital of Mo atoms and the p_z orbital of S atoms. Compared with the in-plane orbitals, these out-of-plane orbitals are more delocalized. For CBM at the K point, it is mainly due to the d_{z^2} orbital of Mo atoms and the p_z orbital of S atoms. The second lowest conduction band minimum is at the Q point, which is mainly contributed by the d_{xy} and $d_{x^2-y^2}$ orbitals of Mo atoms. Note that the energy difference between valleys in the valence band is very small. Thus, external influences such as strain or substrate can easily change their relative positions, leading to a transition from a direct band gap to an indirect band gap [9–14].

The transition can also happen when TMD layers are stacked. As Figure 4.1c shows, from bilayer 2D TMDs onwards, the energy of valence band maximum at the Γ point increases noticeably, and is higher than that at the K points, leading to indirect band gaps. The trend is more pronounced with the increasing thickness of 2D TMDs. This trend can be understood by the orbital characters of valence band at the Γ and K points [8, 15]. Taking MoS_2 as an example, the delocalized out-of-plane d_{z^2} orbital of Mo atom and p_z orbital of S atoms can have considerable spatial overlap with the counterparts in the nearest MoS_2 layer. The stronger inter-layer hopping between the adjacent S atoms increases the band energy at the Γ point. In contrast, due to the localized character of the Mo d_{xy} and $d_{x^2-y^2}$ orbitals, the valence band energy at the K and K′ points remains nearly unchanged. These cause the direct-indirect band gap transition and a reduced band gap in the multilayer MoS_2 and other group-VIB 2D TMDs.

Compared with graphene, spin-orbit coupling (SOC) strength in monolayer TMDs is much stronger, due to the much heavier elements in the TMDs [5, 7, 15, 16]. The strong SOC and breaking of inversion symmetry introduce a notable change to the electronic band structure of monolayer 2D TMDs. As illustrated in Figure 4.1e, the strong SOC in monolayer TMDs induces band slitting in both valence and conduction band edges at the K and K′ points. The valence band splitting for Mo and W based group-VIB 2D TMDs is about 200 and 400 meV, respectively [7, 16]. The spin-splitting at the conduction band edge is much smaller, in a range of several meV, as the Mo d_{z^2} orbital in the conduction band edge contributes to a much-reduced intro-atomic coupling that is responsible for the SOC [17, 18]. It should be noted that the SOC induced band splitting vanishes at the other high symmetry k-points of Γ and M. At the Γ point, the time-reversal symmetry leads to the spin degeneracy, while at the M point, it is due to the time-reversal symmetry and translational symmetry protection [15, 19]. Another important feature induced by the SOC in monolayer group-VIB TMDs is the spin-valley locking effect due to the D_{3h} symmetry of the Bloch state at the inequivalent K and K′ points, in which the spin-polarization direction is opposite in the split bands (see illustration in Figure 4.1e) [5, 7].

Since the direct band gap is in the visible light range, monolayer group-VIB TMDs feature strong light-matter coupling, which is attractive for optoelectronic applications [4, 5, 20–24]. Upon light absorption, an electron in the valence band edge will be excited by the conduction band, leaving a hole in the valence band. The electron-hole pair is bound together through the long-range Coulomb interaction, forming an exciton, as illustrated in Figure 4.2a [4]. In the monolayer group-VIB TMDs, the excitonic effect is more pronounced because of: (i) smaller effective separation between electron and hole in the monolayer limit, (ii) relatively large effective mass of electron and hole at the K and K′ points, and (iii) much reduced dielectric screening to the Coulomb interaction [18, 26]. Thus, electron-hole pairs in monolayer group-VIB TMDs are strongly bound, as evidenced by the localized wavefunction of the exciton (~ several angstrom) and large exciton binding energy (~ several hundred meV) (see Figure 4.2b and c) [25]. This is in contrast with bulk

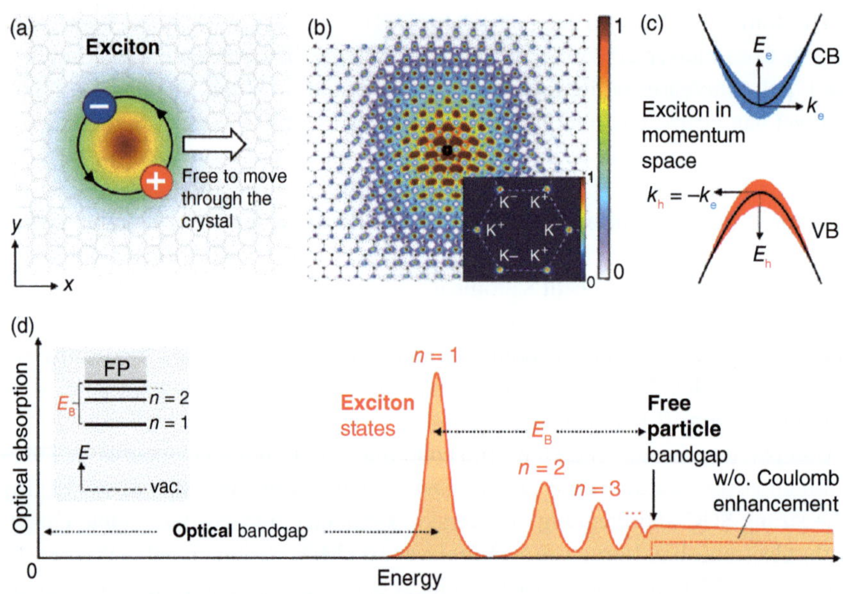

Figure 4.2 **Excitons in monolayer TMDs**. Source: Wang et al. [4]/With permission from American Physical Society. (a) Schematic representation of a strongly bound electron-hole pair (Wannier–Mott exciton). (b) Calculated exciton wavefunction of monolayer MoS_2. Source: Qiu et al. [25] With permission from American Physical Society. (c) Schematic representation of the contribution of electrons and holes in conduction and valence band edge to the exciton in the reciprocal space. (d) An illustration plot of typical optical spectra for monolayer TMD semiconductors, showing the hydrogen-like atomic energy level of the exciton states below the renormalized quasiparticle band gap.

semiconductors, where the exciton wavefunction extends to several nanometres and the exciton binding energy decreases by a factor of four [27]. The exciton in monolayer group-VIB TMDs is often called the Wannier–Mott exciton. The relevant exciton states can be described by hydrogen atom like models and labeled by the principal quantum number n with $n = 1, 2, 3 \ldots$, where $n = 1$ corresponds to the lowest-energy exciton (see Figure 4.2d), and the exciton binding energy can be estimated by $E_B \propto \frac{4R_y\mu}{m_e\varepsilon_{eff}^2}$ (R_y is the Rydberg constant, μ is the reduced mass, m_e is the free electron mass, and ε_{eff} is the effective dielectric constant) [4].

The exciton binding energy can also be estimated from the energy difference between the optical resonant energy and the quasiparticle energy gap [4, 5, 25]. In experiment, the optical resonant energy can be obtained by using the optical techniques such as optical absorption spectra and photoluminescence spectroscopy [28, 29]. The quasiparticle band gap can be measured using scanning tunneling spectroscopy (STS) and angle-resolved photoemission spectroscopy (ARPES) [30–32], which need to be conducted in ultra-high vacuum and are more challenging. The exciton binding energy can also be estimated from the first-principles calculation, where the exciton peak can be obtained by the Bethe–Salpeter equation (BSE) calculated optical spectra and the quasiparticle energy gap can be calculated using the GW method [10, 25, 31]. The optical spectra of monolayer group-VIB TMDs such as MoS_2 and WS_2 feature strong exciton peaks, and the A and B exciton peaks, which are due to SOC-induced splitting of the first exciton peak, can be well resolved, in line with first-principles calculations [25, 29, 33]. Following the first exciton peak, a strong exciton peak at higher energy can be seen in the optical spectra. This enhanced exciton peak is often ascribed to the band nesting effect, in which the band dispersion at the high-symmetry k-point satisfies the condition $\nabla_k(E_c - E_v) = 0$ [34]. In the band nesting region, the valance band is parallel to the conduction band, resulting in the deviation of the joint DOSs from the resonant energy and thus contributing to the substantially enhanced exciton peak. In addition, the aforementioned spin-valley locking effect can lead to the optical selection rules (see Figure 4.1e): the interband transitions at the K and K′ valleys are chiral, where the σ^+ and σ^- circularly polarized light can only couple with electrons at the K and K′ valleys, respectively [5, 7]. This enables the generation and detection of the spin-valley polarization and opens up an exciting venue toward valleytronics [5, 35].

4.1.2 Electronic and Optical Properties of 1T-Phase 2D-TMDs

The atomic planes of 2D TMDs can be arranged in different ways. In addition to A-B-A arrangement that gives rise to the 1H-phase as discussed above, the transition metal atom plane and the chalcogen atom planes can also take A-B-C stacking sequence, as illustrated in Figure 4.3a [2]. This results in the 1T phase of monolayer TMDs, in which the metal atoms and chalcogen atoms are bonded in octahedral coordination. In contrast with the fully filled d orbital in the 1H group-VIB semiconducting TMDs, the d orbital of transition metal atoms in 1T-TMDs is partially filled, leading to the metallic behaviors [2, 38, 39]. This renders metallic

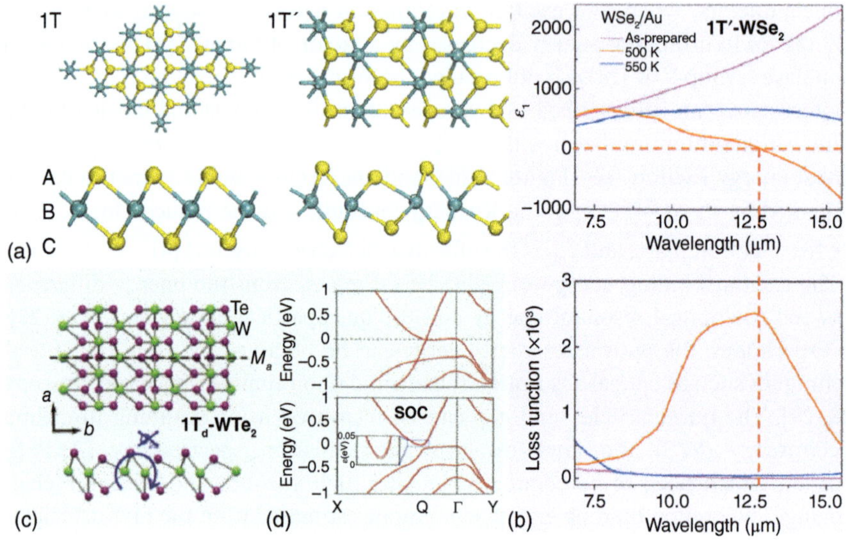

Figure 4.3 Structural and electronic properties of T-TMDs. (a) Atomic structure of 1T and 1T′ TMDs. (b) Temperature dependent dielectric function (real part) and loss function of monolayer 1T′-WSe_2. Source: (a) Tang et al. [36]/With permission from Springer Nature. (b) Molina-Sánchez et al. [6]/John Wiley & Sons / Licensed under CC BY 4.0. The atomic structure (c) and electronic band structure (with/without SOC) (d) of monolayer $1T_d$-WTe_2. Source: Xu et al. [37]/With permission from Springer Nature.

1T-group-VIB TMDs promising for various applications such as electrocatalysis and high-performance metal contact for 2D electronics [40–44].

The free standing1T-group-VIB TMDs are metastable. They will transform into a lower energy state, 1T′-phase, where the neighbouring transition metal atoms will move closer and form metal-metal dimers, as shown in Figure 4.3a [38]. This leads to a (2×1) structural reconstruction with zigzag stripe pattern [45]. The structural reconstruction of monolayer 1T′-TMDs dramatically changes the electronic properties. The formation of metal-metal dimers makes the d orbital of transition metal atoms lower in energy than the *p* orbital of chalcogen atoms. The valence band maximum and conduction band minimum, which are derived from the *p* orbital of chalcogen atoms and the d orbital of transition metal atoms, respectively, touch each other at the Γ point, forming quasi-metallic behavior [46]. By including the strong SOC effect, a band gap opens, accompanied by the band inversion. This is an indication of the phase transition to a topological insulator [47].

Various emerging properties have been found in the monolayer 1T′-TMDs, including anisotropic plasmon excitation, topological properties, anisotropic transport, and magnetoresistance properties [36, 48–52]. For example, mid-infrared plasmon has been observed in monolayer 1T′-WS_2 and MoS_2. As the high-resolution spectroscopic ellipsometry measurement shows in Figure 4.3b, the dielectric function (real part, ε_1) shows a transition from positive to negative values, of which the energy position in the transition point is resonant with the peak in the loss function spectra. This is a hallmark of the plasmon. Further first-principles calculations indicate that the occurrence of the plasmon is due to anisotropic collective charge excitation in the 1T$'$-phase. It only propagates along the direction perpendicular to the zigzag metal-metal stripe [36].

The monolayer monoclinic 1T$'$-TMDs can undergo further structural transition to the orthogonal 1T$_d$-phase [2, 49, 53]. The structure of 1T$_d$-phase is similar to that of 1T$'$ phase, but some deviation is noticeable, where the distortion of the metal atoms is more pronounced as shown in Figure 4.3c. This further reduces the crystal symmetry, and the two-fold screw rotational symmetry is broken, resulting in slightly different electronic properties [37]. Figure 4.3d shows the band structure of monolayer 1T$_d$-WTe_2 without and with SOC. The low energy band structure features 2D massless fermions at the Q and Q$'$ points. With the SOC effect, an inverted quantum-spin-hall gap occurs. Furthermore, the weakly broken rotational symmetry further introduces a small spin splitting at the bottom of the conduction band. Interestingly, due to the non-trivial wavefunction near the inverted gap, monolayer 1T$_d$-WTe_2 shows a Berry curvature dipole and electric field effect, which might provide a useful platform to realize various quantum phenomena such as quantum non-linear Hall effect, and orbital-Edelstein and chiral polaritonic effects [37, 55–58].

4.2 Polaron States of 2D-TMDs

When a charge carrier propagates through a crystal, it interacts with the local atoms that surround it and alters the atomic motion of the latter. Thus, a local potential is induced and in return experienced by the carrier. While propagating, the charge carrier drags the self-induced local potential or virtual phonon cloud, and the whole composite is often envisioned as a quasiparticle termed "polaron" [59, 60]. Recent experiments have confirmed the existence of polarons in real 2D materials and van der Waals heterostructures [61, 62]. Meanwhile, theoretical investigations based on model Hamiltonians have been reported for 2D systems [54, 62, 63]. With appropriate approximations made to the electron-phonon interaction, the theoretical models suggest that the presence of polarons may rationalize some experimental observations and even (partially) resolve discrepancies between experiments and density functional theory (DFT) simulations. This section focuses on polaron effects on the electronic band structures of 2D TMDs.

4.2.1 Holstein Polarons in MoS_2

4.2.1.1 Experimental Characterizations of Holstein Polarons

As an *n*-type semiconductor, the electronic band structure of the 1H MoS_2 around CBM has been thoroughly scrutinized by Mingu Kang et al. [62] Electron doping of the top-most layer of single crystal MoS_2 samples was achieved by the in situ deposition of Rb atoms under 30 K. The large atomic radius prevents deposited Rb atoms from intercalation into MoS_2 layers. As such, the interlayer symmetry would be broken and induce the valence band spin splitting, which is corroborated by the experimentally measured band structure of Rb-deposited MoS_2 using ARPES.

Hence, electrons are mostly donated to the top-most MoS_2 layer, forming 2D dipole layers or electric double layers. The doped electrons fill the conduction band edge of the surface MoS_2 layer and shift it below the Fermi level (E_F), as shown in the 3D ARPES spectra at the electron density $\rho = 5 \times 10^{13}$ cm^{-2} (Figure 4.4a). The CBM of the doped MoS_2 sample also moves to the K/K′ point/valley due to the surface electric-field effect [64].

Figure 4.4b–d display the high-resolution ARPES spectra around the K valley and the spectral analysis using a series of energy-distribution curves (EDCs). It is striking that instead of a continuous parabolic energy band, its profile is significantly reconstructed, and two gaps are opened along the energy scale. EDCs show that there are multiple poles associated with every k-point around the K valley. The energy position of these poles suggests the two small gaps to be uniformly spaced in energy at -26 ± 3 and -53 ± 3 meV below the Fermi level. On the other hand, the dispersive energy bands are connected by flat bands around the small gaps. The two band segments have distinct characters – the dispersive segment is electron-like, whereas the flat band is phonon-like. In particular, the band character abruptly transforms around the region where they are connected (Figure 4.4e), indicative of the strong coupling to the host MoS_2 lattice. Such strong band renormalizations and band character transitions at $-n\Omega_0$ ($\Omega_0 = 26 \pm 3$ meV) manifest the formation of Holstein polarons, where the first sub-band below the Fermi level (red in Figure 4.4d) is the polaron band and others (yellow and green) are the excited states.

The evolution of the polaron state and the band renormalization with respect to the electron density is revealed by controlling the density of Rb atoms deposited on MoS_2 (Figure 4.4f–i). At a low electron density $\rho = 1.1 \times 10^{13}$ cm^{-2}, it is worth noting that the spectra are analogous to those in anatase TiO_2 and acquire the characteristics of the long-range Fröhlich polarons, rather than the short-range Holstein polarons [65, 66]. The broad tail comprises a main peak and a series of shake-off replicas. They are uniformly spaced at $-n\Omega_0$ and their intensity follows a Poisson distribution (Figure 4.4h). However, as the width of the filled conduction band (W) exceeds Ω_0, the long-range electron-phonon coupling is significantly suppressed. Instead, the signature of the short-range Holstein polarons gradually appears, indicating its dominance at $W/\Omega_0 > 1$. With the increase in ρ, The energy positions where the small gaps are opened remain at $-n\Omega_0$, whereas the gap size increases remarkably. The latter suggests a progressive enhancement of the short-range electron-phonon coupling, or the Holstein polaron effects. The enhancement is also evidenced by the increase in the mass associated with the polaron (m^*) as shown in Figure 4.4j. In contrast, the effective mass (m_0) of the bare band almost remains $m_0 = 0.85m_e$ across the electron density range, where m_e is the electron rest mass.

The equally spaced small gaps suggest that the phonon modes associated with the Holstein polaron in electron-doped MoS_2 should have vibrational frequencies of around $\Omega_0 = 26 \pm 3$ meV. These correspond to the longitudinal acoustic (LA) phonons around the Brillouin zone edges (Figure 4.4k). Along the M – K/K′ path, the LA branch is dispersionless and analogous to the optical branches. Note that the unique double degeneracy of the K and K′ valleys endows MoS_2 with an additional intervalley channel for the short-range electron-phonon coupling. In particular,

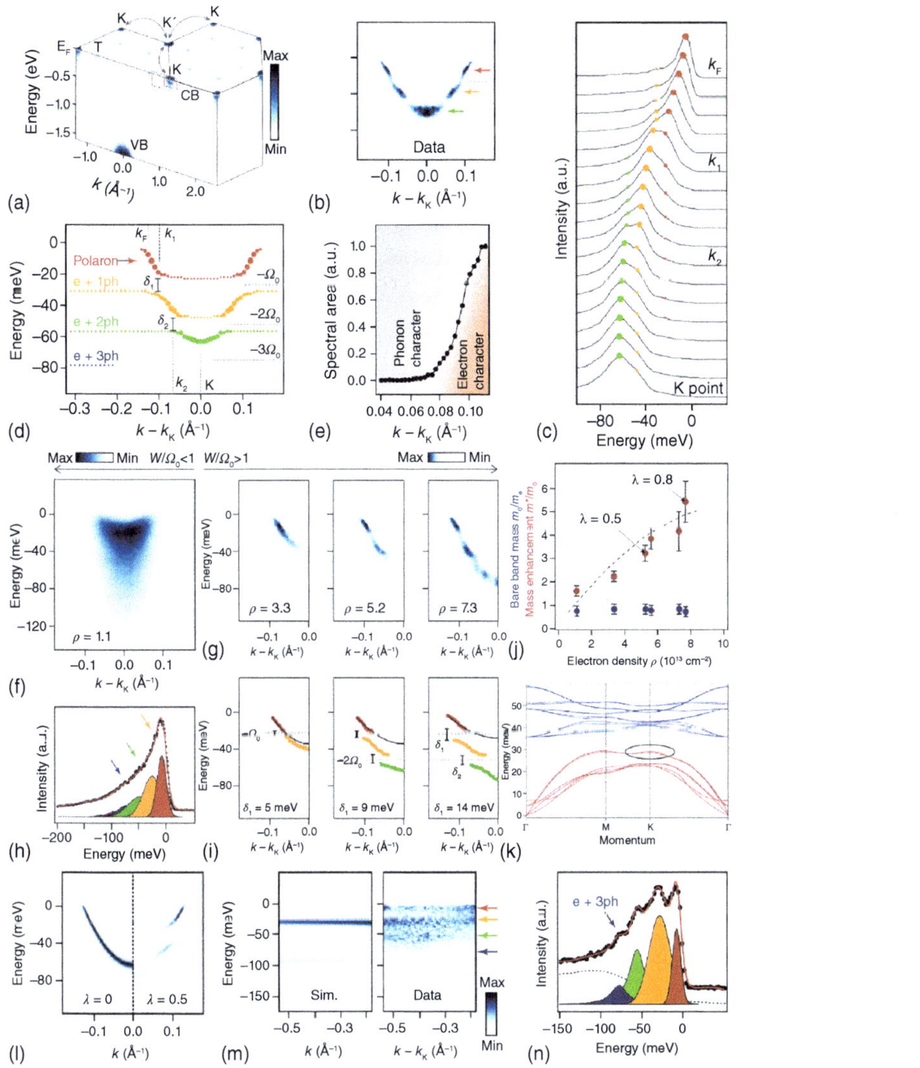

Figure 4.4 (a) The 3D ARPES spectra taken for the Rb-deposited MoS_2 at $\rho = 5 \times 10^{13}$ cm^{-2} and at temperature $T = 80$ K. The photon energy was 54 eV. (b) The high-resolution ARPES data around the region enclosed by the red dashed box in (a). (c) Series of EDCs from the K point to the Fermi momentum (k_F). Filled circles denote the peak positions. (d) Fitted energy position of poles to EDCs in (c) as a function of k. (e) Spectral area of EDC peaks of the polaron band as a function of k. (f, g) The ARPES spectra of MoS_2 at a series of electron density ρ (10^{13} cm^{-2}). (h) EDC at k_F for the data shown in (f). (i) The extracted peak positions from (g). Dotted lines mark the energy positions of small gaps at $-n\Omega_0$. The gap sizes are shown at the bottom of each panel. Overlaid black curves are used to fit the polaron dispersion. (j) Effective mass of polaron bands and bare bands as a function of the electron density. The dashed line is the intervalley electron-phonon coupling matrix for the LA phonons near the K valleys of MoS_2. (k) The theoretical phonon dispersion of MoS_2. The black eclipse marks the longitudinal acoustic phonons around the K valley. (l) The simulated spectral function of Holstein polarons around the K (K′) valley with zero ($\lambda = 0$) and intermediate ($\lambda = 0.5$) electron-phonon coupling. (m) The simulated spectral function of the flat band region at $\lambda = 0.5$ (left panel) and the corresponding experimental spectrum (right panel). (n) k-integrated EDC of the experimental data shown in (m). Source: Kang et al. [62]/With permission from Springer Nature.

the intervalley electron-phonon coupling matrix for LA phonons near the K valley of MoS_2 is very consistent with the measured mass enhancement (Figure 4.4j). Therefore, the dispersionless LA phonons around the Brillouin zone corners (K and K′) are highly likely to be the major phonon modes involved in the Holstein polaron formation in electron-doped MoS_2 or the conduction band renormalizations of electron-doped MoS_2.

4.2.1.2 Theoretical Simulations of the Spectral Functions

The model Hamiltonian, which describes how an electron interacts with a dispersionless phonon mode (Holstein polarons), reads [67–69]

$$\mathcal{H} = \sum_{\vec{k}} \varepsilon_{\vec{k}} a^{\dagger}_{\vec{k}} a_{\vec{k}} + \sum_{\vec{q}} \Omega_0 b^{\dagger}_{\vec{q}} b_{\vec{q}} + \frac{g}{\sqrt{N}} \sum_{\vec{k},\vec{q}} c^{\dagger}_{\vec{k}-\vec{q}} c_{\vec{k}} \left(b^{\dagger}_{\vec{q}} + b_{-\vec{q}} \right) \tag{4.1}$$

The first two terms describe the kinetic energy of the electron and the phonon mode, respectively. $\varepsilon_{\vec{k}}$ is the electronic band energy, and Ω_0 is the phonon energy, which is 26 meV for MoS_2. $a^{\dagger}_{\vec{k}}$ ($a_{\vec{k}}$) and $b^{\dagger}_{\vec{q}}$ ($b_{\vec{q}}$) denote the creation (annihilation) operator for an electron and phonon with the crystal momentum $\vec{k}$ and $\vec{q}$, respectively. The last term accounts for the electron-phonon coupling, where N denotes the total number of lattice sites.

The Green's function of the Holstein polaron can be written as

$$G(\vec{k}, \tau) = -i\Theta(\tau) \left\langle 0 \left| c_{\vec{k}} e^{-i\mathcal{H}\tau} c^{\dagger}_{\vec{k}} \right| 0 \right\rangle \tag{4.2}$$

where $\Theta(\tau)$ is the Heaviside function of time τ. $|0\rangle$ is the ground state of the zero-particle system. This Green's function is only exactly known in two asymptotic limits where either the electron-phonon coupling or the electronic inter-site hoping vanishes. The ratio of the ground-state energies in the two asymptotic limits (λ) is here used to describe the effective electron-phonon coupling

$$\lambda = \frac{g^2}{2dt\Omega_0} \tag{4.3}$$

where d and t denote the dimensionality and the hoping amplitude (~62 meV), respectively. For simplicity, the one-dimensional (1D) model is considered, i.e. $d = 1$. Although the electron-doped MoS_2 surface layer is two-dimensional, the 1D model already allows the key features to be captured, as discussed below. Within the momentum average (MA) approximation [69], the Green's function can be simplified as

$$G(\vec{k}, \omega) = \frac{1}{\hbar\omega - \varepsilon_{\vec{k}} - \Sigma_{\mathrm{MA}}(\omega) + i\eta} \tag{4.4}$$

where $\Sigma_{\mathrm{MA}}(\omega)$ is the self-energy

$$\Sigma_{\mathrm{MA}}(\omega) = \cfrac{g^2 \overline{g}_0(\hbar\omega - \Omega_0)}{1 - \cfrac{2g^2 \overline{g}_0(\hbar\omega - \Omega_0)\overline{g}_0(\hbar\omega - 2\Omega_0)}{1 - \cfrac{3g^2 \overline{g}_0(\hbar\omega - 2\Omega_0)\overline{g}_0(\hbar\omega - 3\Omega_0)}{1 - \dots}}} \tag{4.5}$$

$$\bar{g}_0(\hbar\omega) = \frac{sgn(\hbar\omega)}{\sqrt{(\hbar\omega + i\eta)^2 - 4t^2}} \tag{4.6}$$

The single-particle spectral function $A(\vec{k},\omega)$, which is a function of λ, can be obtained by taking the imaginary part of $G(\vec{k},\omega)$. It is worth noting that $A(\vec{k},\omega)$ directly corresponds to the experimental ARPES spectra.

In the absence of the electron-phonon coupling ($\lambda = 0$), the simulated spectral function shows a parabolic band (Figure 4.4l), which is in stark contrast to the ARPES spectrum. This experimental spectrum can be well reproduced by the simulation at $\lambda = 0.5$. In addition, the flat band segments are also predicted by the same simulation (Figure 4.4m and n). More specifically, the simulated flat bands illustrate a decrease in the spectral weight (the electronic character) from the polaron band to the deeper excited states, in line with the experiments. These agreements further corroborate the formation of the Holstein polaron in the electron-doped MoS_2 and its crucial role in the conduction band renormalization.

4.2.2 Asymmetric Intervalley Polaron Effects on Band Edges of 2D-TMDs

From the ARPES spectra, the electron effective mass can be extracted as $m_0{\sim}0.85m_e$ and has little dependence on the electron density on the order of 10^{13} cm^{-2}(Figure 4.4j) [62]. With the decrease in the electron density, the electron-phonon coupling would gradually turn into the weak coupling regime, as suggested by the mass enhancement in Figure 4.4j. Riccardo Pisoni et al. conducted the magnetotransport experiments for MoS_2 monolayer at $\rho = 1{\sim}5\times10^{12}$ cm^{-2} and measured an electron effective mass of $m_0 = 0.7m_e$ [70]. Interestingly, the effective mass was also found to have no obvious dependence on the electron density on the order of 10^{12} cm^{-2}. This may suggest that in such electron density range the electron-phonon coupling already falls into the weak coupling regime, where the electronic band is less altered. However, this experimental electron effective mass of $0.7m_e$ is remarkably larger than that predicted by DFT calculations ($0.43 - 0.47m_e$), whereas a good agreement is reached for the hole effective mass of MoS_2 (experiment: $0.6m_e$; DFT: $0.54 - 0.61m_e$) [26, 71]. This also holds for $MoSe_2$ monolayer, for which the experimental (DFT-calculated) electron effective mass is $0.8m_e$ ($0.49 - 0.58m_e$), and the experimental (DFT-calculated) spin-resolved hole effective mass is $0.67m_e$ and $0.75m_e$ ($0.59 - 0.64m_e$ and $0.69 - 0.72m_e$) [26, 72]. With regard to the electron effective mass of MoS_2 and $MoSe_2$ monolayers, Glazov et al. demonstrated that the weak-coupling intervalley polaron could be a pathway to partially close the gap between experiments and DFT calculations (Figure 4.5a) [63].

According to the second-order perturbation theory, the general electron self-energy induced by the coupling with phonons can be written as

$$\Sigma_{\vec{k}}(\varepsilon) = \sum_{\vec{q},\alpha;j} \frac{\left|M^j_{\vec{q},\alpha}\right|^2}{\varepsilon - \Omega_{\vec{q},\alpha} - E^j_{\vec{k}-\vec{q}} + i\eta} \tag{4.7}$$

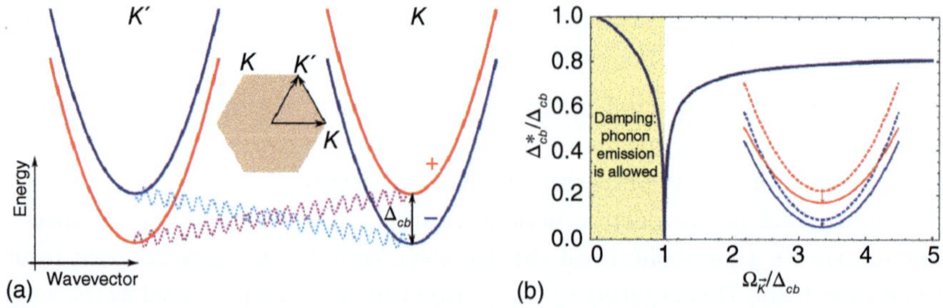

Figure 4.5 (a) Schematic illustration of the spin-resolved conduction band edge around the K and K′ valleys of TMDs. Signs "+" and "−" denote the top and bottom subbands, respectively. (b) The renormalized conduction band spin splitting as a function of the ratio of the phonon energy $\Omega_{\vec{K}}$ to the bare conduction band spin splitting Δ_{cb}. Inset depicts the relative shifts of the top and bottom subbands due to the intervalley polaron effect. Source: Glazov et al. [63]/With permission from American Physical Society.

where $\vec{k}$ and $\vec{q}$ are the wave vector of the electron and the phonon in the crystal, respectively. α is the phonon branch index, j enumerates the electron intermediate states, $\Omega_{\vec{q},\alpha}$ is the phonon energy, $M^j_{\vec{q},\alpha}$ denotes the electron-phonon coupling matrix element, and $E^j_{\vec{k}}$ is the electronic band energy of state j.

The calculation of $\Sigma_{\vec{k}}(\varepsilon)$ involves the integration over the electron intermediate states and the phonons across the whole Brillouin zone. Note that around the K and K′ valleys, the electronic density of states is the largest, and the phonon bands are dispersionless. Therefore, the electronic contributions around the K and K′ valleys are mainly considered, and the phonon dispersions are neglected. In what follows, we concentrate on the electron states in the K valley. In this case, $\Omega_{\vec{q},\alpha}$ and $M^j_{\vec{q},\alpha}$ are replaced by $\Omega_{\vec{k}}$ and $M^j_{\vec{k}}$ (index α is hereafter omitted for brevity). Assuming a parabolic band dispersion ($E_{\vec{k}} = \frac{\hbar^2 k^2}{2m_0}$; m_0: the bare electron effective mass), the intermediate state energy of the top (bottom) spin subband in the K valley can be written as $E^j_{\vec{k}-\vec{q}} \approx \mp\Delta_{cb} + E_{\vec{k}-\vec{q}'}$, where Δ_{cb} is the conduction band spin splitting and $\vec{q}' = \vec{q} - \vec{K}$. Eq. (4.7) for the top (+) and bottom (−) spin subband then can be recast into

$$\Sigma_{\vec{k},\pm}(\varepsilon) = -\frac{\beta_{\vec{K}}(\Delta_{cb} + \Omega_{\vec{K}})}{4\pi} \times \int_0^{E_Q} \frac{dE}{\sqrt{(\varepsilon - E_{\vec{k}} \pm \Delta_{cb} - \Omega_{\vec{K}} - E)^2 - 4E_{\vec{k}}E}} \tag{4.8}$$

$E_Q = \hbar^2 Q^2/2m_0$ is the cut-off energy with the cut-off wave vector $Q \sim |\vec{K}|$. $\beta_{\vec{K}}$ is the effective intervalley coupling constant, which reads

$$\beta_{\vec{K}} = \frac{2Sm_0|M_{\vec{K}}|^2}{\hbar^2(\Delta_{cb} + \Omega_{\vec{K}})} \tag{4.9}$$

where S is the normalization area. It should be noted that the term $i\eta$ is also omitted for brevity.

With $\Sigma_{\vec{k},\pm}(\varepsilon)$, the energy shift of the conduction band edge can be calculated, as illustrated in Figure 4.5b. At $\Omega_{\vec{K}} > \Delta_{cb}$, the intervalley coupling pushes both subbands downwards in energy, between which the top subband shifts more. As a result, the conduction band spin splitting is reduced by $\delta\Delta_{cb} = \lim_{\vec{k}\to 0}[\Sigma_{\vec{k},+}(E_{\vec{k}}) - \Sigma_{\vec{k},-}(E_{\vec{k}})]$. In terms of the bare Δ_{cb}, the reduced spin splitting can be expressed as

$$\Delta^*_{cb} \equiv \Delta_{cb} + \delta\Delta_{cb} = \Delta_{cb}\left[1 - \frac{\beta_{\vec{K}}}{4\pi}\left(1 + \frac{\Omega_{\vec{K}}}{\Delta_{cb}}\right)\ln\left(\frac{\Omega_{\vec{K}} + \Delta_{cb}}{\Omega_{\vec{K}} - \Delta_{cb}}\right)\right] \tag{4.10}$$

For $\Omega_{\vec{K}} < \Delta_{cb}$, however, Δ^*_{cb} becomes complex. The energy shift is given by the real part. In this case, phonon emission is likely to happen during the intervalley transition. The associated electron damping is described by the imaginary part of Δ^*_{cb} (Figure 4.5b). It should be noted that the conduction band spin splitting of MoS_2 ($MoSe_2$) is 0.8 – 3 meV (~20 meV), whereas their optical phonon branches usually have energies of ~20 – 60 meV [26, 73, 74]. Thus, $\Omega_{\vec{K}} > \Delta_{cb}$ is more likely to be the real situation for MoS_2 and $MoSe_2$.

The renormalization of the effective mass can be extracted from the k^2 contribution in the self-energy Eq. (4.8). For the top (+) and bottom (−) spin subband, it reads

$$m^*_{\pm} = m_0\left(1 + \frac{\beta_{\vec{K}}}{4\pi} \times \frac{\Omega_{\vec{K}} + \Delta_{cb}}{\Omega_{\vec{K}} \mp \Delta_{cb}}\right) \tag{4.11}$$

For $\Omega_{\vec{K}} > \Delta_{cb}$, Eq. (4.11) implies that the weak-coupling intervalley polaron effect increases the effective mass of both sub-bands, between which that of the top subband increases more (Figure 4.5b). Note that the same formulation applies to the valence band edge of MoS_2 and $MoSe_2$. However, the spin splitting of the valence band ($\Delta_{vb} \geq 140$ meV for MoS_2 and $\Delta_{vb} \geq 180$ meV for $MoSe_2$) is much larger than the conduction band counterpart and even their typical phonon energies. As a result, the effective intervalley coupling constant significantly decreases, leading to a negligible impact on the hole's effective mass. In a sense, this asymmetric weak-coupling intervalley polaron effect on the electron and hole effective mass is because of the large difference in the band edge spin splitting in MoS_2 and $MoSe_2$.

It is worth noting that the intravalley intermediate states, which correspond to the long-range Fröhlich-like polaron, could also make significant contributions to the electron self-energy. But the asymmetric band edge spin splitting does not enter the formulation. As such, the intravalley polaron effect on both the electron and hole effective mass is likely to be similar. Hence, it cannot be responsible for the discrepancy in the electron effective mass between experiments and DFT calculations.

4.2.3 Polaron Effects on the Band Gap Size of 2D-TMDs

Sections 4.2.1 and 4.2.2 show the significant role of the Holstein or intervalley polarons in the band edge renormalizations, where the phonon modes involved have short wavelengths. Apart from them, the long-wavelength phonon modes could also strongly couple with carriers and modulate the electronic structures. An example is the experimentally observed dependence of TMDs' band gap size on the surrounding dielectric environment. Archana Raja et al. found that engineering the dielectric environment is able to tune the electronic band gap of WS_2 and WSe_2 monolayers by hundreds of meV [75]. Based on theoretical simulations in the Fröhlich polaron model, Yao Xiao et al. reported that the long-wavelength surface optical (SO) phonon modes may play a key role [54].

In the presence of a magnetic field (B), the energies within the conduction and valence bands will be quantized and form Landau levels, as illustrated in Figure 4.6a. A peculiar character of the Landau levels for 2D-TMDs is that the zeroth ($n = 0$) Landau level at the K valley is usually bound to the top of the valence band, whereas it is bound to the bottom of the conduction band at the K′ valley [76]. This allows a direct measurement of the band gap of 2D-TMDs. Considering the presence of a polar substrate, the general model Hamiltonian of a TMD monolayer reads

$$\mathcal{H} = \gamma v_F(\sigma_1\pi_x + \sigma_2\pi_y + \sigma_3 G) + \Sigma_{\vec{q},\alpha}\Omega_\alpha b^\dagger_{\vec{q}} b_{\vec{q}} + \Sigma_{\vec{q},\alpha} M_{\vec{q},\alpha}\left(b^\dagger_{-\vec{q}} + b_{\vec{q}}\right) e^{i\vec{q}\cdot\vec{r}} \quad (4.12)$$

The first term describes the electron ($\gamma = +$) or hole ($\gamma = -$) momentum energy. v_F is the Fermi velocity, which is assumed as 0.5×10^6 m s^{-1} for all TMDs; σ_i ($i = 1, 2, 3$) is the Pauli matrix; $2G$ is the band gap of TMD monolayer. In the symmetry gauge of magnetic field, $\pi_x = P_x - eBy/2$ and $\pi_y = P_y + eBx/2$, where P_j ($j = x, y$) and r_j are the momentum and coordinate, respectively. They can be expressed as $P_j = \frac{\hbar\lambda}{\sqrt{2}} \times (a_j^\dagger + a_j)$ and $r_j = \frac{i}{\sqrt{2}\lambda}(a_j - a_j^\dagger)$ in terms of creation and annihilation operators and the magnetic confinement length $\lambda = \sqrt{eB/2\hbar}$. The second and third terms describe the optical phonon modes and their coupling to the carrier, respectively. According to the Lee-Low-Pines (LLP) theory, applying two consecutive unitary transformations,

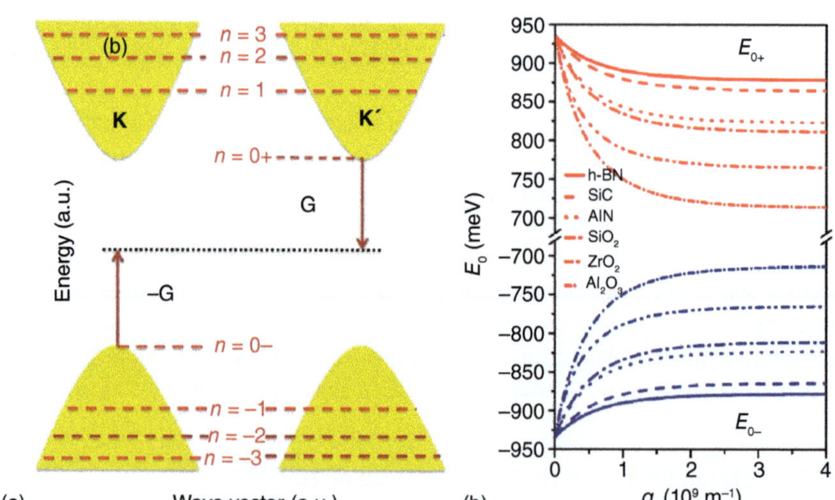

Figure 4.6 (a) Schematic illustration of the Landau levels in the K and K′ valleys of TMD monolayers. 2G denotes the intrinsic band gap. (b) The renormalized zeroth Landau levels (E_{0+} for electron; E_{0-} for hole) of MoS_2 monolayer as a function of the cut-off wave vector q_c in the presence of selected polar substrates. Source: Xiao et al. [54]/With permission from IOP Publishing.

i.e. $U_1 = \exp(-i\sum_{\vec{q}}\vec{q}\cdot\vec{r}b_{\vec{q}}^{\dagger}b_{\vec{q}})$ and $U_2 = \exp(\sum_{\vec{q}}(f_{\vec{q}}b_{\vec{q}}^{\dagger} - f_{\vec{q}}^{*}b_{\vec{q}}))$, and conducting energy minimization with respect to $f_{\vec{q}}$ and $f_{\vec{q}}^{*}$ could give the renormalized zeroth Landau levels due to the polaron effect (E_{0+} for electron; E_{0-} for hole):

$$E_{0\pm} = \pm\sqrt{G^2 + \left(\sum_{\vec{q},\alpha}\hbar q v_{\mathrm{F}}\frac{M_{\vec{q},\alpha}^2}{(v_{\mathrm{F}}\hbar q + \Omega_\alpha)^2}\right)^2}$$

$$\pm\sum_{\vec{q},\alpha}\Omega_\alpha\frac{M_{\vec{q},\alpha}^2}{(v_{\mathrm{F}}\hbar q + \Omega_\alpha)^2} \mp \sum_{\vec{q},\alpha}\frac{2M_{\vec{q},\alpha}^2}{v_{\mathrm{F}}\hbar q + \Omega_\alpha} \tag{4.13}$$

Equation (4.13) shows that both E_{0+} and E_{0-} do not depend on the applied magnetic field, nor does the modulated band gap due to the polaron effect ($E_{0+} - E_{0-}$). $2\Delta G = 2G - (E_{0+} - E_{0-})$ is hereafter introduced to measure the magnitude of the band gap modulation.

Regarding the coupling of the carrier to surface optical phonons, the coupling element in Eq. (4.13) reads

$$M_{\vec{q},\mathrm{SO}} = \sqrt{\frac{e^2\eta\Omega_{\mathrm{SO},\nu}}{2A\varepsilon_0 q}}e^{-qz_0} \tag{4.14}$$

where $\eta = (\kappa_0 - \kappa_\infty)/[(\kappa_\infty + 1)(\kappa_0 + 1)]$ describes the polarizability of the substrate. κ_0 and κ_∞ are the low- and high-frequency dielectric constants, respectively. $\Omega_{\mathrm{SO},\nu}$ ($\nu = 1, 2$) denotes the phonon energy of the νth branch. A, ε_0 and z_0 stand for the interface area, the vacuum permittivity, and the internal distance between the TMD monolayer and the substrate, respectively. The substrate-specific parameters are listed in Table 4.1 for the considered polar substrates and z_0 are assumed to be 0.5 nm. By converting the summation over q into integration in Eq. (4.13) and cutting the wave vector at q_c, A is canceled out and $E_{0\pm}$ can be numerically calculated as a function of q_c.

Figure 4.6b shows the impact of the polaron effect on the zeroth Landau levels of MoS_2 monolayer. With the increase in q_c, E_{0+} (E_{0-}) sharply shifts downwards (upwards) in energy and quickly converges. This indicates that the phonon

Table 4.1 Parameters of the selected polar substrates and the surface optical phonon modes.

	h-BN	SiC	AlN	SiO_2	ZrO_2	Al_2O_3
κ_0 (ε_0)	5.1	9.7	9.1	3.9	24.0	12.5
κ_∞ (ε_0)	4.1	6.5	4.8	2.5	4.0	3.2
$\Omega_{\mathrm{SO},1}$ (meV)	167	146	84	25	94	53
$\Omega_{\mathrm{SO},2}$ (meV)	116	60	105	71	55	19
η	0.032	0.040	0.074	0.082	0.160	0.164

Source: [54]/With permission from IOP Publishing.

Table 4.2 Band gap modulation of TMD monolayers in the presence of the SiO_2 substrate at $z_0 = 0.5$ nm.

	MoS_2	$MoSe_2$	WS_2	WSe_2
$2G$ (meV)	1870	1560	2100	1650
$2\Delta G$ (meV)	124.3	124.1	124.4	124.2
$\Delta G/G$ (%)	13.3	16.0	11.9	15.1

Source: [54]/With permission from IOP Publishing.

modes with small wave vectors dominate the coupling to carriers, in line with the long-wavelength nature of the surface optical phonon modes. Depending on the polar substrate, $E_{0\pm}$ could be reduced in a range from tens to more than 200 meV. It is interesting to note that the resulting modulation of the band gap is on the same order of magnitude as that observed in experiments [75], which confirms the significant role of the long-wavelength surface optical phonon modes. Table 4.2 further tabulates the band gap modulation of different TMDs on top of the SiO_2 substrate. One sees that the magnitudes of the modulations are very similar to one another. This is because only the respective band gap size enters the model Hamiltonian, whereas the detailed electronic structure of the respective TMD monolayer is ignored.

In addition to the surface optical phonons, the intrinsic longitudinal optical (LO) phonons are another type of widely studied long-wavelength phonon modes. The corresponding carrier-phonon coupling element in the Fröhlich polaron model can be written as

$$M_{\vec{q},\mathrm{LO}} = \sqrt{\frac{e^2\eta_0 L_m \Omega_{\mathrm{LO}}}{2A\varepsilon_0}}\, erfc\left(\frac{q\sigma}{2}\right), \tag{4.15}$$

where η_0, L_m, and Ω_{LO} are the dimensionless polarizability, the atomic thickness, and the characteristic phonon energy of TMD monolayers, respectively. σ denotes the effective width of the electronic Bloch states due to the confinement effect between LO phonons and carriers in TMD monolayers. $\eta_0 = 0.2$, $L_m = 0.5$ nm and $\sigma = 0.6$ nm are here assumed for all TMDs. By plugging Eq. (4.15) into Eq. (4.13), the modulated band gap for four TMD monolayers are numerically calculated and summarized in Table 4.3. Compared with $2\Delta G$ induced by the SO phonons, the band gap modulation due to the LO phonons is much weaker ($2\Delta G{\sim}10$ meV). Therefore, coupling to the SO phonons more likely makes a dominant contribution to the band gap modulation of TMD monolayers in varying surrounding dielectric environments. In addition, it is noteworthy that the $2\Delta G$ induced by the LO phonon modes are similar to the experimental spectroscopy linewidth of TMD monolayers [77], which may provide an insight into the temperature dependence of experimental spectra.

This section illustrates the diverse polaron effects on the electronic structure of 2D-TMDs. It should be pointed out that the theoretical simulations discussed here are mainly based on either the Holstein model or the Fröhlich model. In fact, these

Table 4.3 Band gap modulation of TMD monolayers due to the intrinsic longitudinal optical (LO) phonon modes.

	MoS_2	$MoSe_2$	WS_2	WSe_2
Ω_{LO} (meV)	48	34	43	30
$2G$ (meV)	1870	1560	2100	1650
$2\Delta G$ (meV)	14	11	13	10
$\Delta G/G$ (%)	1.5	1.4	1.2	1.2

Source: [54]/With permission from IOP Publishing.

two models to a large extent simplify the electronic structure, the phonon modes and the coupling elements between them for a material of interest. Thus, they are only capable of describing certain types of polarons and are less material-specific [59]. To make the theoretical results more comparable with experimental data, the more general yet complex electron-phonon Hamiltonian needs to be calculated. In this regard, first-principles methods may make a great contribution. However, the DFT simulations of polaron states usually involve a large supercell where an electron is removed or added. In addition, DFT+U methods or hybrid functionals are also needed to alleviate the self-interaction error. Both make the DFT simulations of polaron states very computationally costly. In 2019, Giustino et al. recasted the Landau–Pekar equations in a DFT formalism, which enables polarons to be simulated from first principles without supercells [78]. But efforts are demanded to extend this method to 2D systems. Notwithstanding, recent progress along this direction further demonstrates it as a promising pathway toward the full simulation of polarons in 2D systems at an affordable computational cost [79, 80].

4.3 Valley Properties of 2D-TMDs

Besides the degrees of freedom of charge and spin, electrons also have "valley" degrees of freedom. Valleytronics is an emerging field that promises transformational advances in information processing through the use of a particle's momentum index [81]. Valley specifically refers to the local extremum in the momentum space of crystals, representatively the local region near CBM and VBM. In some materials, the electrons have a valley degree of freedom due to the appearance of inequivalent valleys (band extrema with equal energy located at different k points in the Brillouin zone) such as silicon [82], bismuth [83], graphene [84], and TMDs [7]. For example, graphene has two energy valleys at the inequivalent K and K′ points. However, because graphene is semi-metallic with zero energy gap, its inequivalent energy valleys are difficult to manipulate. Different from the metallic graphene, monolayer 1H-TMD is a typical semiconductor material, this provides the possibility for its practical application [85]. TMDs are the most appropriate materials in valleytronics because of the inversion symmetry breaking in the monolayer. SOC can result in valley-dependent Berry curvature, orbital magnetic moment [86], and optical circular dichroism. The valley properties of monolayer transition metal dichalcogenides have attracted extensive attention owing to its intriguing fundamental physics and potential application for future electronics.

As discussed in Section 4.1, monolayer TMDs, such as MoS_2 and $MoSe_2$, have a direct band gap with CBM and VBM located at both the K and K′ points [87] as shown in Figure 4.7a. The PBE band gap of MoS_2 is 1.60 eV and SOC gives rise to a remarkable spin splitting near the VBM of 150 meV and a negligible spin splitting near the CBM [88]. Near the VBM, the upper (lower) band is dominated by the spin-up (spin-down) component at the K point and the spin-down (spin-up) component at the K′ point, indicating spin-valley coupling. Figure 4.7b shows schematically the K and K′ valleys. As discussed in Section 4.2.1, the spin and valley degrees

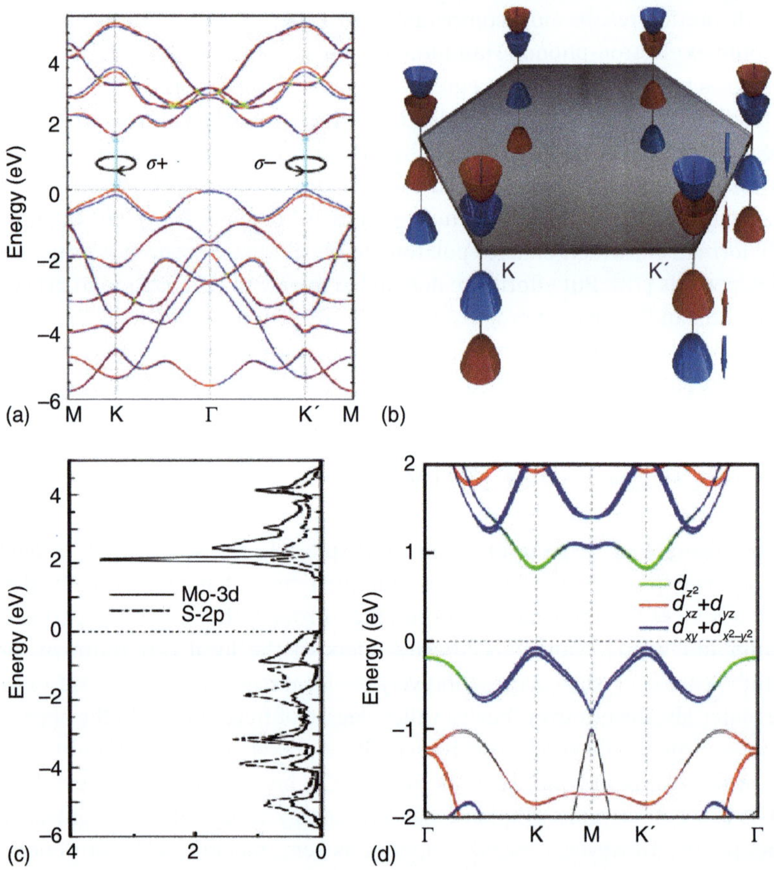

Figure 4.7 (a) Spin polarization band of monolayer MoS_2, Red (blue) color represents the dominance of the spin-up (spin-down) component. (b) Schematic of the valleys near the K and K′ points, (c) partial densities of states of monolayer MoS_2, (d) orbital-projected band structure of monolayer MoS_2. Source: Feng et al. [88]/With permission from American Physical Society.

of freedom are coupled by the time-reversal symmetry, which leads to opposite spin polarization near the VBM at the K and K′ points.

The projected densities of states displayed in Figure 4.7c demonstrate that the valley states are dominated by the d orbitals, which split into a_1 ($d_{3z^2-r^2}$), e_1 ($d_{x^2-y^2}$, d_{xy}), and e_2 (d_{xz}, d_{yz}) groups due to the trigonal crystal field [89]. Specifically, the CBM is mainly due to the Mo a_1 states, and the VBM is mainly due to the Mo e_1 states, as shown in the orbital-projected band structure in Figure 4.7d.

Consequently, the basic functions are defined

$$|\varphi_c\rangle = |d_{3z^2-r^2}\rangle\,;\,|\varphi_v^\tau\rangle = \frac{1}{\sqrt{2}}\left(\left|d_{x^2-y^2}\right\rangle + i\tau\left|\,d_{xy}\right\rangle\right), \tag{4.16}$$

where $c(v)$ denotes the conduction (valence) band and $\tau = +1$ (-1) denotes the K (K′) valley, the Hamiltonian of the two-band effective $k \cdot p$ model without SOC can be expressed as

$$\mathcal{H}_0^\tau = at(\tau k_x\sigma_x + k_y\sigma_y) + \frac{\Delta}{2}\sigma_z \tag{4.17}$$

where a denotes the lattice parameter, t denotes the effective hopping integral, $\sigma_{x/y/z}$ denotes the Pauli matrices, and Δ denotes the band gap [7]. It is well-known that SOC plays a key role in the electronic structure of TMDs, which can directly cause spin splitting of the bands at VBM and CBM. When SOC is taken into account, the effective $k \cdot p$ Hamiltonian can be written as:

$$\mathcal{H}_0^\iota = at(\tau k_x\sigma_x + k_y\sigma_y) + \frac{\Delta}{2}\sigma_z - \lambda\tau\frac{\sigma_z - 1}{2}\hat{s}_z \tag{4.18}$$

where 2λ is the spin splitting at the VBM (induced by SOC) and $\hat{s}_z$ is the Pauli operator. The effective parameters of the $k \cdot p$ model can be extracted from the first-principles band structures.

The exotic electronic structures will induce the valley-contrasting physical properties, such as valley-contrasting Berry curvature. The out-of-plane Berry curvature $\Omega_z(k)$ (z component) of the monolayer TMDs is given by:

$$\Omega_z(\vec{k}) = -\sum_n\sum_{n\neq n'} f_n \frac{2\mathrm{Im}\left\langle \Psi_{n\vec{k}}|v_x|\Psi_{n'\vec{k}}\right\rangle\left\langle \Psi_{n'\vec{k}}|v_y|\Psi_{n\vec{k}}\right\rangle}{\left(E_n - E_n'\right)^2} \tag{4.19}$$

where f_n is the Fermi-Dirac distribution function, $\vec{v}$ is the velocity operator, and E_n is the eigenvalue of Bloch wave function Ψ_{nk}. Figure 4.8a and b display the Berry curvature of MoS_2 along its high symmetry path and the color map in the hexagonal Brillouin zone, respectively [89]. The time-reversal symmetry requires $\Omega_z(-\vec{k}) = -\Omega_z(\vec{k})$. Hence, the Berry curvature $\Omega_z(\vec{k})$ is an odd function of the crystal momentum $\vec{k}$. The peaks of Berry curvatures are located at K and K′ valleys with the same magnitude but opposite sign. Figure 4.8c depicts a schematic diagram of the valley Hall effect driven by the Berry curvature. Because of the finite Berry curvature with opposite signs in the two valleys, when an in-plane longitudinal current passes through the valley system, the carriers at K and K′ valleys have opposite velocities along the direction perpendicular to the longitudinal current, which will accumulate on each side of the Hall bar. The system has an imbalance of spin index and

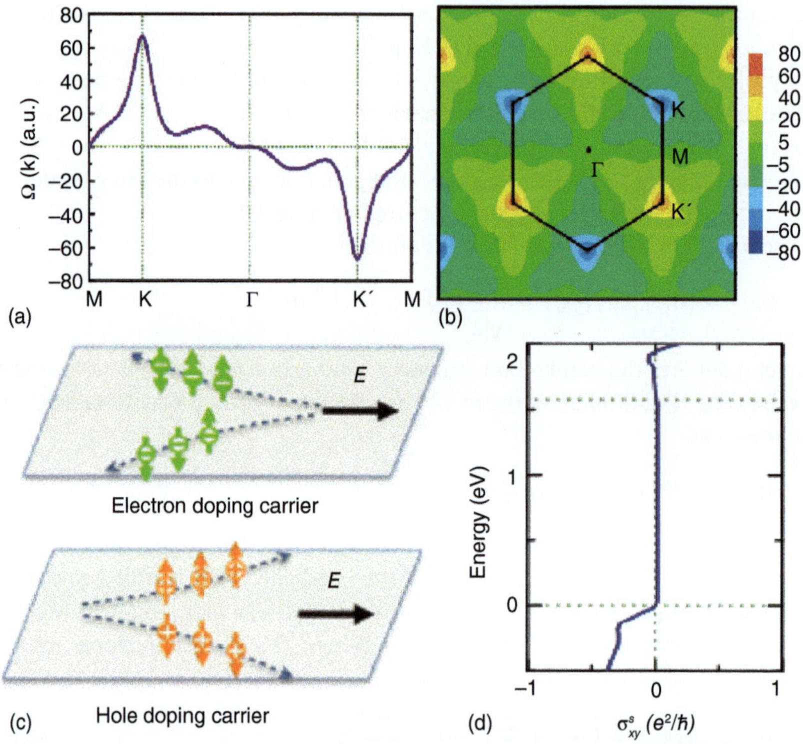

Figure 4.8 The Berry curvatures of monolayer MoS_2 along the high-symmetry lines (a) and in the 2D k plane (b). (c) Schematic of valley Hall effect, (d) the intrinsic spin Hall conductivity as a function of the Fermi energy for monolayer MoS_2. Source: Xiao et al. [7]/With permission from American Physical Society.

valley index at the boundary, forming a transverse spin flow and valley flow, which is the valley Hall effect [90]. Spin Hall effect will appear simultaneously with the valley Hall effect owing to the spin-valley locking relationship [88, 91]. The calculated intrinsic spin Hall conductivity σ^s_{xy} of MoS_2 is shown in Figure 4.8d. It is clear that σ^s_{xy} in the valence band is higher than that in the conduction band owing to the larger spin splitting in the valence band. Under a low hole-doping concentration, the spin Hall conductivity is negative, and the spin Hall conductivity will be reversed in sign under electron doping.

Due to the time-reversal symmetry, currents from the K and K′ valleys are equal, and there is no valley-polarized current. The basic and significant principle to realize the application of valleytronics is the inequivalent distribution of carriers in different valleys (valley polarization) and generation of a valley-polarized current. Valley polarization is able to be controlled in the systems with broken inversion symmetry by an external magnetic field [92], magnetic proximity effects [93], magnetic doping [94], and circularly polarized light [95, 96].

4.3.1 Circularly Polarized Light

As shown in Figure 4.9a, the direct interband transition of carriers in the K and K′ valleys must satisfy not only the conservation of momentum but also the conservation of angular momentum. Therefore, the interband transitions of the K and K′ valleys need to be pumped with different circularly polarized light, where the K valley absorbs the left-handed polarized light, and the K′ valley absorbs the right-handed polarized light, which is the valley optical selection rule [97]. The existence of circularly polarized light dichroism provides the necessary means to selectively excite K or K′ carriers. As shown in Figure 4.9b, Zeng et al. [97] used a right (left) excitation light source to excite a monolayer MoS_2 sample under the condition of $T = 10\,\text{K}$ and measured the right (left) luminescence helicity. The valley polarization efficiency $\rho = \frac{I(\sigma_-)-I(\sigma_+)}{I(\sigma_-)+I(\sigma_+)}$ is 32%. Gao et al. [98] found that valley polarization is achievable via valley-selective circular dichroism arising from its unique symmetry using first principles calculations, and achieved a valley polarization efficiency of 50% in monolayer MoS_2 under the experimental condition of $T = 300$ K (Figure 4.9c). Park et al. [99] encapsulated monolayer MoS_2 with two h-BN flakes to promote the

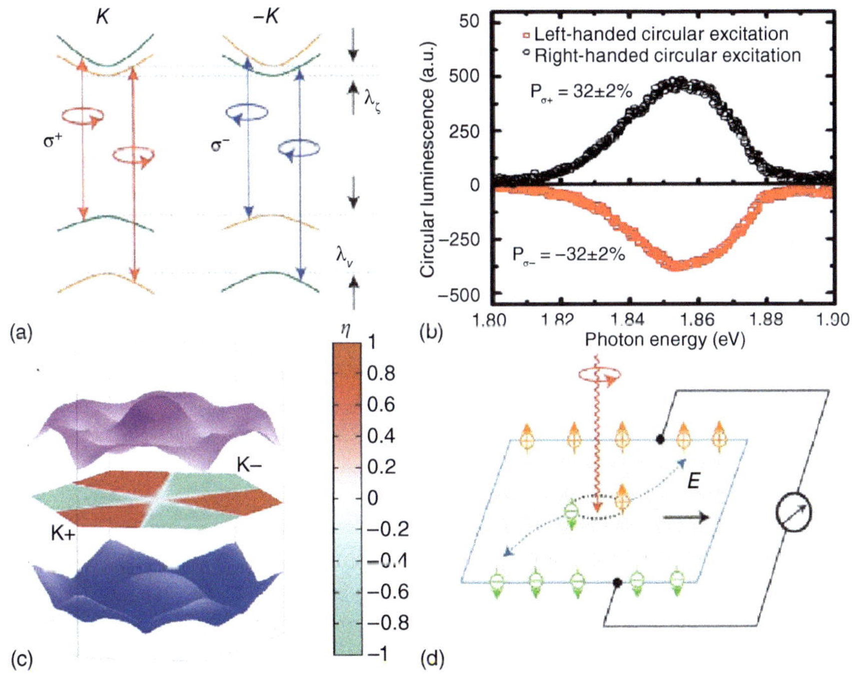

Figure 4.9 (a) Schematic diagram of valley optical selection rule. (b) Polarized luminescence spectrum of MoS_2. (c) First-principles prediction of valley polarization efficiency in MoS_2. (d) Schematic diagram of valley-based optoelectronic applications. Source: Xu et al. [89]/With permission from Springer Nature.

valley polarization application process of monolayer TMDs. The dynamic regulation of energy valleys by circularly polarized light pumping has been proved theoretically and experimentally, which can be used for valley-based optoelectronic applications as shown in Figure 4.9d.

4.3.2 External Field

Breaking the time-reversal symmetry with an external field is an effective way to achieve valley polarization [100, 101]. G. Aivazian et al. [102] observed the Zeeman splitting of WSe_2 under a vertical magnetic field at temperature $T = 30$ K through circularly polarized light measurements. When the magnetic field is 0 T, the phenomenon of polarized emission of excitons in K and K′ valleys is consistent, following the requirement of time-reversal symmetry. When an additional magnetic field of 7 T is applied, the position of the exciton peak of the K valley is higher than that of the K′ valley. When the magnetic field direction is opposite, the position of the K′ valley excitons is higher than that of the K valley, indicating that valley splitting occurs, and a polarization efficiency of 0.11 ± 0.01 meV per Tesla of the magnetic field is achieved. Li et al. [103] also found the same polarization luminescence phenomenon in $MoSe_2$ experiments, and increased the polarization efficiency to 50% by doping, reaching 0.18 ± 0.01 meV T^{-1}. However, a large external magnetic field is difficult to achieve. Compared with manipulating the magnetic field, electronically controlled valley polarization has the advantages of convenient adjustment of equipment, such as easy injection of valley current, continuous adjustment of inversion symmetry, tunable carrier doping density, and tunable exciton energy. Wu et al. [104] used the reversal electric field to continuously tune the magnetic moment of the Dirac valley. The presence of the vertical electric field realizes the non-equilibrium distribution of the interlayer potential in the bilayer MoS_2, breaking the spatial inversion symmetry. As the voltage increases, the inversion symmetry is destroyed, and the valley polarization efficiency is gradually improved.

4.3.3 Magnetic Metal Doping

When magnetic metal atoms are doped or adsorbed, local magnetic moments are introduced. The coupling between magnetic metal atoms and TMD will lift the energy degeneracy of K and K′ valley. However, the choice of the doping element and doping position has a great influence on the electronic structure of the system, and it is difficult to precisely control in experiments. Figure 4.10a and b are a schematic diagram of common doping sites for TMDs. Chen et al. [94] predicted the effect of adsorption of different metal atoms on the valley polarization properties of MoS_2 by first-principles calculations. As shown in Figure 4.10d, upon adsorption of magnetic Sc, Mn, Fe and Cu atoms on monolayer MoS_2, out-of-plane Zeeman field can be induced, and valley polarization could reach tens of meV. But for Ti, V, Cr adsorbed systems, the bands of transition metal atoms and the valence band tops of MoS_2 are hybridized with each other, and the energy valley characteristic disappears. Zhao et al. [105] realized the regulation of valley polarization of Co-doped

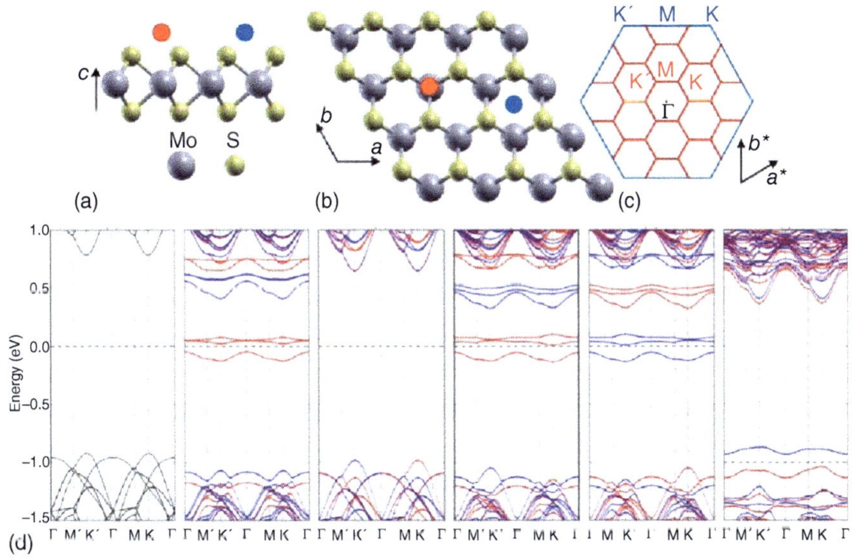

Figure 4.10 (a) and (b) Schematic diagram of TMD doping sites. (c) Brillouin zone corresponding to the 1 × 1 unit cell (blue) and 4 × 4 supercell (red), (d) Band structures of a 4 × 4 supercell of the MoS2 monolayer adsorbed by Sc, Ti, V, Cr, Mn, Fe, Co, and Cu atoms with the spin-orbit coupling. Source: Cheng et al. [94]/With permission from American Physical Society.

WTe_2 in a wide range by stress. With the compressive stress, not only can the valley splitting at the top of the valence band be regulated in a wide range (58–147 meV), but also the electronic properties at CBM are greatly changed, resulting in a valley splitting of 46 meV. The valley splitting at the top of the valence band is much larger than the valley splitting at the bottom of the conduction band. Under an appropriate hole doping and in-plane electric field, pure spin-up valley currents and spin currents can be generated.

4.3.4 Magnetic Substrate

Compared with the above strategies for inducing valley polarization, the introduction of magnetic substrates and the construction of van der Waals heterojunctions have a higher likelihood to realize valley regulation [93]. The proximity effect of the magnetic substrate can significantly change the band structure at the top of the valence band and realize valley polarization. Until now, a large number of theoretical and experimental works have been carried out in this area. For example, Xu et al. [93] constructed WS_2/MnO_2 heterostructures (Figure 4.11a) and successfully induced valley splitting of 15–43 meV as shown in Figure 4.11b. This indicates that the construction of van der Waals heterostructures can effectively induce valley polarization. The Berry curvatures are not equal any more in Figure 4.11c, hence the anomalous valley Hall effect will appear, and the valleytronic device based on WS_2/MnO_2 heterostructures can be used for the generation and transmission of valley information.

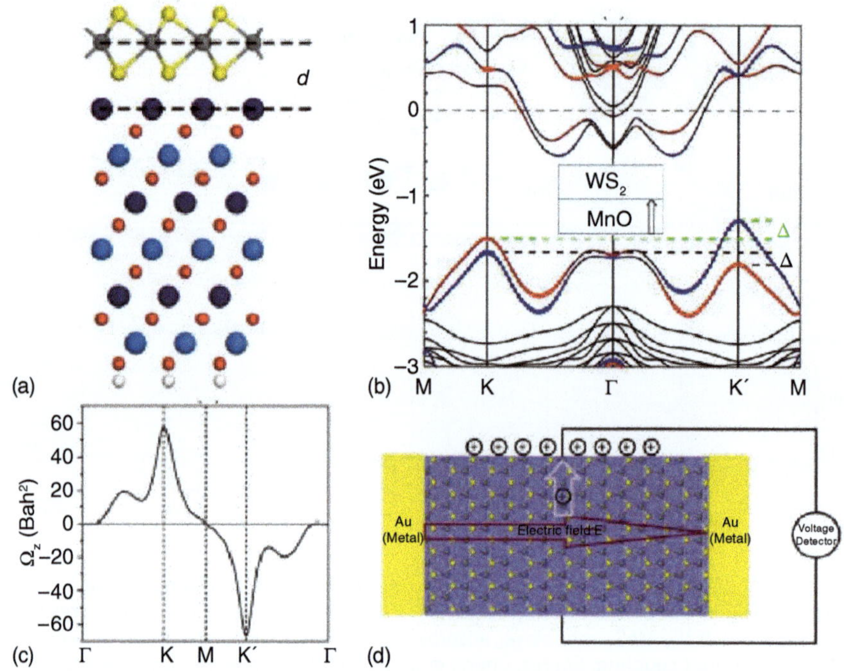

Figure 4.11 (a) side view of the WS_2/MnO heterostructure. (b) Band structure of WS_2/MnO heterostructure with SOC, the spin projections for monolayer WS_2 states along positive (spin up) and negative (spin down) *z* axis are denoted by red and blue weighted solid circles, respectively. (c) Calculated Berry curvature of monolayer WS_2 on the MnO substrate along high symmetry lines. Variation curve of valley splitting and magnetic field. (d) Proposed valleytronic device with intrinsic anomalous Hall effect. Source: Xu et al. [93]/With permission from American Physical Society.

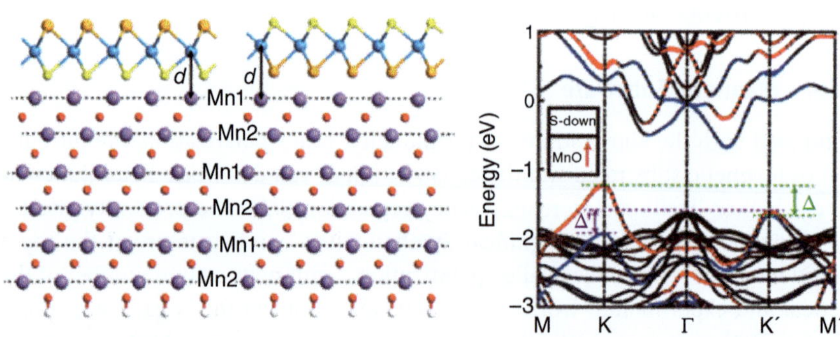

Figure 4.12 (a) Atomic structure and, (b) the band structure of WSSe on MnO substrate. Source: Zhou et al. [106].

Furthermore, Zhou et al. [106] predicted by DFT calculations that WSSe induces a 410 meV valley polarization in the valence band top of the (111) section of MnO and changes the effective Zeeman field by rotating the magnetic moment as shown in Figure 4.12. The size of the valley splitting can be continuously tuned. Zhong et al. [107] detected a valley splitting of 3.5 meV in the WSe_2/CrI_3 heterojunctions that corresponds to an effective exchange field of 13 T. Norden et al. [108] utilized the proximity effect of EuS to achieve valley splitting at 16 meV T^{-1}. A larger separation occurs between the $\sigma+$ and $\sigma-$ spectra, where the $\sigma+$ spectrum shifts to higher energy and $\sigma-$ is transferred to lower energy. Experimentally, Zhao et al. [109] successfully exploited the magnetic exchange field of the EuS substrate to measure a valley splitting of 2.5 meV on WSe_2 by magnetic reflectivity, corresponding to an effective exchange field of 12 T, which may accelerate the development of valleytronics and promote the applications in the field of quantum computing.

4.4 Charge Density Waves of 2D-TMDs

4.4.1 Charge Density Waves in TMDs

Charge density wave (CDW) is a quantum phenomenon that happens below a critical temperature, which features real-space spontaneous periodic charge density, lattice distortions, as well as reciprocal-space band gap opening. The idea of CDW is rooted in the Peierls model of a 1D half-filled atomic chain with a lattice constant a (Figure 4.13). In the 1950s, Peierls reported that such a system has an intrinsic instability (Peierls instability) [110], and its electron density ρ_0 can be spontaneously modulated by:

$$\rho(r) = \rho_0[1 + \rho_1 \cos(\vec{Q} \cdot \vec{r} + \phi)] \tag{4.20}$$

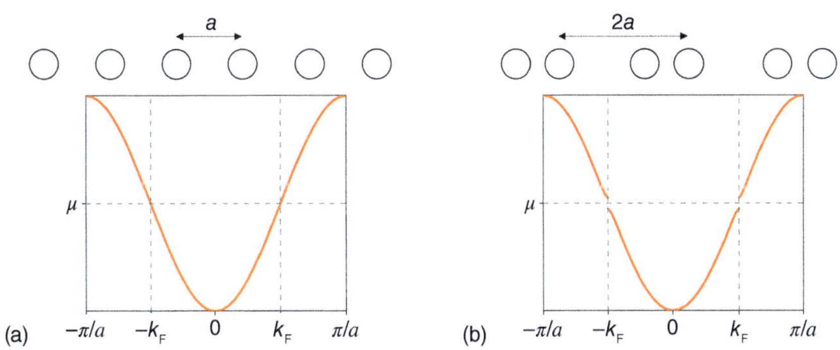

Figure 4.13 Peierls model of a 1D half-filled atomic chain with lattice constant a. The atoms are represented by black balls. (a) Before the CDW transition, the electrons fill up to Fermi energy μ. The Brillouin zone of this system is from $-\pi/a$ to π/a. (b) After the CDW transition, the dimerization of atoms doubles the lattice constant. New electronic gaps open at the new Brillouin zone boundary $\pm k_F$. The system changes from metal to insulator after the CDW transition.

with wavevector $\vec{Q} = 2\vec{k}_F$, amplitude ρ_1 and phase ϕ. Due to the electron-phonon interaction, it is energetically favorable to introduce a lattice distortion

$$u_n = u_0 \sin\left(n\frac{2\pi}{\lambda}a + \phi\right), \tag{4.21}$$

with atomic amplitude u_0 and wavelength

$$\lambda = \frac{2\pi}{|\vec{Q}|} = 2a \tag{4.22}$$

Since the lattice undergoes dimerization between neighboring atoms, the energy gaps appear at the new Brillouin zone boundary $k = \pm\pi/2a$. This gap opening decreases the electronic energy of the system ($\delta E_{\text{ele}} < 0$). When it surpasses the elastic energy generated by lattice distortions ($\delta E_{\text{lat}} > 0$), the dimerized atomic chain becomes the ground state ($\delta E_{\text{ele}} + \delta E_{\text{lat}} < 0$). This is known as a weak-coupling mechanism, as a small interaction can order the system at a low enough temperature [111]. For an arbitrary band filling, depending on the rationality of the wavelength to the lattice constant, CDWs can be categorized into commensurate or incommensurate CDWs.

Experimentally, there are many techniques to identify a CDW formation. For example, a CDW state has periodic modulation of charge density and lattices in real space, which can be easily observed by scanning tunneling microscopy (STM) and X-ray scattering. In the reciprocal space, CDW gap opening can be detected by electronic transport, spectroscopy, and ARPES measurements. Besides, inelastic X-ray scattering, neutron scattering, and Raman scattering can be used to probe phonon behaviors at different temperatures.

With the spirit of Peierls instability, people searched for CDWs in quasi-1D materials. CDW was first experimentally observed in the Krogmann's salt $K_2Pt(CN)_4Be_{0.3}{\cdot}xH_2O$ in the 1970s [112]. Hereafter, CDWs have been observed in many other quasi-1D systems, such as TTF-TCNQ [113], $NbSe_3$ [114], $(TaSe_4)I_2$ [115], and etc.

The discovery of CDW phases in TMDs has strongly broadened the CDW family. The phenomenology of the CDW is rich in 2D TMDs (Table 4.4 and Figure 4.14). Interestingly, the same CDW distortions can be found on the isologue of 1T phase TMDs in group IVB, VIB, and VIIB, while CDWs exist in both 1H and 1T phases in most of group VB TMDs, and their distortion is strongly dependent on the chemical compositions. For example, although 1T-VS_2, 1T-VSe_2, and 1T-VTe_2 have the same transition metal element of V and high-symmetry structure, their CDW distortions are very different from each other [123, 126, 129]. Conversely, CDWs are absent in 1H-VSe_2 and 1H-NbS_2 because of the semiconducting and anharmonic effects [116, 118], respectively. More interestingly, other quantum phenomena in TMDs, including superconductivity, Mottness, magnetism, and nontrivial topology, are closely related to the electronic structure distorted by CDW transition [48, 146–148]. In the following, two prototype CDW materials: 2D 1H-$NbSe_2$ and 1T-VTe_2, are selected as representative CDW examples to introduce their electronic structures.

Table 4.4 Diverse CDWs in the 2D TMDs. In some materials, different CDWs have been observed; we choose the most recognized one. The distorted structures in VIB and VIIB group TMDs are known as T′ and T″ phases with 1×2 and 2×2 supercells, respectively.

		VB			
		V	**Nb**	**Ta**	
H	S	?	No CDW [116]	3×3 [117]	
	Se	Semi [118]	3×3 [119]	3×3 [120]	
	Te	$2\sqrt{3} \times 2\sqrt{3}$ [121]	?	$\sqrt{13} \times \sqrt{13}$ [122]	
T	S	$7 \times \sqrt{3}$ [123]	$\sqrt{13} \times \sqrt{13}$ [124]	$\sqrt{13} \times \sqrt{13}$ [125]	
	Se	$\sqrt{7} \times \sqrt{3}$ [126]	$\sqrt{13} \times \sqrt{13}$ [127]	$\sqrt{13} \times \sqrt{13}$ [128]	
	Te	4×4 [129]	No CDW [130]	$\sqrt{19} \times \sqrt{19}$ [131]	
		IVB	**VIB**		**VIIB**
		Ti	**Mo**	**W**	**Re**
T	S	2×2 [132]	1×2 (T′) [133]	1×2 (T′) [134]	2×2 (T″) [135]
	Se	2×2 [136]	1×2 (T′) [133]	1×2 (T′) [137]	2×2 (T″) [138]
	Te	2×2 [139]	1×2 (T′) [140]	1×2 (T′) [48]	2×2 (T″) [141]

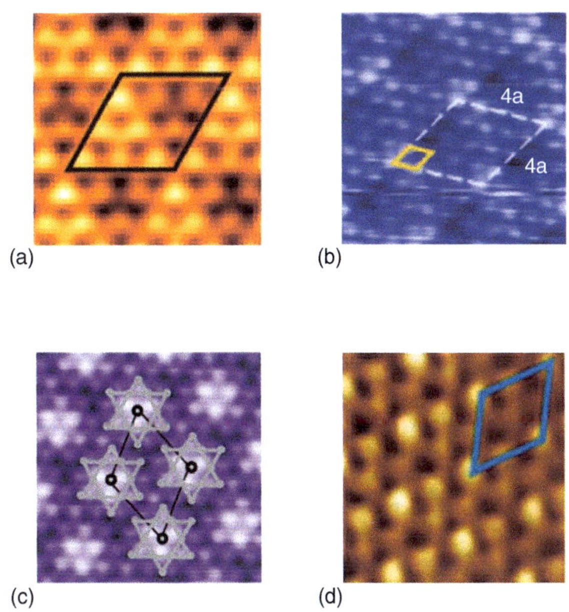

Figure 4.14 STM images reveal (a) 3×3 CDW in 1H-$NbSe_2$, (b) 4×4 CDW in 1T-VTe_2, (c) $\sqrt{13} \times \sqrt{13}$ CDW in 1T-TaS_2, and (d) 2×2 CDW in 1T-$TiSe_2$. Source: Eunseok Oh et al. [142]/Reproduced with permission from American Physical Society. (b) Wong et al. [143]/Reproduced from American Chemical Society. (c) Qiao et al. [144]/Reproduced from American Physical Society/CC BY 4.0. (d) Wang et al. [145]/Reproduced from American Chemical Society.

4.4.2 Effects of CDW on Electronic Properties

The 1H-$NbSe_2$ receives great research interest, partially due to the coexistence of CDW and superconductivity [119, 149–151]. 1H-$NbSe_2$ has the typical electronic structure of 1H-TMDs. It has one electronic band crossing the Fermi energy, which is composed of the Nb d_{z^2}, $d_{x^2-y^2}$ and d_{xy} orbitals (Figure 4.15a). After the CDW transition of $NbSe_2$ at 145 K, a CDW gap will open on the $\Gamma - \mathrm{K}$ path (Figure 4.15b). The full band-gapped sector is at the K pocket (see the cyan arrow in (Figure 4.15c). This $NbSe_2$ CDW phase becomes a superconductor below 3 K. Intriguingly, the calculated superconducting gap function $\Delta_{\mathbf{k}}$ suggests that the remaining Fermi surface is opened by the superconducting transition.

The typical electronic structure of 1T-TMDs can be found in monolayer 1T-VTe_2 [129, 143, 152]. Figure 4.16a is the second derivative of ARPES spectra along the

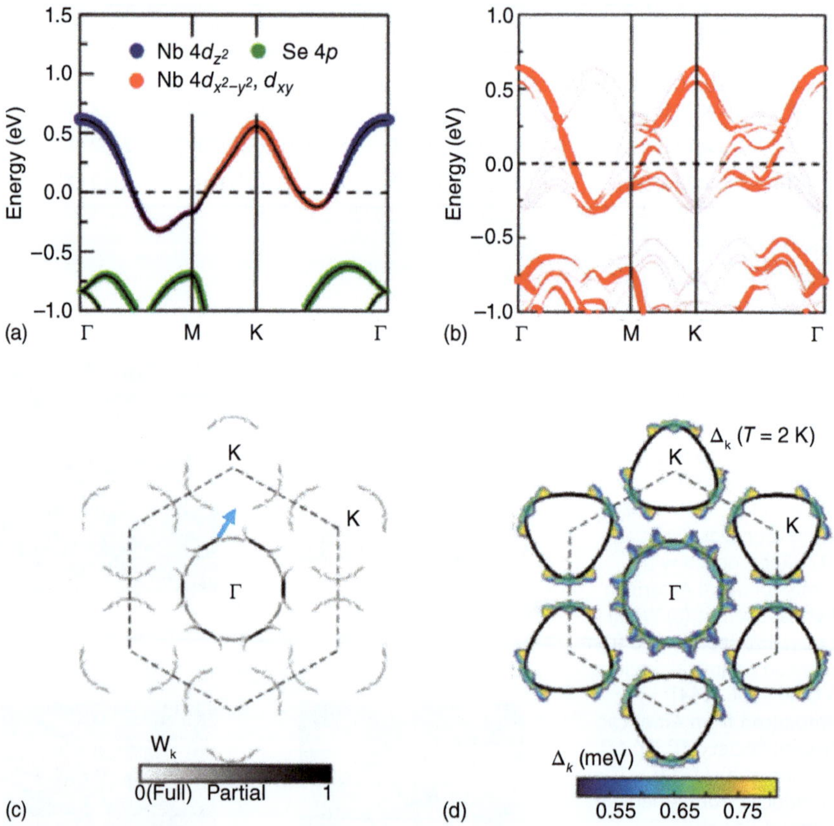

Figure 4.15 (a) Band structure of monolayer 1H-$NbSe_2$. The blue, red, and green dots represent Nb 4d_{z^2}, Nb 4$d_{x^2-y^2}$ + 4d_{xy}, and Se 4p orbitals, respectively. (b) Unfolded band structure and (c) Fermi surface of $NbSe_2$ CDW structure. (d) The distribution of superconducting gaps at 2 K. Source: Figure reproduced from: (a and b) Lian et al. [150], © 2018 American Chemical Society. (c and d) Zheng and Feng [146], ©2019 American Physical Society.

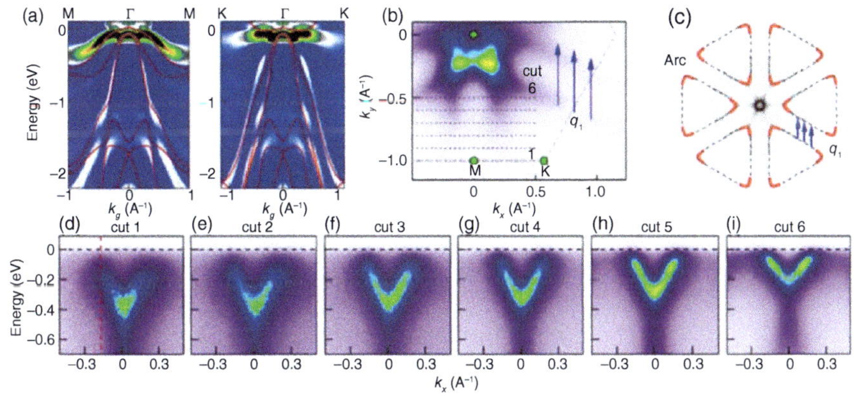

Figure 4.16 (a) The second derivative of ARPES dispersions along ΓM and ΓK paths of monolayer 1T-VTe_2 with the calculated band structure indicated by red lines. (b) ARPES map of Fermi surface of VTe_2 CDW structure. (c) Schematic diagram of Fermi surface with the gapped region (black dashed lines) and the remaining Fermi surface (red solid lines). (d–I) Evolution of ARPES along with cuts 1–6 in (b). Source: Wang et al. [129]/Reproduced with permission from American Physical Society.

$\Gamma - \text{M}$ and $\Gamma - \text{K}$ paths. The red lines are calculated band structures. There is an overall agreement between the calculated and experimental band dispersions. The APRES intensity near the M point is strongly suppressed, indicating the CDW gap opening (Figure 4.16b). Moving toward the Γ point, the spectral intensity at the Fermi surface becomes obvious, which reflects the decreasing of the gap sizes (Figure 4.16d–i). Finally, the CDW gap is completely closed, and the remaining Fermi surface forms a Fermi arc at the apex of the triangle K pocket (Figure 4.16c).

4.4.3 Mechanisms in CDW Transitions

Four models have been applied to explain the formation of CDW phases in TMDs, including the strong distorted T′ and T″ phases, that is, Fermi surface nesting (FSN), electron-phonon coupling (EPC), excitonic insulator (EI), and chemical bonding (CB).

The classical example of FSN is the Peierls instability of 1D atomic chains, which arises from the electronic instabilities [153]. The formation of electron-hole pairs is required at the Fermi surface. Such nesting most likely occurs when the shape of the Fermi surface has large parallel sectors, such as in an ideal 1D case (Figure 4.17a). FSN can be described by static Lindhard susceptibility [154]:

$$\chi'_{\vec{q}} = \sum_{\vec{k}} \frac{f(\varepsilon_{\vec{k}}) - f(\varepsilon_{\vec{k}+\vec{q}})}{\varepsilon_{\vec{k}+\vec{q}} - \varepsilon_{\vec{k}}} \tag{4.23}$$

where f is the Fermi Dirac function, $\varepsilon_{\vec{k}}$ is the electron energy at the $\vec{k}$ point. The criterion for CDW formation is that there would be a divergence in the static Lindhard function at the wave vector $\vec{q}$ where the nesting forms. This divergence will induce an abrupt phonon softening at the CDW vector, which is known as the Kohn

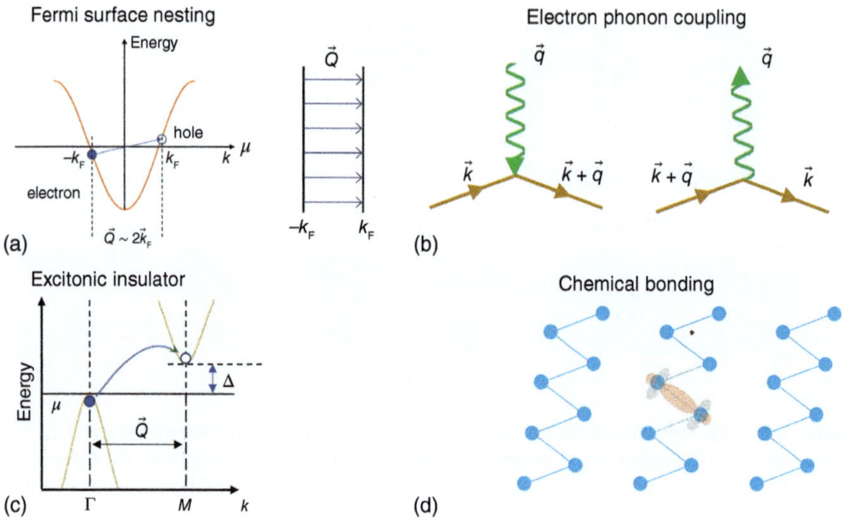

Figure 4.17 Mechanisms for CDW formation in TMDs. (a) Fermi surface nesting. (b) Electron-phonon coupling. (c) An excitonic insulator in a small indirect bandgap semiconductor. (d) Chemical bonding between the transition metal atoms.

anomaly [155]. One should note that FSN only relates to elastic electronic scatterings at the Fermi surface.

Since electronic modulation is always accompanied by lattice modulation, it is reasonable to consider inelastic electronic scatterings mediated by phonons (Figure 4.17b), the Hamiltonian of a coupled electron-phonon system (Fröhlich Hamiltonian [156]) is

$$\mathcal{H} \cong \sum_{\vec{k}} \varepsilon_{\vec{k}} a_{\vec{k}}^{\dagger} a_{\vec{k}} + \sum_{\vec{q}} \hbar\omega_{\vec{q}} b_{\vec{q}}^{\dagger} b_{\vec{q}} + \frac{1}{\sqrt{N}} \sum_{\vec{k},\vec{q}} g_{\vec{k},\vec{k}+\vec{q}} a_{\vec{k}+\vec{q}}^{\dagger} a_{\vec{k}} \left(b_{-\vec{q}}^{\dagger} + b_{\vec{q}} \right) \tag{4.24}$$

Here, $a_{\vec{k}}^{\dagger}$ ($a_{\vec{k}}$) and $b_{\vec{q}}^{\dagger}$ ($b_{\vec{q}}$) are creation (annihilation) operators for an electron with momentum $\vec{k}$ and a phonon with momentum $\vec{q}$, respectively. $\varepsilon_{\vec{k}}$ and $\hbar\omega_{\vec{q}}$ are electronic energy and phonon energy, respectively. $g_{\vec{k},\vec{k}+\vec{q}}$ is the EPC matrix element. The first term is the electronic Hamiltonian, the second term is the phonon Hamiltonian, and the third term is the first-order EPC Hamiltonian.

The solution of this Hamiltonian generates the relationship between the softened phonon frequency $\omega_{\vec{q}}$ and its bare phonon frequency $\Omega_{\vec{q}}$ [157]:

$$\omega_{\vec{q}}^2 = \Omega_{\vec{q}}^2 - 2\Omega_{\vec{q}}\chi_{\vec{q}} \tag{4.25}$$

where

$$\chi_{\vec{q}} = \sum_{\vec{k}} \left| g_{\vec{k},\vec{k}+\vec{q}} \right|^2 \frac{f(\varepsilon_{\vec{k}}) - f(\varepsilon_{\vec{k}+\vec{q}})}{\varepsilon_{\vec{k}+\vec{q}} - \varepsilon_{\vec{k}}} \tag{4.26}$$

is the generalized static electronic susceptibility, which contains both FSN and EPC effects. The imaginary phonon frequency implying CDW formation appears when the $\chi_{\vec{q}}$ is larger than $(1/2)\Omega_{\vec{q}}$. Under the constant matrix element approximation

$$|g_{\vec{k},\vec{k}+\vec{q}}| = 1 \tag{4.27}$$

the $\chi_{\vec{q}}$ is simplified to the $\chi'_{\vec{q}}$, which reflects the pure FSN effect. Similarly, under the constant fraction approximation

$$\frac{f(\varepsilon_{\vec{k}}) - f(\varepsilon_{\vec{k}+\vec{q}})}{\varepsilon_{\vec{k}+\vec{q}} - \varepsilon_{\vec{k}}} = 1 \tag{4.28}$$

the $\chi_{\mathbf{q}}$ is simplified to the momentum-dependent EPC:

$$\overline{g}_{\vec{q}} = \sum_{\vec{k}} \left| g_{\vec{k},\vec{k}+\vec{q}} \right|^2 \tag{4.29}$$

which reflects the pure EPC effect. These two effects are mainly used to understand the CDW formation mechanism in group-VB TMDs [153, 158]. FSN seems to be important in the 1T-TMDs due to their hidden 1D structure [159], while the EPC plays an indispensable role in 1H-TMDs because of poor nesting conditions [160]. Inelastic X-ray scattering experiments of TMDs can probe the phonon behaviors and help to understand the CDW formation mechanism, however, they are rarely reported [161, 162]. Until very recently, researchers could distinguish the contributions of FSN and momentum-dependent EPC to the CDW formation in a theoretical (computational) manner, which indicates that momentum-dependent EPC is a more prevailing CDW formation mechanism in TMDs [111, 152, 163].

When it comes to the origin of CDWs in group-IVB TMDs, EI instability may play an important role [164, 165]. Using a two-band model of 1T-$TiSe_2$ with a small indirect bandgap [166] as an example (Figure 4.17c). The valence band at the Γ point comprises the p orbital of Se atoms, while the conduction band at the M point consists of the d orbital of Ti atoms [165]. In this case, if the binding energy of the electron–hole pair is larger than the bandgap Δ, then the electronic instability occurs due to the formation of excitons at low temperatures [153, 158]. This is a manifestation of an excitonic condensation, leading to the CDW transition in $TiSe_2$ [165].

CDWs in group-VIB are known as 1T$'$ phase, the dimerization of metal atoms form zig-zag chains [48]; while CDWs in group-VIIB are known as 1T$''$ phase (the tetramerization of metal atoms form diamond chains) [135]. They are known as strong coupling CDWs, which are characterized by real space chemical bonding between transition metal atoms [153]. This bonding mechanism can be well described by the Wannier function approach [134]. The strong coupling CDWs have a larger atomic displacement and larger electronic energy gain compared to the weak coupling CDWs [153]. Table 4.5 summarizes the mainstream opinion on CDW formation mechanisms in TMDs. Although both EPC and FSN have been used to understand CDW formation in group-VB TMDs, we emphasize that EPC seems to be more reasonable, according to the recent literature [152, 162, 163, 167]. The minority points also exist, such as a saddle point near the Fermi energy, which supports CDW in 2H-$NbSe_2$ [168], and the combination of momentum-dependent EPC and electron correction in understanding the CDW in 1T-$TiSe_2$ [169].

Table 4.5 A summary of CDW formation mechanisms for the different groups of TMDs.

	IVB	VB	VIB	VIIB
Mechanism	EI	EPC (FSN)	CB	CB

4.4.4 Manipulation of CDWs

Since the ground state of CDW is determined by the delicate balance between electronic energy gain and elastic energy loss, CDW itself and related properties are very sensitive to external stimuli [170]. For example, in intrinsic 1T-TaS_2, it forms an incommensurate CDW (ICCDW) at 550 K, a nearly commensurate CDW (NCCDW) at 350 K, and a commensurate CDW (CCDW) with Mott insulating character at 180 K [171]. However, the external pressure melts CDW phases and induces superconductivity in 1T-TaS_2 (Figure 4.18a). Interestingly, superconductivity also emerges in 1T-$TiSe_2$ when its CDW order is suppressed by either pressure or copper intercalation (Figure 4.18b). On the other hand, the supercell size of CDW can also be modulated. In 1H-TaS_2, electron doping drives the CDW transition from a 3 × 3 to a

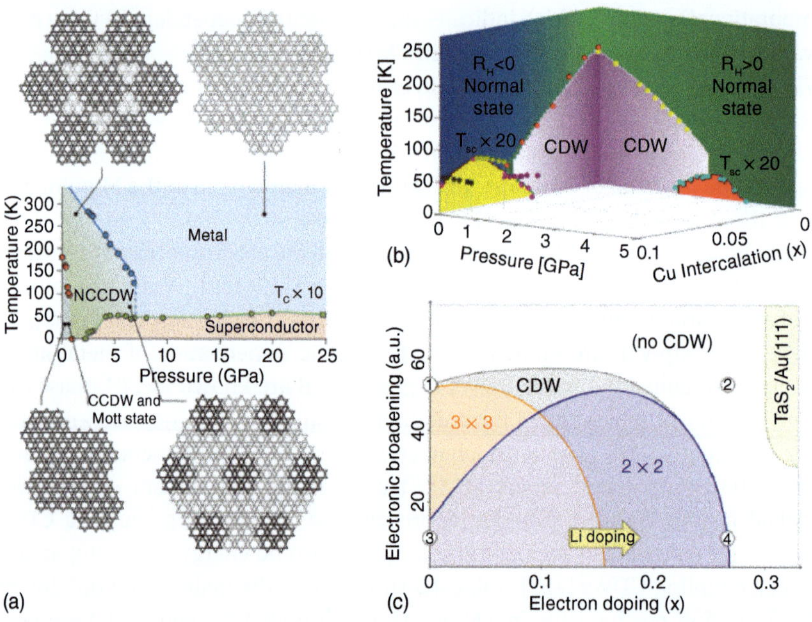

Figure 4.18 (a) The phase diagram of 1T-TaS_2 under pressure. Various CDW patterns are observed. (b) The phase diagram of 1T-$TiSe_2$ under pressure and copper intercalation. (c) The phase diagram of 1H-TaS_2 under electron doping. Source: Figure reproduced from: (a) Manzeli et al. [2] © 2017, Macmillan Publishers Limited, part of Springer Nature. (b) Kusmartseva et al. [172], © 2009 American Physical Society. (c) Ref. [117], © 2019 American Chemical Society.

2 × 2 supercell, and finally, the CDW distortion is completely melted when the doping concentration is high enough (Figure 4.18c).

4.5 Janus Structures of 2D-TMDs

In the other sections, we have shown the interesting properties and potential applications of intrinsic 2D TMDs, including various and tunable bandgaps, high carrier mobility, unique optical properties, and coupled spin valley physics, which are promising for optoelectronic, photovoltaic, electronic, spintronic, and valleytronic devices. Further research has been performed to break their out-of-plane mirror symmetry, aiming to introduce emerging properties to this system, such as the substrate effect, vertical stacking, applying external electric fields, or functionalization. Among them, Janus 2D TMDs with asymmetric structures consisting of two different chalcogen elements (S, Se, or Te) on two sides of the transition metal element (Mo, W, Pt, etc.) atomic layer, resulting in the reduced symmetry from D_{3h} to C_{3v}, are attracting increasing attention [173–176]. The concept of Janus material was first proposed in 1991 by de Gennes in his Nobel Lecture [176]. Although Janus 2D TMDs have been theoretically predicted to be thermally and dynamically stable, experimental investigations remain scarce until recent reports on the experimental synthesis of high-quality Janus TMDs [133, 177–180]. This offers a synthetic strategy to induce vertical asymmetry in TMDs, yielding rich emerging properties, such as out-of-plane piezoelectric polarization, the second harmonic generation response and strong Rashba effects, and excellent exciton properties [174, 175]. Leveraging these unique properties, Janus 2D materials are appealing in various applications ranging from sensors, anode materials for batteries, spintronics, optoelectronics, and catalysis [174, 181–185].

4.5.1 Fabrication Approaches for Janus 2D TMDs

For practical applications, the growth of high-quality samples is the first key step. However, unlike the conventional 2D materials such as graphene that can be exfoliated directly from their parent layered bulk materials (a top-down approach), Janus 2D TMDs are synthetic materials. Hitherto, four different fabrication approaches in a bottom-up manner have been reported to successfully synthesize Janus 2D TMDs and their heterostructures.

(i) Two-step chemical vapor deposition (CVD) process. H_2 plasma was first utilized to strip the top S atomic layer of MoS_2. Then, Se powder was thermally vaporized for the selenization at the optimal temperature (350–450 °C) and Janus MoSSe monolayers could be produced [186]. (ii) One-step direct sulfurization process by CVD. By careful control of the growth temperature (750–850 °C) and pressure, the thermally vaporized gaseous sulfur molecules can sulfurize the top selenium atomic layer of $MoSe_2$ while keeping the bottom selenium layer intact [187]. A similar strategy has been applied to synthesize 2D WSSe and PtSSe [188]. (iii) One-step direct sulfurization process by pulsed laser deposition (PLD).

By tuning the hyperthermal kinetic energy of species inherent within pulsed laser deposition plasma, the selective top layer replacement of sulfur by selenium can be achieved in a ~3–5 eV/atom window at low temperature (300 °C), which leads to the formation of high-quality WSSe Janus monolayers [189]. (iv) Selective epitaxy atomic replacement (SEAR) process. This approach relies on H radicals to efficiently remove and substitute the top chalcogen layer [93]. Compared to the high temperature required in the other three approaches, this is a mild process at room temperature, which enables a variety of high-quality Janus TMD layers and their heterostructures to be fabricated in either vertical or lateral form [190, 191].

4.5.2 Emerging Properties of Janus 2D TMDs

The formation of Janus structures introduces two basic features to the materials: the internal electric field from the nonuniform charge distribution caused by the different electronegativities and atomic radii of the top and bottom chalcogen atoms, and the breaking of mirror symmetry. Based on these two features, emergent interesting properties have been demonstrated in Janus TMDs. For example, from the breaking of mirror symmetry and strong spin-orbital coupling, large Rashba spin slitting, which exceeds the splitting in traditional heterostructures such as $LaAlO_3/SrTiO_3$, has been reported in Janus 2D TMDs, appealing for spintronic applications [192, 193]. On the other hand, from the internal vertical dipole moment, out-of-plane piezoelectricity has been predicted and experimentally demonstrated in Janus 2D TMDs [186, 194].

Besides, the formation of Janus structures can also be utilized as a strategy for band structure engineering. The band gap value of a Janus material is between the magnitudes of its parent TMDs. For example, the band gap of MoSSe is smaller than that of MoS_2, but larger than that of $MoSe_2$. Besides, it can fine tune the location of valence band maximum and/or conduction band minimum, leading to a change of band gap types (direct or indirect) [192]. Another important effect of the Janus structure is the decrease in exciton binding energy due to the internal electric filed. The excitons in Janus 2D TMDs have been found to form about 30% faster, and their radiative recombination lifetime is also longer in the Janus materials than those in pristine TMDs revealed by absorption and photoluminescence (PL) spectroscopic measurements [195]. This, combined with the tunable band gaps, makes Janus 2D TMDs promising for photovoltaic, photodetection, and photocatalytic applications.

4.5.3 Potential Applications of Janus 2D TMDs

The possible applications of Janus 2D TMDs have been discussed above. In this section, we will focus on the potential applications of Janus 2D TMDs based on their modified electronic properties in detail: photocatalyst and sensor.

The photocatalytic performance for water splitting is the result of the holistic effects from photon absorption, exciton dissociation, carrier mobility and redox reactions. Janus 2D TMDs have been systematically scrutinized for these criteria and were found to be ideal for water splitting [196, 197]. The proper band gaps of the Janus 2D TMDs allow a wide solar light absorption range extended to the near-infrared region. The internal electric field and the real-space separation of the conduction band minimum and the valence band maximum suppress the photoexcited electron–hole recombination rate. The different mobilities of the electrons and holes enable efficient carrier generation and transfer separation. The low overpotentials of the oxygen-evolution reaction and the hydrogen-evolution reaction ensure the high solar-to-hydrogen efficiency.

Pristine 2D TMDs have been reported to be promising for hazardous gas sensors, and interestingly, the sensing performance could be even enhanced by applying gate bias [198, 199]. Janus 2D TMDs naturally have an intrinsic internal electric field, which is expected to bring more advantages than the pristine TMDs. By examining the adsorption behaviors of CO, CO_2, NH_3, NO, and NO_2 on a Janus 2D TMD MoSSe monolayer, it has been reported that the interaction strengths between these molecules and the 2D material differ by the adsorption side of the Janus structures. And uniaxial tensile strain can further enhance such a difference. Importantly, the different adsorption behaviors of these gas molecules on both sides of Janus MoSSe lead to distinct electronic variations. Thus, with the intrinsic built-in electric field, Janus MoSSe is appealing as an ideal gas sensing material for its surface selectivity, enhanced gas sensitivity, and strain tunability [183]. Another work demonstrated that the out-of-plane dipole in a Janus 2D TMD could enhance the Raman vibrational intensity of biomolecules such as glucose and dopamine [200]. These findings make Janus 2D TMDs promising for developing novel ultrahigh sensitivity nanodevices for detecting hazardous gases and biomolecules.

4.6 Moiré Superlattices of 2D-TMDs

In 2018, Yuan Cao et al. for the first time realized the magic-angle twisted bilayer graphene (t-BLG) in experiments [201, 202]. It was surprising that Mott insulating states and unconventional superconductivity were observed in such systems. These findings led to a "gold rush" in the exploration of 2D Moiré superlattices in the past few years. 2D Moiré superlattices can be formed by slightly twisting homobilayers or in heterobilayers of which the constituent van der Waals materials differ in the lattice constant. The twist angle can serve as a new degree of freedom to tune the interlayer interaction. As the twist angle approaches zero, the Moiré potential can become so pronounced as to dramatically modulate the electronic structure.

Although computationally expensive, large-scale DFT simulations in combination with theoretical modeling have been conducted to investigate the twisted homobilayer TMDs. Apart from the AA stacking, where the atoms in the top layer are exactly aligned with those of the same species in the bottom layer, the 3-fold rotational symmetry of TMDs leads to a different stacking configuration after one layer is rotated by 60°, denoted as the AB stacking. Taking bilayer MoS_2 as an example, different Moiré superlattices with the same period of 51.8 Å could be generated by twisting the AA and AB stackings by the same angle of 3.5° and are labeled as $M^{3.5}$ and $M^{56.5}$, as shown in Figure 4.19. By examining the local atomic arrangements,

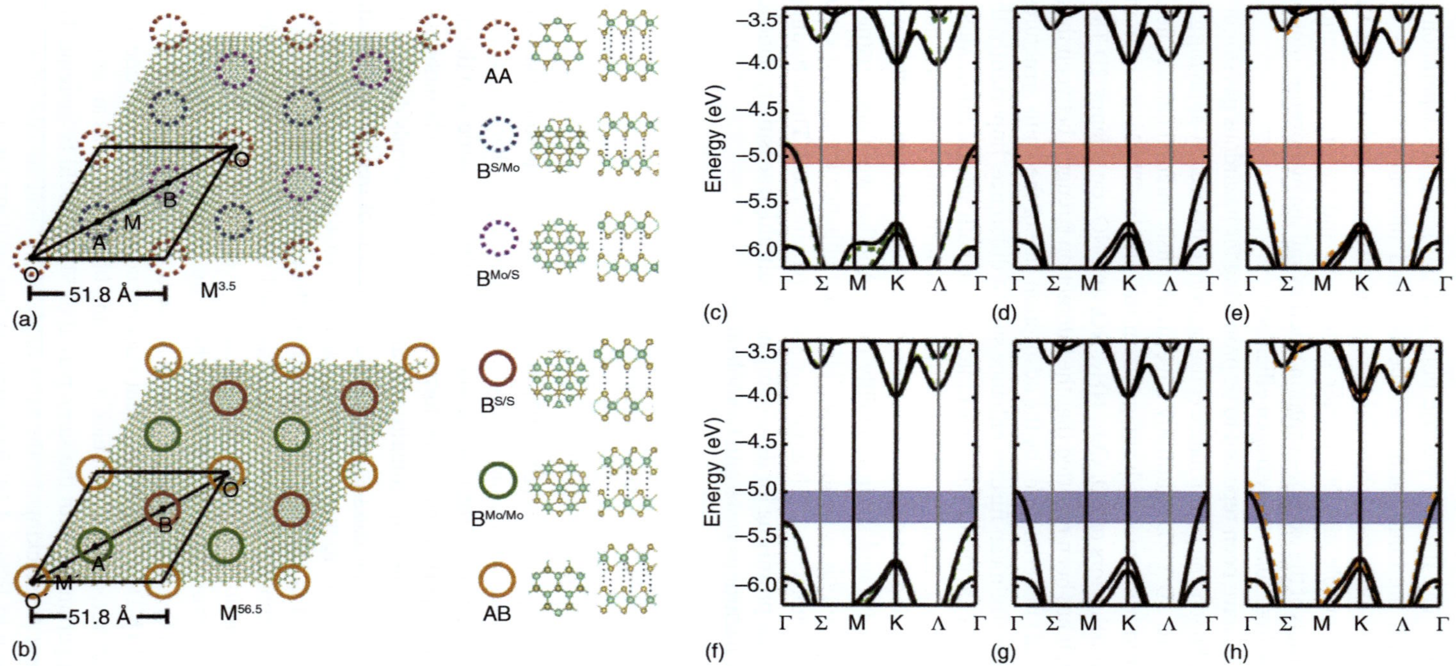

Figure 4.19 (a and b) The Moiré superlattice is formed by twisting bilayer MoS_2 by (a) 3.5° and (b) 56.5°. The high-symmetry stackings highlighted by circles are shown on the right panel. (c–e) Band structure of isolated high-symmetry stackings of bilayer MoS_2 at a fixed interlayer spacing of 5.9 Å. (c) AA using black solid lines and $B^{S/S}$ using green dashed lines; (d) AB; (e) $B^{Mo/S}$ using black sold lines and $B^{Mo/Mo}$ using orange dashed lines. (f–h) The band structure of the same isolated high-symmetry stackings as (c–e) but the interlayer spacing is optimized. Source: Naik and Jain [203]/With permission from American Physical Society.

$M^{3.5}$ is found to consist of three types of high-symmetry stackings, that is, the AA, $B^{S/Mo}$, and $B^{Mo/S}$ stackings. In the $B^{X/Y}$ stacking, the X atoms in the top layer are aligned with the Y atoms in the bottom layer. In contrast, $M^{56.5}$ is composed of the high-symmetry AB, $B^{S/S}$, and $B^{Mo/Mo}$ stackings (Figure 4.19). DFT calculations show significant structural relaxations of the Moiré MoS_2 superlattices. Due to the strong repulsion between the out-of-plane S-p_z orbitals, the AA and $B^{S/S}$ stacking regions have the largest interlayer spacing for $M^{3.5}$ and $M^{56.5}$, respectively. On the other hand, the interlayer distance reaches its smallest in the $B^{Mo/S}$ and $B^{S/Mo}$ stacking regions for $M^{3.5}$ and the $B^{Mo/Mo}$ and AB stacking regions for $M^{56.5}$.

The pronounced structural relaxations significantly affect the electronic structure of the high-symmetry stackings, as shown in Figures 4.19c,d. It is noted that regardless of the stacking type and the structural relaxations, the valence band maximum has been shifted back to the Γ point from the K/K′ point due to the interlayer interaction. However, the increased interlayer spacing weakens the interlayer hybridization in the AA and $B^{S/S}$ stackings and lowers their valence band energies. Consequently, the VBMs of the $B^{S/Mo}$ and $B^{Mo/S}$ stackings of $M^{3.5}$, which are the same as one another, are higher in energy. Therefore, the VBM of $M^{3.5}$ is mainly contributed by the local $B^{S/Mo}$ and $B^{Mo/S}$ stackings, as corroborated by the charge density of its VBM wavefunction in Figure 4.20a. As for $M^{56.5}$, the equilibrium AB stacking has a similar VBM to the equilibrium $B^{Mo/Mo}$ stacking. But the local small tensile strain lowers VBM of the $B^{Mo/Mo}$ region, and the compressive strain raises VBM of the AB region. Therefore, the AB stacking region mainly contributes to VBM of $M^{56.5}$, as evidenced by the localized charge density of its VBM wavefunction in Figure 4.20b.

The localization of the VBM Bloch state is accompanied by flattening of the valence bands in the mini-Brillouin zone of the Moiré superlattices (Figures 4.20c and d). The Moiré bands of both $M^{3.5}$ and $M^{56.5}$ are much flatter than their untwisted counterparts. The bandwidth of the topmost flattened valence band is below 50 meV for both $M^{3.5}$ and $M^{56.5}$, between which the latter is narrower due to a more localized charge distribution (Figure 4.20b). The band dispersion can be further suppressed by decreasing the twist angle [204]. In terms of the two topmost flat bands, a distinct difference between $M^{3.5}$ and $M^{56.5}$ exists. $M^{3.5}$ exhibits Dirac points at the K and K′ points of the Moiré Brillouin zone, whereas for $M^{56.5}$ there is a gap. This is because, in addition to the out-of-plane threefold rotational symmetry, $M^{3.5}$ has additional in-plane twofold rotational axes that bisect the line connecting the neighboring $B^{S/Mo}$ and $B^{Mo/S}$ regions (Figure 4.19a). The D_3 point group protects these Dirac points in $M^{3.5}$. It is worth noting that below the two topmost valence bands, there is another set of interesting valence bands when the AA-stacking bilayer MoS_2 is slightly twisted [204]. More specifically, two dispersionless flat bands touch the top and bottom of a pair of Dirac bands at the Moiré Brillouin zone center. Likewise, these bands also become flatter as the twist angle approaches zero. This set of valence bands has been reported as an intriguing realization of the strongly asymmetric $p_x - p_y$ Hubbard model [204]. Based on the model; the two dispersionless bands instead originate from destructive interference, which is similar to the origin of the dispersionless bands in Lieb or Kagome lattices.

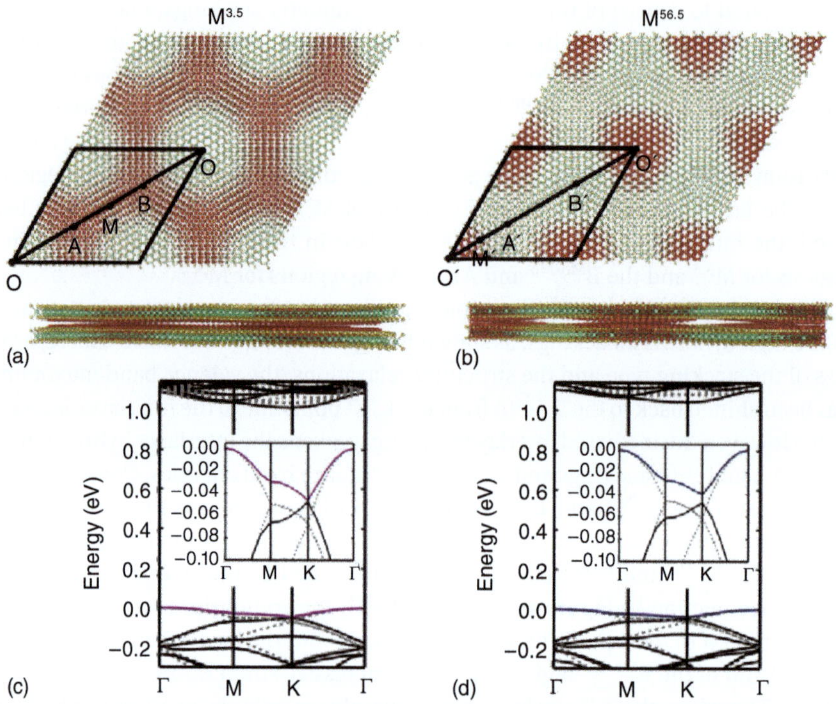

Figure 4.20 (a and b) Charge density of the VBM wave function of (a) $M^{3.5}$ and (b) $M^{56.5}$. The isovalue is set to 3×10^{-4} eV/Å^3. (c and d) The band structure of (c) $M^{3.5}$ and (d) $M^{56.5}$. The topmost valence band is highlighted with magenta in (c) and blue colors in (d), respectively. Inset: the magnified band dispersion near the valence band edges. The dashed lines represent the band structure of the untwisted (c) $B^{Mo/S}$ and (d) AB. Source: Naik and Jain [203]/With permission from American Physical Society.

With the decrease in the twist angle, the electronic kinetic lattice hopping (t) becomes smaller as a result of flattening of Moiré bands. At a critical twist angle, the on-site Coulomb repulsion interaction (U) is comparable to t. When the angle is further decreased, strong electronic correlations come into play. Various fascinating electronic phases have been predicted when the topmost flat valence bands of Moiré TMD superlattices are fractionally filled [203, 205–208]. Among them is the Mott insulator transition at half filling due to large U/t of the Moiré flat bands [205, 206, 208].

The flat valence band from the Γ pocket has been experimentally observed in twisted bilayer WSe_2 [209]. Unlike twisted bilayer MoS_2, the VBM of the local high-symmetry stackings of bilayer WSe_2 is instead located at the K/K′ points. As a result, the topmost Moiré flat band of twisted bilayer WSe_2 originates from the K/K′ pockets. By controlling the voltage bias, nevertheless, the localized charge density associated with the Γ-pocket flat band has been observed in the 3° and 57.5° twisted WSe_2 [209]. They are in good agreement with the DFT predictions of twisted bilayer MoS_2 (Figures 4.20a,b) [203]. In the meantime, the Mott insulator transition in twisted bilayer WSe_2 has also been experimentally confirmed [210]. The resistance against the hole density (n) was measured for the twisted bilayer WSe_2 with the angle ranging from 4° to 5.1°. A resistive peak is observed at large densities and shifts toward a lower density at smaller angles. The hole density associated with the peak is found to coincide with one hole per Moiré unit cell or half filling of the topmost Moiré flat band, i.e. $n/n_s = 0.5$, where n_s is the density to reach full filling of the Moiré band. This finding indicates that the electron correlations are sufficiently strong within the K/K′-pocket flat band to induce a Mott insulating state. Furthermore, two superconductivity domes have also been experimentally measured in the 5.1° twisted bilayer WSe_2 near half filing [210]. The resistance drops to zero at around 3 K, which is higher than the magic-angle twisted bilayer graphene (1.7 K) [201].

Alternatively, the Moiré superlattices can be created by stacking two van der Waals materials with different lattice constants (heterobilayers). Analogous to twisted bilayer TMDs, narrow minibands close to the band edges appear due to the long wavelength Moiré potential [211]. Experiments have demonstrated various correlated electronic states in the hereobilayer Moiré TMD superlattices, such as Mott insulator, Moiré band insulator, generalized Wigner crystals, and charge density waves [212, 213]. It is worth noting that these strong insulating states also occur on the electron-filled Moiré TMD superlattices, and the correlation effects might be even stronger on the electron side than on the hole side [213]. All these discoveries demonstrate that the Moiré TMD superlattices are of great value to the exploration of correlated physics and the simulation of a wealth of quantum many-body problems in reduced dimension.

References

1 Chhowalla, M., Shin, H.S., Eda, G. et al. (2013). The chemistry of two-dimensional layered transition metal dichalcogenide nanosheets. *Nat. Chem.* 5: 263–275.

2 Manzeli, S., Ovchinnikov, D., Pasquier, D. et al. (2017). 2D transition metal dichalcogenides. *Nat. Rev. Mater.* 2: 17033. https://doi.org/10.1038/natrevmats.2017.33.

3 Zhou, J., Lin, J., Huang, X. et al. (2018). A library of atomically thin metal chalcogenides. *Nature* 556: 355–359.

4 Wang, G., Chernikov, A., Glazov, M.M. et al. (2018). Colloquium: excitons in atomically thin transition metal dichalcogenides. *Rev. Mod. Phys.* 90: 021001.

5 Liu, G.-B., Xiao, D., Yao, Y. et al. (2015). Electronic structures and theoretical modelling of two-dimensional group-VIB transition metal dichalcogenides. *Chem. Soc. Rev.* 44: 2643–2663.

6 Yin, X., Tang, C.S., Zheng, Y. et al. (2021). Recent developments in 2D transition metal dichalcogenides: phase transition and applications of the (quasi-) metallic phases. *Chem. Soc. Rev.* 50: 10087–10115.

7 Xiao, D., Liu, G.-B., Feng, W. et al. (2012). Coupled spin and valley physics in monolayers of MoS_2 and other group-VI dichalcogenides. *Phys. Rev. Lett.* 108: 196802. https://doi.org/10.1103/PhysRevLett.108.196802.

8 Cappelluti, E., Roldán, R., Silva-Guillén, J. et al. (2013). Tight-binding model and direct-gap/indirect-gap transition in single-layer and multilayer MoS_2. *Phys. Rev. B* 88: 075409.

9 Espejo, C., Rangel, T., Romero, A. et al. (2013). Band structure tunability in MoS_2 under interlayer compression: a DFT and GW study. *Phys. Rev. B* 87: 245114.

10 Shi, H., Pan, H., Zhang, Y.-W., and Yakobson, B.I. (2013). Quasiparticle band structures and optical properties of strained monolayer MoS_2 and WS_2. *Phys. Rev. B* 87: 155304.

11 Qiu, Z., Trushin, M., Fang, H. et al. (2019). Giant gate-tunable bandgap renormalization and excitonic effects in a 2D semiconductor. *Sci. Adv.* 5: eaaw2347.

12 Yang, Y., Song, T., Zhang, X. et al. (2020). Substrate mediated electronic and excitonic reconstruction in a MoS_2 monolayer. *J. Mater. Chem. C* 8: 11778–11785.

13 Chang, C.-H., Fan, X., Lin, S.-H., and Kuo, J.-L. (2013). Orbital analysis of electronic structure and phonon dispersion in MoS_2, $MoSe_2$, WS_2, and WSe_2 monolayers under strain. *Phys. Rev. B* 88: 195420.

14 Yang, T., Song, T.T., Callsen, M. et al. (2019). Atomically thin 2D transition metal oxides: Structural reconstruction, interaction with substrates, and potential applications. *Adv. Mater. Interfaces* 6: 1801160.

15 Zhu, Z.Y., Cheng, Y.C., and Schwingenschlögl, U. (2011). Giant spin-orbit-induced spin splitting in two-dimensional transition-metal dichalcogenide semiconductors. *Phys. Rev. B* 84: 153402.

16 Ramasubramaniam, A. (2012). Large excitonic effects in monolayers of molybdenum and tungsten dichalcogenides. *Phys. Rev. B* 86: 115409.

17 Cheiwchanchamnangij, T. and Lambrecht, W.R. (2012). Quasiparticle band structure calculation of monolayer, bilayer, and bulk MoS_2. *Phys. Rev. B* 85: 205302.

18 Liu, G.-B., Shan, W.-Y., Yao, Y. et al. (2013). Three-band tight-binding model for monolayers of group-VIB transition metal dichalcogenides. *Phys. Rev. B* 88: 085433. https://doi.org/10.1103/PhysRevB.88.085433.

19 Koroteev, Y.M., Bihlmayer, G., Gayone, J.E. et al. (2004). Strong spin-orbit splitting on Bi surfaces. *Phys. Rev. Lett.* 93: 046403.

20 Mak, K.F., Lee, C., Hone, J. et al. (2010). Atomically thin MoS_2: a new direct-gap semiconductor. *Phys. Rev. Lett.* 105: 136805.

21 Wang, Q.H., Kalantar-Zadeh, K., Kis, A. et al. (2012). Electronics and optoelectronics of two-dimensional transition metal dichalcogenides. *Nat. Nanotechnol.* 7: 699–712.

22 Tang, C.S., Yin, X., Yang, M. et al. (2019). Three-dimensional resonant exciton in monolayer tungsten diselenide actuated by spin–orbit coupling. *ACS Nano* 13: 14529–14539.

23 Wang, Z., Yuan, G., Yang, M. et al. (2020). Exciton-enabled meta-optics in two-dimensional transition metal dichalcogenides. *Nano Lett.* 20: 7964–7972.

24 Pan, H., Chu, H., Pan, Z. et al. (2020). Large-scale monolayer molybdenum disulfide (MoS_2) for mid-infrared photonics. *Nanophotonics* 9: 4703–4710.

25 Qiu, D.Y., Felipe, H., and Louie, S.G. (2013). Optical spectrum of MoS_2: many-body effects and diversity of exciton states. *Phys. Rev. Lett.* 111: 216805.

26 Kormányos, A., Burkard, G., Gmitra, M. et al. (2015). *k·p* theory for two-dimensional transition metal dichalcogenide semiconductors. *2D Mater.* 2: 022001. https://doi.org/10.1088/2053-1583/2/2/022001.

27 Ivchenko, E.L. (2005). *Optical Spectroscopy of Semiconductor Nanostructures.* (Alpha Science Int'l Ltd.)

28 Zhao, W., Ribeiro, R.M., and Eda, G. (2015). Electronic structure and optical signatures of semiconducting transition metal dichalcogenide nanosheets. *Acc. Chem. Res.* 48: 91–99.

29 Zhang, C., Wang, H., Chan, W. et al. (2014). Absorption of light by excitons and trions in monolayers of metal dichalcogenide MoS_2: experiments and theory. *Phys. Rev. B* 89: 205436.

30 Zhang, C., Johnson, A., Hsu, C.-L. et al. (2014). Direct imaging of band profile in single layer MoS_2 on graphite: quasiparticle energy gap, metallic edge states, and edge band bending. *Nano Lett.* 14: 2443–2447.

31 Ugeda, M.M., Bradley, A.J., Shi, S.F. et al. (2014). Giant bandgap renormalization and excitonic effects in a monolayer transition metal dichalcogenide semiconductor. *Nat. Mater.* 13: 1091–1095.

32 Bussolotti, F., Chai, J., Yang, M. et al. (2018). Electronic properties of atomically thin MoS_2 layers grown by physical vapour deposition: band structure and energy level alignment at layer/substrate interfaces. *RSC Adv.* 8: 7744–7752.

33 Eda, G., Yamaguchi, H., Voiry, D. et al. (2011). Photoluminescence from chemically exfoliated MoS_2. *Nano Lett.* 11: 5111–5116.

34 Carvalho, A., Ribeiro, R., and Neto, A.C. (2013). Band nesting and the optical response of two-dimensional semiconducting transition metal dichalcogenides. *Phys. Rev. B* 88: 115205.

35 Schaibley, J.R., Yu, H., Clark, G. et al. (2016). Valleytronics in 2D materials. *Nat. Rev. Mater.* 1: 1–15.

36 Tang, C.S., Yin, X., Yang, M. et al. (2020). Anisotropic collective charge excitations in quasimetallic 2D transition-metal dichalcogenides. *Adv. Sci.* 7: 1902726.

37 Xu, S.-Y., Ma, Q., Shen, H. et al. (2018). Electrically switchable Berry curvature dipole in the monolayer topological insulator WTe_2. *Nat. Phys.* 14: 900–906.

38 Voiry, D., Mohite, A., and Chhowalla, M. (2015). Phase engineering of transition metal dichalcogenides. *Chem. Soc. Rev.* 44: 2702–2712.

39 Yin, X., Wang, Q., Cao, L. et al. (2017). Tunable inverted gap in monolayer quasi-metallic MoS_2 induced by strong charge-lattice coupling. *Nat. Commun.* 8: 1–9.

40 Liu, Y., Huang, Y., and Van der Duan, X. (2019). Waals integration before and beyond two-dimensional materials. *Nature* 567: 323–333.

41 Kappera, R., Voiry, D., Yalcin, S.E. et al. (2014). Phase-engineered low-resistance contacts for ultrathin MoS_2 transistors. *Nat. Mater.* 13: 1128–1134.

42 Acerce, M., Voiry, D., and Chhowalla, M. (2015). Metallic 1T phase MoS_2 nanosheets as supercapacitor electrode materials. *Nat. Nanotechnol.* 10: 313–318.

43 Chai, J.W., Yang, M., Callsen, M. et al. (2017). Tuning contact barrier height between metals and MoS_2 monolayer through interface engineering. *Adv. Mater. Interfaces* 4: 1700035.

44 Yang, T., Zhou, J., Song, T.T. et al. (2020). High-throughput identification of exfoliable two-dimensional materials with active basal planes for hydrogen evolution. *ACS Energy Lett.* 5: 2313–2321.

45 Hu, T., Li, R., and Dong, J. (2013). A new (2×1) dimerized structure of monolayer 1T-molybdenum disulfide, studied from first principles calculations. *J. Chem. Phys.* 139: 174702.

46 Lai, Z., He, Q., Tran, T.H. et al. (2021). Metastable 1T′-phase group VIB transition metal dichalcogenide crystals. *Nat. Mater.* 20: 1113–1120.

47 Ugeda, M.M., Pulkin, A., Tang, S. et al. (2018). Observation of topologically protected states at crystalline phase boundaries in single-layer WSe_2. *Nat. Commun.* 9: 1–7.

48 Tang, S., Zhang, C., Wong, D. et al. (2017). Quantum spin Hall state in monolayer 1T'-WTe_2. *Nat. Phys.* 13: 683–687.

49 Sun, Y., Wu, S.-C., Ali, M.N. et al. (2015). Prediction of Weyl semimetal in orthorhombic $MoTe_2$. *Phys. Rev. B* 92: 161107.

50 Ali, M.N., Xiong, J., Flynn, S. et al. (2014). Large, non-saturating magnetoresistance in WTe_2. *Nature* 514: 205–208.

51 Zheng, F., Cai, C., Ge, S. et al. (2016). On the quantum spin hall gap of monolayer 1T′-WTe_2. *Adv. Mater.* 28: 4845–4851.

52 Wu, S., Fatemi, V., Gibson, Q.D. et al. (2018). Observation of the quantum spin Hall effect up to 100 kelvin in a monolayer crystal. *Science* 359: 76–79.

53 Dawson, W. and Bullett, D. (1987). Electronic structure and crystallography of $MoTe_2$ and WTe_2. *J. Phys. C: Solid State Phys.* 20: 6159.

54 Xiao, Y., Li, Z.-Q., and Wang, Z.-W. (2017). Polaron effect on the bandgap modulation in monolayer transition metal dichalcogenides. *J. Phys. Condens. Matter* 29: 485001. https://doi.org/10.1088/1361-648x/aa94fb.

55 Moore, J.E. and Orenstein, J. (2010). Confinement-induced Berry phase and helicity-dependent photocurrents. *Phys. Rev. Lett.* 105: 026805.

56 Yoda, T., Yokoyama, T., and Murakami, S. (2018). Orbital Edelstein effect as a condensed-matter analog of solenoids. *Nano Lett.* 18: 916–920.

57 Basov, D., Fogler, M., and García de Abajo, F. (2016). Polaritons in van der Waals materials. *Science* 354: aag1992.

58 Low, T., Chaves, A., Caldwell, J.D. et al. (2017). Polaritons in layered two-dimensional materials. *Nat. Mater.* 16: 182–194.

59 Franchini, C., Reticcioli, M., Setvin, M., and Diebold, U. (2021). Polarons in materials. *Nat. Rev. Mater.* https://doi.org/10.1038/s41578-021-00289-w.

60 Emin, D. (2012). *Polarons*. Cambridge University Press.

61 Chen, C., Avila, J., Wang, S. et al. (2018). Emergence of interfacial polarons from electron–phonon coupling in graphene/h-BN van der Waals heterostructures. *Nano Lett.* 18: 1082–1087. https://doi.org/10.1021/acs.nanolett.7b04604.

62 Kang, M., Jung, S.W., Shin, W.J. et al. (2018). Holstein polaron in a valley-degenerate two-dimensional semiconductor. *Nat. Mater.* 17: 676–680. https://doi.org/10.1038/s41563-018-0092-7.

63 Glazov, M.M., Semina, M.A., Robert, C. et al. (2019). Intervalley polaron in atomically thin transition metal dichalcogenides. *Phys. Rev. B* 100: 041301. https://doi.org/10.1103/PhysRevB.100.041301.

64 Saito, Y., Nakamura, Y., Bahramy, M.S. et al. (2016). Superconductivity protected by spin–valley locking in ion-gated MoS_2. *Nat. Phys.* 12: 144–149. https://doi.org/10.1038/nphys3580.

65 Moser, S., Moreschini, L., Jaćimović, J. et al. (2013). Tunable polaronic conduction in anatase TiO_2. *Phys. Rev. Lett.* 110: 196403. https://doi.org/10.1103/PhysRevLett.110.196403.

66 Verdi, C., Caruso, F., and Giustino, F. (2017). Origin of the crossover from polarons to Fermi liquids in transition metal oxides. *Nat. Commun.* 8: 15769. https://doi.org/10.1038/ncomms15769.

67 Holstein, T. (1959). Studies of polaron motion: part I. The molecular-crystal model. *Ann. Phys.* 8: 325–342. https://doi.org/10.1016/0003-4916(59)90002-8.

68 Holstein, T. (1959). Studies of polaron motion: part II. The “small” polaron. *Ann. Phys.* 8: 343–389. https://doi.org/10.1016/0003-4916(59)90003-X.

69 Goodvin, G.L., Berciu, M., and Sawatzky, G.A. (2006). Green’s function of the Holstein polaron. *Phys. Rev. B* 74: 245104. https://doi.org/10.1103/PhysRevB.74.245104.

70 Pisoni, R., Kormányos, A., Brooks, M. et al. (2018). Interactions and magnetotransport through spin-valley coupled landau levels in monolayer MoS_2. *Phys. Rev. Lett.* 121: 247701. https://doi.org/10.1103/PhysRevLett.121.247701.

71 Eknapakul, T., King, P.D.C., Asakawa, M. et al. (2014). Electronic structure of a quasi-freestanding MoS_2 monolayer. *Nano Lett.* 14: 1312–1316. https://doi.org/10.1021/nl4042824.

72 Larentis, S., Movva, H.C., Fallahazad, B. et al. (2018). Large effective mass and interaction-enhanced Zeeman splitting of K-valley electrons in $MoSe_2$. *Phys. Rev. B* 97: 201407. https://doi.org/10.1103/PhysRevB.97.201407.

73 Marinov, K., Avsar, A., Watanabe, K. et al. (2017). Resolving the spin splitting in the conduction band of monolayer MoS_2. *Nat. Commun.* 8: 1938. https://doi.org/10.1038/s41467-017-02047-5.

74 Horzum, S. et al. (2013). Phonon softening and direct to indirect band gap crossover in strained single-layer $MoSe_2$. *Phys. Rev. B* 87: 125415. https://doi.org/10.1103/PhysRevB.87.125415.

75 Raja, A., Chaves, A., Yu, J. et al. (2017). Coulomb engineering of the bandgap and excitons in two-dimensional materials. *Nat. Commun.* 8: 15251. https://doi.org/10.1038/ncomms15251.

76 Rose, F., Goerbig, M.O., and Piéchon, F. (2013). Spin- and valley-dependent magneto-optical properties of MoS_2. *Phys. Rev. B* 88: 125438. https://doi.org/10.1103/PhysRevB.88.125438.

77 Tongay, S., Zhou, J., Ataca, C. et al. (2012). Thermally driven crossover from indirect toward direct bandgap in 2D semiconductors: $MoSe_2$ versus MoS_2. *Nano Lett.* 12: 5576–5580. https://doi.org/10.1021/nl302584w.

78 Sio, W.H., Verdi, C., Poncé, S., and Giustino, F. (2019). Polarons from first principles, without supercells. *Phys. Rev. Lett.* 122: 246403. https://doi.org/10.1103/PhysRevLett.122.246403.

79 Lafuente-Bartolome, J., Lian, C., Sio, W.H. et al. (2022). Unified approach to polarons and phonon-induced band structure renormalization. *Phys. Rev. Lett.* 129: 076402. https://doi.org/10.1103/PhysRevLett.129.076402.

80 Lafuente-Bartolome, J., Lian, C., Sio, W.H. et al. (2022). Ab initio self-consistent many-body theory of polarons at all couplings. *Phys. Rev. B* 106: 075119. https://doi.org/10.1103/PhysRevB.106.075119.

81 Vitale, S.A., Nezich, D., Varghese, J.O. et al. (2018). Valleytronics: opportunities, challenges, and paths forward. *Small* 14: 1801483. https://doi.org/10.1002/smll.201801483.

82 Goswami, S., Slinker, K.A., Friesen, M. et al. (2007). Controllable valley splitting in silicon quantum devices. *Nat. Phys.* 3: 41–45. https://doi.org/10.1038/nphys475.

83 Küchler, R., Steinke, L., Daou, R. et al. (2014). Thermodynamic evidence for valley-dependent density of states in bulk bismuth. *Nat. Mater.* 13: 461–465. https://doi.org/10.1038/nmat3909.

84 Xiao, D., Yao, W., and Niu, Q. (2007). Valley-contrasting physics in graphene: magnetic moment and topological transport. *Phys. Rev. Lett.* 99: 236809. https://doi.org/10.1103/PhysRevLett.99.236809.

85 Mak, K.F., He, K., Shan, J., and Heinz, T.F. (2012). Control of valley polarization in monolayer MoS_2 by optical helicity. *Nat. Nanotechnol.* 7: 494–498. https://doi.org/10.1038/nnano.2012.96.

86 Son, J., Kim, K.-H., Ahn, Y.H. et al. (2019). Strain engineering of the berry curvature dipole and valley magnetization in monolayer MoS_2. *Phys. Rev. Lett.* 123: 036806. https://doi.org/10.1103/PhysRevLett.123.036806.

87 Splendiani, A., Sun, L., Zhang, Y. et al. (2010). Emerging photoluminescence in monolayer MoS_2. *Nano Lett.* 10: 1271–1275.

88 Feng, W., Yao, Y., Zhu, W. et al. (2012). Intrinsic spin Hall effect in monolayers of group-VI dichalcogenides: a first-principles study. *Phys. Rev. B* 86: 165108. https://doi.org/10.1103/PhysRevB.86.165108.

89 Xu, X., Yao, W., Xiao, D., and Heinz, T.F. (2014). Spin and pseudospins in layered transition metal dichalcogenides. *Nat. Phys.* 10: 343–350.

90 Mak, K.F., McGill, K.L., Park, J., and McEuen, P.L. (2014). The valley Hall effect in MoS_2 transistors. *Science* 344: 1489–1492. https://doi.org/10.1126/science.1250140.

91 Li, X., Zhang, F., and Niu, Q. (2013). Unconventional quantum Hall effect and tunable spin Hall effect in dirac materials: application to an isolated MoS_2

trilayer. *Phys. Rev. Lett.* 110: 066803. https://doi.org/10.1103/PhysRevLett.110.066803.

92 Arora, A., Schmidt, R., Schneider, R. et al. (2016). Valley Zeeman splitting and valley polarization of neutral and charged excitons in monolayer $MoTe_2$ at high magnetic fields. *Nano Lett.* 16: 3624–3629. https://doi.org/10.1021/acs.nanolett.6b00748.

93 Xu, L., Yang, M., Shen, L. et al. (2018). Large valley splitting in monolayer WS_2 by proximity coupling to an insulating antiferromagnetic substrate. *Phys. Rev. B* 97: 041405. https://doi.org/10.1103/PhysRevB.97.041405.

94 Cheng, Y.C., Zhang, Q.Y., and Schwingenschlögl, U. (2014). Valley polarization in magnetically doped single-layer transition-metal dichalcogenides. *Phys. Rev. B* 89: 155429. https://doi.org/10.1103/PhysRevB.89.155429.

95 Seyler, K.L., Zhong, D., Huang, B. et al. (2018). Valley manipulation by optically tuning the magnetic proximity effect in WSe_2/CrI_3 heterostructures. *Nano Lett.* 18: 3823–3828. https://doi.org/10.1021/acs.nanolett.8b01105.

96 Li, X., Cao, T., Niu, Q. et al. (2013). Coupling the valley degree of freedom to antiferromagnetic order. *Proc. Natl. Acad. Sci.* 110: 3738–3742. https://doi.org/10.1073/pnas.1219420110.

97 Zeng, H., Dai, J., Yao, W. et al. (2012). Valley polarization in MoS_2 monolayers by optical pumping. *Nat. Nanotechnol.* 7: 490–493. https://doi.org/10.1038/nnano.2012.95.

98 Cao, T., Wang, G., Han, W. et al. (2012). Valley-selective circular dichroism of monolayer molybdenum disulphide. *Nat. Commun.* 3: 887. https://doi.org/10.1038/ncomms1882.

99 Park, S., Arscott, S., Taniguchi, T. et al. (2022). Efficient valley polarization of charged excitons and resident carriers in Molybdenum disulfide monolayers by optical pumping. *Commun. Theor. Phys.* 5: 73. https://doi.org/10.1038/s42005-022-00850-1.

100 Lu, J.-D. and Chen, X.-S. (2021). Strain-controlled valley polarization in a graphene under the modulation of a realistic magnetic field. *J. Supercond. Novel Magn.* 34: 2545–2550. https://doi.org/10.1007/s10948-021-05883-1.

101 Wu, Y.-C., Taniguchi, T., Watanabe, K., and Yan, J. (2022). Negative valley polarization in doped monolayer $MoSe_2$. *Phys. Chem. Chem. Phys.* 24: 191–196. https://doi.org/10.1039/D1CP03490A.

102 Aivazian, G., Gong, Z., Jones, A.M. et al. (2015). Magnetic control of valley pseudospin in monolayer WSe_2. *Nat. Phys.* 11: 148–152. https://doi.org/10.1038/nphys3201.

103 Li, Y., Ludwig, J., Low, T. et al. (2014). Valley splitting and polarization by the Zeeman effect in monolayer $MoSe_2$. *Phys. Rev. Lett.* 113: 266804. https://doi.org/10.1103/PhysRevLett.113.266804.

104 Wu, S., Ross, J.S., Liu, G.B. et al. (2013). Electrical tuning of valley magnetic moment through symmetry control in bilayer MoS_2. *Nat. Phys.* 9: 149–153. https://doi.org/10.1038/nphys2524.

105 Zhao, X.W., Li, Y., Liang, R.D. et al. (2020). Enhanced valley polarization at valence/conduction band in transition-metal-doped WTe_2 under strain force. *Appl. Surf. Sci.* 504: 144367. https://doi.org/10.1016/j.apsusc.2019.144367.

106 Zhou, W., Yang, Z., Li, A. et al. (2020). Spin and valley splittings in Janus monolayer WSSe on a MnO(111) surface: large effective Zeeman field and opening of a helical gap. *Phys. Rev. B* 101: 045113. https://doi.org/10.1103/PhysRevB.101.045113.

107 Zhong, D., Seyler, K.L., Linpeng, X. et al. (2017). Van der Waals engineering of ferromagnetic semiconductor heterostructures for spin and valleytronics. *Sci. Adv.* 3: e1603113. https://doi.org/10.1126/sciadv.1603113.

108 Norden, T., Zhao, C., Zhang, P. et al. (2019). Giant valley splitting in monolayer WS_2 by magnetic proximity effect. *Nat. Commun.* 10: 4163. https://doi.org/10.1038/s41467-019-11966-4.

109 Zhao, C., Norden, T., Zhang, P. et al. (2017). Enhanced valley splitting in monolayer WSe_2 due to magnetic exchange field. *Nat. Nanotechnol.* 12: 757–762. https://doi.org/10.1038/nnano.2017.68.

110 Peierls, R.E. (1955). *Quantum Theory of Solids*. Clarendon Press.

111 Flicker, F. and van Wezel, J. (2015). Charge order from orbital-dependent coupling evidenced by $NbSe_2$. *Nat. Commun.* 6: 7034. https://doi.org/10.1038/ncomms8034.

112 Comès, R., Lambert, M., Launois, H., and Zeller, H.R. (1973). Evidence for a peierls distortion or a Kohn anomaly in one-dimensional conductors of the Type $K_2Pt(CN)_4Br_{0.30}{\cdot}xH_2O$. *Phys. Rev. B* 8: 571–575. https://doi.org/10.1103/PhysRevB.8.571.

113 Mook, H.A. and Watson, C.R. (1976). Neutron inelastic scattering study of tetrathiafulvalene tetracyanoquinodimethane (TTF-TCNQ). *Phys. Rev. Lett.* 36: 801–803. https://doi.org/10.1103/PhysRevLett.36.801.

114 Nicholson, C.W., Berthod, C., Puppin, M. et al. (2017). Dimensional crossover in a charge density wave material probed by angle-resolved photoemission spectroscopy. *Phys. Rev. Lett.* 118: 206401. https://doi.org/10.1103/PhysRevLett.118.206401.

115 Shi, W., Wieder, B.J., Meyerheim, H.L. et al. (2021). A charge-density-wave topological semimetal. *Nat. Phys.* 17: 381–387. https://doi.org/10.1038/s41567-020-01104-z.

116 Lin, D., Li, S., Wen, J. et al. (2020). Patterns and driving forces of dimensionality-dependent charge density waves in 2H-type transition metal dichalcogenides. *Nat. Commun.* 11: 2406. https://doi.org/10.1038/s41467-020-15715-w.

117 Hall, J., Ehlen, N., Berges, J. et al. (2019). Environmental control of charge density wave order in monolayer 2H-TaS_2. *ACS Nano* 13: 10210–10220. https://doi.org/10.1021/acsnano.9b03419.

118 Li, D., Wang, X., Kan, C.M. et al. (2020). Structural phase transition of multilayer VSe_2. *ACS Appl. Mater. Interfaces* 12: 25143–25149. https://doi.org/10.1021/acsami.0c04449.

119 Ugeda, M.M., Bradley, A.J., Zhang, Y. et al. (2016). Characterization of collective ground states in single-layer $NbSe_2$. *Nat. Phys.* 12: 92–97. https://doi.org/10.1038/nphys3527.

120 Ryu, H., Chen, Y., Kim, H. et al. (2018). Persistent charge-density-wave order in single-layer $TaSe_2$. *Nano Lett.* 18: 689–694. https://doi.org/10.1021/acs.nanolett.7b03264.

121 Zhu, Z.-L., Liu, Z.L., Wu, X. et al. (2022). Charge density wave states in phase-engineered monolayer VTe_2. *Chin. Phys. B* 31: 077101. https://doi.org/10.1088/1674-1056/ac6739.

122 Miller, D.C., Mahanti, S.D., and Duxbury, P.M. (2018). Charge density wave states in tantalum dichalcogenides. *Phys. Rev. B* 97: 045133. https://doi.org/10.1103/PhysRevB.97.045133.

123 van Efferen, C., Berges, J., Hall, J. et al. (2021). A full gap above the Fermi level: the charge density wave of monolayer VS_2. *Nat. Commun.* 12: 6837. https://doi.org/10.1038/s41467-021-27094-x.

124 Tresca, C. and Calandra, M. (2019). Charge density wave and spin 1/2 insulating state in single layer 1T-NbS_2. *2D Mater.* 6: 035041. https://doi.org/10.1088/2053-1583/ab23c0.

125 Lee, J., Jin, K.-H., and Yeom, H.W. (2021). Distinguishing a Mott insulator from a trivial insulator with atomic adsorbates. *Phys. Rev. Lett.* 126: 196405. https://doi.org/10.1103/PhysRevLett.126.196405.

126 Chen, P., Pai, W.W., Chan, Y.H. et al. (2018). Unique gap structure and symmetry of the charge density wave in single-layer VSe_2. *Phys. Rev. Lett.* 121: 196402. https://doi.org/10.1103/PhysRevLett.121.196402.

127 Liu, L., Yang, H., Huang, Y. et al. (2021). Direct identification of Mott Hubbard band pattern beyond charge density wave superlattice in monolayer 1T-$NbSe_2$. *Nat. Commun.* 12: 1978. https://doi.org/10.1038/s41467-021-22233-w.

128 Chen, Y., Ruan, W., Wu, M. et al. (2020). Strong correlations and orbital texture in single-layer 1T-$TaSe_2$. *Nat. Phys.* 16: 218–224. https://doi.org/10.1038/s41567-019-0744-9.

129 Wang, Y., Ren, J., Li, J. et al. (2019). Evidence of charge density wave with anisotropic gap in a monolayer VTe_2 film. *Phys. Rev. B* 100: 241404. https://doi.org/10.1103/PhysRevB.100.241404.

130 Wu, R., Tao, Q., Dang, W. et al. (2019). van der Waals epitaxial growth of atomically thin 2D metals on dangling-bond-free WSe_2 and WS_2. *Adv. Funct. Mater.* 29: 1806611. https://doi.org/10.1002/adfm.201806611.

131 Hwang, J., Jin, Y., Zhang, C. et al. (2022). A novel $\sqrt{19}\times\sqrt{19}$ superstructure in epitaxially grown 1T-$TaTe_2$. *Adv. Mater.* 34: 2204579. https://doi.org/10.1002/adma.202204579.

132 Liao, J.-H., Zhao, Y.-C., Zhao, Y.-J. et al. (2020). Doping induced charge density wave in monolayer TiS_2 and phonon-mediated superconductivity. *J. Appl. Phys.* 127: 044301. https://doi.org/10.1063/1.5135615.

133 Duerloo, K.-A.N., Li, Y., and Reed, E.J. (2014). Structural phase transitions in two-dimensional Mo- and W-dichalcogenide monolayers. *Nat. Commun.* 5: 4214. https://doi.org/10.1038/ncomms5214.

134 Pasquier, D. and Yazyev, O.V. (2019). Unified picture of lattice instabilities in metallic transition metal dichalcogenides. *Phys. Rev. B* 100: 201103. https://doi.org/10.1103/PhysRevB.100.201103.

135 Yu, W., Wang, Z., Zhao, X. et al. (2020). Domain engineering in ReS_2 by coupling strain during electrochemical exfoliation. *Adv. Funct. Mater.* 30: 2003057. https://doi.org/10.1002/adfm.202003057.

136 Singh, B., Hsu, C.-H., Tsai, W.-F. et al. (2017). Stable charge density wave phase in a 1T-$TiSe_2$ monolayer. *Phys. Rev. B* 95: 245136. https://doi.org/10.1103/PhysRevB.95.245136.

137 Chen, W., Hu, M., Zong, J. et al. (2021). Epitaxial growth of single-phase 1T′-WSe_2 monolayer with assistance of enhanced interface interaction. *Adv. Mater.* 33: 2004930. https://doi.org/10.1002/adma.202004930.

138 Jiang, S., Hong, M., Wei, W. et al. (2018). Direct synthesis and in situ characterization of monolayer parallelogrammic rhenium diselenide on gold foil. *Commun. Chem.* 1: 17. https://doi.org/10.1038/s42004-018-0010-6.

139 Sky Zhou, J., Bianco, R., Monacelli, L. et al. (2020). Theory of the thickness dependence of the charge density wave transition in 1 T-$TiTe_2$. *2D Mater.* 7: 045032. https://doi.org/10.1088/2053-1583/abae7a.

140 Keum, D.H., Cho, S., Kim, J.H. et al. (2015). Bandgap opening in few-layered monoclinic $MoTe_2$. *Nat. Phys.* 11: 482–486. https://doi.org/10.1038/nphys3314.

141 YAĞMURCUKARDEŞ, M. (2020). Stable anisotropic single-layer of $ReTe_2$: a first principles prediction. *Turk. J. Phys.* 44: 450–457. https://doi.org/10.3906/fiz-2004-17.

142 Oh, E., Gye, G., and Yeom, H.W. (2020). Defect-selective charge-density-wave condensation in 2H-$NbSe_2$. *Phys. Rev. Lett.* 125: 036804. https://doi.org/10.1103/PhysRevLett.125.036804.

143 Wong, P.K.J., Zhang, W., Zhou, J. et al. (2019). Metallic 1T phase, $3d^1$ electronic consnd charge density wave order in molecular beam epitaxy grown monolayer vanadium ditelluride. *ACS Nano* 13: 12894–12900. https://doi.org/10.1021/acsnano.9b05349.

144 Qiao, S., Li, X., Wang, N. et al. (2017). Mottness collapse in 1T-$TaS_{2-x}Se_x$ transition-metal dichalcogenide: an interplay between localized and itinerant orbitals. *Phys. Rev. X* 7: 041054. https://doi.org/10.1103/PhysRevX.7.041054.

145 Wang, J., Zheng, H., Xu, G. et al. (2016). Controlled synthesis of two-dimensional 1T-$TiSe_2$ with charge density wave transition by chemical vapor transport. *J. Am. Chem. Soc.* 138: 16216–16219. https://doi.org/10.1021/jacs.6b10414.

146 Zheng, F. and Feng, J. (2019). Electron-phonon coupling and the coexistence of superconductivity and charge-density wave in monolayer $NbSe_2$. *Phys. Rev. B* 99: 161119. https://doi.org/10.1103/PhysRevB.99.161119.

147 Calandra, M. (2018). Phonon-assisted magnetic Mott-insulating state in the charge density wave phase of single-layer 1T-$NbSe_2$. *Phys. Rev. Lett.* 121: 026401. https://doi.org/10.1103/PhysRevLett.121.026401.

148 Coelho, P.M., Nguyen Cong, K., Bonilla, M. et al. (2019). Charge density wave state suppresses ferromagnetic ordering in VSe_2 monolayers. *J. Phys. Chem. C* 123: 14089–14096. https://doi.org/10.1021/acs.jpcc.9b04281.

149 Xi, X., Zhao, L., Wang, Z. et al. (2015). Strongly enhanced charge-density-wave order in monolayer $NbSe_2$. *Nat. Nanotechnol.* 10: 765–769. https://doi.org/10.1038/nnano.2015.143.

150 Lian, C.-S., Si, C., and Duan, W. (2018). Unveiling charge-density wave, superconductivity, and their competitive nature in two-dimensional $NbSe_2$. *Nano Lett.* 18: 2924–2929. https://doi.org/10.1021/acs.nanolett.8b00237.

151 Liu, X., Chong, Y.X., Sharma, R., and Davis, J.C.S. (2021). Discovery of a Cooper-pair density wave state in a transition-metal dichalcogenide. *Science* 372: 1447–1452. https://doi.org/10.1126/science.abd4607.

152 Wang, Z., C Chen, J Mo et al. (2022). Decisive role of electron-phonon coupling for phonon and electron instabilities in transition metal dichalcogenides. arXiv preprint arXiv:2208.07121.

153 Rossnagel, K. (2011). On the origin of charge-density waves in select layered transition-metal dichalcogenides. *J. Phys. Condens. Matter* 23: 213001. https://doi.org/10.1088/0953-8984/23/21/213001.

154 Johannes, M.D. and Mazin, I.I. (2008). Fermi surface nesting and the origin of charge density waves in metals. *Phys. Rev. B* 77: 165135. https://doi.org/10.1103/PhysRevB.77.165135.

155 Kohn, W. (1959). Image of the Fermi surface in the vibration spectrum of a metal. *Phys. Rev. Lett.* 2: 393–394. https://doi.org/10.1103/PhysRevLett.2.393.

156 Fröhlich, H. (1954). On the theory of superconductivity: the one-dimensional case. *Proc. R. Soc. London, Ser. A* 223: 296–305. https://doi.org/10.1098/rspa.1954.0116.

157 Flicker, F. and van Wezel, J. (2016). Charge order in $NbSe_2$. *Phys. Rev. B* 94: 235135. https://doi.org/10.1103/PhysRevB.94.235135.

158 Xu, Z., Yang, H., Song, X. et al. (2021). Topical review: recent progress of charge density waves in 2D transition metal dichalcogenide-based heterojunctions and their applications. *Nanotechnology* 32: 492001. https://doi.org/10.1088/1361-6528/ac21ed.

159 Whangbo, M.H. and Canadell, E. (1992). Analogies between the concepts of molecular chemistry and solid-state physics concerning structural instabilities. Electronic origin of the structural modulations in layered transition metal dichalcogenides. *J. Am. Chem. Soc.* 114: 9587–9600. https://doi.org/10.1021/ja00050a044.

160 Calandra, M., Mazin, I.I., and Mauri, F. (2009). Effect of dimensionality on the charge-density wave in few-layer 2H-$NbSe_2$. *Phys. Rev. B* 80: 241108. https://doi.org/10.1103/PhysRevB.80.241108.

161 Weber, F., Rosenkranz, S., Castellan, J.P. et al. (2011). Extended phonon collapse and the origin of the charge-density wave in 2H-$NbSe_2$. *Phys. Rev. Lett.* 107: 107403. https://doi.org/10.1103/PhysRevLett.107.107403.

162 Diego, J., Said, A.H., Mahatha, S.K. et al. (2021). van der Waals driven anharmonic melting of the 3D charge density wave in VSe_2. *Nat. Commun.* 12: 598. https://doi.org/10.1038/s41467-020-20829-2.

163 Wang, Z., J You, C Chen et al. (2022). Emergent Topological Superconductor by Charge Density Wave Transition. arXiv preprint arXiv:2210.07633.

164 Kogar, A., Rak, M.S., Vig, S. et al. (2017). Signatures of exciton condensation in a transition metal dichalcogenide. *Science* 358: 1314–1317. https://doi.org/10.1126/science.aam6432.

165 Chen, C., Singh, B., Lin, H., and Pereira, V.M. (2018). Reproduction of the charge density wave phase diagram in 1T-$TiSe_2$ exposes its excitonic character. *Phys. Rev. Lett.* 121: 226602. https://doi.org/10.1103/PhysRevLett.121.226602.

166 Zhou, J.S., Monacelli, L., Bianco, R. et al. (2020). Anharmonicity and doping melt the charge density wave in single-layer $TiSe_2$. *Nano Lett.* 20: 4809–4815. https://doi.org/10.1021/acs.nanolett.0c00597.

167 Zhu, X., Cao, Y., Zhang, J. et al. (2015). Classification of charge density waves based on their nature. *Proc. Natl. Acad. Sci.* 112: 2367–2371. https://doi.org/10.1073/pnas.1424791112.

168 Rice, T.M. and Scott, G.K. (1975). New mechanism for a charge-density-wave instability. *Phys. Rev. Lett.* 35: 120–123. https://doi.org/10.1103/PhysRevLett.35.120.

169 Novko, D., Torbatian, Z. and Lončarić, I. (2022). Electron correlations rule phonon-driven instability in single layer $TiSe_2$. arXiv preprint arXiv:2203.10658.

170 Wang, Z., Zhou, J., Loh, K.P., and Feng, Y.P. (2021). Controllable phase transitions between multiple charge density waves in monolayer 1T-VSe_2 via charge doping. *Appl. Phys. Lett.* 119: 163101. https://doi.org/10.1063/5.0068241.

171 Sipos, B., Kusmartseva, A.F., Akrap, A. et al. (2008). From Mott state to superconductivity in 1T-TaS_2. *Nat. Mater.* 7: 960–965. https://doi.org/10.1038/nmat2318.

172 Kusmartseva, A.F., Sipos, B., Berger, H. et al. (2009). Pressure induced superconductivity in pristine 1T-$TiSe_2$. *Phys. Rev. Lett.* 103: 236401. https://doi.org/10.1103/PhysRevLett.103.236401.

173 Pang, X., Wan, C., Wang, M., and Lin, Z. (2014). Strictly biphasic soft and hard Janus structures: synthesis, properties, and applications. *Angew. Chem. Int. Ed.* 53: 5524–5538. https://doi.org/10.1002/anie.201309352.

174 Zhang, L., Yang, Z., Gong, T. et al. (2020). Recent advances in emerging Janus two-dimensional materials: from fundamental physics to device applications. *J. Mater. Chem. A* 8: 8813–8830. https://doi.org/10.1039/D0TA01999B.

175 Li, R., Cheng, Y., and Huang, W. (2018). Recent progress of Janus 2D transition metal chalcogenides: from theory to experiments. *Small* 14: 1802091. https://doi.org/10.1002/smll.201802091.

176 de Gennes, P.G. (1992). Soft matter. *Science* 256: 495–497. https://doi.org/10.1126/science.256.5056.495.

177 Wang, Z. (2018). 2H → 1T′ phase transformation in Janus monolayer MoSSe and MoSTe: an efficient hole injection contact for 2H-MoS_2. *J. Mater. Chem. C* 6: 13000–13005. https://doi.org/10.1039/C8TC04951C.

178 Tang, X., Li, S., Ma, Y. et al. (2018). Distorted Janus transition metal dichalcogenides: stable two-dimensional materials with sizable band gap and ultrahigh carrier mobility. *J. Phys. Chem. C* 122: 19153–19160. https://doi.org/10.1021/acs.jpcc.8b04161.

179 Li, Y., Duerloo, K.-A.N., Wauson, K., and Reed, E.J. (2016). Structural semiconductor-to-semimetal phase transition in two-dimensional materials induced by electrostatic gating. *Nat. Commun.* 7: 10671. https://doi.org/10.1038/ncomms10671.

180 Zhang, C., Kc, S., Nie, Y. et al. (2016). Charge mediated reversible metal–insulator transition in monolayer $MoTe_2$ and $W_xMo_{1-x}Te_2$ alloy. *ACS Nano* 10: 7370–7375. https://doi.org/10.1021/acsnano.6b00148.

181 Ma, X., Wu, X., Wang, H., and Wang, Y. (2018). A Janus MoSSe monolayer: a potential wide solar-spectrum water-splitting photocatalyst with a low carrier recombination rate. *J. Mater. Chem. A* 6: 2295–2301. https://doi.org/10.1039/C7TA10015A.

182 Er, D., Ye, H., Frey, N.C. et al. (2018). Prediction of enhanced catalytic activity for hydrogen evolution reaction in Janus transition metal dichalcogenides. *Nano Lett.* 18: 3943–3949. https://doi.org/10.1021/acs.nanolett.8b01335.

183 Jin, C., Tang, X., Tan, X. et al. (2019). A Janus MoSSe monolayer: a superior and strain-sensitive gas sensing material. *J. Mater. Chem. A* 7: 1099–1106. https://doi.org/10.1039/C8TA08407F.

184 Zhang, L., Xia, Y., Li, X. et al. (2022). Janus two-dimensional transition metal dichalcogenides. *J. Appl. Phys.* 131: 230902. https://doi.org/10.1063/5.0095203.

185 Tang, X. and Kou, L. (2022). 2D Janus transition metal dichalcogenides: properties and applications. *Phys. Status Solidi(b)* 259: 2100562. https://doi.org/10.1002/pssb.202100562.

186 Lu, A.-Y., Zhu, H., Xiao, J. et al. (2017). Janus monolayers of transition metal dichalcogenides. *Nat. Nanotechnol.* 12: 744–749. https://doi.org/10.1038/nnano.2017.100.

187 Zhang, J., Jia, S., Kholmanov, I. et al. (2017). Janus monolayer transition-metal dichalcogenides. *ACS Nano* 11: 8192–8198. https://doi.org/10.1021/acsnano.7b03186.

188 Sant, R., Gay, M., Marty, A. et al. (2020). Synthesis of epitaxial monolayer Janus SPtSe. *npj 2D Mater. Appl.* 4: 41. https://doi.org/10.1038/s41699-020-00175-z.

189 Lin, Y.-C., Liu, C., Yu, Y. et al. (2020). Low energy implantation into transition-metal dichalcogenide monolayers to form Janus structures. *ACS Nano* 14: 3896–3906. https://doi.org/10.1021/acsnano.9b10196.

190 Trivedi, D.B., Turgut, G., Qin, Y. et al. (2020). Room-temperature synthesis of 2D Janus crystals and their heterostructures. *Adv. Mater.* 32: 2006320. https://doi.org/10.1002/adma.202006320.

191 Guo, Y., Lin, Y., Xie, K. et al. (2021). Designing artificial two-dimensional landscapes via atomic-layer substitution. *Proc. Natl. Acad. Sci.* 118: e2106124118. https://doi.org/10.1073/pnas.2106124118.

192 Xia, C., Xiong, W., Du, J. et al. (2018). Universality of electronic characteristics and photocatalyst applications in the two-dimensional Janus transition metal dichalcogenides. *Phys. Rev. B* 98: 165424. https://doi.org/10.1103/PhysRevB.98.165424.

193 Caviglia, A.D., Gabay, M., Gariglio, S. et al. (2010). Tunable Rashba spin-orbit interaction at oxide interfaces. *Phys. Rev. Lett.* 104: 126803. https://doi.org/10.1103/PhysRevLett.104.126803.

194 Dong, L., Lou, J., and Shenoy, V.B. (2017). Large in-plane and vertical piezoelectricity in Janus transition metal dichalchogenides. *ACS Nano* 11: 8242–8248. https://doi.org/10.1021/acsnano.7b03313.

195 Zheng, T., Lin, Y.C., Yu, Y. et al. (2021). Excitonic dynamics in Janus MoSSe and WSSe monolayers. *Nano Lett.* 21: 931–937. https://doi.org/10.1021/acs.nanolett.0c03412.

196 Ju, L., Bie, M., Tang, X. et al. (2020). Janus WSSe monolayer: an excellent photocatalyst for overall water splitting. *ACS Appl. Mater. Interfaces* 12: 29335–29343. https://doi.org/10.1021/acsami.0c06149.

197 Gao, X., Shen, Y., Liu, J. et al. (2021). Boosting the photon absorption, exciton dissociation, and photocatalytic hydrogen- and oxygen-evolution reactions by built-in electric fields in Janus platinum dichalcogenides. *J. Mater. Chem. C* 9: 15026–15033. https://doi.org/10.1039/D1TC03628A.

198 Kuc, A. and Heine, T. (2015). The electronic structure calculations of two-dimensional transition-metal dichalcogenides in the presence of external electric and magnetic fields. *Chem. Soc. Rev.* 44: 2603–2614. https://doi.org/10.1039/C4CS00276H.

199 Late, D.J., Huang, Y.K., Liu, B. et al. (2013). Sensing behavior of atomically thin-layered MoS_2 transistors. *ACS Nano* 7: 4879–4891. https://doi.org/10.1021/nn400026u.

200 Jia, S., Bandyopadhyay, A., Kumar, H. et al. (2020). Biomolecular sensing by surface-enhanced Raman scattering of monolayer Janus transition metal dichalcogenide. *Nanoscale* 12: 10723–10729. https://doi.org/10.1039/D0NR00300J.

201 Cao, Y., Fatemi, V., Fang, S. et al. (2018). Unconventional superconductivity in magic-angle graphene superlattices. *Nature* 556: 43–50. https://doi.org/10.1038/nature26160.

202 Cao, Y., Fatemi, V., Demir, A. et al. (2018). Correlated insulator behaviour at half-filling in magic-angle graphene superlattices. *Nature* 556: 80–84. https://doi.org/10.1038/nature26154.

203 Naik, M.H. and Jain, M. (2018). Ultraflatbands and shear solitons in Moiré patterns of twisted bilayer transition metal dichalcogenides. *Phys. Rev. Lett.* 121: 266401. https://doi.org/10.1103/PhysRevLett.121.266401.

204 Xian, L., Claassen, M., Kiese, D. et al. (2021). Realization of nearly dispersionless bands with strong orbital anisotropy from destructive interference in twisted bilayer MoS_2. *Nat. Commun.* 12: 5644. https://doi.org/10.1038/s41467-021-25922-8.

205 Zhang, Y., Liu, T., and Fu, L. (2021). Electronic structures, charge transfer, and charge order in twisted transition metal dichalcogenide bilayers. *Phys. Rev. B* 103: 155142. https://doi.org/10.1103/PhysRevB.103.155142.

206 Devakul, T., Crépel, V., Zhang, Y., and Fu, L. (2021). Magic in twisted transition metal dichalcogenide bilayers. *Nat. Commun.* 12: 6730. https://doi.org/10.1038/s41467-021-27042-9.

207 Wu, F., Lovorn, T., Tutuc, E. et al. (2019). Topological insulators in twisted transition metal dichalcogenide homobilayers. *Phys. Rev. Lett.* 122: 086402. https://doi.org/10.1103/PhysRevLett.122.086402.

208 Wu, F., Lovorn, T., Tutuc, E., and MacDonald, A.H. (2018). Hubbard model physics in transition metal dichalcogenide Moiré bands. *Phys. Rev. Lett.* 121: 026402. https://doi.org/10.1103/PhysRevLett.121.026402.

209 Zhang, Z., Wang, Y., Watanabe, K. et al. (2020). Flat bands in twisted bilayer transition metal dichalcogenides. *Nat. Phys.* 16: 1093–1096. https://doi.org/10.1038/s41567-020-0958-x.

210 Wang, L., Shih, E.M., Ghiotto, A. et al. (2020). Correlated electronic phases in twisted bilayer transition metal dichalcogenides. *Nat. Mater.* 19: 861–866. https://doi.org/10.1038/s41563-020-0708-6.

211 Ruiz-Tijerina, D.A. and Fal'ko, V.I. (2019). Interlayer hybridization and moiré superlattice minibands for electrons and excitons in heterobilayers of transition-metal dichalcogenides. *Phys. Rev. B* 99: 125424. https://doi.org/10.1103/PhysRevB.99.125424.

212 Regan, E.C., Wang, D., Jin, C. et al. (2020). Mott and generalized Wigner crystal states in WSe_2/WS_2 moiré superlattices. *Nature* 579: 359–363. https://doi.org/10.1038/s41586-020-2092-4.

213 Xu, Y., Liu, S., Rhodes, D.A. et al. (2020). Correlated insulating states at fractional fillings of moiré superlattices. *Nature* 587: 214–218. https://doi.org/10.1038/s41586-020-2868-6.

5

Magnetism and Spin Structures of Polymorphic 2D TMDs

Meizhuang Liu[1], Zuxin Chen[1], Jingbo Li[1], Yuli Huang[2], Kuan Eng Johnson Goh[3,4], and Andrew T. S. Wee[3]

[1] *South China Normal University, School of Semiconductor Science and Technology, Taoyuan East Road, Nanhai District, Foshan, 528225, China*
[2] *International Campus of Tianjin University, Joint School of National University of Singapore and Tianjin University, Binhai New City, Fuzhou, 350207, China*
[3] *National University of Singapore, Department of Physics, 2 Science Drive 3, 117542, Singapore*
[4] *Institute of Materials Research and Engineering (IMRE), Agency for Science Technology and Research (A*Star), 2 Fusionopolis Way, Innovis, Singapore 138634, Singapore*

Long-range ferromagnetic order has previously been believed to hardly survive in two-dimensional (2D) systems due to the enhanced thermal fluctuations, according to the Mermin–Wagner theorem [1]. The introduction of magnetic anisotropy in 2D systems opens up a spin-wave excitation gap, which can resist the thermal fluctuation and stabilize the long-range ferromagnetic order, resulting in finite Curie temperatures [2–4]. The observation of intrinsic 2D magnetism in atomically thin CrI_3 and $Cr_2Ge_2Te_6$ has stimulated much research interest in low-dimensional magnets [5–9]. The subsequent studies have revealed many other 2D magnetic materials such as $CrBr_3$ [10, 11], $CrCl_3$ [12, 13], $CrPS_4$ [14], Fe_3GeTe_2 [15], $FePS_3$ [16], $MnBi_2Te_4$ [17], VSe_2 [18–20], Cr_mX_n (X=S, Se and Te) [21–30], FeTe [31], and $MnSe_x$ [32] with exotic electrical, magnetic and optical properties, including magnetoelectric effect, giant tunneling magnetoresistance and magnetic second harmonic generation [33–36]. The magnetic properties can be tuned by different layer thicknesses, stacking order, external electric field, electrostatic doping [37] and external perturbations such as strain, light, and proximity coupling [38, 39], creating opportunities for realization of new functional devices. The atomically-thin 2D magnets also make it possible for the seamless integration with other van der Waals materials and the combination of distinct physical properties, which could give rise to unprecedented interfacial phenomena such as multiferroicity, topological superconductivity, and the quantum anomalous Hall effect [40, 41].

2D TMDs, as a family of 2D materials, have polymorphic structures with unique physical properties such as quantum spin Hall effect [42, 43], superconductivity [44, 45], charge density wave [46, 47], ferroelectricity [48], and ferromagnetism. The magnetic properties of TMDs can vary with phase and different stoichiometric

Two-Dimensional Transition-Metal Dichalcogenides: Phase Engineering and Applications in Electronics and Optoelectronics, First Edition. Edited by Chi Sin Tang, Xinmao Yin, and Andrew T. S. Wee.

ratios. The magnetic property of MoS_2 changes from diamagnetism to paramagnetism as the crystal structure transforms from 2H to 1T phase [49]. The magnetic moment in monolayer VSe_2 is theoretically calculated to be enhanced from 0.44 to 1.00 μB per unit cell with the phase transition from 1T to 1H phase [50]. The chromium chalcogenides (Cr_mX_n) usually possess diverse crystal structures with different stoichiometric ratios, such as CrX, CrX_2, Cr_2X_3, Cr_3X_4, and Cr_5X_8, which leads to the occurrence of ferrimagnetic, antiferromagnetic, and ferromagnetic behaviors. The chemical vapor deposition (CVD) grown Cr_2S_3 is ferrimagnetic semiconductor with Néel temperature of 120 K [51]. The CVD-grown CrSe flakes have a ferromagnetic property with the transition temperature close to 280 K [21]. The MBE-grown $CrSe_2$ monolayer is antiferromagnetic and Cr_2Se_3 monolayer has a weak ferromagnetism with Currie temperature of ~200 K [23]. Room-temperature ferromagnetism has been observed in $CrTe_2$ ultrathin films [25–27]. Antiferromagnetic spin texture was revealed in the epitaxial monolayer $CrTe_2$ by spin-polarized scanning tunneling microscopy [28]. Thickness-dependent magnetic behavior has also been reported in Cr_2Te_3, Cr_3Te_4 and Cr_5Te_8 films [24, 29, 30]. The polymorphic structures endow TMD materials with the abundant magnetic properties.

5.1 Two-dimensional Ferromagnetism

In 2D magnets, exchange interaction and magnetic anisotropy are fundamental parameters. The exchange interactions originate from the electron's antisymmetric wave function and are governed by electrostatic interactions under the Pauli exclusion principle, which is the driving force of long-range magnetic ordering of electron spin moments (Figure 5.1a) [2, 6]. Depending on the magnetic ion species and the corresponding crystalline structures, different exchange mechanisms, including direct exchange interaction, superexchange interaction, double exchange interaction, itinerant electrons, and Ruderman–Kittel–Kasuya–Yosida (RKKY) mechanisms, have been utilized to interpret the magnetism in 2D systems [52]. According to the different spin dimensionality, the magnetic anisotropy in 2D magnets can be described by the Ising, XY, or Heisenberg models. The Ising model features an out-of-plane easy axis and strong uniaxial magnetic anisotropy energies. In the XY model, there is an easy-plane anisotropy whereby spins lie in a given plane and have no energetic barrier to rotate in the whole plane. For the isotropic Heisenberg model, the spins are free to rotate with no constraint on the directions. The dimensionality effect plays a crucial role in determining the impact of thermal fluctuations on 2D magnetism. For 2D systems without magnetic anisotropy, the spin-wave excitation gap diminishes and long-range magnetic order collapses. The presence of uniaxial magnetic anisotropy (UMA) can open up a spin-wave excitation gap to resist the thermal fluctuation, leading to the finite Curie temperatures. As the dimensionality increases from 2D to 3D, the magnon spectrum evolves from a step function to a gradually increasing function (Figure 5.1c). When the ferromagnetic exchange interactions and magnetic anisotropy are strong enough, the intrinsic 2D magnets with long-range magnetic order at high temperatures can be achieved.

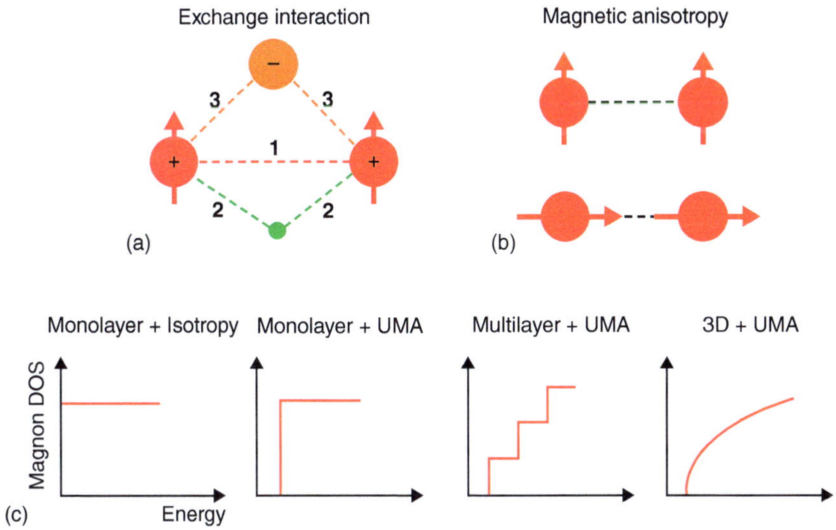

Figure 5.1 (a) Schematic diagram of the exchange interaction between spins directly established (red line 1) and indirectly mediated by conduction electrons (green line 2) or intermediate anions (orange line 3) [6]. (b) Magnetic anisotropy with spins aligned in a preferred orientation [6]. (c) The evolution of magnon density of states with dimensionality and magnetic anisotropy. Source: Gong and Zhang [6]/with permission from American Association for the Advancement of Science.

5.2 Cr-based Magnetic Materials and Device Applications

CrI_3. Bulk CrI_3 is a layered van der Waals ferromagnet with a Curie temperature of 61 K, a band gap of about 1.2 eV, and an out-of-plane easy axis [53]. Few-layer CrI_3 is found to have a distinct magnetic property from the bulk, which is a layered antiferromagnet with intralayer ferromagnetic interaction. The different magnetic properties between bulk and few-layer CrI_3 are ascribed to the different layer stacking orders. Bulk CrI_3 has a phase transition from high-temperature monoclinic structure to low-temperature rhombohedral structure at ~200 K, whereas few-layer CrI_3 can retain the monoclinic structure at low temperature, which leads to a different magnetic configuration [53–56].

Monolayer CrI_3 can be prepared by mechanical exfoliation from bulk crystals or molecular beam epitaxy (MBE) growth [57], in which Cr^{3+} ions are coordinated to six I^- ions with a hexagonal honeycomb structure and a high spin state of $S = 3/2$ with three electrons at the t_{2g} levels (Figure 5.2a) [58]. Huang et al. demonstrated that long-range ferromagnetic order can exist in the mechanically exfoliated single-layer CrI_3 with a Curie temperature of 45 K [3]. Magneto-optical Kerr effect (MOKE) measurements allow non-contact and non-destructive magnetic characterization and describe the polarization change of a light reflected from a magnetic surface. The MOKE measurements of monolayer CrI_3 show a single hysteresis loop

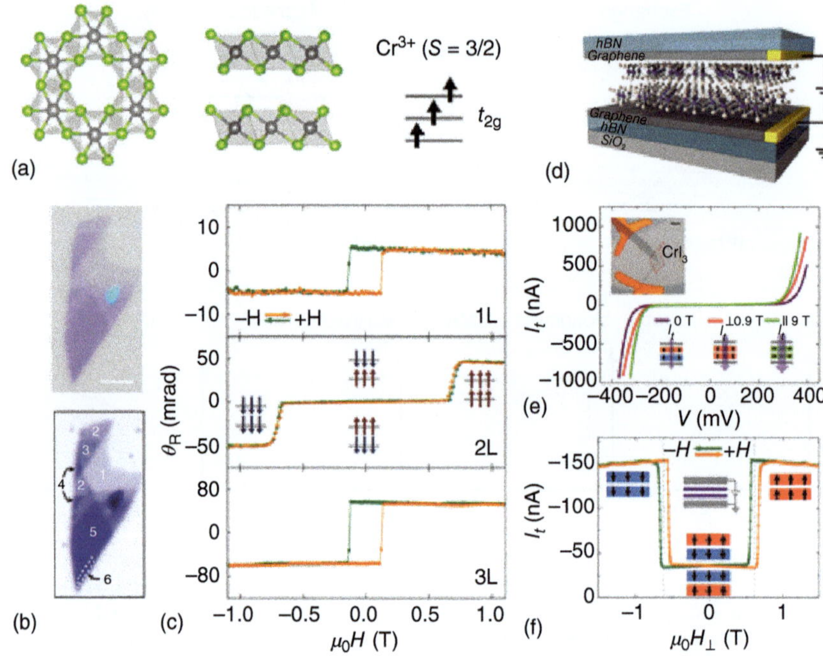

Figure 5.2 (a) Atomic structure of CrI_3. Source: Huang et al. [3]/with permission from Springer Nature. (b) the optical micrograph of the exfoliated CrI_3 flake. Source: Huang et al. [3]/with permission from Springer Nature. (c) Different layers and MOKE signals of monolayer, bilayer, and trilayer CrI_3 flakes. Source: Huang et al. [3]/with permission from Springer Nature. (d) Schematic of a bilayer CrI_3 spin-filter magnetic tunnel junction (sf-MTJ). Source: Song et al. [35]/with permission from American Association for the Advancement of Science. (e) Tunneling current of a sf-MTJ as a function of bias voltage at selected magnetic fields. Source: Song et al. [35]/with permission from American Association for the Advancement of Science. (f) Tunneling current as a function of magnetic field at a fixed voltage (−290 mV). Source: Song et al. [35]/with permission from American Association for the Advancement of Science.

with a non-zero remanent magnetization at μ_0H = 0 T (Figure 5.2b), demonstrating ferromagnetic order with out-of-plane spin polarization. Ferromagnetic order in few-layer CrI_3 demonstrates a layer-dependent behavior: mono- and tri-layer CrI_3 are ferromagnetic, but bilayer CrI_3 is antiferromagnetic (AFM) due to the intralayer ferromagnetic interactions and interlayer antiferromagnetic interactions. The net magnetization in bilayer CrI_3 is close to zero below a critical field of about 0.65 T at 15 K (Figure 5.2c). As the magnetic field increases in the out-of-plane direction, a metamagnetic phase transition from AFM state to FM state occurs with all the spins flipping in the direction of the external magnetic field and a sharp jump in the MOKE signal. The character of layer-by-layer antiferromagnetic order in few-layer CrI_3 can be exploited for fabricating spin-filter magnetic tunnel junctions (sf-MTJ).

Prototypical magnetic tunnel junctions generally consist of two ferromagnetic layers separated by a thin nonmagnetic insulating layer, and the tunnel magnetoresistance (TMR) is defined as $(R_{AP} - R_P)/R_P$ and depends on the relative magnetization direction (antiparallel or parallel) of the two ferromagnetic layers. The layered antiferromagnetic structure in bilayer CrI_3 naturally forms the interlayer spin filter, which can avoid the usage of nonmagnetic spacers. The basic structure of the vdW heterostructure sf-MTJ is sketched in Figure 5.2d, which consists of bilayer CrI_3 tunnel barrier sandwiched by few-layer graphene contacts and hexagonal boron nitride (hBN) protection layers [35]. In the magnetic tunnel junction of bilayer CrI_3, the tunneling current is suppressed at the layered-antiferromagnetic ground state under zero magnetic field and significantly enhanced when the spin states of bilayer CrI_3 are uniaxially polarized under an out-of-plane or in-plane magnetic field. With the tunneling current at $\mu_0 H = 0\,T$ much smaller than the current under an out-of-plane magnetic field or in-plane magnetic field, it indicates the existence of spin-polarized currents in the sf-MTJ device (Figure 5.2e). The high-resistance state or low-resistance state of tunnel magnetoresistance corresponds to the different magnetic states of bilayer CrI_3, which can be revealed by the measured tunneling current (Figure 5.2f) as a function of the out-of-plane magnetic field at a selected bias voltage (−290 mV). The tunneling current will jump sharply to a higher plateau from the lower plateau when the external magnetic field exceeds the critical value by switching the magnetic states in bilayer CrI_3. The tunneling magnetoresistances in bilayer CrI_3 sf-MTJ are 310% and 530% for magnetization aligned perpendicular and parallel to the plane respectively. As the layer thickness of CrI_3 increases, a series of spin filters form through the different interlayer antiferromagnetic coupling structures. The tunneling magnetoresistance can be drastically enhanced due to the layer-by-layer antiferromagnetic order and is measured to be a record-breaking 19 000% in four-layer sf-MTJ devices at 2K [35].

Breaking the time-reversal and spatial-inversion symmetries is required for the induction of magnetization by an electric field [59]. The non-centrosymmetric bilayer CrI_3 with interlayer antiferromagnetic coupling meets the condition, which provides the possibility for the electric-field control of magnetism through the magneto-electric effect. Compared with the single-gated field-effect devices, the dual-gated ones enable the independent control of the electric field and electrostatic doping by gate voltages (Figure 5.3a) [37]. The vertical electric fields can be applied in the dual-gated field-effect devices by setting the top and back gate voltages with the same amplitude but opposite signs. A linear magnetoelectric effect with non-zero net magnetization as large as ~30% of the saturation magnetization can be induced in the bilayer CrI_3 by the applied electric fields [34]. In the AFM phase, the induced change in magnetization ΔM depends linearly on E and changes sign at $E = 0\,\text{Vnm}^{-1}$. In contrast, the saturation magnetization is independent of E because the linear ME effect is not allowed due to the existing spatial-inversion symmetry in the FM state [34]. The occurrence of a magneto-electric effect in bilayer CrI_3 is attributed to the existence of an interlayer potential difference accompanied by charge transfer. As for monolayer CrI_3, the externally applied electric field has a negligible effect on the magnetization. However, gate-induced doping can significantly modify the hysteresis loop of monolayer CrI_3 with increasing (decreasing) saturation magnetization M_S and coercive force H_C through the

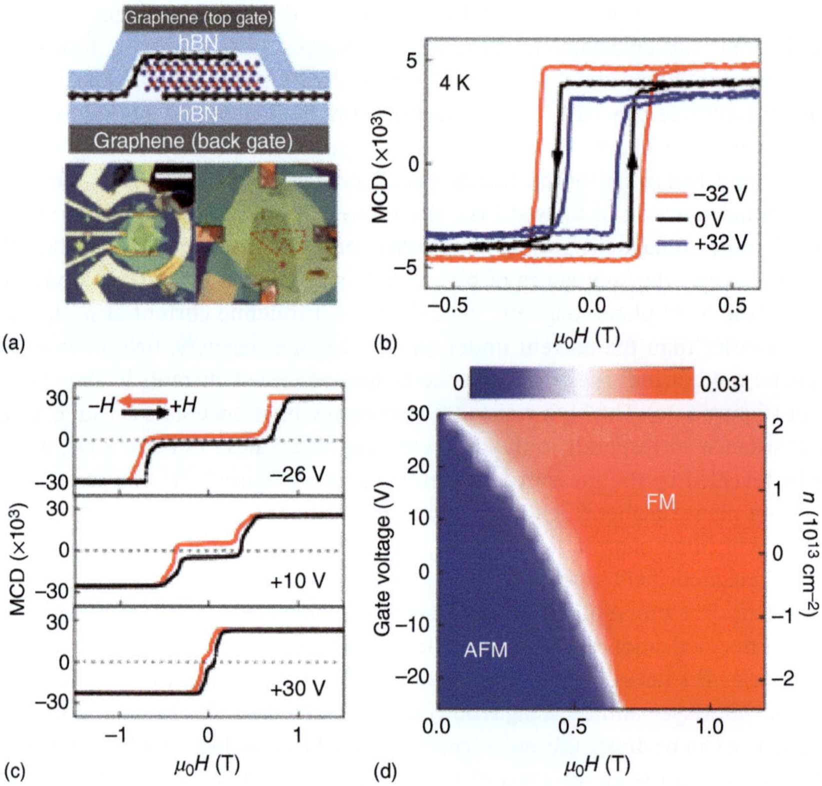

Figure 5.3 (a) Schematic of a dual-gate field-effect device based on bilayer CrI_3. (b) MCD signals of monolayer CrI_3 as a function of out-of-plane magnetic field at gate voltages of 0 and ±32 V. (c) MCD signals versus magnetic field at different gate voltages. (d) Gate-induced doping density-magnetic field Phase diagram at 4 K. Source: Jiang et al. [37].

hole (electron) doping [37]. The Curie temperature T_C can also be modified over ~20% by the gate-induced doping. Hole doping can strengthen the magnetic order, and conversely, electron doping weakens the magnetic order in monolayer CrI_3, the microscopic mechanisms of which are still not well understood. When the electrostatic doping is considered in bilayer CrI_3, the applied magnetic field the converting AFM phase to the FM phase decreases as the gate-induced electron doping density increases (Figure 5.3c). The AFM phase finally vanishes at a gate voltage of +30 V corresponding to a critical electron density of $\sim 2.5 \times 10^{13}\,cm^{-2}$ (Figure 5.3d). The AFM-FM transition under zero magnetic field is ascribed to the modified interlayer exchange coupling by electron doping in bilayer CrI_3 [37]. The MOKE measurements reported by Huang et al. also exhibited gate-controlled magnetism in bilayer CrI_3 by electrostatic doping [33]. Voltage-controlled switching between AFM and FM states can be realized at a selected magnetic field near the metamagnetic transition [33]. The efficient control of magnetism in monolayer and bilayer CrI_3 holds promise for future voltage-controlled magnetic switching applications of 2D magnets.

$CrBr_3$: The other analogues of CrI_3, monolayer $CrBr_3$ and $CrCl_3$, have also been demonstrated to be ferromagnetic semiconductors with the Curie temperatures of 34 and 13 K respectively [10–13]. For $CrBr_3$, the monolayer samples can be prepared by MBE growth or mechanical exfoliation. The atomic structure of the MBE-grown high-quality monolayer $CrBr_3$ is revealed by scanning tunneling microscopy (STM) with periodically arranged triangular clusters formed by three top Br atoms [10]. Spin-polarized STM measurements, a powerful technique for investigating the magnetism at the atomic scale [60], were carried out to measure the local magnetic properties of monolayer $CrBr_3$. The spin flip of monolayer $CrBr_3$ under the reversal of magnetic field results in the difference of spin-polarized dI/dV curves (Figure 5.4a). The rectangular hysteresis loop formed by a series of dI/dV spectra demonstrates the ferromagnetic property of monolayer $CrBr_3$ with out-of-plane easy axis (Figure 5.4b) [10]. Intrinsic ferromagnetism of monolayer $CrBr_3$ has also been revealed by polarization resolved magneto-photoluminescence. The hysteresis loop progressively diminishes as the temperature is increased above the Curie temperature of 34 K [11]. When the layer thickness of $CrBr_3$ increases to bilayer, there are two different stacking orders of H-type and R-type determined by the orientation of triangular Br clusters (Figure 5.4c,d). The H-type and R-type stacked bilayer $CrBr_3$ exhibit FM and AFM interlayer coupling, respectively. The H-type stacked bilayer $CrBr_3$ has a rectangular magnetic hysteresis loop, similar to that of the monolayer $CrBr_3$. In contrast, the R-type stacked bilayer $CrBr_3$ has

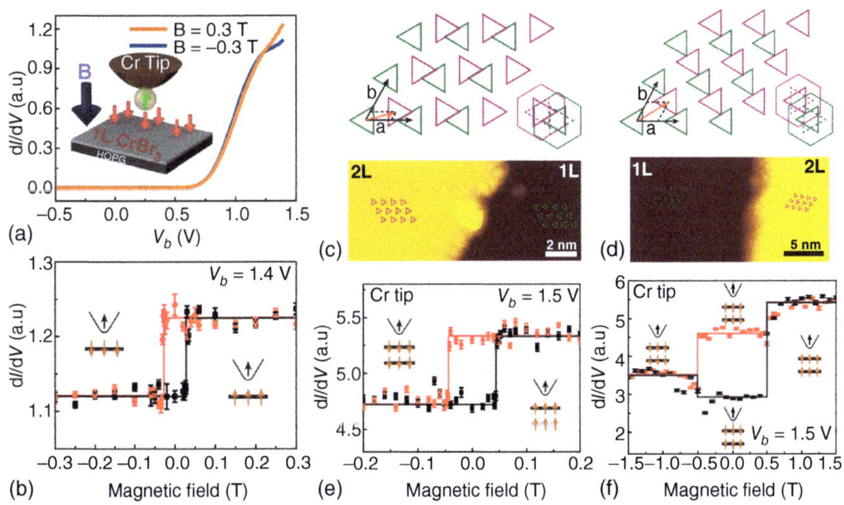

Figure 5.4 (a) Spin-polarized tunneling spectra of monolayer $CrBr_3$ under out-of-plane magnetic fields of ± 0.3 T. The inset illustrates the experimental geometry. (b) dI/dV signal as a function of the applied magnetic field at a selected bias voltage of 1.4 V. (c) Atomic structure of the H-type stacked bilayer $CrBr_3$ and the corresponding STM image across the step edge. (d) Atomic structure of the R-type stacked bilayer $CrBr_3$ and the corresponding STM image with a monolayer region and bilayer island. (e) Spin-polarized dI/dV signal measured on H-type stacked bilayer $CrBr_3$. (f) Spin-polarized dI/dV signal measured on R-type stacked bilayer $CrBr_3$. Source: Chen et al. [10]/with permission from American Association for the Advancement of Science.

four plateaus in the hysteresis loop. Two middle plateaus correspond to two new magnetization configurations (Figure 5.4f) differing from the antiferromagnetic ground states in bilayer $CrBr_3$ [10].

$CrCl_3$: Bulk $CrCl_3$ is a weakly coupled, layered antiferromagnet with small XY anisotropy and Néel temperature of ~14 K [12, 13]. Ferromagnetic order exists in monolayer $CrCl_3$ with the easy-plane normal to the c-axis (Figure 5.5a) [61]. Compared to the band gap of 3.0 eV in bulk samples, the monolayer $CrCl_3$ has a smaller band gap of 1.6 eV extracted from the dI/dV spectrum (Figure 5.5b) [12]. The element-specific X-ray magnetic circular dichroism (XMCD) signal reveals that the MBE-grown monolayer $CrCl_3$ possesses a magnetic moment of 2.9 μ_B/Cr at 3.5 K. The magnetic anisotropy in monolayer $CrCl_3$ favors an in-plane easy axis, as revealed from the XMCD hysteresis loops (Figure 5.5c). Ferromagnetic-to-paramagnetic transition in monolayer $CrCl_3$ can be confirmed by the vanishing of remanence and coercive fields in temperature-dependent XMCD measurements (Figure 5.5d) [12]. With regards to device applications, Cai et al. fabricated a magnetic tunnel junction based on the bilayer $CrCl_3$ [13]. Vertical tunneling measurements indicate the tunneling current can be suppressed at zero field by a spin-filter effect, and the spin polarizations of two layers become increasingly aligned in an external field. Based on the tunneling current as a function of temperature and magnetic fields, the magnetic phase diagram of bilayer $CrCl_3$ exhibits transitions between AFM, a fully spin-polarized phase, and paramagnetic (PM) phases (Figure 5.5h).

$Cr_2Ge_2Te_6$: $Cr_2Ge_2Te_6$ is a nearly ideal 2D Heisenberg ferromagnet with small magnetic anisotropy [6]. As a ternary vdW ferromagnetic material, the crystal structure of $Cr_2Ge_2Te_6$ consists of edge-sharing $CrTe_6$ octahedrons with Ge-dimers located in the center of the octahedron honeycomb (Figure 5.6a) [62]. The mechanically exfoliated bilayer $Cr_2Ge_2Te_6$ strip can be clearly distinguishable from the surrounding substrate by the scanning Kerr microscope as the temperature decreases to 4.7 K. The temperature of ferromagnetic phase transition in bilayer $Cr_2Ge_2Te_6$ is confirmed by the disappearance of the Kerr rotation signal at higher temperatures [4]. The thickness-dependent Kerr measurements reveal a strong dimensionality effect on the Curie temperature of $Cr_2Ge_2Te_6$, which increases from 30 K of bilayer $Cr_2Ge_2Te_6$ to 68 K of the bulk. From STM imaging, the surface of $Cr_2Ge_2Te_6$ bulk shows a well-ordered honeycomb lattice in which bright spots correspond to the outermost Te atoms (inset in Figure 5.6b). The dI/dV spectra measured on the $Cr_2Ge_2Te_6$ surface at 50 and 77 K exhibit different energy gaps of about 0.74 and 0.86 eV respectively (Figure 5.6b), which is contrary to the behavior of conventional semiconductors whose band gap decreases with increasing temperature. The formation of ferromagnetic order at 50 K renormalizes the electronic structure with e_g peak downshift toward E_F and reduces the band gap [62]. The magnetic domains of bulk $Cr_2Ge_2Te_6$ have been revealed by a magnetic force microscope (MFM), which exhibits the long-range magnetic order in real space (Figure 5.6c) [63]. The fractal-like domain structure transforms into a stripe-shaped magnetic structure when the temperature increases from 60 to 62 K (Figure 5.6d).

The magnetic order in few-layer $Cr_2Ge_2Te_6$ can be controlled via electrostatic gating. In the Si-gated field-effect transistors (FETs) of ~3.5 nm $Cr_2Ge_2Te_6$, the

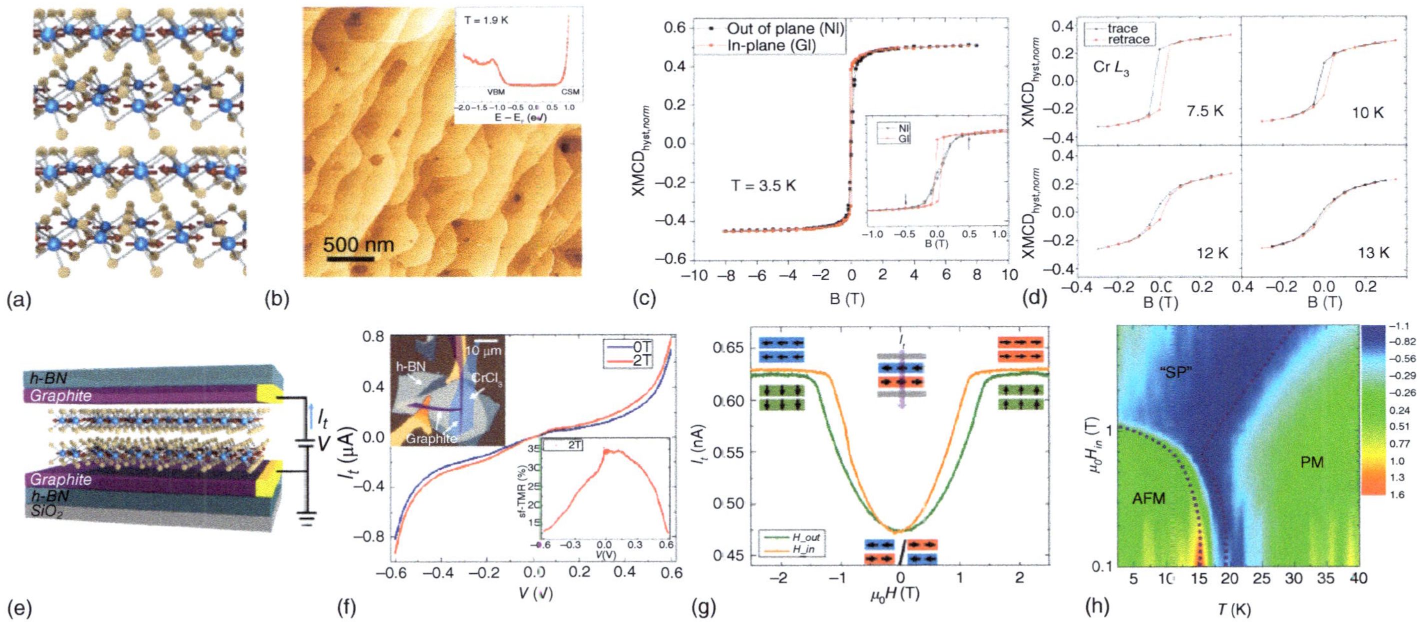

Figure 5.5 (a) Crystal structure of $CrCl_3$. Source: Cai et al. [13], (2019) © 2019 American Chemical Society. (b) (Science AAAS) STM image of monolayer $CrCl_3$ grown on graphene (Inset: dI/dV spectrum at $CrCl_3$ surface). Source: Pinto et al. [12]/with permission from American Association for the Advancement of Science. (c) (Science AAAS) Field-dependent XMCD hysteresis loops taken in normal and grazing incidence directions at 3.5 K. Source: Pinto et al. [12]/with permission from American Association for the Advancement of Science. (d) (Science AAAS) XMCD hysteresis loops taken at various temperatures. Source: Pinto et al. [12]/with permission from American Association for the Advancement of Science. (e) Schematic of bilayer $CrCl_3$ spin-filter magnetic tunnel junction device with graphite contacts encapsulated by h-BN. Source: Cai et al. [13]. (f) Different tunneling currents with and without an in-plane magnetic field of $\mu_0 H = 2$ T. Insets: optical microscope of the bilayer $CrCl_3$ spin-filter magnetic tunnel junction and TMR ratio. Source: Cai et al. [13]/American Chemical Society. (g) Field-dependent tunneling current under in-plane and out-of-plane magnetic fields at $V = 1$ mV. Source: Cai et al. [13]. (h) Magnetic phase diagram of bilayer $CrCl_3$. Source: Cai et al. [13].

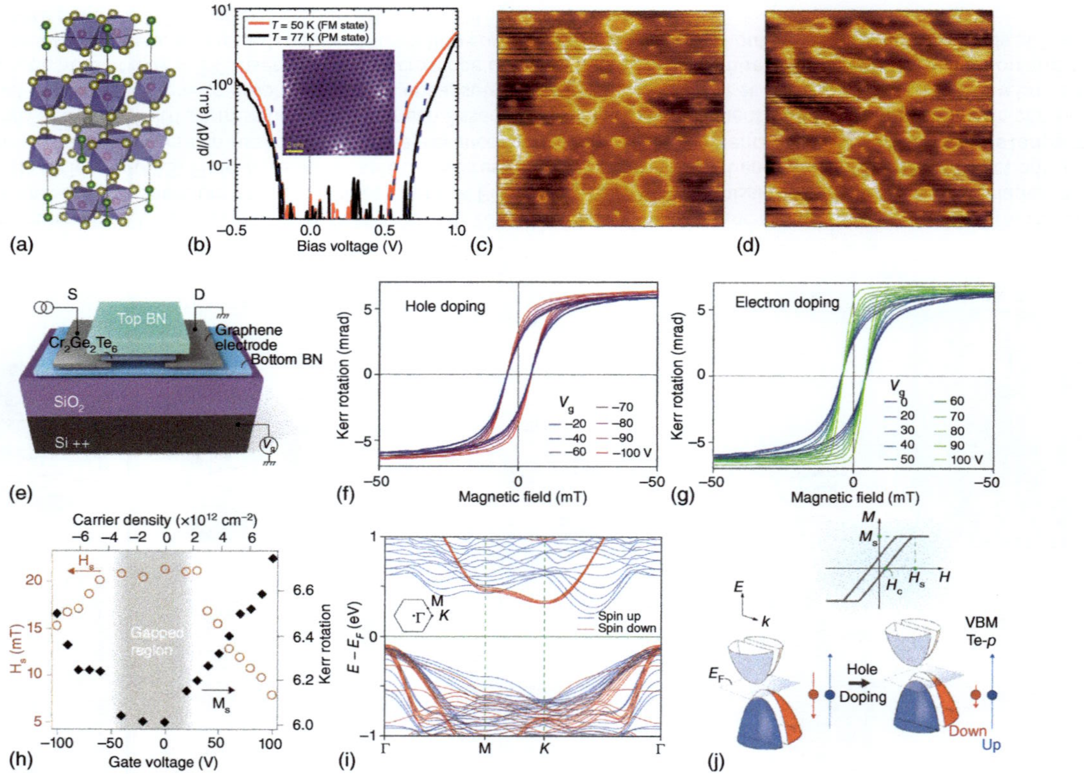

Figure 5.6 (a) Crystal structure of $Cr_2Ge_2Te_6$. Source: Hao et al. [62]/with permission from ELSEVIER. (b) dI/dV spectra on $Cr_2Ge_2Te_6$ surface taken at 50 and 77 K. Inset: STM image of cleaved $Cr_2Ge_2Te_6$ surface. Source: Hao et al. [62]/reproduced with permission from Elsevier. Magnetic force microscope image of bulk $Cr_2Ge_2Te_6$ at 60 K (c) and 62 K (d), respectively. Source: Wang et al. [63]/with permission from Springer Nature. (e) Schematic of few-layered $Cr_2Ge_2Te_6$ FETs with Si gate. Source: Wang et al. [63]/with permission from Springer Nature. Kerr measurement of 3.5 nm $Cr_2Ge_2Te_6$ sample with Si gate at 40 K for negative (f) and positive (g) gate voltages, respectively. (h) Bipolar gate tenability of H_s and M_s as a function of gate voltage at 40 K. (i) Calculated spin-polarized band structure of few-layer $Cr_2Ge_2Te_6$. (j) Schematic of spin realignment model after hold doping [63].

enhanced saturation magnetization (M_S), reduced saturation field (H_S), and doping-insensitive coercivity (H_c) can be observed by the Kerr measurements regardless of hole or electron doping (Figure 5.6f,g) [63]. The bipolar gate tunability of H_s and M_s with a stronger gating effect on the electron side and a gapped region can be seen in Figure 5.6h. Hole and electron doping in semiconducting few-layered $Cr_2Ge_2Te_6$ can shift the Fermi level into the valence band and conduction band by depleting Te-*p* and filling Cr-*d* orbitals, respectively. As the valence band maximum (VBM) and conduction band minimum (CBM) are contributed by spin minority (spin down) and spin majority (spin up) states (Figure 5.6i), the shift of the Fermi level could result in an increased net magnetic moment for mild doping of either electrons or holes (Figure 5.6j).

$CrPS_4$: An air-stable vdW material, $CrPS_4$ (Figure 5.7a), reported by Son et al. has a layer-dependent magnetic property [14]. The magnetic susceptibilities of bulk $CrPS_4$ have a maximum at the Néel temperature, indicating its antiferromagnetic nature (Figure 5.7b). The layer-dependent magnetization can be observed by 2D scanning MOKE microscopy, whereby the 1ML and 3ML flakes have uniformly positive signals of Kerr rotation angles below the Curie temperature of ~23 K and no net magnetization is detected on the 2ML flakes (Figure 5.7c,d). The ferromagnetism in odd-layer $CrPS_4$ is further confirmed by field-sweep MCD measurements, whereby clear hysteresis curves in 1 and 3 ML flakes (Figure 5.7e,f) are observed [14].

Another ternary Cr-based magnetic material, CrOCl, has been demonstrated to be a stable vdW antiferromagnetic material with the Néel temperature of ~13.6 K. Monolayer CrOCl has a staggered CrO double layer sandwiched between two halogen atomic layers [64, 65]. The atomic structure of the (001) plane of the mechanically exfoliated CrOCl flake was obtained by scanning transmission electron microscopy (STEM) where the Cr atoms have a higher brightness than Cl/O atoms. The antiferromagnetic property of CrOCl flakes was confirmed by field-dependent magnetic measurements, which show that the net magnetization is close to zero below the critical field and the hysteresis loop occurring at ±3–5 T gradually shrinks with increasing temperature [64]. It has also been theoretically predicted that monolayer CrOCl is a ferromagnetic semiconductor with a high Curie temperature of up to 160 K, but this has not been demonstrated in experiments to date [66].

5.3 Polymorphic 2D Cr-based Magnetic TMDs

2D Cr-based TMDs have also attracted much interest for their intrinsic ferromagnetism and good environmental stability. Zhang et al. utilized the CVD method to synthesize ultrathin CrSe flakes with a non-layered structure as shown in Figure 5.8a,b [21]. The ferromagnetic order was demonstrated by a temperature-dependent and field-dependent magnetic moment, which shows a ferromagnetic-paramagnetic transition at a temperature of ~280 K (Figure 5.8c,d) [21]. The $CrSe_2$ crystal has been synthesized through deintercalation of lithium or potassium from a stable parent compound ($LiCrSe_2$ or $KCrSe_2$) [67, 68]. Furthermore, Li *et al.* reported CVD-grown $CrSe_2$ nanosheets with tunable thickness

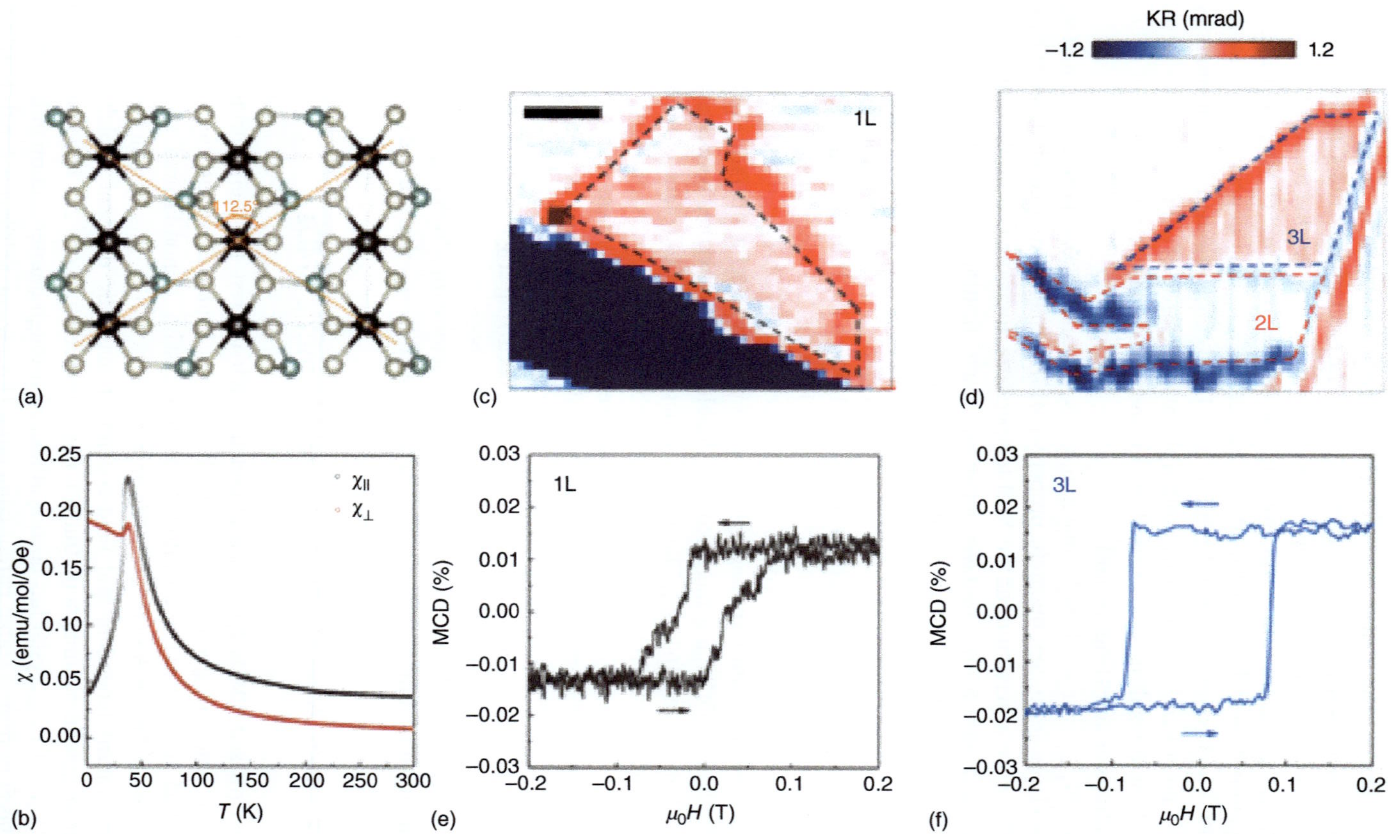

Figure 5.7 (a) Crystal structure of $CrPS_4$. Source: Son et al. [14] Reproduced from *ACS Nano 15*, 16904–16912 (2021). © 2021 American Chemical Society. (b) Magnetic susceptibilities of bulk $CrPS_4$ [14]. 2D scanning MOKE microscope image of 1 monolayer (c), 2 ML, and 3 ML (d) $CrPS_4$ [14]. MCD hysteresis loops of 1 ML (e) and 3 ML (f) samples at 3.5 K [14].

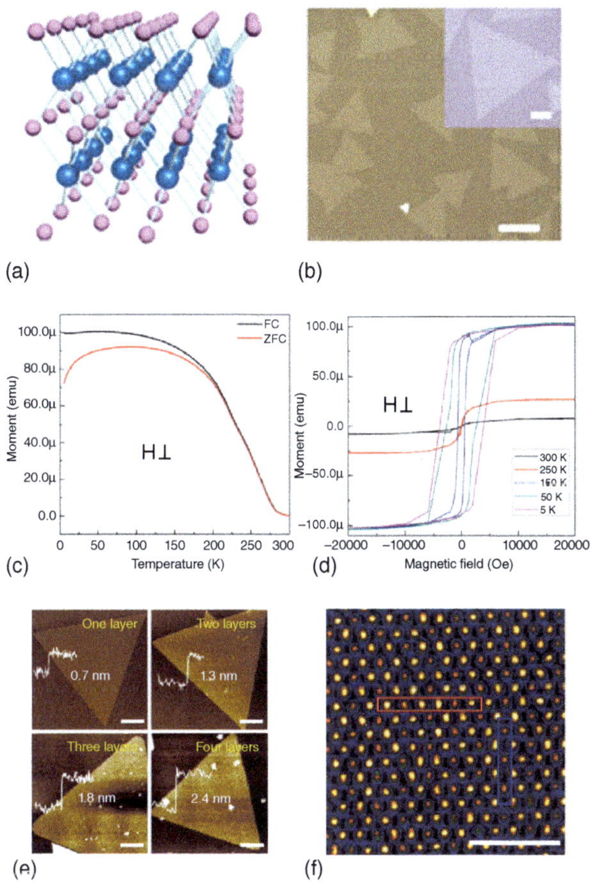

Figure 5.8 (a) Atomic model of CrSe crystal. Source: Zhang et al. [21]/with permission from John Wiley & Sons. (b) Optical micrograph of CVD-grown CrSe flakes on mica substrate. Source: Zhang et al. [21]/with permission from John Wiley & Sons. (c) Temperature-dependent magnetic moments of CVD-grown CrSe flake under a magnetic field of 1000 Oe. Source: Zhang et al. [21]/with permission from John Wiley & Sons. (d) Magnetic hysteresis loops of CrSe flake at various temperatures under a vertical magnetic field. Source: Zhang et al. [21]/with permission from John Wiley & Sons. (e) Atomic force microscope images of $CrSe_2$ nanosheets with various thicknesses [22]. (f) STEM image of $CrSe_2$ nanosheets. Source: Li et al. [22]/with permission from Springer Nature. (g) Polar RMCD signals of monolayer $CrSe_2$ nanosheet at different temperatures. Source: Li et al. [22]/with permission from Springer Nature. (h) RMCD signals of $CrSe_2$ with different layer numbers at 15 K. Source: Li et al. [22]/with permission from Springer Nature. (i) STM image of MBE-grown monolayer $CrSe_2$ at 200 °C with a CrSe thin layer on the islands. Source: Liu et al. [23] Reproduced from *J. Phys. Chem. Lett. 12*, 7752–7760 (2021). © 2021 American Chemical Society. (j) STM image of MBE-grown monolayer Cr_2Se_3 at 300 °C. Source: Liu et al. [23]. (k) Differential conductance spectrum taken on Cr_2Se_3 surface. Inset: the valence band photoelectron spectrum of monolayer Cr_2Se_3. Source: Liu et al. [23]. (l) XMCD spectra of monolayer Cr_2Se_3 taken at different temperatures. Source: Liu et al. [23].

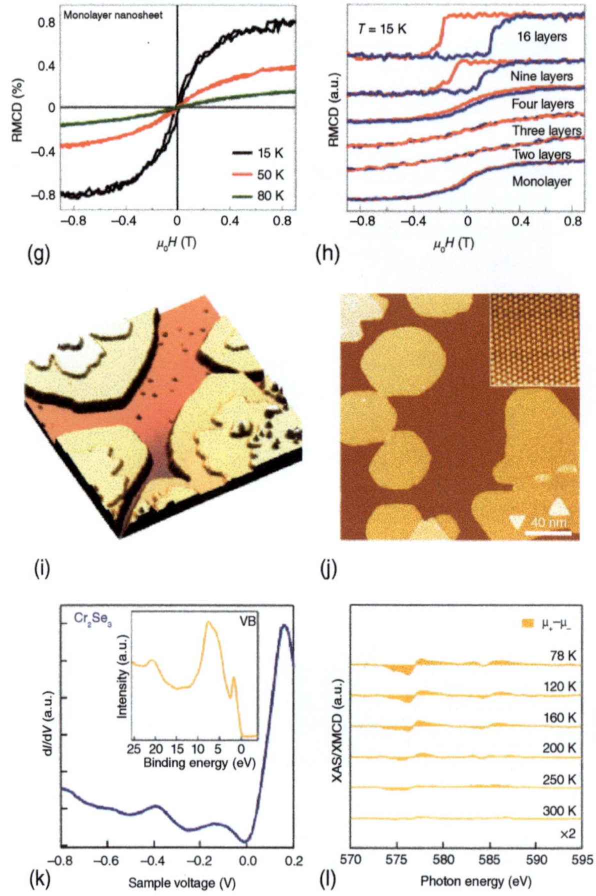

Figure 5.8 *(Continued)*

down to monolayer on WSe_2 substrates [22]. In their experiments, a series of $CrSe_2$ samples with different thicknesses, down to monolayer thickness (~0.7 nm) can be obtained by controlling the growth temperature (Figure 5.8e). The atomically resolved STEM image shows that the Cr atom is surrounded by six Se atoms with a lattice constant of 0.363 nm (Figure 5.8f), but the critical cross-sectional image supporting the layered structure in the $CrSe_2$ sample was lacking. For monolayer $CrSe_2$, the reflective magnetic circular dichroism (RMCD) signal shows a negligible hysteresis, indicating weak FM behavior (Figure 5.8g). As the thickness increases from monolayer to dozens of layers, the hysteresis loops become more obvious, and T_c increases from 65 K for the monolayer to 110 K for the 16-layer $CrSe_2$ (Figure 5.8h) [22].

In our recent work, monolayer $CrSe_2$ and Cr_2Se_3 were grown by the MBE method [23]. A CrSe thin layer on $CrSe_2$ islands with apparent height of 0.3 nm was observed in the experiments (Figure 5.8i), which indicates a non-layered growth mechanism. Monolayer $CrSe_2$ has a 1T phase with lattice constant of 3.3 Å and a coexisting 1T″ phase with 2 × 2 superstructure. The metastable monolayer $CrSe_2$ can be transformed into Cr_2Se_3 by annealing at 300 °C, with a lattice constant of 3.6 Å (inset in Figure 5.8j). The STS and photoelectron spectroscopy measurements revealed the metallic character of 1T-$CrSe_2$ and 1T-Cr_2Se_3. (Figure 5.8k). In the XMCD measurements, no obvious ferromagnetic signal was observed in monolayer $CrSe_2$ at 78 K, and only a weak ferromagnetic signal with T_c of ~200 K was found in monolayer Cr_2Se_3 (Figure 5.8l). The weak ferromagnetism in monolayer Cr_2Se_3 might be due to the dimensionality and interfacial effects [23].

Chromium tellurides are expected to generate strong magnetic anisotropy to stabilize the long-range magnetic order in 2D systems owing to the strong spin-orbit coupling of telluride [70]. Several stable stoichiometries have been reported, such as CrTe, $CrTe_2$, Cr_2Te_3, Cr_3Te_4, and Cr_5Te_8 [24–30, 71–74]. Among them, $CrTe_2$ is a vdW layered ferromagnetic material with T_c up to 300 K [25–27]. MBE and CVD methods have been utilized by several different research groups in the growth of atomically thin $CrTe_2$. The temperature-dependent ferromagnetic properties of the MBE grown 1T-$CrTe_2$ samples were confirmed by SQUID and XMCD measurements (Figure 5.9c–e) [26]. The out-of-plane and in-plane magnetization-magnetic field (M-H) hysteresis loops of 7 ML $CrTe_2$ at different temperatures revealed an out-of-plane anisotropy and T_c up to 300 K. XMCD characterization of the 7 ML $CrTe_2$ samples further confirms the temperature-dependent ferromagnetism with atomic magnetic moment of ~3 μ_B/Cr at 5 K and 0.21 μ_B at 300 K (Figure 5.9e). The band structure of $CrTe_2$ with splitting of majority and minority bands (~0.2 eV at Γ point) (Figure 5.9f) was observed by angle-resolved photoemission spectroscopy (ARPES), which corroborates the interplay of ferromagnetism and electronic structure in $CrTe_2$ [26]. In the MBE grown thin-layer $CrTe_2$, T_c gradually decreases from 300 to 200 K as the film thickness is reduced to monolayer $CrTe_2$ (Figure 5.9g). The MBE-grown monolayer $CrTe_2$ has a 2 × 1 structure with the in-plane isosceles lattice constants of $a_1 = 3.4$ Å, $a_2 = 3.7$ Å, $a_3 = 3.7$ Å (Figure 5.9h) [28]. The spin-polarized STM study revealed a zigzag spin contrast in 1 ML 1T-$CrTe_2$ which reverses its contrast under the opposite field (Figure 5.9i,j). The zigzag spin contrast originates from the magnetic configuration of zigzag AFM ground state in 1 ML 1T-$CrTe_2$ (Figure 5.9k) [28]. Antiferromagnetic monolayer $CrTe_3$ with band gap of ~0.9 eV can be obtained by MBE at a specific temperature range, which gives rise to a lateral $CrTe_2$-$CrTe_3$ heterojunction with an atomically sharp and seamless interface (Figure 5.9l–o) [69]. As the substrate temperature is elevated to above 350 °C, the $CrTe_2$ single-layer films are reported to transform into more three-dimensional Cr_xTe_y islands with step heights of 1.2 nm [75].

The multiple polymorphs of chromium telluride, such as $CrTe_2$, Cr_2Te_3, Cr_3Te_4, and Cr_5Te_8, have been obtained by tuning the ratio of Cr and Te precursors in the CVD growth [24, 27, 30, 74]. The crystal structures could be viewed as self-intercalated compounds of layered $CrTe_2$ (Figure 5.10a,e,i, and m). They have similar structural features except possessing different percentages of intercalated Cr atoms. The CVD-grown $CrTe_2$ has a 1T layered structure with AA stacking order revealed by the cross-sectional STEM images (Figure 5.10b) [27]. The different thicknesses of CVD-grown samples can be controlled by precisely

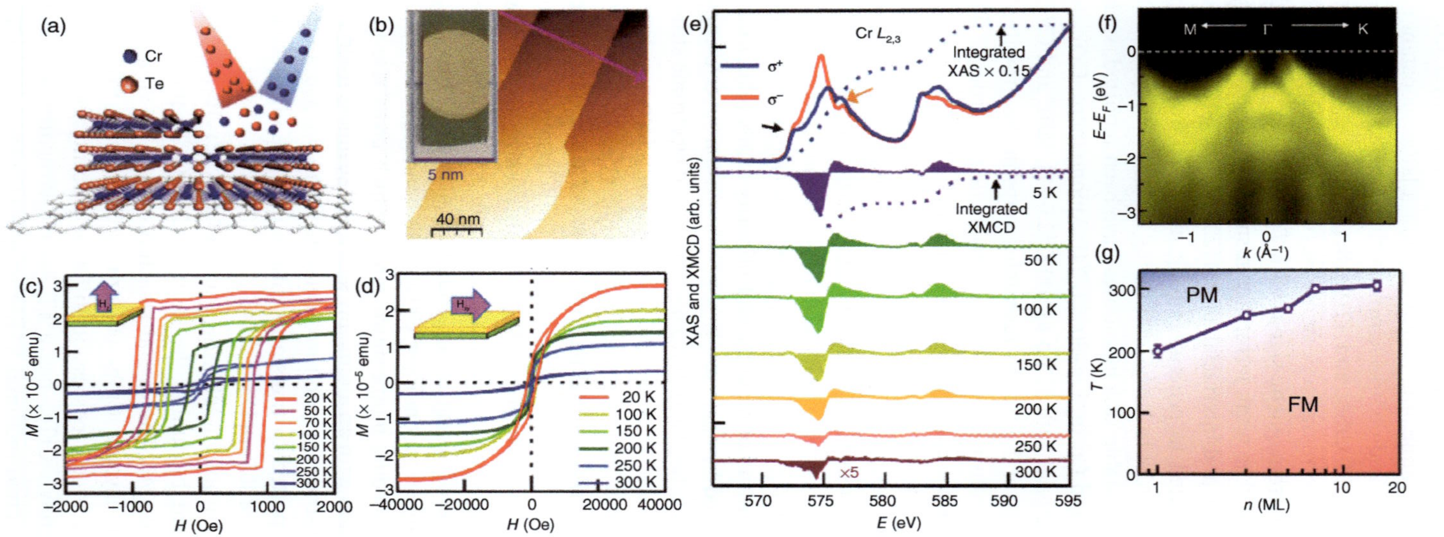

Figure 5.9 (a) Schematic of MBE growth of $CrTe_2$ films on graphene. Source: Zhang et al. [26]/Springer Nature/Licensed under CC BY 4.0. (b) Atomic-resolution STM image of $CrTe_2$ surface with a hexagonal structure. Source: Zhang et al. [26]/Springer Nature/Licensed under CC BY 4.0. Temperature-dependent superconducting quantum interface device (SQUID) measurements of 7 ML $CrTe_2$ with magnetic fields along the perpendicular (c), and parallel orientations (d). Source: Zhang et al. [26]/Springer Nature/Licensed under CC BY 4.0. (e) X-ray absorption spectroscopy (XAS) and XMCD spectra of 7 ML $CrTe_2$ measured from 5 to 300 K. Source: Zhang et al. [26]/Springer Nature/Licensed under CC BY 4.0. (f) Plot of valence-band dispersion of 7 ML $CrTe_2$ along high symmetry direction μ-Γ-K. Source: Zhang et al. [26]/Springer Nature/Licensed under CC BY 4.0. (g) Thickness-dependent T_C obtained from XMCD and SQUID measurements. Source: Zhang et al. [26]/Springer Nature/Licensed under CC BY 4.0. (h) Atomic-resolution STM image of 1ML $CrTe_2$. Source: Xian et al. [28]/Springer Nature/Licensed under CC BY 4.0 (i) Spin-polarized STM image of 1ML $CrTe_2$ at the magnetic field of +1T (i) and −1T (j), respectively. Source: Xian et al. [28]/Springer Nature/Licensed under CC BY 4.0 (k) DFT-calculated zigzag AFM order in 1ML $CrTe_2$. Source: Xian et al. [28]/Springer Nature/Licensed under CC BY 4.0 (l) In-plane $CrTe_2$-$CrTe_3$ metal-semiconductor heterostructure. Source: Yao et al. [69]/with permission from John Wiley & Sons. (m) High-resolution STM image of 1 ML $CrTe_3$. Source: Yao et al. [69]/with permission from John Wiley & Sons. (n) NC-AFM image of 1ML $CrTe_3$. Source: Yao et al. [69]/with permission from John Wiley & Sons. (o) dI/dV conductance maps taken across the $CrTe_2$/$CrTe_3$ heterojunction. Source: Yao et al. [69]/with permission from John Wiley & Sons.

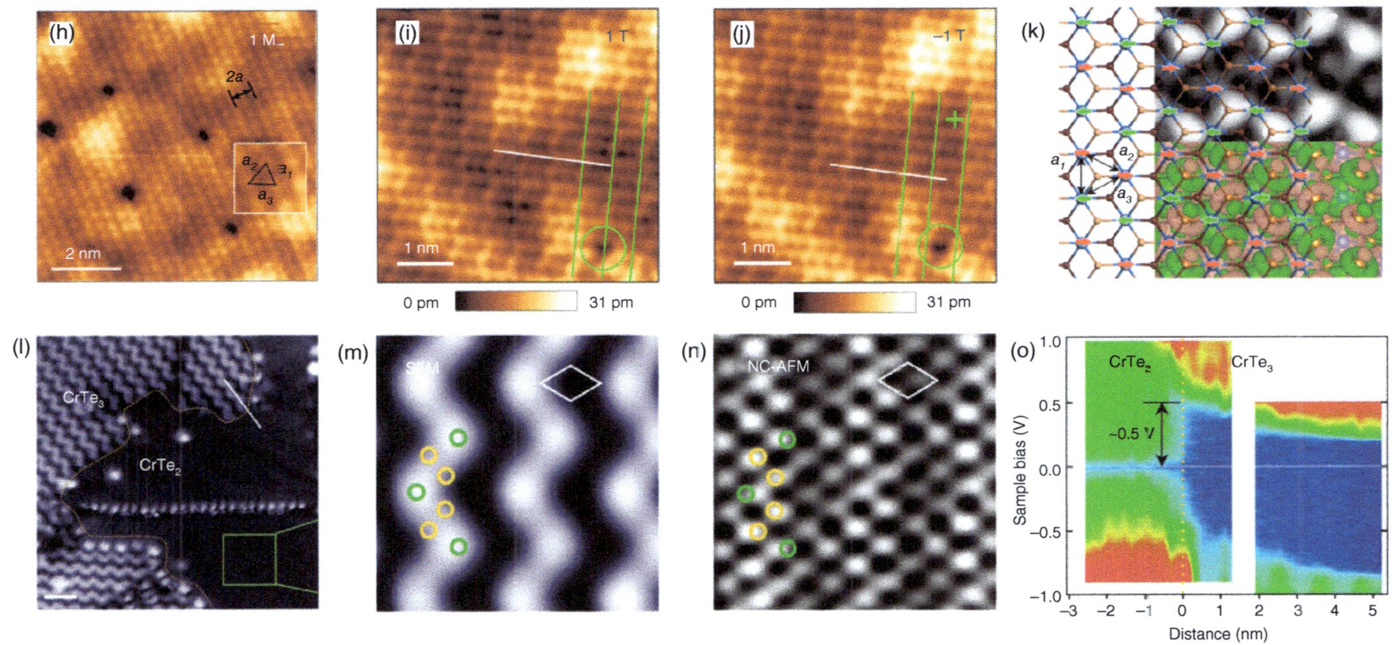

Figure 5.9 (*Continued*)

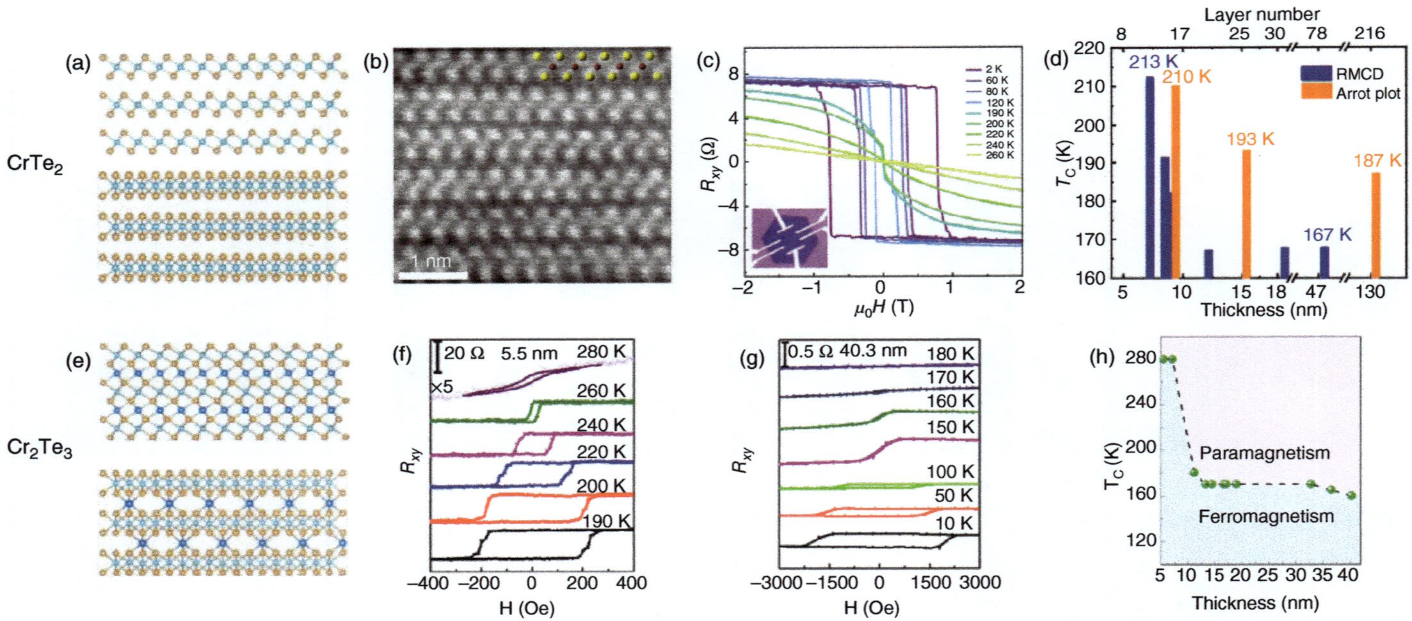

Figure 5.10 Atomic structures (side views) of $CrTe_2$ (a), Cr_2Te_3 (e), Cr_3Te_4 (i), Cr_5Te_8 (m), respectively. (b) Atomic-resolution cross-section STEM image of a CVD-grown $CrTe_2$ sample. (c) Out-of-plane Hall resistance hysteresis loops and magneto-resistance of a 10 nm 1T-$CrTe_2$ sample. (d) T_C of 1T-$CrTe_2$ as a function of thickness. *H*-dependent Hall electrical resistivity of Cr_2Te_3 with thickness of 5.5 nm (f) and 40.3 nm (g) at different temperatures. (h) Thickness-dependent T_c of Cr_2Te_3. (j) Atomic-resolution cross-section STEM image of CVD-grown Cr_3Te_4 sample (k) 2D magnetic component maps of Lorentz TEM image of Cr_3Te_4 nanosheet under 100 K. (l) Thickness-temperature phase diagram of Cr_3Te_4. (n) STEM image of Cr_5Te_8 nanosheet. (o) Hysteresis loops of Cr_5Te_8 nanosheets with different thicknesses at 50 K. (p) Thickness-temperature phase diagram of Cr_5Te_8. Source: (a, e, l, m–p) Chen et al. Yao et al. [30]/with permission from John Wiley & Sons. Meng et al. (b–d) [27]/Springer Nature/Licensed under CC BY 4.0. (f–h) Wen et al. [24]/American Chemical Society. (j, k) Li et al. [74]/with permission of ELSEVIER.

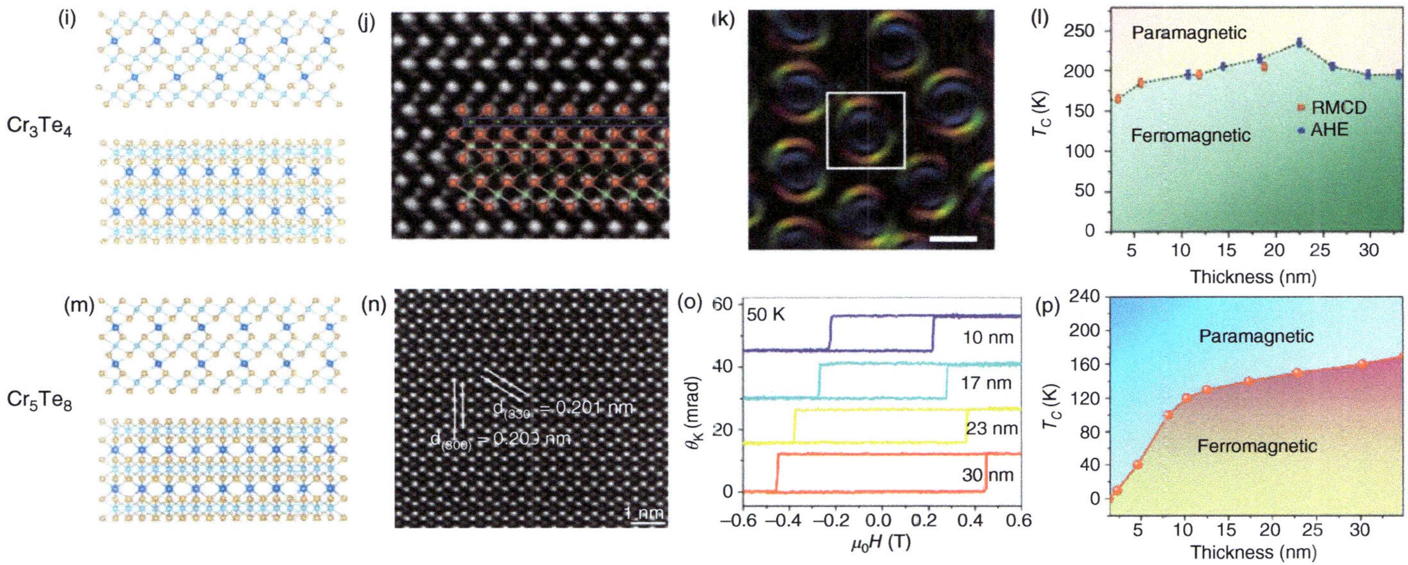

Figure 5.10 (*Continued*)

tuning the growth temperature. The thickness-dependent magnetic properties were revealed by the magneto-transport and RMCD measurements, whereby Curie temperature increases monotonically from ~187 to ~212 K with decreasing thickness from ~130.0 to ~7.6 nm, and the magnetic easy axis changes from in-plane to out-of-plane at a critical thickness of 10.0 nm (Figure 5.10c,d) [27]. The CVD-grown Cr_2Te_3 samples have similar thickness-dependent magnetic behaviors, whereby T_C increases from 160 to 280 K as thickness decreases from 40.3 to 5.5 nm (Figure 5.10f–h) [24]. The increases of T_c in $CrTe_2$ and Cr_2Te_3 thinner flakes are ascribed to the weakening of Coulomb screening and possible reconstruction in the 2D limit, respectively [24, 27]. Cr_3Te_4 can be viewed as 50% Cr atoms intercalated in $CrTe_2$ layers (Figure 5.10i). The intercalated Cr atoms located between two individual $CrTe_2$ layers can be directly observed in the cross-sectional STEM images (Figure 5.10j) [74]. Topologically nontrivial magnetic-spin states, biskymionic bubbles with the tunable size and density, can be realized in the Cr_3Te_4 nanosheets by tuning the external magnetic field and thickness of the nanosheets (Figure 5.10k) [74]. The magnetic phase diagram of Cr_3Te_4 nanosheets shows that T_C first gradually increases with increasing thickness and has a maximum value of 235 K at the thickness of 22.5 nm (Figure 5.10l). A tube-in-tube CVD growth strategy was used to ensure a Te-rich growth atmosphere during the synthesis of Cr_5Te_8 nanosheets. The MOKE measurements of Cr_5Te_8 nanosheets revealed that the coercive field (H_C) values decrease with decreasing thickness and T_C gradually increases with increasing thickness, reaching to 160 K at the thickness of 30 nm (Figure 5.10o,p) [30]. The polymorphic structures and superior environmental stability of chromium telluride bring great opportunities to explore the distinctive 2D magnetic properties and spintronic device applications.

5.4 Magnetism in 2D Vanadium, Ion, Manganese Chalcogenides

Monolayer VSe_2 is notable for its room-temperature ferromagnetism, which was first reported experimentally by Bonilla *et al.* [18] The room-temperature ferromagnetism in monolayer VSe_2 grown on HOPG and MoS_2 substrates by MBE method was revealed by using SQUID and vibrating sample magnetometry (VSM) [18]. There are some controversies about this work [20, 76]. The extremely high magnetic moments (15 μ_B per V atom) are much larger than the theoretically calculated result (~0.6 μ_B) [50, 79]. Furthermore, the observed easy-plane anisotropy can hardly sustain the strong ferromagnetism at room temperature [6, 18]. Subsequent angle-resolved photoemission (ARPES) and XMCD studies revealed the absence of long-range ferromagnetic order in MBE-grown monolayer VSe_2, as no exchange-split V electronic bands and ferromagnetically coupled V 3d state were observed (Figure 5.11a,b) [76]. Ferromagnetism in monolayer VSe_2 can be promoted through interfacial hybridization with an Co atomic overlayer. The emergence of magnetic polarization with a magnetic moment of ~0.4 μ_B per V atom in monolayer VSe_2 is confirmed by XMCD hysteresis loops (Figure 5.11c).

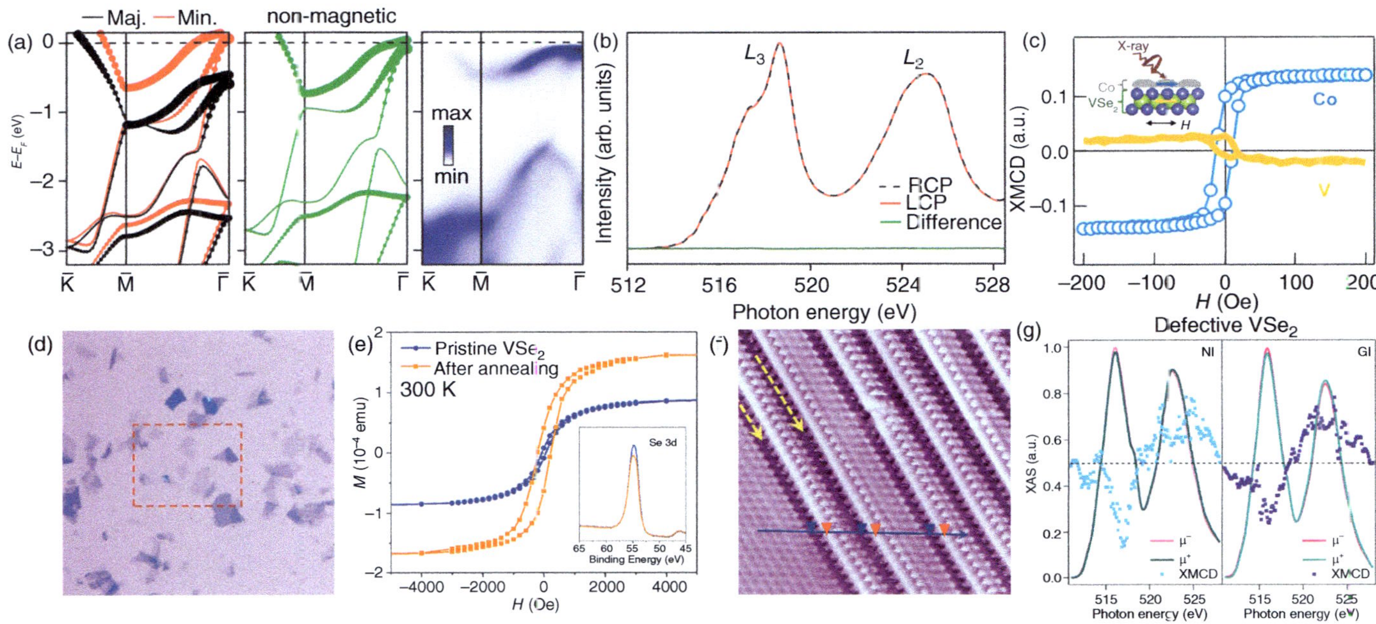

Figure 5.11 (a) DFT calculated and experimental electronic structures of monolayer VSe_2. Source: Feng et al. [76] Reproduced from *Nano Lett. 18*, 4493–4499 (2018). © 2018 American Chemical Society. (b) XAS and XMCD spectra of monolayer VSe_2 under a magnetic field of 9 T at 100 K. Source: Feng et al. [76]. (c) XMCD hysteresis loops of monolayer VSe_2 with deposited ~3 ML Co at 65 K. Source: Zhang et al. [77] Reproduced from *ACS Nano 13*, 8997–9004 (2019). © 2019 American Chemical Society. (d) Optical microscopy image of electrochemically exfoliated 1T-VSe_2 flakes on SiO_2 substrate. Source: Yu et al. [19]/with permission from John Wiley & Sons. (e) M–H hysteresis loops of pristine and thermally annealed VSe_2 samples. Source: Yu et al. [19]/with permission from John Wiley & Sons. (f) NC-AFM image of thermally annealed monolayer VSe_2. Source: Chua et al. [78]/with permission from John Wiley & Sons. (g) XMCD spectra of defective VSe_2 taken in normal and grazing incidence directions at room temperature. Source: Chua et al. [78]/with permission from John Wiley & Sons.

The magnetic transition in the Co/VSe_2 interface provides further proof for the absence of intrinsic ferromagnetic ordering in monolayer VSe_2 [77]. Coelho et al. performed DFT calculations by considering the CDW effect in monolayer VSe_2 and concluded that the existing CDW distortion can quench the intrinsic ferromagnetism in the monolayer limit [80]. Similar results that the CDW instability not only competes with but also suppresses the ferromagnetic ground state have been demonstrated in monolayer VTe_2 [47, 81, 82]. Monolayer VSe_2 flakes are extremely unstable in air, making it challenging to isolate monolayer flakes by mechanical exfoliation. Yu et al. reported monolayer VSe_2 obtained by an organic-solvent based electrochemical exfoliation approach, in which Se vacancies can enhance the ferromagnetism that is protected by thiol molecule layers (Figure 5.11d,e) [19]. In MBE-grown monolayer VSe_2 samples, 1D Se-deficient line defects can be initiated by thermal annealing above 400 °C which are determined to be 8-member-rings by non-contact atomic force microscopy (Figure 5.11f) [78]. The room-temperature ferromagnetism in 2D VSe_2 can be induced by the introduction of 1D Se-deficient line defects, which was confirmed by the XMCD measurements (Figure 5.11g).

Iron chalcogenides (FeX, X=S, Se, and Te) have attracted much attention for their polymorphic structures, superconductivity, and magnetism [83–85]. Different magnetic properties, tunable with the structural phases, have been observed in CVD-grown ultrathin FeTe nanoplates [31]. Controlled synthesis of 2D FeTe with the layered tetragonal and nonlayered hexagonal structures was realized by tuning the growth temperature (Figure 5.12a). The obvious structural difference between the tetragonal and hexagonal phases was revealed by atomic-resolution STEM, in which a P_{4g} wallpaper group symmetry and an in-plane sixfold symmetry can be directly observed in tetragonal- and hexagonal-shaped FeTe crystals, respectively (Figure 5.12b,c) [31]. Magneto transport measurements exhibited a clear change in resistance of tetragonal FeTe devices, which corresponds to a PM to AFM phase transition. The Néel temperature of antiferromagnetic tetragonal FeTe changes from 70 to 45 K as the thickness decreases from 32 to 5 nm (Figure 5.12d). Magnetoelectric and Hall measurements demonstrate that hexagonal FeTe has a ferromagnetic behavior with a thickness-dependent T_C dropping from 220 K (30 nm) to 170 K (4 nm) (Figure 5.12e). The observed ferromagnetism in hexagonal FeTe is calculated to originate from the structural distortion and Stoner instability [31].

Manganese chalcogenides have also been studied, as they have a maximum magnetic moment of 5 μ_B in their elemental form. 1T-$MnSe_2$ is theoretically predicted to be ferromagnetic with a T_C of about 250 K and $\mu = 3\ \mu_B$ [86, 87]. O'Hara et al. reported that MBE-grown monolayer $MnSe_x$ films on GaSe and $SnSe_2$ substrates have ferromagnetic ordering up to room temperature with a large saturation magnetization of ~4 μ_B per Mn atom [32]. As the thickness increases to ~40 nm, the magnetic moment is enhanced to be twice as that of the monolayer sample. The ferromagnetism in the $MnSe_x$ film was attributed to intrinsic ferromagnetism of a likely 1T-$MnSe_2$ monolayer in the ML limit and interfacial magnetism between 1T-$MnSe_2$ and α-MnSe in the thick film, which needs to be further explored. The growth

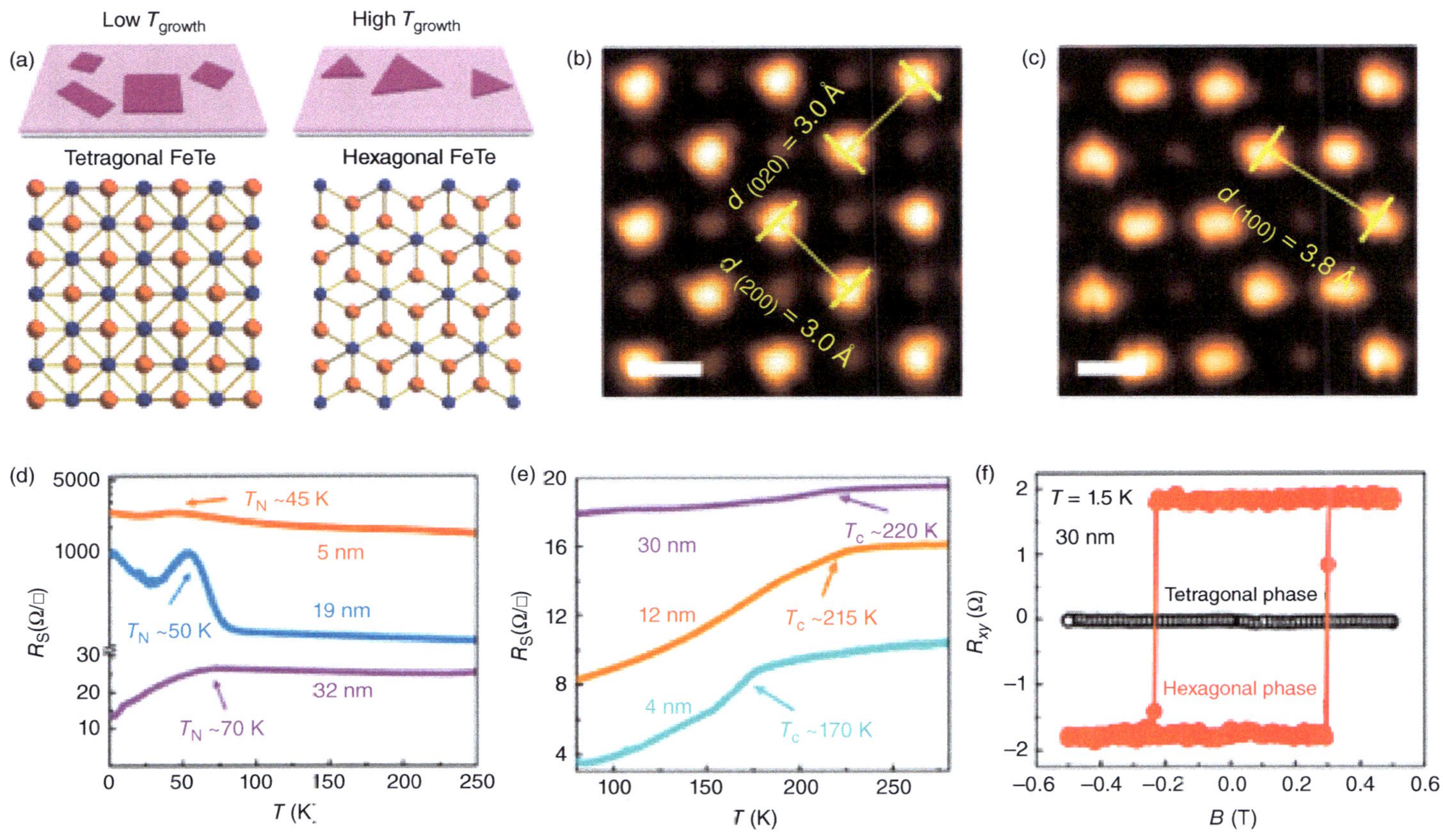

Figure 5.12 (a) Schematic view and crystal structures of tetragonal and hexagonal FeTe. Source: Reproduced with permission from Kang et al. [31] under the Creative Commons Attribution 4.0 International License. Atomic-resolution STEM image of tetragonal (b) and hexagonal (c) FeTe samples, respectively. Temperature-dependent longitudinal sheet resistance (R_S) of tetragonal (d) and hexagonal (e) FeTe Hall-bar devices with different thicknesses, respectively. (f) The magnetic field dependence of Hall resistance (R_{xy}) of tetragonal and hexagonal FeTe samples at 1.5 K. Source: (a–f) Kang et al. [31]/Springer Nature/Licensed under CC BY 4.0.

mechanism of manganese chalcogenides is not very clear, and different magnetic properties significantly affected by the structural phases need more efforts to explore. Besides the intrinsic magnetism in 2D materials, doping the magnetic atoms into 2D nonmagnetic materials is an alternative method to induce ferromagnetism in 2D materials [88–92].

5.5 Conclusions and Outlook

In summary, 2D magnetic materials have received much attention, and significant research progress has been made with an in-depth understanding of magnetism in 2D limit over the past few years. A variety of methods, including CVD, MBE, and CVT techniques, have been developed to fabricate high-quality 2D magnetic materials [17–24]. Van der Waals layered and non-layered magnetic materials have exhibited promising magneto-electric and magneto-optical applications, which offer opportunities for novel spintronic and magnetic storage devices. The study of 2D ferromagnets and novel magnetic phenomena in 2D materials is still in the preliminary stage, accompanied by both challenges, and opportunities. Exploring ferromagnetic semiconductors with high T_C, good environmental stability, and tunable electronic and magnetic properties remains a challenge. Most of the recently discovered 2D ferromagnets suffer from rapid degradation and dissolution through hydration or oxidation in ambient conditions. The poor stability of air to some extent restricts its property characterization and device applications. The relatively low Curie temperature also prevents the potential spintronic applications of 2D magnetic materials. The synthesis of air-stable magnetic semiconductors with T_C higher than room temperature is desired for spintronic applications.

Tuning 2D magnetism by external perturbations and exploring exotic magnetic phenomena in the heterostructures provide exciting opportunities in field of 2D materials. The atomically thick 2D magnets make it possible for integration with other van der Waals materials, which can produce unprecedented properties and exotic functionalities. Exotic spin textures and topologically protected spin configurations in 2D magnets remain open for further investigation, which could lead to important advances in quantum computation and communication. The understanding and control of 2D magnets will open possibilities for numerous applications ranging from flexible spintronics to quantum communication.

Acknowledgements

This work was supported by Guangdong Basic and Applied Basic Research Foundation (2022A1515111001), "The Pearl River Talent Recruitment Program" (2019ZT08X639), National Natural Science Foundation of China (Grant No. 62104073), China Postdoctoral Science Foundation (2021M691088) and Guangdong Provincial Key Laboratory of Chip and Integration Technology, Guangzhou 510631, P.R. China.

References

1 Mermin, N.D. and Wagner, H. (1966). Absence of ferromagnetism or Antiferromagnetism in one- or two-dimensional isotropic Heisenberg models. *Phys. Rev. Lett.* 17: 1133–1136.

2 Gibertini, M., Koperski, M., Morpurgo, A.F., and Novoselov, K.S. (2019). Magnetic 2D materials and heterostructures. *Nat. Nanotechnol.* 14: 408–419.

3 Huang, B., Clark, G., Navarro-Moratalla, E. et al. (2017). Layer-dependent ferromagnetism in a van der Waals crystal down to the monolayer limit. *Nature* 546: 270–273.

4 Gong, C., Li, L., Li, Z. et al. (2017). Discovery of intrinsic ferromagnetism in two-dimensional van der Waals crystals. *Nature* 546: 265–269.

5 Burch, K.S., Mandrus, D., and Park, J.G. (2018). Magnetism in two-dimensional van der Waals materials. *Nature* 563: 47–52.

6 Gong, C. and Zhang, X. (2019). Two-dimensional magnetic crystals and emergent heterostructure devices. *Science* 363: 6428.

7 Zhang, W., Wong, P.K.J., Zhu, R., and Wee, A.T.S. (2019). Van der Waals magnets: wonder building blocks for two-dimensional spintronics? *InfoMat* 1: 479–495.

8 Huang, Y.L., Chen, W., and Wee, A.T.S. (2021). Two-dimensional magnetic transition metal chalcogenides. *SmartMat* 2: 139–153.

9 Soriano, D., Katsnelson, M.I., and Fernández-Rossier, J. (2020). Magnetic two-dimensional chromium Thihalides: a theoretical perspective. *Nano Lett.* 20: 6225–6234.

10 Chen, W., Sun, Z., Wang, Z. et al. (2019). Direct observation of van der Waals stacking–dependent interlayer magnetism. *Science* 366: 983–987.

11 Zhang, Z., Shang, J., Jiang, C. et al. (2019). Direct photoluminescence probing of ferromagnetism in monolayer two-dimensional $CrBr_3$. *Nano Lett.* 19: 3138–3142.

12 Bedoya-Pinto, A., Ji, J.-R., Pandeya, A. et al. (2021). Intrinsic 2D-XY ferromagnetism in a van der Waals monolayer. *Science* 374: 616–620.

13 Cai, X., Song, T., Wilson, N.P. et al. (2019). Atomically thin $CrCl_3$: an in-plane layered antiferromagnetic insulator. *Nano Lett.* 19: 3993–3998.

14 Son, J., Son, S., Park, P. et al. (2021). Air-stable and layer-dependent ferromagnetism in atomically thin van der Waals $CrPS_4$. *ACS Nano* 15: 16904–16912.

15 Deng, Y., Yu, Y., Song, Y. et al. (2018). Gatetunable room-temperature ferromagnetism in two-dimensional Fe_3GeTe_2. *Nature* 563: 94–99.

16 Lee, J.U., Lee, S., Ryoo, J.H. et al. (2016). Ising-type magnetic ordering in atomically thin $FePS_3$. *Nano Lett.* 16: 7433–7438.

17 Deng, Y., Yu, Y., Shi, M.Z. et al. (2020). Quantum anomalous hall effect in intrinsic magnetic topological insulator $MnBi_2Te_4$. *Science* 367: 895–900.

18 Bonilla, M., Kolekar, S., Ma, Y. et al. (2018). Strong room-temperature ferromagnetism in VSe_2 monolayers on van der Waals substrates. *Nat. Nanotechnol.* 13: 289–293.

19 Yu, W., Li, J., Herng, T.S. et al. (2019). Chemically exfoliated VSe_2 monolayers with room-temperature ferromagnetism. *Adv. Mater.* 31: 1903779.

20 Wong, P.K.J., Zhang, W., Bussolotti, F. et al. (2019). Evidence of spin frustration in vanadium diselenide monolayer magnet. *Adv. Mater.* 31: 1901185.

21 Zhang, Y., Chu, J., Yin, L. et al. (2019). Ultrathin magnetic 2D single-crystal CrSe. *Adv. Mater.* 31: 1900056.

22 Li, B., Wan, Z., Wang, C. et al. (2021). Van der Waals epitaxial growth of air-stable $CrSe_2$ nanosheets with thickness-tunable magnetic order. *Nat. Mater.* 20: 818–825.

23 Liu, M., Huang, Y.L., Gou, J. et al. (2021). Diverse structures and magnetic properties in nonlayered monolayer chromium selenide. *J. Phys. Chem. Lett.* 12: 7752–7760.

24 Wen, Y., Liu, Z., Zhang, Y. et al. (2020). Tunable room-temperature ferromagnetism in two-dimensional Cr_2Te_3. *Nano Lett.* 20: 3130–3139.

25 Sun, X., Li, W., Wang, X. et al. (2020). Room temperature ferromagnetism in ultra-thin van der Waals crystals of 1T-$CrTe_2$. *Nano Res.* 13: 3358–3363.

26 Zhang, X., Lu, Q., Liu, W. et al. (2021). Room-temperature intrinsic ferromagnetism in epitaxial $CrTe_2$ ultrathin films. *Nat. Commun.* 12: 2492.

27 Meng, L., Zhou, Z., Xu, M. et al. (2021). Anomalous thickness dependence of curie temperature in air-stable two-dimensional ferromagnetic 1T-$CrTe_2$ grown by chemical vapor deposition. *Nat. Commun.* 12: 1–8.

28 Xian, J., Wang, C., Nie, J. et al. (2022). Spin mapping of intralayer antiferromagnetism and field-induced spin reorientation in monolayer $CrTe_2$. *Nat. Commun.* 13: 257.

29 Chua, R., Zhou, J., Yu, X. et al. (2021). Room temperature ferromagnetism of monolayer chromium telluride with perpendicular magnetic anisotropy. *Adv. Mater.* 33: 2103360.

30 Chen, C., Chen, X., Wu, C. et al. (2022). Air-stable 2D Cr_5Te_8 nanosheets with thickness-tunable ferromagnetism. *Adv. Mater.* 34: 2107512.

31 Kang, L., Ye, C., Zhao, X. et al. (2020). Phase-controllable growth of ultrathin 2D magnetic FeTe crystals. *Nat. Commun.* 11: 3729.

32 O'Hara, D.J., Zhu, T., Trout, A.H. et al. (2018). Room temperature intrinsic ferromagnetism in epitaxial manganese selenide films in the monolayer limit. *Nano Lett.* 18: 3125–3131.

33 Huang, B., Clark, G., Klein, D.R. et al. (2018). Electrical control of 2D magnetism in bilayer CrI_3. *Nat. Nanotechnol.* 13: 544–548.

34 Jiang, S., Shan, J., and Mak, K.F. (2018). Electric-field switching of two-dimensional van der Waals magnets. *Nat. Mater.* 17: 406–410.

35 Song, T., Cai, X., Tu, M.W.-Y. et al. (2018). Giant tunneling magnetoresistance in spin-filter van der Waals heterostructures. *Science* 360: 1214–1218.

36 Sun, Z., Yi, Y., Song, T. et al. (2019). Giant nonreciprocal second-harmonic generation from antiferromagnetic bilayer CrI_3. *Nature* 572: 497–501.

37 Jiang, S., Li, L., Wang, Z. et al. (2018). Controlling magnetism in 2D CrI_3 by electrostatic doping. *Nat. Nanotechnol.* 13: 549–553.

38 Zhong, D., Seyler, K.L., Linpeng, X. et al. (2017). Van der |Waals engineering of ferromagnetic semiconductor heterostructures for spin and valleytronics. *Sci. Adv.* 3: e1603113.

39 Zhong, D., Seyler, K.L., Linpeng, X. et al. (2020). Layer-resolved magnetic proximity effect in van der Waals heterostructures. *Nat. Nanotechnol.* 15: 187–191.

40 Kezilebieke, S., Huda, M.N., Vaňo, V. et al. (2020). Topological superconductivity in a van der Waals Heterostructure. *Nature* 588: 424–428.

41 Kezilebieke, S., Silveira, O.J., Huda, M.N. et al. (2021). Electronic and magnetic characterization of epitaxial $CrBr_3$ monolayers on a superconducting substrate. *Adv. Mater.* 33: 2006850.

42 Qian, X., Liu, J., Fu, L., and Li, J. (2014). Quantum spin hall effect in two dimensional transition metal dichalcogenides. *Science* 346: 1344–1347.

43 Tang, S., Zhang, C., Wong, D. et al. (2017). Quantum spin hall state in monolayer 1T′-WTe_2. *Nat. Phys.* 13: 683–687.

44 Ugeda, M.M., Bradley, A.J., Zhang, Y. et al. (2016). Characterization of collective ground states in single-layer $NbSe_2$. *Nat. Phys.* 12: 92–97.

45 Xing, Y., Zhao, K., Shan, P. et al. (2017). Ising superconductivity and quantum phase transition in macro-size monolayer $NbSe_2$. *Nano Lett.* 17: 6802–6807.

46 Ma, L., Ye, C., Yu, Y. et al. (2016). A metallic mosaic phase and the origin of Mott-insulating state in 1T-TaS_2. *Nat. Commun.* 7: 10956.

47 Liu, M., Wu, C., Liu, Z. et al. (2020). Multimorphism and gap opening of charge-density-wave phases in monolayer VTe_2. *Nano Res.* 13: 1733–1738.

48 Chang, K., Liu, J.W., Lin, H.C. et al. (2016). Discovery of robust in-plane ferroelectricity in atomic-thick SnTe. *Science* 353: 274–278.

49 Yan, S., Qiao, W., He, X. et al. (2015). Enhancement of magnetism by structural phase transition in MoS_2. *Appl. Phys.Lett.* 106: 012408.

50 Li, F., Tu, K., and Chen, Z. (2014). Versatile electronic properties of VSe_2 bulk, few-layers, monolayer, nanoribbons, and nanotubes: a computational exploration. *J. Phys. Chem. C* 118: 21264–21274.

51 Cui, F.F., Zhao, X.X., Xu, J.J. et al. (2020). Controlled growth and thickness-dependent conduction-type transition of 2D Ferrimagnetic Cr_2S_3 semiconductors. *Adv. Mater.* 32: 1905896.

52 Jiang, X., Liu, Q., Xing, J. et al. (2021). Recent progress on 2D magnets: fundamental mechanism, structural design and modification. *Appl. Phys. Rev.* 8: 031305.

53 McGuire, M.A., Dixit, H., Cooper, V.R., and Sales, B.C. (2015). Coupling of crystal structure and magnetism in the layered, ferromagnetic insulator CrI_3. *Chem. Mater.* 27: 612.

54 Sivadas, N., Okamoto, S., Xu, X. et al. (2018). Stacking-dependent magnetism in bilayer CrI_3. *Nano Lett.* 18: 7658–7664.

55 Niu, B., Su, T., Francisco, B.A. et al. (2020). Coexistence of magnetic orders in two-dimensional magnet CrI_3. *Nano Lett.* 20: 553–558.

56 Liu, Z., Deng, L., and Peng, B. (2021). Ferromagnetic and ferroelectric two-dimensional materials for memory application. *Nano Res.* 14: 1802–1813.

57 Li, P., Wang, C., Zhang, J. et al. (2020). Single-layer CrI_3 grown by molecular beam epitaxy. *Sci. Bull.* 65: 1064–1071.

58 Tartaglia, T.A., Tang, J.N., Lado, J.L. et al. (2020). Accessing new magnetic regimes by tuning the ligand spin-orbit coupling in van der Waals magnets. *Sci. Adv.* 6, No. eabb9379.

59 Matsukura, F., Tokura, Y., and Ohno, H. (2015). Control of magnetism by electric fields. *Nat. Nanotechnol.* 10: 209–220.

60 Kubetzka, A., Bode, M., Pietzsch, O., and Wiesendanger, R. (2002). Spin-polarized scanning tunneling microscopy with antiferromagnetic probe tips. *Phys. Rev. Lett.* 88: 057201.

61 Wang, Z., Gibertini, M., Dumcenco, D. et al. (2019). Determining the phase diagram of atomically thin layered antiferromagnet $CrCl_3$. *Nat. Nanotechnol.* 14: 1116–1122.

62 Hao, Z., Li, H., Zhang, S. et al. (2018). Atomic scale electronic structure of the ferromagnetic semiconductor $Cr_2Ge_2Te_6$. *Sci. Bull.* 63: 825–830.

63 Wang, Z., Zhang, T., Ding, M. et al. (2018). Electric-field control of magnetism in a few-layered van der Waals ferromagnetic semiconductor. *Nat. Nanotechnol.* 13: 554–559.

64 Zhang, T., Wang, Y., Li, H. et al. (2019). Magnetism and optical anisotropy in van der Waals antiferromagnetic insulator CrOCl. *ACS Nano* 13: 11353–11362.

65 Zhang, M., Hu, Q., Hua, C. et al. (2021). Metamagnetic transitions in few-layer CrOCl controlled by magnetic anisotropy flipping. *arXiv:2108.02825* .

66 Miao, N., Xu, B., Zhu, L. et al. (2018). 2D intrinsic ferromagnets from van der Waals antiferromagnets. *J. Am. Chem. Soc.* 140: 2417–2420.

67 van Bruggen, C.F., Haange, R.J., Wiegers, G.A., and de Boer, D.K.G. (1980). $CrSe_2$, a new layered dichalcogenide. *Physica B+C* 99: 166–172.

68 Kobayashi, S., Katayama, N., Manjo, T. et al. (2019). Linear trimer formation with antiferromagnetic ordering in 1T-$CrSe_2$ originating from peierls-like instabilities and interlayer se–se interactions. *Inorg. Chem.* 58: 14304–14315.

69 Yao, J., Wang, H., Yuan, B. et al. (2022). Ultrathin Van der Waals antiferromagnet $CrTe_3$ for fabrication of in-plane $CrTe_3/CrTe_2$ monolayer magnetic heterostructures. *Adv. Mater.* 34: 2200236.

70 Huang, X., Guan, J., Lin, Z. et al. (2017). Epitaxial growth and band structure of Te film on graphene. *Nano Lett.* 17: 4619–4623.

71 Zhao, D., Zhang, L., Malik, I.A. et al. (2018). Observation of unconventional anomalous hall effect in epitaxial CrTe thin films. *Nano Res.* 11: 3116–3121.

72 Roy, A., Guchhait, S., Dey, R. et al. (2015). Perpendicular magnetic anisotropy and spin glass-like behavior in molecular beam epitaxy grown chromium telluride thin films. *ACS Nano* 9: 3772–3779.

73 Li, H., Wang, L., Chen, J. et al. (2019). Molecular beam epitaxy grown Cr_2Te_3 thin films with tunable curie temperatures for spintronic devices. *ACS Appl. Nano Mater.* 2: 6809–6817.

74 Li, B., Deng, X., Shu, W. et al. (2022). Air-stable ultrathin Cr_3Te_4 nanosheets with thickness-dependent magnetic biskyrmions. *Mater. Today* 57: 66–74.

75 Lasek, K., Coelho, P.M., Zberecki, K. et al. (2020). Molecular beam epitaxy of transition metal (Ti-, V-, and Cr-) tellurides: from monolayer ditellurides to multilayer self-intercalation compounds. *ACS Nano* 14: 8473–8484.

76 Feng, J., Biswas, D., Rajan, A. et al. (2018). Electronic structure and enhanced charge-density wave order of monolayer VSe_2. *Nano Lett.* 18: 4493–4499.

77 Zhang, W., Zhang, L., Wong, P.K.J. et al. (2019). Magnetic transition in monolayer VSe_2 *via* Interface hybridization. *ACS Nano* 13: 8997–9004.

78 Chua, R., Yang, J., He, X. et al. (2020). Can reconstructed se-deficient line defects in monolayer VSe_2 induce magnetism? *Adv. Mater.* 32: 2000693.

79 Ma, Y., Dai, Y., Guo, M. et al. (2012). Evidence of the existence of magnetism in pristine VX2 monolayers (X = S, se) and their strain-induced tunable magnetic properties. *ACS Nano* (6): 1695–1701.

80 Coelho, P.M., Nguyen Cong, K.N., Bonilla, M. et al. (2019). Charge density wave state suppresses ferromagnetic ordering in VSe_2 monolayers. *J. Phys. Chem. C* 123: 14089–14096.

81 Coelho, P.M., Lasek, K., Cong, K.N. et al. (2019). Monolayer modification of VTe_2 and its charge density wave. *J. Phys. Chem. Lett.* 10: 4987–4993.

82 Wong, P.K.J., Zhang, W., Zhou, J. et al. (2019). Evidence for metallic 1T phase, 3d1 electronic configuration and charge density wave order in molecular-beam epitaxy grown monolayer VTe_2. *ACS Nano* 13: 12894–12900.

83 Guo, J., Jin, S., Wang, G. et al. (2010). Superconductivity in the Iron selenide $K_xFe_2Se_2$ $(0 \leq x \leq 1.0)$. *Phys. Rev. B* 82: 180520.

84 Liu, H. and Xue, Y. (2021). Van Der Waals epitaxial growth and phase transition of layered $FeSe_2$ nanocrystals. *Adv. Mater.* 33: 2008456.

85 Chen, S., Liu, H., Chen, F. et al. (2020). Synthesis, transfer, and properties of layered $FeTe_2$ nanocrystals. *ACS Nano* 14: 11473–11481.

86 Ataca, C., Sahin, H., and Ciraci, S. (2012). Stable, single-layer MX_2 transition-metal oxides and dichalcogenides in a honeycomb-like structure. *J. Phys. Chem. C* 116: 8983–8999.

87 Kan, M., Adhikari, S., and Sun, Q. (2014). Ferromagnetism in MnX2 (X = S, se) monolayers. *Phys. Chem. Chem. Phys.* 16: 4990–4994.

88 Yun, S.J., Duong, D.L., Ha, D.M. et al. (2020). Ferromagnetic order at room temperature in monolayer WSe_2 semiconductor via vanadium dopant. *Adv. Sci.* 7: 1903076.

89 Duan, H., Guo, P., Wang, C. et al. (2019). Beating the exclusion rule against the coexistence of robust luminescence and ferromagnetism in chalcogenide monolayers. *Nat. Commun.* 10: 1584.

90 Li, B., Xing, T., Zhong, M. et al. (1958). A two-dimensional Fe-doped SnS_2 magnetic semiconductor. *Nat. Commun.* 2017: 8.

91 Zhang, K., Feng, S., Wang, J. et al. (2015). Manganese doping of monolayer MoS_2: the substrate is critical. *Nano Lett.* 15: 6586–6591.

92 Yang, L., Wu, H., Zhang, L. et al. (2021). Tunable and robust near-room-temperature intrinsic ferromagnetism of a van der Waals layered Cr-doped 2H-$MoTe_2$ semiconductor with an out-of-plane anisotropy. *ACS Appl. Mater. Interfaces* 13: 31880–31890.

6

Recent Progress of Mechanical Exfoliation and the Application in the Study of 2D Materials

Yunyun Dai[1,2], Xinyu Huang[1,2], Xu Han[1,2], Jiangang Guo[3], Xiangfan Xu[4], Lei Wang[5], Luqi Liu[6], Ningning Song[7], Yeliang Wang[1,2], and Yuan Huang[1,2]

[1] *Beijing Institute of Technology, Advanced Research Institute of Multidisciplinary Science, Beijing, 100081, China*
[2] *Beijing Institute of Technology, School of Integrated Circuits and Electronics, MIIT Key Laboratory for Low-Dimensional Quantum Structure and Devices, Beijing, 100081, China*
[3] *Institute of Physics, Chinese Academy of Sciences, Beijing, 100190, China*
[4] *Tongji University, Center for Phononics and Thermal Energy Science, School of Physical Science and Engineering, Shanghai 200092, China*
[5] *Nanjing University, National Laboratory of Solid-State Microstructures, School of Physics, Nanjing 210093, China*
[6] *National Center for Nanoscience and Nanotechnology, Beijing 100190, China*
[7] *Beijing University of Chemical Technology, College of Mathematics and Physics, Beijing 100029, China*

6.1 Introduction

In the development of material sciences, a new type of carbon materials often plays the role of milestones. In 1985, the discovery of "football" structured C_{60} attracted worldwide attention immediately [1], which opened a new gate for zero-dimensional (0D) nanomaterials and nanotechnology. In 1991, a new one-dimensional (1D) carbon material, carbon nanotube (CNT), was observed by transmission electron microscope (TEM) [2]. Although the CNTs had been first synthesized in a carbon fiber generator in 1976 by M. Endo et al. [3, 4], the high-resolution atomic structure of CNT was still challenging to detect by TEM. After the discovery of C_{60} and CNTs, various 0D and 1D materials were synthesized by many physical or chemical methods, which makes nano sciences more and more promising beyond just a concept. Graphene was still a mysterious material at that time since most researchers in nano science field were focusing on 0D and 1D materials.

As early as 1930s, one of the most famous theoretical physicists in the Former Soviet Union, Lev Landau, demonstrated that 2D crystals cannot be stable in principle [5]. Before 2000, researchers in condensed matter physics mainly focused on bulk crystals, while 2D-related studies were still a small branch. However, some experimental scientists who always like to challenge inherent knowledge. At the end of the twentieth century, with the fast development of advanced instruments,

* Yunyun Dai and Xinyu Huang contributed equally.

Two-Dimensional Transition-Metal Dichalcogenides: Phase Engineering and Applications in Electronics and Optoelectronics, First Edition. Edited by Chi Sin Tang, Xinmao Yin, and Andrew T. S. Wee.

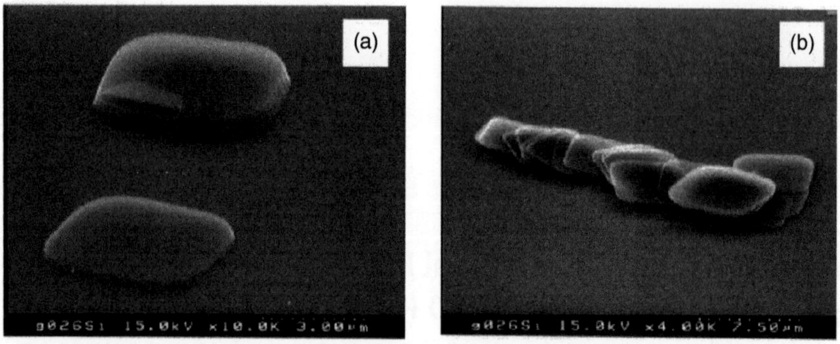

Figure 6.1 (a and b) Ruoff et al. from the University of Washington in the United States used AFM tip to push graphene away on a graphene column. Source: Lu et al. [6]/Reproduced with permission from IOP Publishing.

such as scanning tunneling microscope (STM) and atomic force microscope (AFM), some researchers started to think about how to obtain thinner flakes from bulk layered materials. In 1999, R. Ruoff's team created an array of small graphite islands by reactive ion etching of the HOPG surface with the oxygen plasma [6]. These islands were then manipulated by AFM tip and tailored into few layer flakes (Figure 6.1a,b). Limited by the fabrication method, monolayer graphene was not obtained by thinning-down graphite islands. However, this pioneering work has made a huge step toward 2D materials. Recently, nano-scale monolayer graphene islands were manipulated by STM tip and folded into bilayer graphene with an arbitrary twist angle, as reported by Hong-Jun Gao's team in 2019 [7].

In the year 2004, Andre Geim et al. reported the successful exfoliation of thin graphene flakes by scotch tape and fabricated field-effect transistor devices, which demonstrated that the electrical performance of thin graphene films can be tuned by electrical field [8]. The simple tape method, also known as ME, has been widely used for preparing high-quality monolayer and few layer 2D materials. Among many other techniques for preparing 2D materials, including chemical vapor deposition (CVD), molecular beam epitaxy (MBE), and liquid exfoliation methods, ME method shows unique advantages for its simplicity, low cost, and high quality. The 2D materials obtained by ME method can avoid liquid contamination, and the exfoliated 2DACs have been proven to be the most ideal samples for studying their intrinsic properties. Many interesting phenomena have been discovered on exfoliated 2D materials, such as the quantum Hall effect (QHE), mass-less Fermions in graphene [9, 10], and the indirect-to-direct band gap transition in MoS_2 [11]. This series of progresses indicate that ME method is critical to seize the opportunity in the study of 2D materials. In this review, we will introduce the development and challenge of ME method in the field of 2D materials. Besides, recent progress of some typical 2D materials is also summarized here, such as graphene, MoS_2 et al. We also present some applications of exfoliated 2D materials in basic research, including the mechanical and thermal properties. An outlook on the ME method in the study of 2D materials is addressed at the end of this review.

6.2 Different Ways for Preparing 2D Materials

After more than ten years of development, several preparation methods for obtaining 2D materials have been reported, which can be divided into two strategies: top-down and bottom-up. CVD and MBE techniques are typical bottom-up methods, which are involving the self-assembling of small atoms and molecules. The 2D materials grown by these methods often have strong interactions with the substrate underneath and many grain boundaries since the nuclear-sites are difficult to control; thus, the intrinsic properties of 2D materials are more or less affected by substrates. The ME and chemical exfoliation techniques are two typical top-down methods [12], which make bulk layered materials thin down to monolayer or few layer counterparts. We will briefly summarize some progress on the two strategies, respectively.

6.2.1 Chemical Vapor Deposition (CVD)

Although graphene has many fascinating properties, the small-size exfoliated graphene flakes cannot meet the requirements of large-area device integration. To get large-area graphene, researchers tried different metal substrates to prepare CVD graphene, such as Ru and Ni [13–15]. In 2010, X. S. Li et al. grew monolayer CVD graphene on Cu foil [16]. The CVD method is also widely used for preparing h-BN and other transition-metal dichalcogenides (TMCs, MoS_2, WS_2, WSe_2, et al.) [17–20]. By controlling the components of precursor, in-plane and interlayer heterostructures can be synthesized by CVD method [21, 22]. Even though 2D materials grown by CVD method have large sizes and facilitate integration, there are still many challenges in controlling the defects and quantity of nuclear sites [23, 24]. Besides, the CVD growth method for 2D materials shows strong substrate dependence, and many parameters cannot be well controlled in the growth process [25]. Therefore, it's hard to grow high-quality 2D materials using the CVD method on a large scale.

6.2.2 Mechanical Exfoliation (ME)

ME method is famous for the discovery of graphene. Thanks to the weak interlayer van der Waals (vdWs) interaction, many layered materials can be exfoliated into monolayer and transferred onto other substrate by scotch tape or some other organic films. After the first successful exfoliation of graphene by scotch tape in 2004, many TMD materials (MoS_2, WSe_2, et al.) and other layered materials (e.g. black phosphorous, Fe_3GeTe_2) were subserviently discovered by ME method. Many breakthroughs in 2D materials were unveiled on exfoliated samples, such as QHEs in graphene, band gap transitions in TMD monolayers. Even though the exfoliated 2D samples show high-performance in the intrinsic property study, some drawbacks are also obvious, like small size, low yield, and poor repeatability.

Researchers have tried different ways to improve the yield ratio and flake size of 2D materials prepared by exfoliation method. Jayasena et al. devised a lathe-like experimental setup to cleave HOPG samples for generating graphene flakes [26], but the thinnest resulting flakes typically have a thickness of tens of nanometers.

Coleman and co-workers first reported the high-yield production of graphene by sonication assisted liquid phase exfoliation of graphite [27], but the relatively small size of monolayer flakes, along with low quality, limits its use in both research and applications. Shukla et al. reported a different exfoliation approach based on anodic bonding of graphite to insulating (Pyrex, borosilicate glass) supports [28]. Moldt and coworkers later reported a detailed survey of the process parameter space in the anodic bonding process, and identified conditions for large-yield production of graphene flakes on glass [29]. After transferring the flakes to SiO_2/Si substrates, it is demonstrated that the carrier mobility is as high as 6000 $cm^2/(V\,s)$ in electrical measurements on back-gated devices. While capable of substantially increasing the yield and flake size over conventional exfoliation methods, anodic bonding suffers from several practical difficulties: the need to apply a high voltage (0.6–1.2 kV) between the support and a graphite crystal at high temperatures (~180–250 °C) and, most importantly, the limitation to certain classes of materials, e.g. electrically conducting layered crystals (ruling out important systems, such as h-BN) and substrates with relatively mobile ions, such as borosilicate glass. The latter not only makes the identification of thin graphene sections difficult due to low optical contrast but also complicates device fabrication (e.g. for back-gated transport measurements) by requiring additional transfer steps to other substrates. There were other reports of very large graphene flakes in the literature. For example, Geim showed a large (~0.9 × 0.5 mm) exfoliated graphene flake [30], but no explanation was provided about the way such flakes were obtained and the reproducibility of the method used.

Generally, the small size of exfoliated samples hindered the progress of the study of 2D materials. In recent years, more and more characterization instruments require high quality and large-area 2D samples, such as ultrared spectrum, and angle-resolved photoemission spectroscopy (ARPES). Therefore, optimizing the exfoliation method and getting large-area/high-quality sample is critical for promoting the development of 2D materials. In the past five years, we have been devoted to exploring new exfoliation methods, and have highly improved the yield ratio and sizes of monolayer and bilayer samples, such as graphene, $Bi_2Sr_2CaCu_2O_x$, MoS_2, WSe_2, et al. We will briefly summarize the oxygen-plasma-enhanced method and Au-film-enhanced exfoliation methods below.

6.3 New Mechanical Exfoliation Methods

6.3.1 Oxygen Plasma Enhanced Exfoliation

As the first discovered 2D material, graphene shows many unique properties, such as high mobility, high Young's modulus, and transmittance. Among different strategies for preparing graphene, bottom-up methods, like CVD and MBE, have been widely studied. However, the top-down methods for preparing graphene and other 2D materials were rarely reported. One reason is the small size of bulk crystals (around cm scale), which makes the exfoliated 2D crystals too small to fabricate integrated devices. Another main reason may be attributed to the simplicity of top-down methods, such as fewer processes and few adjustable parameters. In other words, there are not too many parameters that can be tuned to improve the yield ratio and 2D flake size.

The interaction between substrate and layered crystal is the main factor affecting the yield ratio and flake size of mechanical exfoliation, which shows huge differences depending on various layered crystals and different substrates. For example, graphene has a strong interaction with the oxide substrates, such as SiO_2, Al_2O_3, etc., because the C atoms have a high affinity with O atoms. However, the interaction between most of the metal substrates is relatively weak, which is not suitable for exfoliating graphene. In 2015, Huang and Sutter et al. reported a modified exfoliation method that can be effectively used for the exfoliation of graphene and copper based superconductors (e.g. Bi2212) [31]. It is found that oxide substrates are helpful to get large size monolayers, such as quartz, sapphire, and SiO_2/Si. Graphene is usually exfoliated onto a Si wafer with one layer of SiO_2. The contrast of monolayer graphene depends on the thickness of the SiO_2 layer. Both theoretical calculations and experimental results demonstrate that the 90 and 300 nm SiO_2 layers show high contrast for graphene [32].

Figure 6.2 shows the exfoliation process. SiO_2/Si was used as the substrate and ordinary adhesive tape as the transfer medium. Contact with the tape and transfer thick graphite flakes from a highly oriented pyrolitic graphite (HOPG) or natural graphite crystal (Figure 6.2).

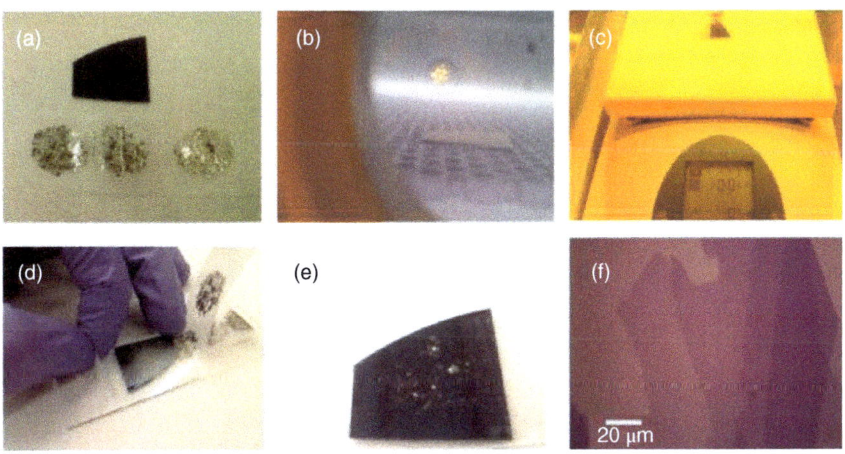

Figure 6.2 Illustration of the modified exfoliation process for layered crystals (shown here for graphene): (a) Optical image of the SiO_2/Si substrate and adhesive tape with graphite flakes, which adhere after contact with a HOPG crystal; (b) Oxygen plasma cleaning of the SiO_2/Si substrate; (c) Contact between the graphite decorated tape and the substrate surface, followed by heating of the substrate (with tape) on a hot plate in air at ~100 °C for 2 minutes; (d) Removal of the substrate from the hot plate and peeling off of the tape; (e) Optical image of the substrate after graphene exfoliation; (f) Optical micrograph of one of the graphene flakes on the substrate in (e). The flake has a thickness varying in steps between 14 layers. Source: Yuan et al. [31]/Reproduced from American Chemical Society.

Actually, except for the interaction between the substrate and graphite surface, another interaction exists in the exfoliating process, which is generated between the graphite layers. Both interactions are vdWs forces, and the competition between them plays the most important role in the results of mechanical exfoliation. Only when the interaction between substrate and graphite surface is stronger than that between graphite layers, the graphite layers can be transferred to substrate. To increase the interaction between substrate and graphite surface, a modified exfoliation method has been used to prepare a high yield ratio and large size of graphite layers. As a typical modified exfoliation method, prior to exfoliating thin graphene from these flakes, the SiO_2/Si substrate is ultrasonically cleaned in acetone, 2-propanol, and deionized (DI) water, and then subjected to oxygen plasma to remove ambient adsorbates from its surface (Figure 6.2b). Following the plasma cleaning step, the graphite-loaded tape is brought in contact with the substrate. Instead of immediately removing it to complete the exfoliation, the substrate with the attached tape is annealed for 2–5 minutes at ~100 °C in air on a conventional laboratory hot plate (Figure 6.2c). After the sample is cooled to room temperature, the adhesive tape is removed (Figure 6.2d), which completes the exfoliation. Inspection by optical microscopy shows the successful transfer of a few-layer graphene to the SiO_2/Si substrate. The linear dimensions of areas with uniform thickness are routinely in the range from 20 μm to well above 100 μm, and the thin (few-layer) sections of the flakes are up to several hundreds of micrometers in size (Figure 6.2f).

Figure 6.3 shows optical micrographs comparing graphene flakes prepared by the standard exfoliation method and the modified approach. The size of the monolayer

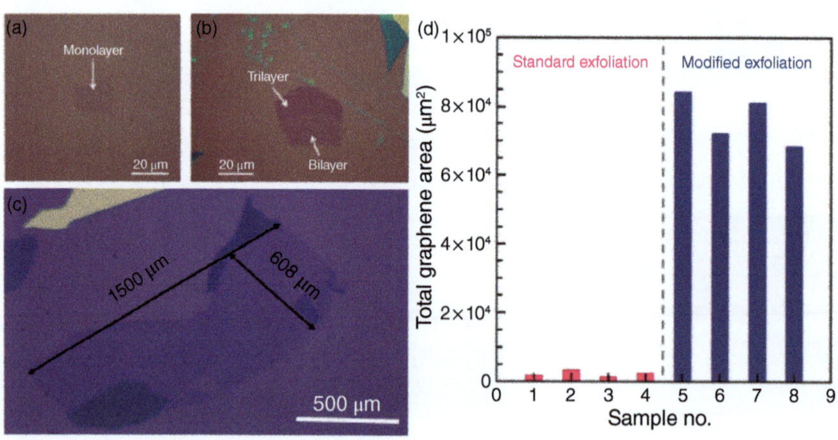

Figure 6.3 Optical images of graphene flakes prepared by the standard exfoliation method and our modified method: (a and b) Optical microscopy images of typical monolayer to trilayer graphene prepared by the standard method, including a solvent wash and O_2 plasma cleaning of the substrate followed by graphene transfer; (c) Optical microscopy images of millimeter size graphene flake prepared by the new modified method, with O_2 plasma clean of the SiO_2/Si surface, followed by contact with graphite-loaded tape, annealing to 100 °C, cooling to room temperature, and peel-off; (d) Total area of exfoliated graphene on eight different samples. Samples 1–4 were prepared by the standard exfoliation process; samples 5–8 were prepared by the method introduced here. The size of all substrates was 2×2 cm^2. Source: Yuan et al. [31]/Reproduced from American Chemical Society.

sample exfoliated by the standard method is around $20 \times 20\,\mu m^2$, a typical large size prepared by this approach. Previous reports show that the size of graphene flakes prepared by this approach is normally smaller than $\sim 20 \times 20\,\mu m^2$ [9, 10]. Figure 6.3 is an optical image of a bilayer and trilayer flake with a size of $\sim 40 \times 40\,\mu m^2$, slightly larger than the previously reported bilayer and trilayer graphene devices [33]. Figure 6.3c shows typical graphene flakes prepared by modified exfoliation method. The flake size in Figure 6.3c is $\sim 1500 \times 600\,\mu m^2$, more than 2250 times larger than that of the flake shown in Figure 6.3a. These results show that it is easier for the modified exfoliation methods to get larger size monolayer or few layer 2D samples.

The effect of pre-heating time and temperature on the 2D samples prepared by ME method has also been investigated. No obvious or orderly size-changing has been observed in the graphene flakes with increasing heating time from 2 to 30 minutes or temperature from 80 to 140 °C. The samples prepared by modified exfoliation method show higher production than that of standard exfoliation method. The highest sample production shows when the heating temperature is 100 °C and heating time is 2 minutes, which is 60% higher than the lowest sample production heated at 140 °C for 2 minutes. With further increasing heating time or temperature, excess glue would remain on the surface of the substrate. This means the best condition to prepare graphene flakes of large size by the modified exfoliation method is to heat the substrate at 100 °C for 2 minutes before standard exfoliation.

Except for various layered crystals, different substrates, pre-treat temperature and exfoliating time are other important factors affecting the yield ratio and flake size of mechanical exfoliation. It is important to know that the graphite flakes are exfoliated from the graphite crystal but not the adhesive tape. Therefore, it is hard to get graphene of large size after too many times of exfoliation. Normally, three to four exfoliations by adhesive tape from bulk graphite are enough; after that, the graphite crystal adhered on tape is thick enough before contacting to the substrate.

Apart from graphite, $Bi_2Sr_2CaCu_2O_x$ (Bi2212) as one of the layered superconducting materials has also been successfully prepared by our modified exfoliation method (Figure 6.4) [31]. The size of as-exfoliated Bi2212 layer is around 500 µm

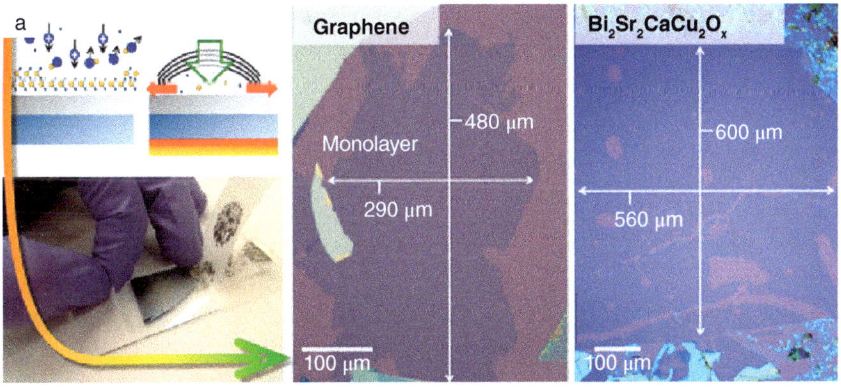

Figure 6.4 Monolayer graphene and Bi2212 samples prepared by an oxygen plasma-assisted exfoliation method. Source: Yuan et al. [31]/Reproduced from American Chemical Society.

and the yield ratio is extremely high. Compared to the reported Bi2212 layer with the size of only a few microns prepared by the standard exfoliation method in 2005 [34], this result is more exciting. In 2019, Yuanbo Zhang et al. successfully exfoliated large size Bi2212 by combing oxygen plasma enhanced exfoliation method and low-temperature preparation condition [35], and found superconducting properties in the as-prepared single-layer Bi2212, which is consistent with that of bulk Bi2212.

6.3.2 Gold Film Enhanced Exfoliation

Although oxygen plasma enhanced exfoliation has proven useful for preparing graphite and Bi2212 layer, Y. Huang et al. found that this method does not work well for TMDs. In 2015, the Tapasztó group successfully prepared single-layer MoS_2 and WSe_2 with large size by evaporation of 100 nm gold film on bulk MoS_2 and WSe_2. But in the following gold film exfoliating step, both the gold film and single-layer TMDs would be heavily damaged. A new exfoliation method called gold film enhanced exfoliation method has been reported in the following investigation to prepare 2D TMDs materials. In this method, gold film is evaporated first on TMDs, then transformed to another substrate combing with heat release tape, and removed by iodine solution at last [36]. This method is very useful for preparing TMDs materials, but no other layered crystal system has been reported by this method. Nevertheless, the detailed mechanism of this method is still unclear.

In 2020, Y. Huang et al. reported a universally applicable exfoliation method [37], which has triggered great attention in the studies of 2D materials and their physics properties (Figure 6.5). In this work, the interaction between different substrates and 58 kinds of layered crystals have been calculated. According to the rules of interaction between different elements in the periodic table, it is found that the interaction between substrate and the outermost element in the layered crystal plays the most important role in exfoliation. The gold film enhanced exfoliation proved useful to prepare layered crystals containing outermost elements in V A, VI A or VII A main groups. More than 40 kinds of large size 2D material have been successfully exfoliated by the gold film enhanced exfoliation. The size of the monolayer sample with a single crystal structure is over a millimeter magnitude, and the as-prepared monolayer material can be further stacked to heterostructure.

Theoretical calculations show that quasi-covalent bond can be formed between gold and many kinds of 2D materials, and this interaction is much stronger than vdWs force between layers in 2D materials [37]. That's why a large, single-layer sample can be prepared by the gold film enhanced exfoliation method. What's more, the gold film enhanced exfoliation method can realize multifaceted regulation of 2D materials. Firstly, the folded film used in the gold film enhanced exfoliation method does not need to be continuous, and suspended samples can be efficiently prepared, which provides an ideal system for investigating intrinsic optical and transport properties. Secondly, the electrical properties of the substrate can be tuned from insulation to electrical conductivity to satisfy different experimental requirements. For example, for STM and ARPES measurements, the 2D sample can be exfoliated directly onto a conductive substrate by increasing the thickness of gold film. While

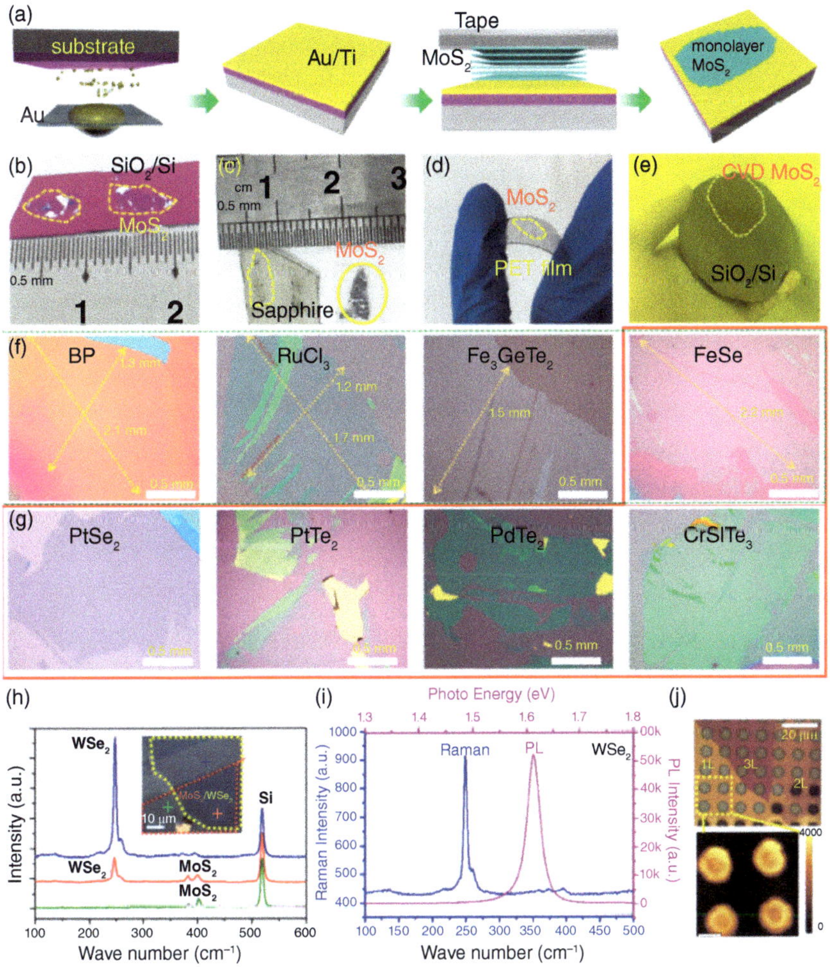

Figure 6.5 Mechanical exfoliation of different monolayer materials with macroscopic size. (a) Illustration of the modified exfoliation process; (b–e) Optical images of large-area MoS_2 exfoliated on different substrates; (h–j) Raman and PL spectra of heterostructure and suspended 2D material. Source: Huang et al. [37]/Reproduced from Springer Nature/

for fluorescence spectra and electrical measurements, an insulating metal island can be obtained when the thickness of gold film is thinner than 3 nm. FeSe, $PtSe_2$, $PdTe_2$, and $CrSiTe_3$ have been exfoliated by the gold film enhanced exfoliation method for the first time, which provides a new approach to exploit novel materials properties. This gold film enhanced exfoliation method can also be used to exfoliate single-layer material of large size on transparent or flexible substrates, which opens a new avenue to exploit further optical research and flexible devices.

Though the recently discovered Au-enhanced exfoliation technique provides an effective strategy for preparation of large-scale 2D crystals, the high cost of gold hinders this method from being widely adopted in industrial applications. In addition, direct Au contact can significantly quench photoluminescence (PL) emission in 2D semiconductors. It is therefore crucial to find alternative metals that can replace gold to achieve efficient exfoliation of 2D materials. In 2022, Y. Huang et al. presented a one-step Ag-assisted method that can efficiently exfoliate many large-area 2D monolayers [38], where the yield ratio is comparable to Au-enhanced exfoliation method (Figure 6.6). In contrast to Au film, however, the surface roughness of as-prepared Ag films on SiO_2/Si substrate is much higher, which facilitates the generation of surface plasmons resulting from the nanostructures formed on the

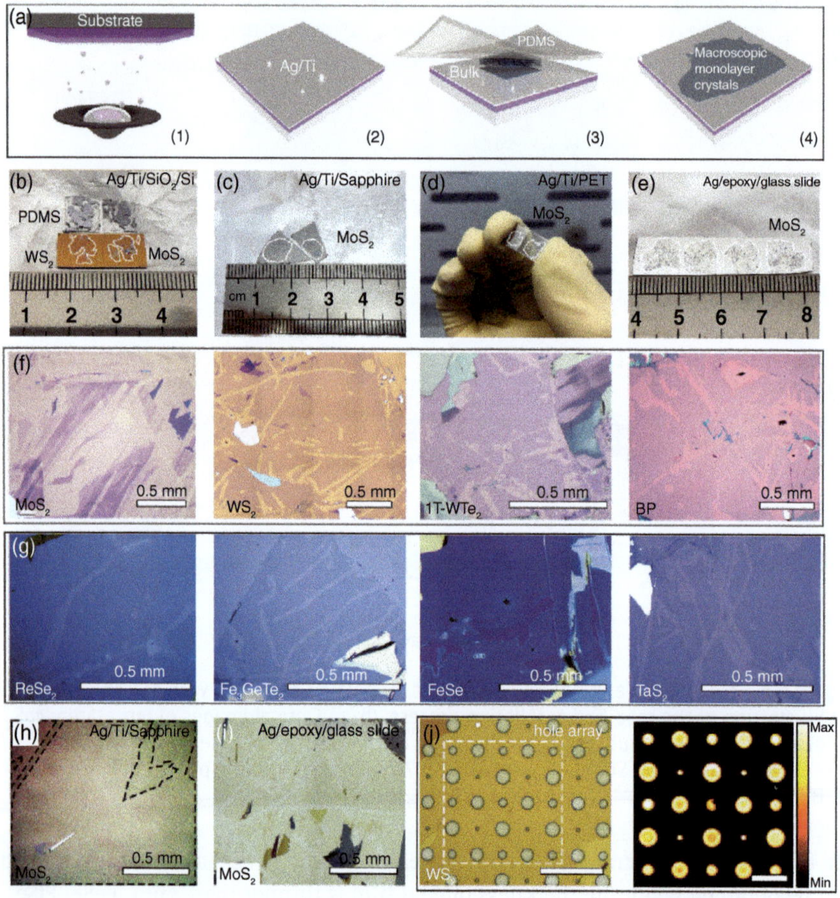

Figure 6.6 Ag-assisted exfoliation procedures and optical characterization of exfoliated samples. (a) Schematic illustration of the exfoliation procedures. (b) Exfoliated macroscopic MoS_2 and WS_2 on 15 nm Ag film supported by SiO_2/Si substrates, and bulk crystals on PDMS tapes. (c) Exfoliated macroscopic MoS_2 supported by sapphire substrates. (d) Exfoliated MoS_2 supported by elastic PET substrates. (e) Exfoliated MoS_2 on Ag/epoxy/glass slide substrate. (f) Optical microscope images of some 2D crystals exfoliated on 15 nm Ag film, including MoS_2, WS_2, 1T-WTe_2, and BP. (g) Optical microscope images of exfoliated millimeter size 2D crystals on 5 nm Ag film, including $ReSe_2$, Fe_3GeTe_2, FeSe, and TaS_2. (h and i) Optical microscope images of exfoliated millimeter size MoS_2 on sapphire substrate and Ag, respectively. (j) Optical microscope and PL mapping images of exfoliated monolayer WS_2 on 15 nm Ag film with hole array, the scale bars in the two images are 40 and 20 μm, respectively. Source: Fu et al. [38]/Reproduced from John Wiley & Sons / CC BY 4.0.

rough Ag surface. More interestingly, the strong coupling between 2D semiconductor crystals (e.g. MoS_2, $MoSe_2$) and Ag film leads to a significant PL enhancement that has not been observed in other ME techniques, which can be mainly attributed to enhanced light-matter interaction as a result of surface plasmonic polariton (SPP). This work provides a lower-cost and universal Ag-assisted exfoliation method, while at the same time offering enhanced SPP-matter interactions.

2D materials are highly sensitive to substrates, interfaces, and the surrounding environment. Suspended 2D materials are free from substrate-induced effects, thus an ideal approach to study their intrinsic properties. However, it is very challenging to prepare large-area suspended 2D materials with high efficiency. Recently, large-area suspended mono- and few-layer 2D materials were prepared on pretreatments of densely patterned hole array substrates with either oxygen-plasma or gold film deposition (Figure 6.7) [39]. Multiple structural, optical, and electrical

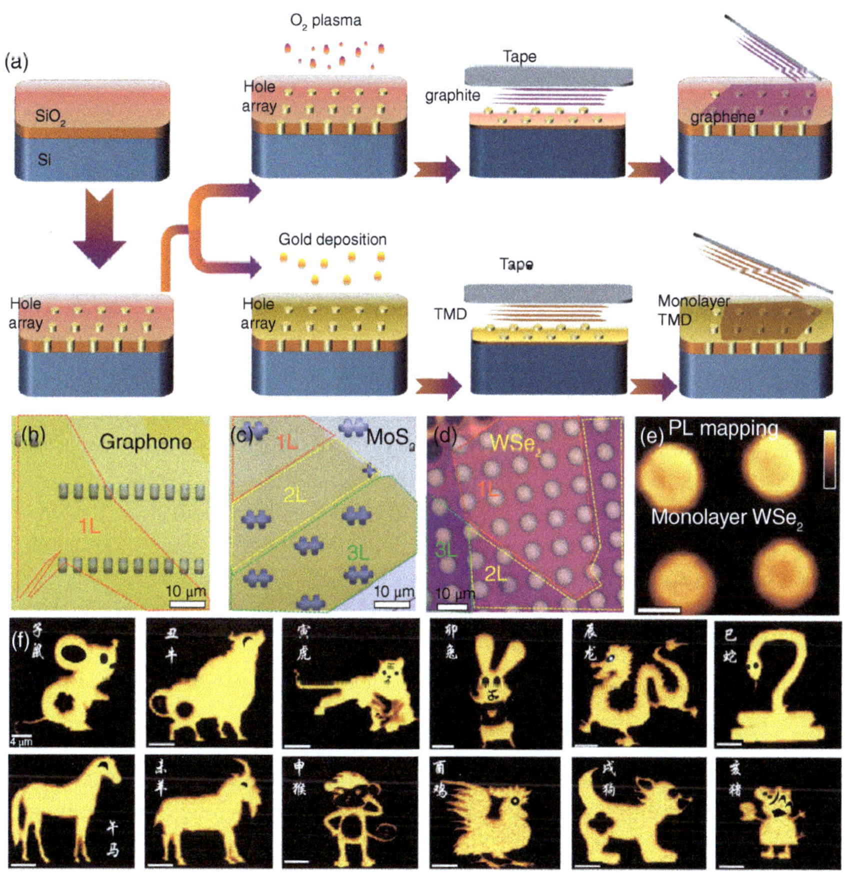

Figure 6.7 Fabrication process and characterization of suspended 2D materials. (a) Schematic images for preparing suspended samples. (b and d) Optical images of exfoliated graphene, MoS_2, and WSe_2 on different patterned substrates, including rectangle, Hall bar, and circular hole structures. (e) PL mapping image of suspended monolayer WSe_2. (f), PL mapping images of Chinese zodiac signs, which were collected on suspended WSe_2 flakes. Some details are added artificially, such as eyes and mouths. The scale bar is 4 μm. Source: Yuan Huang et al. [39]/Reproduced from John Wiley & Sons/.

characterization tools were used to fully evaluate the improved performance of various suspended 2D layers. The new discoveries reported in this study mainly include: (i) Observation of a new Raman low frequency mode for the suspended MoS_2; (ii) Significantly stronger PL and second harmonic generation (SHG) signals of suspended WSe_2, which enables the study of new optical transition processes; (iii) The low energy electron diffraction (LEED) pattern on suspended MoS_2 also exhibits much sharper spots than that on the supported area; and (iv) The mobility of suspended graphene device approaches 300 000 $cm^2/(V\,s)$, which is desirable to explore the intrinsic properties of graphene. This work provides an innovative and efficient route for fabricating suspended 2D materials, which will be broadly used for studying the intrinsic properties of 2D materials and in applications of hybrid active nanophotonic and electronic devices.

6.4 Application of Mechanical Exfoliation Method

The results of high-throughput calculations show that there are more than 1600 kinds of materials with layered structures that can be exfoliated [40]. While, up to now, few materials have been investigated, and most of the materials still need to be further studied. Based on the energy band structure, the family of 2D materials is classified as insulators, semiconductors, semi-metals, and superconductors. The flexible and transparent properties make 2D materials promising candidates for application in flexible, transparent, and wearable electronic devices. The two representative fabrication strategies include top-down and bottom-up methods for synthesizing 2D materials. It is easier to get high-quality 2D single crystals by top-down method, but the size of as-prepared 2D samples is too small for integration. Meanwhile, large size 2D samples can be obtained by bottom-up methods such as CVD and MBE without sacrificing the quality of crystal structure. Based on the significant progress in the exfoliation methods large-size 2D materials will inspire more practical applications in the fields of electronics and photonics in the future. Herein, we concentrate on the recent progress of mechanical exfoliation method for the physical characteristic investigation of 2D materials (e.g. electrical, optical, magnetic, and thermal properties), as well as the related electronic, photonic, and memory applications.

6.4.1 Electrical Properties and Devices

Current semiconductor technology can be roughly divided into several basic processes, such as thin film deposition, photolithography, etching, doping, and metallization. However, since 2D materials are extremely sensitive to the surrounding environment, most of the above-mentioned microfabrication techniques cannot be directly adopted, and need to be specially modified before implementation. At present, there are several issues facing the preparation of high-quality electronic devices based on 2D materials. (i) Due to the extreme sensitivity to intrinsic and external disorders, preserving the intrinsic properties of 2D materials after the transfer assembly and device processing is the primary question. (ii) Because of its atomic thickness and lack of surface angling bonds, universal methods to achieve ohmic contacts for 2D semiconductors are also elusive. (iii) Based on the above discussion, we will also address the recent progress on moiré superlattice devices in terms of fabrication processes.

6.4.1.1 Screening of Disorders

The so-called "disorder" herein includes various intrinsic disorders caused by crystal defects and external disorders induced by the surrounding environment [41]. For graphene, mechanical exfoliation can indeed provide high-quality 2D crystals with few defects, so external disorders are one of the main factors. We will choose graphene as a model to analyze the influence of external disorders on the electrical transport and how to reduce them through device processing. Since the QHE is closely connected to the disorders in the sample, we take the QHE as the criterion for examining the disorders (Figure 6.8). Although the intrinsic defect density of exfoliated graphene is very low, the carrier mobility on SiO_2/Si at low temperature is 10^3–10^4 cm^2/(V s) [8]. Only a fourfold degeneracy of the zero-energy Landau was observed on graphene directly exfoliated onto SiO_2/Si [9, 10]. The suspended device constructed by etching the underlying SiO_2 layer can greatly reduce the charge inhomogeneity in graphene, increasing the carrier mobility to 200 000 cm^2/(V s) at cryogenic temperatures [43] and thus enabling the observation of fractional quantum Hall effect (FQHE) [44, 45]. However, such suspended structure is quite fragile, and the channel carrier concentration cannot be effectively tuned over a large range.

Compared with SiO_2/Si substrate, stacking graphene onto h-BN (or graphene/h-BN heterostructure) using PMMA transfer can significantly improve the device quality, which can be identified by the lift of the fourfold degeneracy of the zero-energy Landau level [46]. The devices prepared by PMMA transfer are still plagued by the residual polymer during the transferring and patterning, so the carrier mobility can reach 60 000–100 000 cm^2/(V s) at 4 K, which is still lower than that of the suspended devices. Lei Wang et al. used pick-up technology to prepare graphene devices encapsulated by h-BN, which can effectively isolate graphene from the external environment [47]. The encapsulated graphene devices exhibit ballistic transport, with an electron mean free path (MFP) of several micronmeters and the mobility close to the theoretical value of 100 000 cm^2/(V s) at room temperature.

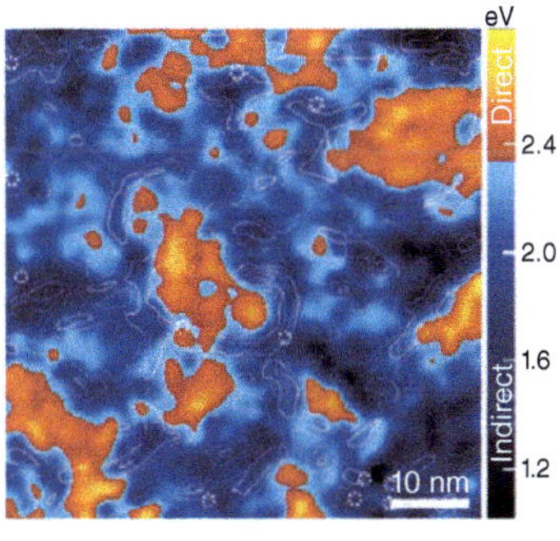

Figure 6.8 Variation of energy band structure of MoS_2 on SiO_2/Si substrate, deduced from STM measurements. Source: Shin et al. [42]/with permission from John Wiley & Sons.

The electron MFP at low temperature is even larger than 20 μm, and the mobility is higher than 1 000 000 $cm^2/(V s)$, which lays a solid foundation for high-quality graphene devices.

The encapsulation of stack significantly helps to reduce the disorder in the devices. However, it has been found that metal gates above or below the encapsulated stack will lead to fluctuations in electrostatic potential, which in turn affect the electrical transport. To solve this issue, employing an ultrathin graphite layer as a gate using the pick-up transfer method can nearly diminish such potential fluctuation. In 2014, Lei Wang et al. used ultrathin graphite as a bottom-gate in an encapsulated bilayer graphene device for the first time, and observed FQH states of bilayer graphene [48]. Recently, even denominator FQH states and other topological states have been experimentally observed using such device geometry [49, 50]. Nevertheless, the Landau level broadening confirms that the disorder of graphene devices is about two orders of magnitude higher than that of GaAs/AlGaAs quantum wells [41]. The origin of the disorder is still unclear, which may be due to the presence of some environmental impurities (such as water or hydrocarbon molecules) at the stack interface, or may be caused by the trace of oxygen or carbon contained in h-BN. To date, almost all QHE phenomena of GaAs heterojunction devices can be observed in graphene devices, apart from the striped phase [51]. With the advancement of the device quality of graphene, one can expect that more QHE physics can be studied in the future.

For the 2D semiconductors such as TMDs, their crystallinity is not as high as that of graphene, and their average defect density lies in the range of 10^{11}–10^{12} cm^{-2}. Therefore, both intrinsic disorder and external disorder should be considered for their electrical transport. In particular, we focus on the impact of external disorders on its transport property and how to lower the effects through processing. Here, we take the typical TMD, MoS_2, as an example. Similar to graphene, substrate-induced disorders in monolayer MoS_2 can also be visualized in STM images, as shown in Figure 6.8 [42]. At room temperature, the mobility of monolayer MoS_2 on SiO_2/Si substrate is typically ~10 $cm^2/(V s)$, which is much lower than the predicted phonon scattering limit value (~400 $cm^2/(V s)$) [52–55]. This disorder can be reduced by using either high-κ dielectric passivation (such as HfO_2) or h-BN encapsulation, which significantly increases the room temperature carrier mobility of MoS_2 [56]. At low temperatures, MoS_2 devices on SiO_2/Si substrates even appear to be near-insulating at low carrier densities, suggesting a disorder-induced localization. HfO_2 passivation can avoid such insulating behavior, but the mobility at high carrier densities is still low. In contrast, the h-BN encapsulation can increase the mobility of monolayer MoS_2 by about one order of magnitude [54]. Since electrical transport measurements are usually performed at cryogenic temperatures under strong magnetic fields where h-BN encapsulation is more applicable for fundamental studies. Note that the lowering degree of disorder in TMDs by h-BN encapsulation is much less than that of graphene, which is largely because the crystal quality of TMDs is poorer than that of graphene crystals [41]. Overall, however, h-BN encapsulation has enabled the electrical transport measurements for TMD devices.

6.4.1.2 Electrical Contacts of 2D Materials

The electrical contact is also one of the key factors determining the device quality. Experimentally, the contact resistance (R_C) is usually used to evaluate the contact characteristics, and the ideal R_C is to obtain a low-resistance ohmic contact. Generally speaking, the 1D edge contact structure can generate low contact resistances for graphene devices. The major target of electrical transport for graphene is how to further reduce disorder. For 2D semiconductors, good electrical contact is more challenging for electrical transport measurements. Here, two types of electrical contacts are introduced below.

1D Edge Contact of Graphene Devices The h-BN encapsulation is currently the most effective method to reduce the external disorders of 2D materials, and it also brings additional complexity for the electrode process. It is very difficult to ensure that the process of etching outside h-BN layers does not destroy the inside 2D material layer in the stack by using the slotting process. To this end, Lei Wang et al. developed a unique 1D edge contact process [47]. The stack is first etched to only expose the edge of the graphene. After which, metal is deposited to naturally form a 1D edge contact. The edge contact process is perfectly matched to the stack encapsulated by h-BN, so the intrinsic properties of 2D materials can be well preserved. This is due to the fact that better interfacial bonding and a lower tunneling barrier can be achieved in 1D edge contact, which facilitates the injection of carriers from the electrode into graphene. The R_C of graphene devices can be as low as 100 Ω μm, and such a contact structure is widely used in the study of electrical transport of graphene. 1D edge contacts are truly suitable for graphene but are not suitable for 2D semiconductors. For 2D semiconductors, a trade-off method is to construct a hybrid structure using graphene. That is, the graphene is used to form a planar vdWs contact with a 2D semiconductor inside the stack, and then use the graphene as electrodes to make 1D edge contacts. Cui et al. prepared such a hybrid edge contact by transferring graphene electrodes onto MoS_2 to achieve better ohmic contact and to observe the SdH oscillations [54].

Electrode Contacts of 2D Semiconductors A low R_C is a prerequisite for electrical transport measurements for 2D semiconductors. Unfortunately, a large contact resistance always forms between 2D semiconductors and metal electrodes. The origin of Fermi level pinning at the 2D materials/metal interface is inconclusive up to date, but regardless of the mechanism, the Fermi level pinning effect leads to large R_C [55]. So far, ideal electrical contacts for a 2D semiconductor can be achieved by constructing vdWs type 2D materials/metal contacts through stacking process. The preparation of top contacts by conventional metal deposition processes has also been reported in electrical transport studies, but it is difficult to obtain ideal 2D materials/metal interfaces. Taking the metal/MoS_2 contact as an example, the commonly used deposition processes (evaporation, sputtering, etc.) generate a large number of defects, strains, and metal diffusion at the interface [56]. These disorders will induce various energy gap states in 2D semiconductors, resulting in serious Fermi level pinning.

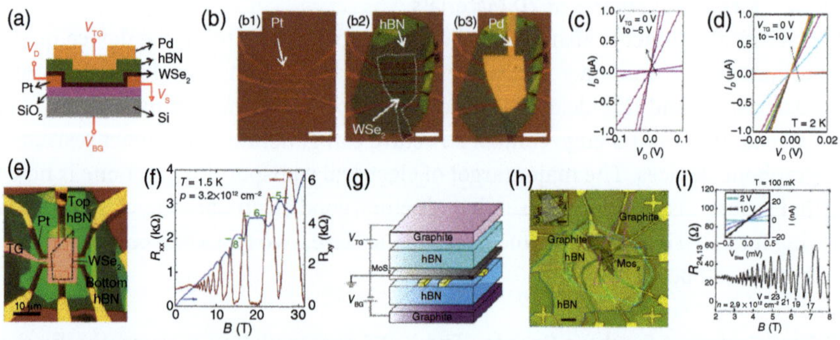

Figure 6.9 Schematic of vdWs contact between bottom electrodes and TDACs. (a) Optical sketch of the device; (b) steps of fabrication process. Source: Movva et al. [57], Reproduced from American Chemical Society. (c and d) I–V curves indicating ohmic contact. Electrical transport measurements of WSe_2 (e and f), and MoS_2 (g–i) using bottom electrode contacts. Source: (e) Fallahazad et al. [58]/with permission from American Physical Society. (f) Pisoni et al. [59]/with permission from American Physical Society.

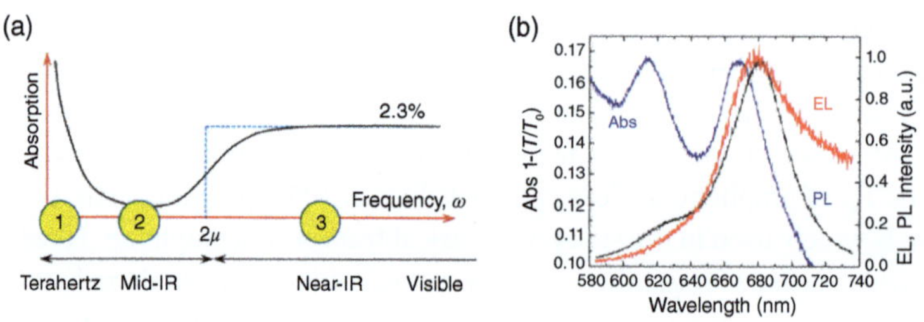

Figure 6.10 (a) Illustration of a typical absorption spectrum of doped graphene. Source: Low and Avouri [60] © 2014 American chemical Chemical Society. (c) Absorption (Abs), EL, and PL spectra of monolayer MoS_2. Source: Sundaram et al. [61] © 2013 American chemical Chemical Society.

At present, the research on electrical transport in 2D semiconductors depends on the bottom electrode (vdWs contact process). Compared with the top electrode process, the bottom electrode contact is more suitable with the encapsulated stacking. For example, Movva et al. fabricated a pre-deposited Pt electrode on SiO_2/Si substrate and then transferred WSe_2 onto the Pt electrode to form vdWs contacts (Figure 6.9a and b) [57]. As shown in Figure 6.10c and d, the contact shows excellent ohmic properties at both room temperature and cryogenic temperature. Fallahazad et al. further deposited the bottom electrode directly on h-BN, thereby fabricating an encapsulated WSe_2 device with bottom electrode contact. Using a graphite gate to further reduce the disorder, the SdH oscillation in WSe_2 was observed (Figure 6.10e,f) [58]. Through the device structure combining bottom electrode and dual graphite gates, the electrical transport measurement of $MoSe_2$ was also carried out (Figure 6.10g–i) [59].

6.4.2 Optical Properties and Photonic Devices

2D materials have recently enabled new scientific and technological breakthroughs across a variety of disciplines in nanophotonics and quantum optics [62–66]. Indeed, these 2D materials have exhibited many remarkable optical properties (e.g. large optical nonlinearities [65, 67–70], and strong excitonic effects [71]), which have already enabled diverse new conceptual photonic devices fundamentally different from those based on traditional bulk materials [72–75]. The mechanical exfoliation method has been adopted to fabricate the photonic devices composed of high-quality 2D materials for excellent performance.

For example, mechanically exfoliated graphene has been widely used for numerous photonic applications [76–78]. Since the energy-momentum of charge carriers shows a linear relationship near the Dirac point, electrons and holes behave as massless Dirac fermions [79–81], which gives rise to the unique optical properties of graphene [65, 82]. For instance, monolayer graphene has a broadband absorption of ~2.3% extending from the ultraviolet, visible, and near-infrared to the mid-infrared, far-infrared, and even to the terahertz and microwave regions [76, 77, 83]. This demonstrated photonic device includes photodetectors [65, 82], saturable absorbers [68, 84, 85], and optical modulators [70].

Apart from graphene, monolayer TMDs and black phosphorus are direct-bandgap semiconductors, offering properties complementary to graphene. The direct bandgap results in efficient and robust PL from monolayer TMDs, which opens the possibility for numerous optoelectronic applications, such as light emitters [75]. Due to the strong interaction between electrons and holes, excitons have a relatively large binding energy in TMDs [75]. As a result, monolayer TMD materials have excitonic resonances [72, 75] and the PL behavior of TMDs is dominated by excitons (Figure 6.10a) [11, 86]. Since monolayer TMDs have no inversion symmetry, second-order (e.g. SHG) and other even-order nonlinear optical effects are allowed in these TMD crystals, which are not present in graphene [87].

A single-layer hBN has a large band gap of approximately ~5.5 eV, which is high compared to other 2D layered materials [88]. As a result, being an insulator with high chemical stability enables single-layer hBN to be widely utilized as an atomically flat and ultraclean substrate, insulation, or protection layer for 2D hybrid structures.

Different atomically thin 2D materials can be readily stacked together by mechanical exfoliation to make 2D heterostructures, offering a flexible and easy approach to design desired physical properties. 2D materials have been demonstrated to be integrated with different photonic structures by mechanical exfoliation and transfer processes, such as well-developed fiber, silicon devices (e.g. waveguides), and photonic resonators (e.g. photonic crystals, plasmonic nanostructures) [64].

6.4.2.1 Photodetectors

Photodetectors that convert incident light into electrical signals are the fundamental optoelectronic components in the development of the innovative technologies, including remote controls, biomedical imaging, and optical communications. Graphene absorbs from the ultraviolet to terahertz range, as a result of which graphene-based photodetectors work over a wide wavelength range. However, given that monolayer graphene absorbs only 2.3% of light in the visible wavelength range, other 2D materials, including TMDs, are adopted for light detection [89]. Lopez-Sanchez et al. developed the ultrasensitive photodetectors based on mechanically exfoliated monolayer MoS_2. Because of the direct bandgap of monolayer MoS_2, devices based on it exhibit a very high photoresponsivity, reaching 880 AW^{-1} [90].

Nevertheless, due to the atomic thickness, the light absorption of 2D materials is still not satisfying. To further enhance the light-matter interactions, several approaches, such as plasmonic techniques, integration with waveguides, and nanocavities, have been developed for higher detectivity in photodetectors made of 2D materials [91–93].

6.4.2.2 Optical Modulators

Optical modulators play a crucial role in optical information processing systems, including optical communication, interconnect and computing [66, 82]. 2D materials have been demonstrated for the compact and efficient optical modulators with superior performance [66, 82]. The modulation mechanism includes electro-optic, magneto-optic, thermo-optic, and all-optical effects.

Electro-optic modulators based on 2D materials revealed the significant advantages owing to their broad operation bandwidth, compactness, low operation voltage, ultrafast modulation speed, and CMOS compatibility. Indeed, light modulation with graphene has been demonstrated to cover the visible, infrared [94] and terahertz range [95]. Graphene modulators working in the terahertz region have been demonstrated through the modulation of intraband absorption, giving good modulation performance, such as >94% modulation depth [95].

Apart from electro-optic modulations, all-optical modulation using 2D materials, such as saturable absorbers, optical limiters, and wavelength convertors, exploits the strong nonlinear optical response of 2D materials [66, 96, 97]. The all-optical modulators based on 2D materials allow ultrafast and broadband optical signal processing [70]. For example, saturable absorbers based on 2D materials operate as self-amplitude modulators that enable ultrafast pulse generations in mode-locking and Q-switching of lasers [98–100]. Given that TMDs have saturable absorption in the visible wavelength range, they operate as saturable absorbers for light modulation in all-fiber pulsed lasers (Figure 6.11) [101].

6.4.2.3 Single Photon Emitters

A single-photon emitter is a few-level quantum system with an optical transition that emits only one photon at a time. Zero-dimensional quantum confinement in bulk semiconductors is typically achieved by either a point defect in the crystal lattice (e.g. electrons bound to donor impurities in bulk GaAs [102], diamond [103], or SiC [104]), or a change in material composition at the nanometer-scale, often accompanied by local strain. SPEs in 2D materials potentially originate from localized

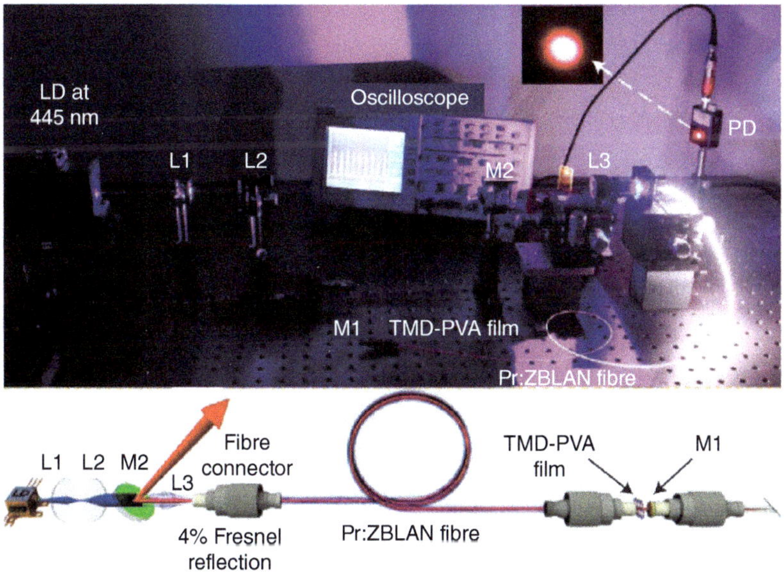

Figure 6.11 Photograph of a TMD-based saturable absorber modulated by visible fiber laser. Bottom: Schematic of the laser set-up. LD, laser diode; L1, L2, and L3, lens; M1 and M2, mirror; PD, photodetector; Pr:ZBLAN, praseodymium-doped ZBLAN fiber; PVA, poly(vinyl alcohol). Source: Luo et al. [101]/Reproduced from Royal Society of Chemistry / CC BY 3.0.

excitons or an electronic perturbation in the crystal environment. The atom-like defects thus result in stable and robust single photon emissions that can be tuned by strains [105] and enhanced by plasmonics [106] or photonic crystal cavities [107].

SPEs in TMDs were first discovered in WSe_2 in 2015 [108–113] and later in monolayer WS_2 [114]. It has been reported that artificially creating anionic vacancies in monolayer MoS_2, $MoSe_2$, and WSe_2 introduces another emission peak in the PL spectrum at an energy below the one of the free-exciton PL, which is defined as the localized exciton [115]. The localized exciton peak in WSe_2 gives rise to a single-photon emission given by their measured second-order correlation function [108–113], and the particular crystalline defects are responsible for exciton binding and deserve further understanding. The SPEs in TMDs can have interesting applications in quantum computation as they have shown similar strong valley degree freedom as the free excitons [111].

Room-temperature single photon emission has also been discovered in monolayer and few-layer hBN [116]. Single photon emissions based on hBN are the brightest that have ever been realized, and they operate mostly in the visible and the ultraviolet ranges [117]. The origin of single-photon emission in hBN differs from other 2D materials. Single-photon emissions with hBN are mainly associated with crystallographic defects, and the emission energy of these emitters falls far below the bandgap, thus preventing the interpretation of recombining excitons. Single-photon emissions with hBN are typically activated with a plasma treatment and by annealing [118, 119]. Other methods, including electron beam [120] and laser irradiation [121] have also been demonstrated. A typical defect proposed to be the origin of single-photon emission is anti-site nitrogen vacancy N_BV_N, in which there is a vacant nitrogen and the neighboring boron atom is substituted by a nitrogen atom [120]. Alternatively, the carbon substitutional impurity at nitrogen site (C_N) [117] and carbon anti-site (C_BV_N) [122] defects have been suggested.

6.4.3 Moiré Superlattice and Devices

Stacking different 2D layered materials on top of each other via the mechanical exfoliation method leads to the construction of a multitude of vdWs heterostructures [123]. This platform has hosted numerous novel physical phenomena and offers the possibility for band-structure engineering (e.g. Hofstadter butterfly effect in graphene-hBN heterostructure) [124–126]. The freedom of combining a rich variety of different 2D materials (e.g. metals, semiconductors, superconductors, topological insulators, etc.) in different-type vdWs heterojunctions and then their stacked superlattices leads to even more exciting discoveries of novel optical properties [123, 126]. Especially, when two layers of 2D materials in a stack have very close lattice constants, or have the same lattice constants but are twisted with each other, periodically modulated moiré superstructures are generated. The period of the superlattice is much larger than the original atomic lattice of 2D materials. Moiré superlattice has become one of the hot topics in the field of 2D materials in recent years, starting a new field of "Twistronics" [127].

6.4.3.1 Graphene/h-BN Moiré Superlattice

The aligned graphene/h-BN stack is the first moiré superlattice system [128]. Due to the lattice mismatch, the lattice alignment between graphene and h-BN can form a moiré superlattice, which can induce secondary Dirac cones. The study of aligned graphene/h-BN focuses on interactions between the moiré potential and magnetic fields at a similar scale. Theories predict that a series of self-similar energy levels can be produced in a moiré superlattice, which is called the fractal spectrum of the Hofstadter butterfly [129]. However, it is extremely challenging to directly design dielectric superlattices that enable the fractal energy spectrum in GaAs/AlGaAs quantum wells. Although van Hove singularities of graphene/h-BN moiré superlattices were observed in STM, the disorder in graphene hindered the electrical transport properties in earlier days. It was only in 2013 that technological advances based on pick-up transferring demonstrated the fractal energy spectrum of graphene/h-BN moiré superlattices using electrical transport experiments [125].

The graphene/h-BN moiré superlattice generates a rich sequence of QHE states under a strong magnetic field at low temperatures, and its trajectory can be traced using the following formula:

$$\frac{n}{n_0} = t\frac{\varphi}{\phi_0} + s$$

where n/n_0 is the carrier density per superlattice cell, s is an integer or rational fraction representing the Bloch band filling factor, and t is an integer or rational fraction related to the energy gap structure. Depending on whether *t* and *s* are integers or fractions and whether they are zero or non-zero, five different electronic states can be observed, which are generally considered due to the competition between the moiré periodic potential and the Coulomb interaction [125, 128]. In a word, to understand the many body ground states in the Hofstadter butterfly energy spectrum, more systematic and in-depth theoretical and experimental studies are both essential.

6.4.3.2 Twisted Graphene Moiré Superlattice

The twisted magic angle graphene is another moiré superlattice system that has been intensively studied in the last few years. The coupling of 2D materials and moiré superlattices will also strongly change the energy bands of electrons. Therefore, the magic angle system can also serve as the platform for flat bands, which are different from the intrinsic flat band materials [130, 131]. In the flat band, electronic interactions dominate the phases, accompanied by a series of emergent correlated states. In recent years, thanks to the improvement of transferring methods, many imperative breakthroughs have been made in magic angle graphene. A series of correlated states have been observed as a result of interlayer hybridization and twisted angle-dependent band shift in magic angle bilayer graphene [132, 133]. In ABC trilayer graphene/h-BN heterostructures [134, 135] and twisted bilayer graphene-bilayer graphene [136], a bandgap is induced by applying a vertical displacement field, enriching the band tuning knobs. Similar correlation behaviors were also found in twisted trilayer graphene [137]. In magic-angle graphene aligned with the underlying h-BN, orbital magnetism has been observed due to the breaking of its valley degeneracy [138]. There have been several specialized reviews on the physics of moiré superlattices [139, 140], and this field has attracted more and more attention.

6.4.3.3 Twisted TMD Moiré Superlattice

The isolated flat bands are realized in a variety of twisted graphene moiré superlattices. However, the tunability of the twisted graphene system is still greatly limited. For instance, in twisted bilayer graphene-bilayer graphene, there effectively remains a narrowly defined magic angle, owing to additional displacement-field effects on the overall band structure. Whereas in the ABC trilayer graphene/h-BN moiré superlattice, the maximal moiré wavelength is fixed by the mismatch. Theoretical work on TMD homo- and hetero-bilayers predicts the existence of flat bands and van Hove singularities in the moiré Brillouin zone that can support emergent electronic phases [141, 142]. Due to coupled spin and valley degrees of freedom, the reduced degeneracy in TMD compared to that in graphene, together with material properties not present in graphene, such as strong spin-orbit coupling, suggest that twisted bilayer TMD can provide an idealized simulator for a single-band Hubbard model on a triangular lattice, in which several correlated states may be realized.

In terms of device process, the tear-and-stack pick-up transfer method and the 1D edge contact transport enable the transport research of twisted graphene moiré superlattices. Yet, studies on correlated states in a large number of 2D semiconductors, such as twisted TMDs, still depend on optical methods or local electrical

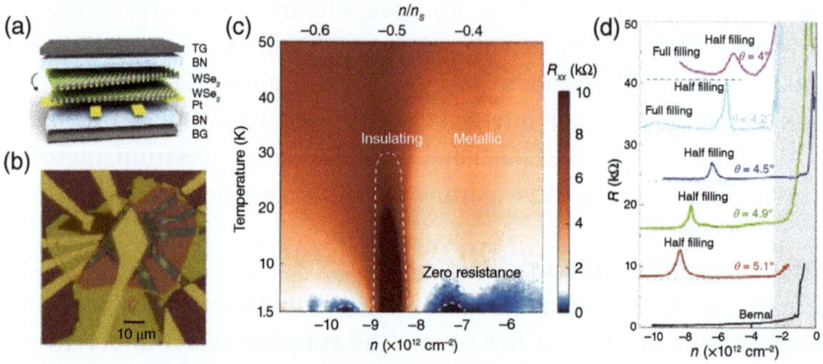

Figure 6.12 The correlated electronic states in a twisted WSe_2 device. Schematic (a) and optical image (b) of the device with bottom electrode. (c) Curves of the longitudinal resistance versus the carrier concentration for devices with different twisted angles under zero magnetic field. The resistance of the half-filling state is marked. (d) The phase diagram of the twisted WSe_2 device. Source: Wang et al. [143]/Reproduced with permission from Springer Nature.

measurements based on scanning probes. Overall, the electrical transport of twisted TMD is seriously impeded by the large contact resistance. Lei Wang et al. have made a breakthrough in electrical transport in twisted TMD, using the Pt bottom electrode described above to achieve better low temperature ohmic contact, and revealed the correlated states in twisted bilayer WSe_2 [143], as shown in Figure 6.12a and b. It was found that in twisted WSe_2 with angles of 4°–5.1°, a resistance peak appeared at the half filling of the flat band. And when the twisted angle was 4° and 4.2°, the resistance peak corresponding to the full filling of the moiré minibands was observed, confirming flat bands produced by the moiré superlattice (Figure 6.12c). For the device with a twisted angle of 5.1°, the strongest correlation states can occur by adjusting the electric field, and zero resistance is observed on both sides of the half filling when the temperature is lower than 3 K (Figure 6.12d). The significant advantage of the twisted TMD is that the bandwidth and doping in the system can be varied independently. And the material diversity of TMD family provides a potential platform for discovering other correlated states, such as exciton condensation, spin liquids, and ferromagnetic ordering.

6.4.4 Magnetic Properties and Memory Devices

It is well known that memory units play an indispensable role in modern electronics, in which bigger storage capacity and faster read/write speeds are consistently being pursued. Many emerging memory technologies, such as resistive random-access memories (RRAM), magnetic random-access memories (MRAM), phase-change memories (PCM), and ferroelectric random-access memories (FRAM), have been developed in the past several decades. Besides advanced device structures, various emerging materials with modulable resistance also play a critical role in promoting future memory technologies. Therefore, TMDs have attracted tremendous interest in fabricating advanced memory devices due to many advantages, such as low-power switching capability, electrostatic gate tunability, and mechanical flexibility.

2D spintronics based on vdW crystals with long-range ferromagnetic order is crucial for the emerging functional devices (for example, ultrafast photodetectors, broadband optical modulators, and excitonic semiconductor lasers), which have been derived primarily from the electron-charge degree of freedom [144]. Besides, the intrinsic ferromagnetism originating from the 2D lattice is significant for understanding the underlying physics of electronic and spin processes. Indeed, some theoretical efforts had been made to demonstrate that such 2D magnetic crystals exhibit a finite ferromagnetic transition temperature T_c [145–147]. Despite all major electronic classes in 2D vdW materials ranging from semiconductors and superconductors to highly correlated materials, the class of ferromagnetic phases was missing until the first experimental observation of low-temperature magnetism in Cr halides and chalcogenides 2D crystals in 2017 [148, 149]. Currently, a good dozen 2D magnetic crystals have been studied in exfoliated materials [150–153]. It has a higher possibility of retaining a magnetic state in an exfoliated 2D magnets because the magnetic anisotropy is important for realizing 2D magnetism [154, 155], and most vdW magnets have an intrinsic magneto-crystalline anisotropy owing to the reduced crystal symmetry of their layered structures [156]. Besides, compared with the magnetic films grown directly on a substrate whose magnetic properties strongly depend on the underlying substrate [157], the exfoliated 2D materials can be regarded as isolated single crystals regardless of the effect of the substrate on the ferromagnetic phenomena.

In a 3D crystal system, it is well known that there are five basic magnetic states, which are diamagnetism, paramagnetism, ferromagnetism, antiferromagnetism, and ferrimagnetism, as shown in Figure 6.13. The crystal shows diamagnetism when it is stacked by atoms or ions without a magnetic moment. Although the paramagnetic crystal is made up of the atoms or ions with the magnetic moment, the spontaneous magnetization is prevented, and the crystal exhibits no collective

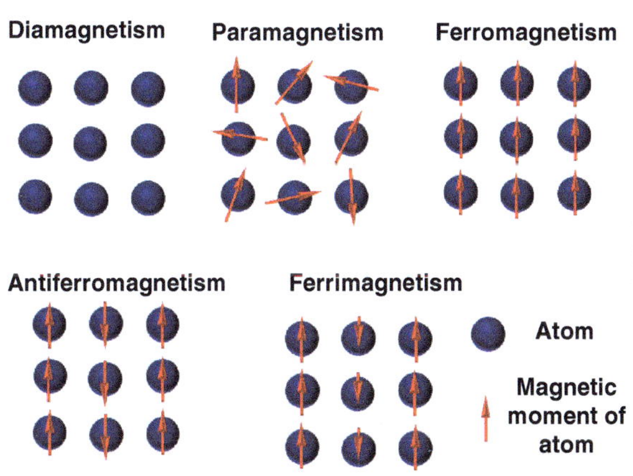

Figure 6.13 Schematic diagram of five basic magnetic states in a 3D crystal system.

magnetic moment. Ferromagnetism, antiferromagnetism, and ferrimagnetism exhibit long-range magnetic order with the collective magnetic moments over the macroscopic scale and spontaneous breaking of time-reversal symmetry below a certain temperature. This is typically driven by the interaction between the neighboring spins (exchange coupling) that tends to favor specific relative orientations between them [158]. With increasing T, the long-range order is not preserved above T_c for ferromagnetism (T_N, Néel temperature, for antiferromagnetism and ferrimagnetism) due to thermal fluctuations. Ferromagnetism refers to the phenomenon that the magnetic moments of adjacent atoms or ions in a substance are roughly arranged in the same direction below the temperature T_c, and the magnetic moment increases with the increasing applied magnetic field until a certain limit. When the magnetic moments of adjacent atoms or ions line up neatly but are antiparallel staggered, no macroscopic magnetic moment can be observed, which is called antiferromagnetism. The above scenarios can be well described by Heisenberg Spin Hamiltonian: [158]

$$H = -\frac{1}{2}\sum_{i,j}\left(J\vec{S}_i \cdot \vec{S}_j + \Lambda \vec{S}_i^z \cdot \vec{S}_j^z\right) - \sum_i A\left(\vec{S}_i^z\right)^2$$

where J is the exchange coupling between spins $\vec{S}_i$ and $\vec{S}_j$ on neighboring sites (favoring either ferromagnetic, $J > 0$, or antiferromagnetic (ferrimagnetic), $J < 0$), A and Λ are the "on-site" and "intersite" (or exchange) magnetic anisotropies, respectively. In some cases, more sophisticated models are needed, including either further neighbors or different kinds of magnetic coupling, such as Dzyaloshinsky–Moriya [156, 159], Kitaev [160, 161] or higher-order (for example, bi-quadratic) interactions. A wide range of physical phenomena is expected depending on the values of J, Λ, and A or on the presence of additional magnetic interactions.

When exfoliating layered magnetic compounds from their 3D counterparts to monolayers, a non-zero magnetization at a finite temperature is greatly determined by the suppression of thermal fluctuations by intrinsic magnetic anisotropy. For CrI_3 and $FeGeTe_2$, magnetism survived in the monolayer limit at a finite temperature due to the magneto-crystalline anisotropy from a strong spin-orbit coupling [162, 163]. While in the case of $CrFeTe_3$, the magnetism is destroyed in the 2D limit, and the critical temperature goes to zero due to the absence of intrinsic anisotropy [144]. Different anisotropy in 2D magnetic crystals leads to different kinds of interlayer spin alignment, as well as different interlayer exchange coupling J_L and intralayer exchange coupling J_i. It is interesting that ferromagnetic states have been reported in VSe_2 monolayers but not in the bulk, which was explained using band structure theory [151]. A metallic layered material leads to a change in the density of states (DOS) at the Fermi level, which may enhance the sensitivity to the interaction effects (J) and lead to the magnetic order. Given the crucially different properties of 2D magnetic crystals, we schematically show different magnetic phenomena in 2D magnetic crystals by the difference in interlayer exchange coupling J_L and intralayer exchange coupling J_i in Figure 6.14. And we will illustrate two experimentally reported magnetic systems in mechanically exfoliated 2D magnetic crystals: the ferromagnetic system (CrI_3, Fe_3GeTe_2) and antiferromagnetic system ($MnPS_3$).

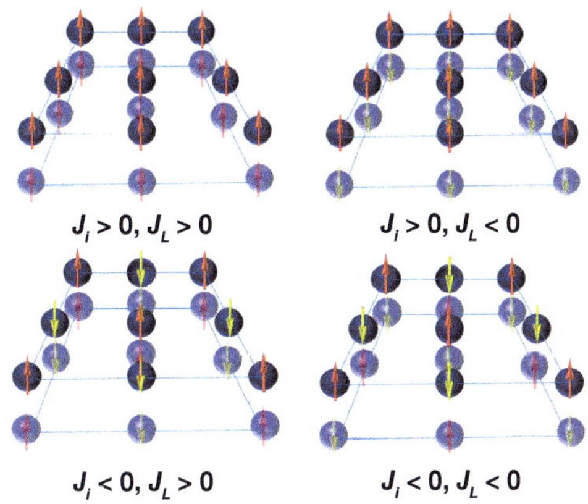

Figure 6.14 The scheme of the reported different magnetic states in a 2D crystal system. The magnetic anisotropy of different magnetic states is not illustrated in the scheme.

6.4.4.1 Ferromagnetism in 2D Materials

In 1960s, bulk CrI_3 was experimentally investigated [164–166]. The magnetic behavior of bulk CrI_3 is a typical soft ferromagnet with its out-of-plane easy axis below the Curie temperature (T_c = 61 K). The first experiments on the magnetic transition in atomically thin crystals were also reported in CrI_3 [149]. In this work, the atomically thin CrI_3 flakes were obtained by ME, and the magnetic order was probed by the polar magneto-optical Kerr effect (MOKE) measurements at different applied external magnetic fields perpendicular to the sample plane. All measured monolayer and trilayer samples consistently show ferromagnetic behavior with a single hysteresis loop centered at $\mu_0 H = 0$ T as shown in Figure 6.15a–c. The intrinsic magneto-crystalline anisotropy in CrI_3 monolayers counteracts thermal fluctuations and preserves the 2D long-range ferromagnetic order. A further observation is that the bilayer CrI_3 shows a markedly different magnetic behavior from the monolayer. The MOKE signal is strongly suppressed at a small field (±0.65T), which implies compensation for the out-of-plane magnetization. Upon crossing a critical field, the MOKE signal shows a sharp jump, depicting a sudden recovery of the out-of-plane co-parallel orientation of the spins. These observations suggest that each individual layer is ferromagnetically ordered (out-of-plane) while the interlayer coupling is antiferromagnetic. The strength of the interlayer coupling determines the field at which jumps between different plateaus occur, but the detailed mechanism of this coupling remains unclear. Although a change in the magnetic state associated with a structural transition has been reported in few-layer CrI_3 [167], and calculations based on antiferromagnetic interlayer exchange coupling suggest that the structure of few-layer CrI_3 corresponds to that of high-temperature bulk CrI_3 [168–170], the crossover from antiferromagnetic to ferromagnetic interlayer coupling as the increase of the number of layers in 2D magnetic materials is still an open question.

Fe_3GeTe_2 is another typical 2D ferromagnet with high T_c (ranging from 150 to 220 K depending on Fe occupancy) [152, 171, 172], large out-of-plane anisotropy [152, 171, 172], and strong electron correlation effects [173] that are expected to stabilize the long-range ferromagnetic order in Fe_3GeTe_2 monolayers. The itinerant

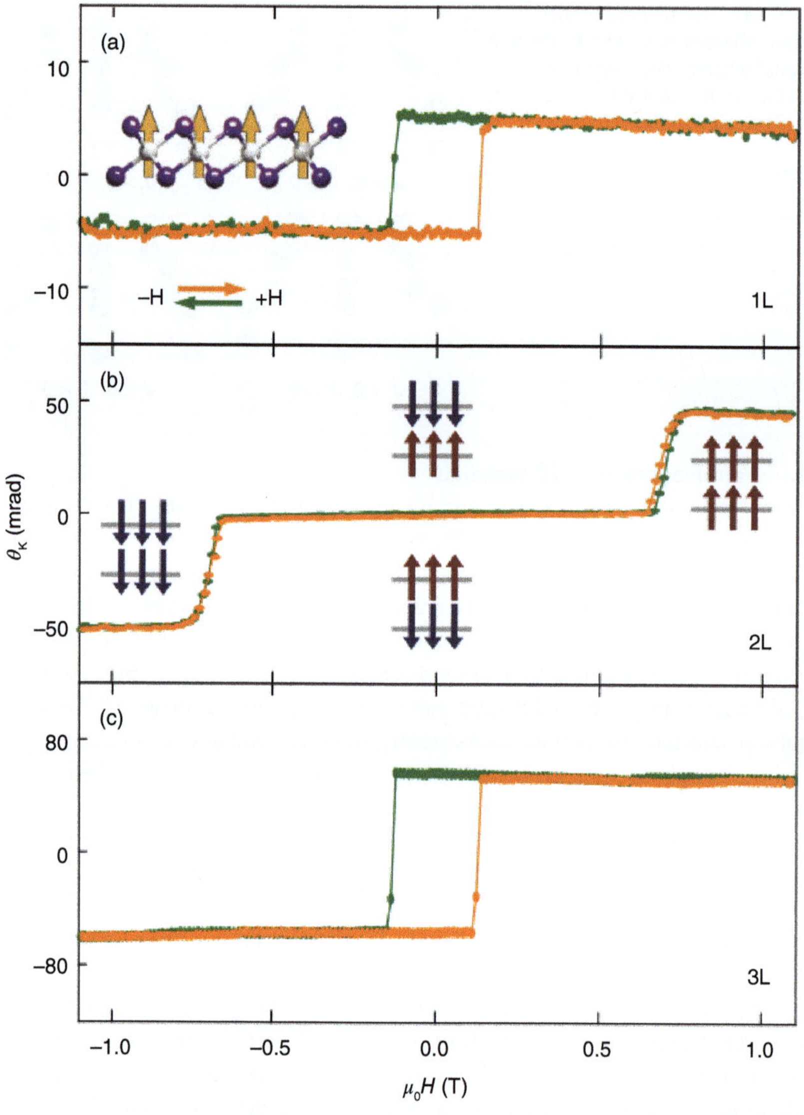

Figure 6.15 Layer-dependent magnetic ordering in atomically-thin CrI_3. (a) MOKE signal on a monolayer (1L) CrI_3 flake, showing hysteresis in the Kerr rotation as a function of applied magnetic field, indicative of ferromagnetic behavior. The inset is a view of the in-plane atomic lattice of a single CrI_3 layer. Grey and purple balls represent Cr and I atoms, respectively. The out-of-plane view of the CrI_3 structure depicts the Ising spin orientation. (b) MOKE signal from a bilayer CrI_3 showing vanishing Kerr rotation for applied fields ±0.65 T, suggesting antiferromagnetic behavior. Insets depict bilayer (2L) magnetic ground states for different applied fields. (c) MOKE signal on a trilayer (3L) flake, showing a return. to ferromagnetic behavior. Source: Reproduced with permission from Huang et al. [149].

vdW structure enables us to elucidate the evolution of the magnetic order from bulk down to monolayer Fe_3GeTe_2. Although bulk Fe_3GeTe_2 cleaves along the vdWs gap, the intralayer bonding is not strong enough for a thin flake to survive conventional mechanical exfoliation processes [150]. Al_2O_3- and gold-assisted exfoliation was performed to increase the chances of obtaining monolayer Fe_3GeTe_2 flakes. As a metallic material, Fe_3GeTe_2 enables the interplay of both spin and charge degree of freedom, which lies at the heart of various spintronic architectures [174]. Apart from Kerr measurements, magntotransport measurements, such as the anomalous Hall effect, can be applied to examine the magnetic properties of Fe_3GeTe_2 thin flakes. Yujun Deng et. al [150] studied the ferromagnetism in 2D Fe_3GeTe_2 by probing the Hall resistance under an external magnetic field applied perpendicular to the vdWs plane. As reducing the layer number, the samples become progressively more insulating, but ferromagnetism persists down to the monolayer. T_c as a function of layer number from the temperature dependence of the anomalous Hall effect was precisely determined to fully elucidate the dimensionality effect. As the samples thin down, T_c decreases monotonically, from about 180 K in the bulk limit to 20 K in a monolayer. The strong dimensionality effect on the ferromagnetism in Fe_3GeTe_2 stems from the fundamental role of thermal fluctuation in two dimensions. According to the classical Heisenberg model with Ruderman–Kittel–Kasuya–Yosida (RKKY) exchange [175], the magnetocrystalline anisotropy gives rise to an energy gap in the magnon dispersion. The gap suppresses low-frequency, long-wavelength magnon excitations, and protects the magnetic order below a finite T_c. Besides, it was demonstrated that an ionic gate can drastically modulate the ferromagnetism in Fe_3GeTe_2 thin flakes, and T_c boost them up to room temperature (Figure 6.16). Although the gate-tuned ferromagnetism in atomically thin Fe_3GeTe_2 can be explained by the Stoner model, a more comprehensive characterization of the gate-voltage induced magnetic properties is required to further establish the definite conclusions, such as the effect of gate-voltage on the electronic structure, the exchange interactions, and magnetocrystalline anisotropy of Fe_3GeTe_2 crystal structure.

6.4.4.2 Antiferromagnetism in 2D Materials

The MPX_3 (M = Mn, Fe, Co, Ni; X = S, Se) family compounds are good candidates for 2D honeycomb magnets for exploring intrinsic magnetic orders (Figure 6.17) [177]. The magnetic transition metal atoms are held together by weak vdWs forces in the *ab* planes, so the coupling between the planes is weak both magnetically and atomically. As one of the MPX_3 family compounds, the main structural and magnetic properties of $MnPS_3$ have been known since 1980s. $MnPS_3$ is a highly resistive broadband semiconductor with a gap close to 3 eV [178]. The valence of manganese atoms is 2+, which means they have a half-filled 3*d* shell and 5/2 spin. Therefore, the classical theory can be adequate to describe the spin dynamics, and orbital contribution doesn't exist [177]. The magnetic susceptibility of $MnPS_3$ is isotropic at high temperatures [179, 180], however, 2D MPS_3 orders antiferromagnetically below T_N = 78 K. Theory shows that a dipole-dipole anisotropy, if strong enough, would be sufficient to induce long-range order in a 2D Heisenberg antiferromagnet

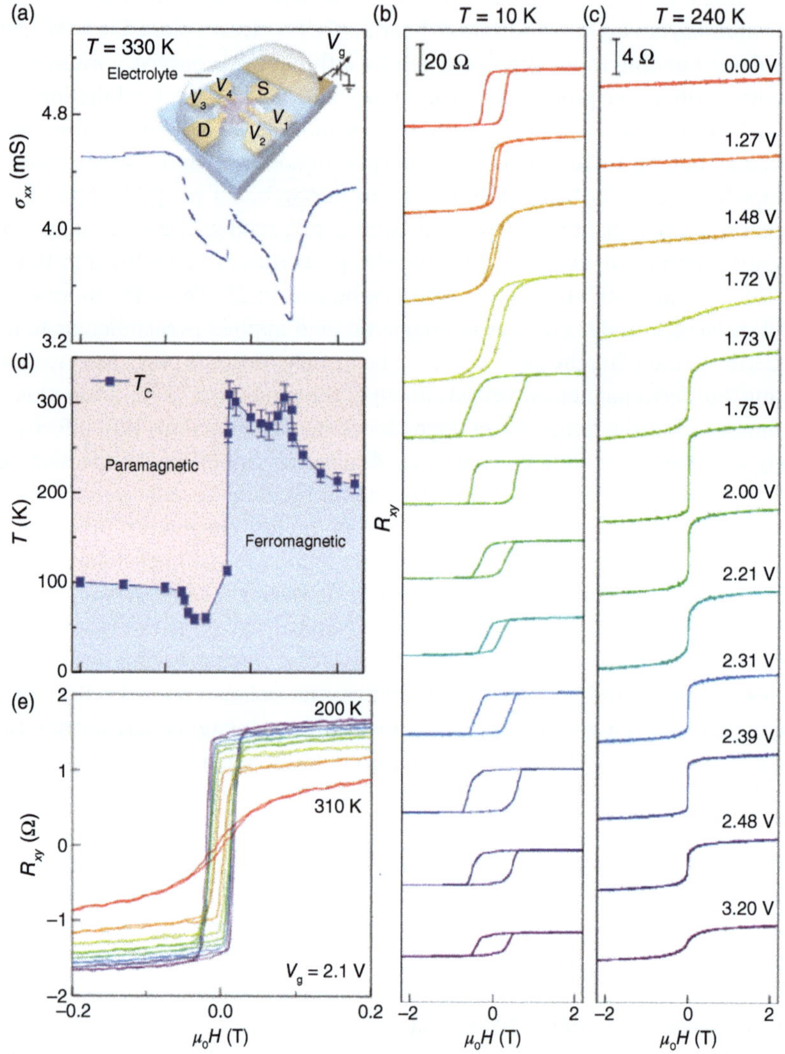

Figure 6.16 Ferromagnetism in an atomically thin Fe_3GeTe_2 flake modulated by an ionic gate. (a) Conductance as a function of gate voltage v_g measured in a trilayer Fe_3GeTe_2 device. Data were obtained at $T = 330$ K. The inset shows a schematic of the Fe_3GeTe_2 device structure and measurement setup. S and D label the source and drain electrodes, respectively, and V_1, V_2, V_3, and V_4 label the voltage probes. (b and c) R_{xy} as a function of external magnetic field recorded at representative gate voltages, obtained at $T = 10$ K (b) and $T = 240$ K (c). (d) Phase diagram of the trilayer Fe_3GeTe_2 sample as the gate voltage and temperature are varied. The transition temperature was determined by temperature-dependent anomalous Hall resistance extrapolating to zero. Vertical error bars represent the uncertainties in determining the onset of non-zero R_{xy}. (e) R_{xy} of a four-layer FGT flake under a gate voltage of $V_g = 2.1$ V. Hysteresis in R_{xy} persists up to $T = 300$ K, providing unambiguous evidence for room-temperature ferromagnetism in the sample. Source: Reproduced with permission from Deng et al. [150]. Copyright © 2018, Springer Nature Limited.

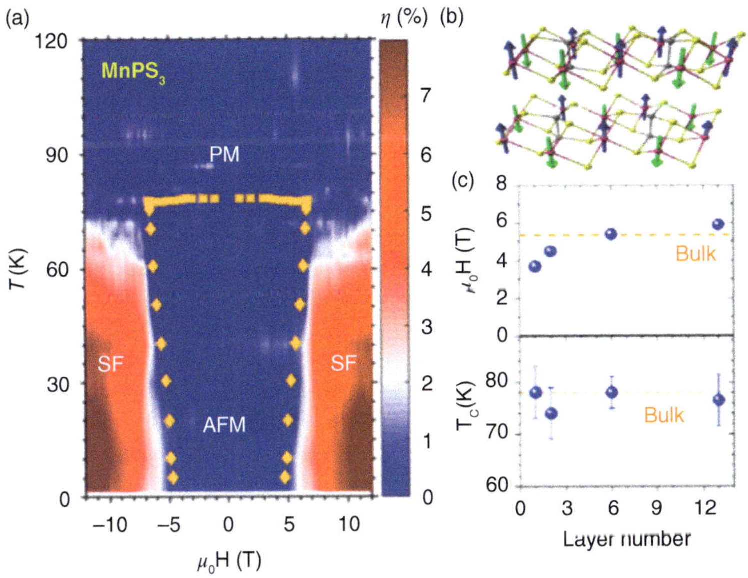

Figure 6.17 The tunnel transport in Atomically Thin MnPS3 crystals. (a) Color plot of the tunneling magnetoresistance $\eta(H, T)$ of 6-layer $MnPS_3$, as a function of applied perpendicular magnetic field (from −12 to +12 T) and temperature (from 1.5 to 120 K). The values of the spin-flop field H_{sf} as a function of T (orange diamonds) and of the T_N as a function of $\mu_0 H$ (orange circles) were extracted from measurements on $MnPS_3$ crystalsenen and overlaid on the same plot. (b) $MnPS_3$ crystal structure, illustrating the antiferromagnetically aligned spins on the Mn atoms (note: past neutron studies have shown that the spin direction is nearly, but not exactly, perpendicular to the layers). (c) The characteristic magnetic fields (the position of the peak in mono and bilayers, and the value of H^e_{sf} in 13L and 6L devices) and the temperature at which magnetoresistance vanishes for all devices, as a function of layer number. The horizontal orange dashed line in the two panels indicates the bulk T_N and the spin-flop in bulk $MnPS_3$, respectively. Source: Reproduced with permission from Long et al. [176]. Copyright © 2020 American Chemical Society.

[181]. $MnPS_3$ is reported to be a 2D system with a weak interplanar exchange and a weak anisotropy that appears to be a combination of both dipolar (out-of-plane) and single-ion (in-plane) contributions [177]. Kerr effect or direct magnetometry has been demonstrated to be useful in a ferromagnetic monolayer, while it would not work in the absence of a net antiferromagnetization. Temperature-dependent Raman measurements have been used to investigate the critical magnetic behaviors of the monolayer $MnPS_3$ crystal. Three phase transition temperatures at around 55, 80, and 120 K are associated with unbinding of spin vortices, magnetic phase transition from the antiferromagnetic to paramagnetic phase, and a second magnetic phase transition due to 2D spin critical fluctuations, respectively. The magnetic state of atomically thin semiconducting layered antiferromagnets has been studied by measuring their resistance as a function of magnetic field and temperature. The tunneling magnetoresistance depends on the relative orientation of the magnetization

in adjacent layers when the spins within each individual layer are ferromagnetically aligned. To study the 2D antiferromagnetic systems, G. Long et al. investigated the tunnel transport properties in atomically thin $MnPS_3$ crystals [176]. For thick multilayers below $T\sim78$ K, a T-dependent magnetoresistance sets in at $\mu_0 H\sim5$ T and the boundary between the antiferromagnetic and the spin-flop phases known from bulk measurements is tracked, which confirmed that antiferromagnetic system survived down to individual monolayers. It is also found that the transition temperature between the paramagnetic and antiferromagnetic states is nearly independent of H. The electron tunneling process in antiferromagnetic tunnel barriers needs further investigation in the future by a great deal of theoretical work to extract microscopic information about the magnetic states of atomically thin $MnPS_3$ layers from the dependency on the tunneling magnetoresistance.

Despite the large amount of work devoted to the 2D magnetic crystal system and various long range magnetic orders that have been reported, so far, no ferrimagnetism in the 2D crystal system has emerged. Ferrimagnets provide an ideal platform to explore magnetic switching devices [182]. The key characteristic of the materials with ferrimagnetic ordering is the strong negative exchange interaction compared to ferromagnetism and the non-zero residual magnetization compared to antiferromagnetism, which causes magnetic features distinct from their ferromagnetic and antiferromagnetic counterparts. A much lower time scale of magnetization dynamics and reversal and a faster demagnetization are observed in the ferrimagnets compared to that of ferromagnets [183, 184]. To evaluate the effect of exchange interaction in ferrimagnetics, studies are required across ferrimagnetic 2D.

6.4.5 Thermal Conduction

Unique thermal properties have been observed in 2D materials, including the anomalous behavior of thermal conduction [185–188]., For instance, the Fourier's law, a fundamental law in understanding thermal transport in bulk materials, is invalid in 2D materials [189–192]. Investigating the thermal conduction in 2D materials is beneficial to further understand the physical essence of thermal transport and explore potential applications [185, 189, 193, 194].

Graphene has been verified to hold high thermal conductivity, reaching 2000–3000 W m^{-1} K^{-1} at room temperature. Results from different research groups show significant variations, ranging from ~680 to 5300 W m^{-1} K^{-1} for single-layer suspended graphene [195–201]. Besides the discrepancy of experimental details (e.g. the power absorbed by graphene, the shape of suspended part), a primary cause is the quality of grapheme, which closely relates to the preparation process. Defects in CVD graphene induce extra scattering resulting in reduced thermal conductivity, compared with that in exfoliated graphene. Residual polymer adhering to surfaces during the sample transfer process induces extra phonon scattering [202, 203].

At low temperatures, ballistic phonon transport, i.e. phonon transport without scattering, can be generated in a short graphene sheet with a clean surface and infinite width. Two independent experiments about quasi-phonon transport in suspended and supported graphene were reported by Xu et al. and Bae et al., respectively

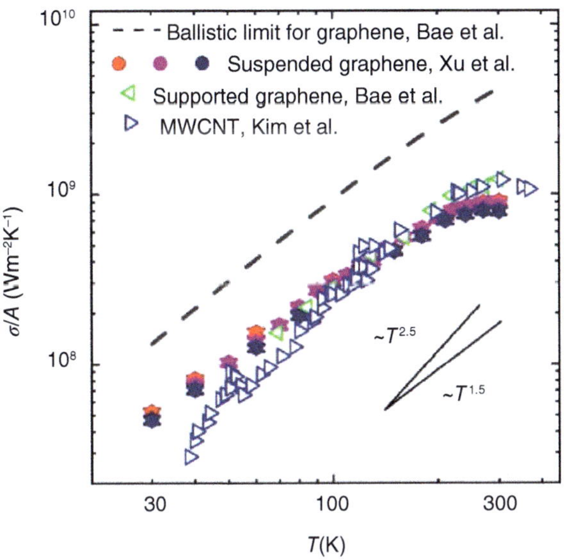

Figure 6.18 Experimental observation of quasi-ballistic phonon transport in suspended graphene [201], supported graphene [204]. Experimental data of individual multi-walled carbon nanotube [208] and calculated ballistic limit value for graphene [204] are plotted for comparison. Source: Reproduced with permission from Xu et al. [189]. © 2016 IOP Publishing Ltd.

[204, 205]. Thermal conductance per unit cross-section (σ/A) is more accurate to analyze this kind of transport [204, 206, 207]. In ballistic regime, the calculated limit value of σ/A should follow $[1/(4.4\times10^5 T^{1.68})+1/(1.2\times10^{10})]^{-1}$. The experimental values reached 40% of the above limit value at $T = 30$ K, as shown in Figure 6.18.

Analyzing the behavior of acoustic phonons in graphene can be favorable for comprehending the physics behind ultrahigh thermal conductivity. Thanks to the anomalously large DOS of flexural phonons, at $T = 300$ K, the out-of-plane phonon modes contributed over 70% heat transport in monolayer graphene, theoretically calculated by Mingo et al. [209] In experiment, Xu et al. first reported that at low temperatures, the thermal conductivity of suspended single-layer graphene followed $\sim T^{1.5\pm0.18}$, which is compelling evidence to demonstrate the prominent contribution of out-of-plane phonon modes [205]. Therefore, when graphene is supported by a substrate, the out-of-plane acoustic vibrations are strongly suppressed, and its thermal conductivity undergoes a dramatic reduction. Thermal conductivity of ~600 W m^{-1} K^{-1} was observed in single-layer graphene supported by SiO_2 [210].

Besides graphene, some physical properties of h-BN resemble those of graphene, such as high temperature stability and superior thermal conductivity. Based on solving the Boltzmann transport equation, Lindsay et al. claimed that, compared with high-quality bulk h-BN (~390 W m^{-1} K^{-1}), monolayer h-BN holds a higher room-temperature thermal conductivity (~600 W m^{-1} K^{-1}) [211]. Limited by preparation technology, the above prediction can't be tested in early experiments where many organic residuals adhered to the surface of single/few-layer h-BN [212–214]. Until recently, the dry-transfer method was applied by Wang et al., and the room-temperature thermal conductivity of suspended bilayer h-BN was measured to be ~484 (+141/−24) $W\,m^{-1}\,K^{-1}$ [215]. Subsequently, Li et al. reported that room-temperature thermal conductivity is as high as ~751 (±340) $W\,m^{-1}\,K^{-1}$ in single-layer h-BN, and decreases to 646 (±242) $W\,m^{-1}\,K^{-1}$ and 602 (±247) $W\,m^{-1}\,K^{-1}$ in the bilayer and trilayer h-BN, respectively [216].

It's well accepted that thermal conduction in nanostructures distinguishes from that in bulk; finding out the transition from bulk to 2D is significantly important. Ghosh et al. experimentally reported that the room-temperature thermal conductivity of suspended graphene dramatically varies with the number of layers [217]. It is as high as ~4000 $W\,m^{-1}\,K^{-1}$ in monolayer graphene, and then gradually reduces with the number of layers, as a result of strong cross-plane coupling and severe phonon Umklapp scattering in thicker samples. When the thickness of graphene is greater than four/five-atoms-layer, the thermal conductivity of graphene approaches to that in graphite (see Figure 6.19a). However, in supported graphene, thermal conductivity shows barely changes or slight increases with thickness due to the suppression of out-of-plane phonon modes, which have been discussed above [210, 218–220]. Similar thickness-dependent thermal conductivity has been theoretically predicted and experimentally observed in suspended h-BN [209, 215, 216, 221]. Xu et al. carried out a study of bilayer and four-layer suspended h-BN, and found that the room-temperature thermal conductivity of four-layer h-BN is lower [215, 221]. What's more, at low temperatures, a relationship, $\kappa \sim T^{2.5}$,

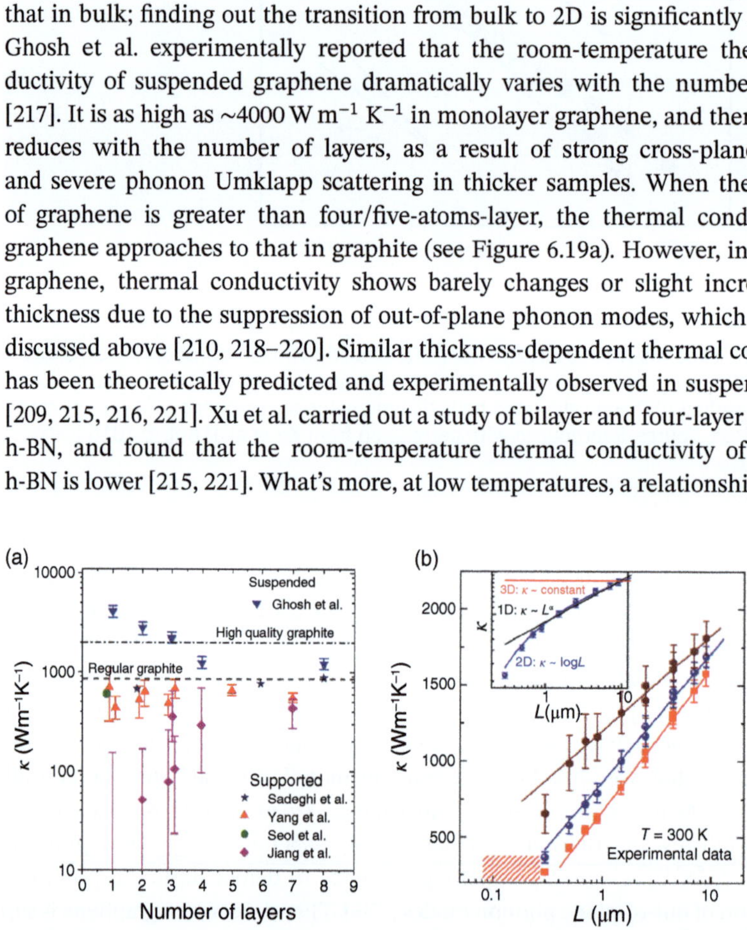

Figure 6.19 (a) Experimental observation of thickness-dependent thermal conductivity in suspended graphene [217] and supported graphene [210, 218–220]. The thermal conductivity of graphene approaches to that in graphite. Length dependence of the extracted intrinsic thermal conductivity when assuming R_c contributes negligible (red squares), 5% (blue circles) and 11.5% (brown diamonds) to the total measured thermal resistance in a 9-μm-long sample. The linear, solid lines are guides to the eyes. Inset: illustration of $\log\kappa \sim \log L$ scaling behavior for 1D, 2D and 3D systems, where thermal conductivity scales as $\sim L^{0.33}$, $\sim \log L$ and constant, respectively. (b) Experimental observation of length-dependent thermal conductivity in monolayer suspended graphene. Source: Reproduced with permission from Xu et al. [201].

is observed in four-layer h-BN, which indicates that its thermal conduction behavior is close to bulk h-BN. Further verification was supported by Li et al., and they claimed that the Grüneisen parameter and the frequencies of out-of-plane phonon modes increased with h BN layers [216].

While for other 2D materials, e.g. black phosphorous and telluride zirconium, their thermal conductivity decrease as thickness decrease because of severe phonon boundary scattering [222, 223]. Note that experimental evidence is necessary to confirm whether this variation tendency still holds with the thickness below 10 nm.

In bulk materials, the size is much larger than the phonon MFP, and phonons propagate diffusively; and therefore, a size-independent thermal conductivity is acknowledged. For 2D materials, the phonon MFP can be close to the thickness of material, or even smaller than that. Thus, quasi-ballistic phonon transport and an anomalous size-dependent thermal conductivity emerge. The first theoretical research on length-dependent thermal conductivity in a 2D lattice was conducted on the Fermi–Pasta–Ulam lattice [224]. A logarithmic divergence between length of materials and thermal conductivity, $\kappa \sim \log(L)$, was predicted. Graphene provides an ideal platform to study size-dependent thermal conductivity. Nika et al. found that the room-temperature thermal conductivity of single-layer graphene is significantly impacted by sample length when $L < 30\,\mu m$ [225]. Through Monte Carlo simulation, Knezevic et al. reported that in grapheme, the MFP of ~20% phonons is longer than 100 μm at $T = 300\,K$ [226]. Suspended single-layer graphene with a length ranging from 300 nm to 10 μm was prepared by Xu et al. [201] The measured thermal conductivity increases with length and follows a logarithmic divergence trend (see Figure 6.19b). What's more, width-dependent thermal conductivity was also verified in this experiment. The narrower graphene (width ~1.5 μm) possesses lower thermal conductivity compared with the wider one (width ~4 μm), due to severe phonon boundary scattering. Recently, experiments about width dependent thermal conductivity in single-layer graphene and size-dependent thermal conductivity in single/few-layer molybdenum sulfide were carried out [227, 228].

2D materials hold wide application prospects in various fields, such as field effect transistors [229, 230], thermoelectric devices [231, 232], and thermal management [233, 234]. Hotspots with high temperatures usually appear in high-power electronic devices. The operating efficiency and service life of devices are seriously damaged if hotspots exist for a long time. Thus, the challenge of heat dissipation becomes a bottleneck for the electronic industry. It's a promising approach that utilizes the excellent thermal conduction properties of 2D materials, represented by graphene and h-BN, to realize rapid heat dissipation and avoid hotspot damage.

In micro-/nano-scale devices, single/few-layer graphene can be qualified as a heat spreader directly. Balandin et al. transferred exfoliated few-layer graphene to a SiN_x substrate, on which an AlGaN/GaN transistor was fabricated [235]. Compared with no graphene device, the temperature of hotspots cooled down about 20 °C when the transistor operated at a power density of ~13 W mm^{-1}, and they predicted that the life time of this device was extended about ten times.

The cooling efficiency of CVD graphene is slightly worse, but it's more suitable for large-scale industrial production [236]. In recent years, high-quality and large-size CVD graphene has been prepared in laboratory, making it worthy to anticipate graphene as a heat dissipation material [237, 238].

When encountering continuous high-density heat flow, single-/few-layer graphene is likely to be invalid for cooling. In this case, the advantages of graphite thin films assembled from graphene become obvious. Through strictly controlling pressure and temperature, Liu et al. achieved a thermal conductivity as high as $\sim$3200 W m^{-1} K^{-1} in graphite film, exceeding that in HOPG [239]. Graphene not only directly serves as a heat spreader but can also be mixed with thermal interface materials to modify their heat conduction properties. Shahil et al. found that the room-temperature thermal conductivity of commercial thermal grease is increased from $\sim$5.8 to $\sim$14 W m^{-1} K^{-1} with only 2 vol% graphene loading, and mechanical properties are hardly impacted [240].

In addition to graphene, h-BN also shows excellent performance in heat dissipation. In certain situations, h-BN is more applicable because of its insulation and smooth surface. With the assistance of h-BN, significant temperature drops were realized in WSe_2 devices, and the saturated power intensity of devices were improved, indicating that h-BN can make heat spread quickly and avoid forming hotspots [241].

6.4.6 Superconductors

2D superconducting materials have received great attention in recent years because of their excellent physical and chemical properties, which are completely different from their bulk. We survey the recent developments in the field of 2D superconductivity. Several typical 2D superconductors and their properties are thoroughly described below, including carbon-based unconventional superconductivity, cuprate-based high-temperature superconductors, mechanically exfoliated 2D crystals and ion-gated superconductors, and highly crystalline atomic layers grown by MBE.

6.4.6.1 2D Superconductors and Their Characteristics

In 2018, Jarillo-Herrero and Yuan Cao reported the Mott insulating state and superconducting state in "magic angle" graphene. The first article [132] mainly described that when the twisted bilayer graphene has a turning angle closing to the 'magic' angle (approximately 1.1° under normal conditions), the electronic band structure near Fermi energy becomes flat, owing to strong interlayer coupling. This may result in a metal-insulator transition that a Mott-like insulator states emerges when the energy band is half-filled. It can arise from electrons being localized in the superlattice that is induced by the moiré pattern. The main significance of this work is to provide a novel and tunable platform for studying the more exotic correlated systems, such as unconventional superconductors and quantum spin liquids.

Upon electrostatic doping, a tunable zero-resistance state with a critical temperature of 1.7 K can be observed [133]. Taking the extended cell formed by moiré as a unit, the carrier concentrations vary from $+4 \times 10^{12}$ to -4×10^{12} cm^{-2} continuously by gating. Because of the existence of the superlattice, there will be an energy gap near $\pm 2.7 \times 10^{12}$ cm^{-2}, as a result, the conductivity drops to zero and presents a band insulator state with the energy band fully filled. However, when it is half-filled (two electrons per moiré unit), that is, near $n_s/2 = \pm 1.4 \times 10^{12}$ cm^{-2}, another platform with zero conductivity appears magically, very similar to the behavior of Mott insulator of cuprates. The discovery of superconductivity in "magic" angle graphene has opened a new door for the low-dimensional world to control electronic states. It is also the first time that superconductivity has been achieved in a pure carbon-based two-dimensional material.

Recently, Yuanbo Zhang and Xianhui Chen's groups reported the high-temperature superconductivity in monolayer $Bi_2Sr_2CaCu_2O_{8+\delta}$ (Bi-2212) [35]. The researchers developed a new manufacturing process to fabricate samples on a cold stage kept at −40 °C inside an Ar-filled glove box with water and oxygen content below 0.1 ppm. They succeeded in obtaining high-quality, intrinsic monolayer Bi-2212 (a monolayer refers to a half-unit cell that contains two CuO_2 planes). Their investigations of superconductivity, the pseudogap, charge order, and the Mott state at various doping concentrations reveal that the phases are indistinguishable from those in the bulk. Therefore, the single-layer Bi-2212 shows all the basic physical principles of high-temperature superconductivity. The results of this study establish a new platform for single-layer copper oxide to study two-dimensional high-temperature superconductivity and other strongly related phenomena.

WTe_2 is a layered material crystallizing in a distorted version of the common MoS_2 structure type, and its single-layer material can be obtained by ME. WTe_2 is predicted to possess nontrivial topological phases in its bulk counterpart as type-II Weyl fermions [242]. Due to the discovery of unsaturated magnetoresistance [243], WTe_2 has attracted widespread attention in condensed matter. Jarillo-Herrero and coworkers [244] used the top and bottom electrostatic gating to tune the superconductivity in the quantum spin Hall insulator monolayer WTe_2 [245]. The monolayer's ground state can be continuously gate-tuned from the topologically insulating state to the superconducting state. $T_c \sim 1$ K is achieved by applying different V_{bg} as the V_{tg} is kept fixed at 5 V. Their results establish monolayer WTe_2 as a material platform for engineering nanodevices that combine superconducting and topological phases of matter.

Li and coworkers successfully synthesized the K-intercalated T_d-WTe_2 by using the liquid ammonia method [246]. K_xWTe_2 undergoes the superconducting transition at ~2.6 K due to the electron doping effect resulting from K intercalation (Figure 6.20). The superconductivity exhibits an evident anisotropic behavior. There is no change observed in the crystal lattice and no prominent suppression of the positive magnetoresistance, it is reasonable to expect that the topology of WTe_2 as the type II Weyl semimetal persists. These results indicate that the K-intercalated WTe_2 may be a promising candidate to explore the topological superconductor.

Qi-Kun Xue and coworkers reported the type-II Ising pairing in few-layer stanene [247]. Trilayer stanene with high-quality atomic structure was grown on PbTe substrates by low-temperature MBE. The temperature dependence of

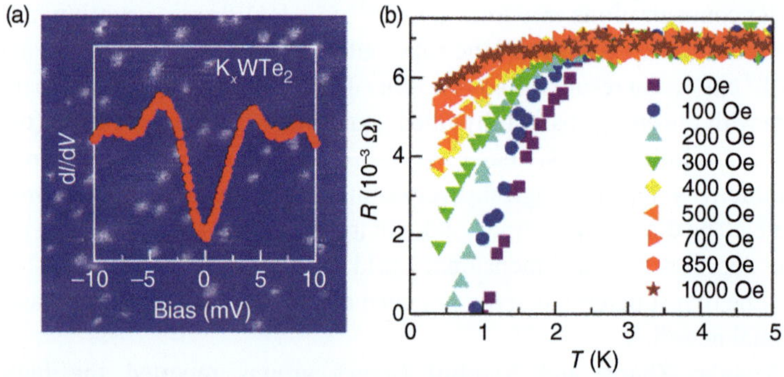

Figure 6.20 (a) dI/dV spectra measured on a bias voltage scale from −10 to +10 mV on the surface of K_xWTe_2. (b) Evolution of R vs T under different magnetic fields perpendicular to the ab plane. Source: Reproduced from Zhu et al. [246]. © 2018 American Chemical Society.

sheet resistance shows that the sample undergoes a superconducting transition at 1.1 K. They observed an anomalous increase of upper critical field $B_{c2}//$ by 30% over the conventional behavior in a temperature window as narrow as 200 mK. A new "type-II Ising pairing" was proposed to elucidate the mechanism of the upper critical field enhancement in stanene. The bands are split as a result of spin-orbit locking without the participation of inversion symmetry breaking. This phenomenon is different from Ising type-I pairing mechanism, which relies on inversion symmetry–breaking. This work points to a broader range of materials hosting such pairing mechanisms without the participation of inversion symmetry–breaking.

Based on the self-developed field-effect transistor with a solid ion conductor as the gate dielectric (SIC-FET) technology, Xianhui Chen and co-authors successfully observed the reversible transition from electric field-controlled superconductors to ferromagnetic insulators in (Li,Fe)OHFeSe thin-layer samples, which expanded SIC-FET technology in the field of state control [248] (Figure 6.21). The origin of the ferromagnetism is ascribed to the order of the interstitial Fe ions expelled from the (Li,Fe)OH layers by gating-controlled Li injection. When the Li ions are initially driven into the thin flake, Li ions replace the Fe ions in the (Li,Fe)OH layer, and the free Fe ions move to a new structural position, thereby realizing a new metastable structure. More importantly, this process is reversible. When a negative gate voltage is applied, Fe ions at the interstitial sites can re-occupy the original positions in (Li,Fe)OH layer, leading to the disappearance of ferromagnetic state and the recovery of superconductivity. These surprising findings offer a unique platform to study the relationship between superconductivity and ferromagnetism in Fe-based superconductors.

Recently, Devarakonda and coauthors developed a bulk superlattice $Ba_6Nb_{11}S_{28}$, which consists of the 2H-NbS_2 layers and a commensurate block layer that yields enhanced two-dimensionality 2D superconductivity [249]. In $Ba_6Nb_{11}S_{28}$, the quantum mobility is on the order of 10^3 cm^2 V^{-1} s^{-1}, which greatly exceeds the 2H-NbS_2 bulk single crystals (1 cm^2 V^{-1} s^{-1}) [250]. By analyzing the quantum oscillations

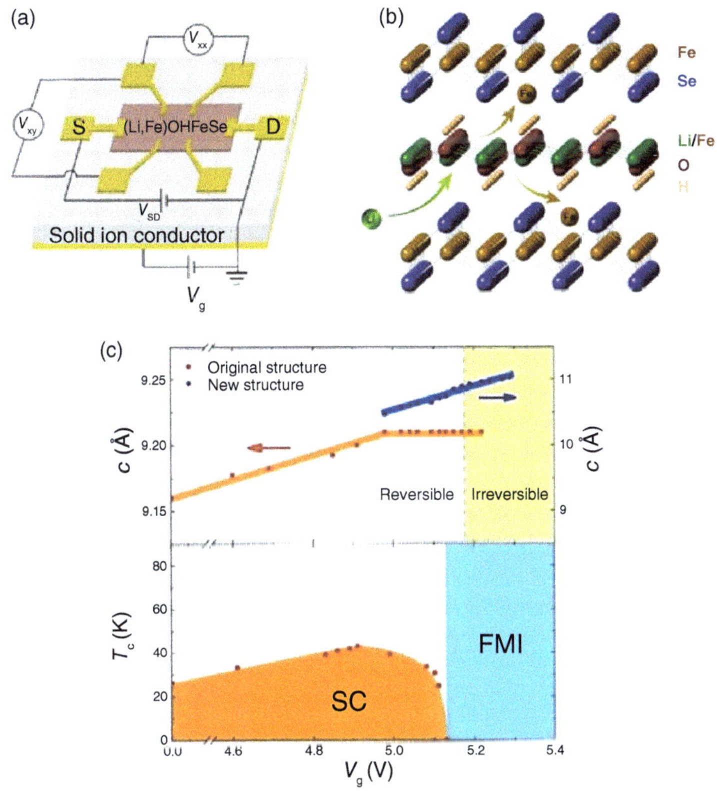

Figure 6.21 (a) A schematic view of the (Li,Fe)OHFeSe based SIC-FET device. (b) The schematic diagram of (Li,Fe)OHFeSe superconducting state-ferromagnetic insulation state reversible phase transition process. (c) The phase diagram of the gate-tuned (Li,Fe)OHFeSe thin flake. Source: Reproduced with permission from Ma et al. [248]. 2019 Elsevier.

and low field magnetoresistance, it is shown that the associated transport MFP is 1.21 mm, which greatly exceeds the Pippard coherence length (254 nm). The profile of $H_{c2}(\theta)$ following the 2D Tinkham form has shown that $Ba_6Nb_{11}S_{28}$ is a 2D superconductor. These observations taken together indicate that $Ba_6Nb_{11}S_{28}$ is a high-quality electronic and clean inorganic 2D superconductor.

6.4.6.2 Regulation Methods

Strain engineering can be used to understand how the physical properties of 2D materials can be tuned by controlling the elastic strain fields applied to them [251]. Strain can be applied in a variety of ways. Recently, Yuan Liu and a co-author reported the efficient strain modulation of MoS_2 via polymer encapsulation (Figure 6.22a) [252] and observed a higher bandgap modulation $\Delta E_g \sim 300$ meV and $S_{\Delta Eg} \sim 136$ meV/% ($S_{\Delta Eg}$, defined as the slope between ΔE_g and applied strain) of monolayer MoS_2, which is approximately two times enhancement compared to previous best. This spin-coated encapsulation method can generate a strong interaction

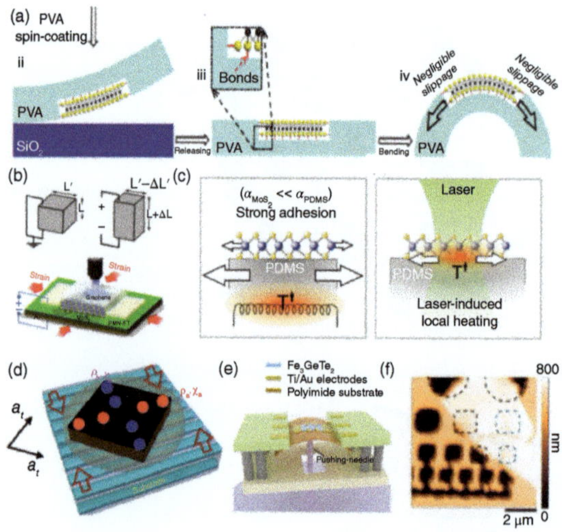

Figure 6.22 (a) Schematical illustration of fabrication processes of spin-coated encapsulation method. Source: Li et al. [252]/Springer Nature / Licensed under CC BY 4.0. (b) Uniform biaxial straining by using piezoelectric substrates. Source: Reproduced from hui et al. [253] © 2013 American Chemical Society. (c) Uniform biaxial straining by exploiting the thermal expansion of the substrate. Source: Reproduced with permission from Plechinger et al. [254] © 2015 IOP Publishing Ltd. (d) Illustration of the symmetry-breaking straining setup. The crystal is glued on top of a glass-fiber-reinforced plastic substrate using epoxy with the [108] direction parallel to fibers. Source: He et al. [255]/Springer Nature/Licensed under CC BY 4.0. (e) Schematic diagram of a device in the strain experimental set-up. Source: Wang et al. [251]/With permission from John Wiley & Sons. (f) AFM topography image of the suspended 2H $MoTe_2$ flake obtained from non-contact mode. Circles (squares) outline the cavities partially (fully) covered by $MoTe_2$. Source: Reproduced from Song et al. [256]. © 2016 American Chemical Society.

force in PVA and MoS_2, and avoid the large slippage between them. This simple strategy can be well extended to other 2D materials, leading to enhanced bandgap modulation. Yeung Yu Hui and coauthors used a piezoelectric substrate, which is capable of providing controllable compressive strain up to 0.2% to apply uniform and controllable biaxial compressive strain in trilayer MoS_2 [253] (Figure 6.22b). PL and Raman characterizations show that the direct band gap can be blue-shifted for ~300 meV per 1% strain, revealing unprecedentedly large strain tunability. Gerd Plechinger [254] exploited the large thermal expansion of polydimethylsiloxane (PDMS) substrates to apply controlled biaxial strain to a single-layer MoS_2 as shown in Figure 6.22c.

Mingquan He and the coworkers applied a large symmetry-breaking in-plane strain based on differential thermal expansion to determine the in-plane susceptibility and resistivity anisotropies of the $BaFe_2As_2$ [255] (Figure 6.22d). Their results imply that lattice and orbital degrees of freedom play a subdominant role in these materials. Wei Ji and Feng Miao investigated the change in magnetic properties of Fe_3GeTe_2 thin flakes caused by tensile strain and a dramatic increase in the coercive field of more than 150% with an applied strain of 0.32% [251]. The schematic diagram of the device in the strain experiment is shown in Figure 6.22e. Young

Hee Lee and coauthors used an AFM tip to apply mechanical strain to control the phase transition temperature of $MoTe_2$ [256] (Figure 6.22f). They demonstrated a reversible phase transition at room temperature under ambient conditions.

Although 2D material research has made exciting and continuous progress, challenges still exist. The exfoliated monolayer of 2D materials is usually extremely sensitive to its environment. The first challenge in single-layer characterization is to obtain large and high-quality single-layer samples. In addition, how to construct a stable and controllable structures on a large scale and how to further increase the concentration and uniformity of carrier doping to tune the superconducting transition temperature are the critical factors in investigating the 2D superconductivity. In the future, it is a trend to combine new types of exfoliation techniques and different tuning methods to prepare new 2D superconducting materials and discover new superconductivity-related properties.

6.5 Summary and Outlook

In conclusion, the exfoliation method opens a new path for controllable synthesis of 2D materials and forecasts a vital potential application of 2D materials. Although CVD, MBE, and many other methods have also been widely used to prepare 2D materials, the exfoliation method still shows great vitality in exploring new single-crystal 2D materials. With the development of investigations about 2D materials, the enhanced exfoliation method has gained significant attention, the size of as-prepared 2D materials has reached over a millimeter scale, and the efficiency of production has greatly increased. Especially in last two years, the optimized methods have shown great potential for realizing wafer-scale monolayer exfoliation. 2D materials have been explored in diverse research areas, including electronics, optoelectronics, magnetism, superconductivity, and many more. They can create a revolutionary effect on the industry in the next few years. We have discussed each research area in detail with the help of recently reported work. This progress greatly expanded the species and application scale of 2D materials by covering a wide spectrum range. Due to their outstanding quality, the exfoliated 2D crystals will still play an important role in basic research and practical applications of 2D materials.

Acknowledgments

This work was supported by the National Key Research and Development Program of China (Grant Nos. 2019YFA0308000, 2021YFA1401800, 2018YFA0704200, 2022YFA1403302), the National Natural Science Foundation of China (Grant Nos. 62022089, 11874405, 52272135, 51872021), Chongqing Outstanding Youth Fund (Grant No. 2021ZX0400005), and the Strategic Priority Research Program (B) of the Chinese Academy of Sciences (Grant Nos. XDB33000000). The authors thank Dr. Norman N. Shi for helpful discussions.

References

1 Kroto, H.-W., Heath, J.R., O'Brien, S.C. et al. (1985). C_{60} buckminsterfullerene. *Nature* 14: 162–163.

2 Iijima, S. (1991). Helical microtubules of graphitic carbon. *Nature* 354: 56–58.

3 Oberun, A. and Endo, M. (1976). Filmentous growth of carbon filamentous growth of carbon through benzene decompisition. *J. Cryst. Growth* 32: 335–349.

4 Endo, M. and Kroto, H.W. (1992). Formation of carbon nanofibers. *J. Phys. Chem.* 96: 6941–6944.

5 Landau, L. and Peierls, R. (1931). *Z. Angew. Phys.* 56.

6 Lu, X., Yu, M., Huang, H., and Ruoff, R.S. (1999). Tailoring graphite with the goal of achieving single sheets. *Nanotechnology* 10: 269–272.

7 Chen, H., Zhang, X.L., Zhang, Y.Y. et al. (2019). Atomically precise, custom-design origami graphene nanostructures. *Science* 365: 1036–1040.

8 Novoselov, K.S., Geim, A.K., Morozov, S.V. et al. (2004). Electric field effect in atomically thin carbon films. *Science* 306: 666–669. https://doi.org/10.1126/science.1102896.

9 Novoselov, K.S., Geim, A.K., Morozov, S.V. et al. (2005). Two-dimensional gas of massless Dirac fermions in graphene. *Nature* 438: 197–200. https://doi.org/10.1038/nature04233.

10 Zhang, Y., Tan, Y.W., Stormer, H.L., and Kim, P. (2005). Experimental observation of the quantum Hall effect and Berry's phase in graphene. *Nature* 438: 201–204. https://doi.org/10.1038/nature04235.

11 Mak, K.F., Lee, C., Hone, J. et al. (2010). Atomically thin MoS_2: a new direct-gap semiconductor. *Phys. Rev. Lett.* 105: 136805. https://doi.org/10.1103/PhysRevLett.105.136805.

12 Tan, C., Xiehong, C., Xue-Jun, W. et al. (2017). Recent advances in ultrathin two-dimensional nanomaterials. *Chem. Rev.* 117: 6225–6331. https://doi.org/10.1021/acs.chemrev.6b00558.

13 Sutter, P.W., Flege, J.I., and Sutter, E.A. (2008). Epitaxial graphene on ruthenium. *Nat. Mater.* 7: 406–411. https://doi.org/10.1038/nmat2166.

14 Pan, Y., Shi, D.-X., and Gao, H.-J. (2007). Formation of graphene on Ru(0001) surface. *Chin. Phys. B* 16: 3151.

15 Yu, Q., Lian, J., Siriponglert, S. et al. (2008). Graphene segregated on Ni surfaces and transferred to insulators. *Appl. Phys. Lett.* 93: https://doi.org/10.1063/1.2982585.

16 Li, X., Cai, W., An, J. et al. (2009). Large-area synthesis of high-quality and uniform graphene films on copper foils. *Science* 324: 1312–1314. https://doi.org/10.1126/science.1171245.

17 Bhaviripudi, S., Jia, X., Dresselhaus, M.S., and Kong, J. (2010). Role of kinetic factors in chemical vapor deposition synthesis of uniform large area graphene using copper catalyst. *Nano Lett.* 10: 4128–4133. https://doi.org/10.1021/nl102355e.

18 Liu, Q., Yu, C., He, Z. et al. (2018). Chemical vapor deposition graphene of high mobility by gradient growth method on an 4H-SiC (0 0 0 1) substrate. *Appl. Surf. Sci.* 454: 68–73. https://doi.org/10.1016/j.apsusc.2018.05.131.

19 Yu, J., Li, J., Zhang, W., and Chang, H. (2015). Synthesis of high quality two-dimensional materials via chemical vapor deposition. *Chem. Sci.* 6: 6705–6716. https://doi.org/10.1039/c5sc01941a.

20 Xu, C., Wang, L., Liu, Z. et al. (2015). Large-area high-quality 2D ultrathin Mo_2C superconducting crystals. *Nat. Mater.* 14: 1135–1141. https://doi.org/10.1038/nmat4374.

21 Zhang, Z., Chen, P., Duan, X. et al. (2017). Robust epitaxial growth of two-dimensional heterostructures, multiheterostructures, and superlattices. *Science* 10: 1126.

22 Liu, L., Park, J., Siegel, D.A. et al. (2014). Heteroepitaxial growth of two-dimensional hexagonal boron nitride templated by graphene edges. *Science* 343: 163–167.

23 Virojanadara, C., Syväjarvi, M., Yakimova, R. et al. (2008). Homogeneous large-area graphene layer growth on6H-SiC(0001). *Phys. Rev. B* 78: https://doi.org/10.1103/PhysRevB.78.245403.

24 Colombo, L., Li, X., Han, B. et al. (2010). Growth kinetics and defects of CVD graphene on Cu. *ECS Trans.* 28: 109–114.

25 Huang, P.Y., Ruiz-Vargas, C.S., Van Der Zande, A.M. et al. (2011). Grains and grain boundaries in single-layer graphene atomic patchwork quilts. *Nature* 469: 389–392. https://doi.org/10.1038/nature09718.

26 Jayasena, B. and Subbiah, S. (2011). A novel mechanical cleavage method for synthesizing few-layer graphenes. *Nanoscale Res. Lett.* 6: 95.

27 Hernandez, Y., Nicolosi, V., Lotya, M. et al. (2008). High-yield production of graphene by liquid-phase exfoliation of graphite. *Nat. Nanotechnol.* 3: 563–568. https://doi.org/10.1038/nnano.2008.215.

28 Shukla, A., Kumar, R., Mazher, J., and Balan, A. (2009). Graphene made easy: High quality, large-area samples. *Solid State Commun.* 149: 718–721. https://doi.org/10.1016/j.ssc.2009.02.007.

29 Moldt, T., Eckmann, A., Klar, P. et al. (2011). High-yield production and transfer of graphene flakes obtained by anodic bonding. *ACS Nano* 5: 7700–7706.

30 Geim, A.K. (2009). Graphene: status and prospects. *Science* 324: 1530–1534.

31 Yuan, H., Sutter, E., Shi, N.N. et al. (2015). Reliable exfoliation of large-area high-quality flakes of graphene and other two-dimensional materials. *ACS Nano* 9: 10612–10620.

32 Blake, P., Hill, E.W., Neto, A.H.C. et al. (2007). Making graphene visible. *Appl. Phys. Lett.* 91: https://doi.org/10.1063/1.2768624.

33 Zhang, Y., Tang, T.T., Girit, C. et al. (2009). Direct observation of a widely tunable bandgap in bilayer graphene. *Nature* 459: 820–823. https://doi.org/10.1038/nature08105.

34 Novoselov, K.S., Jiang, D., Schedin, F. et al. (2005). Two-dimensional atomic crystals. *PNAS* 102: 10451–10453.

35 Yu, Y., Ma, L., Cai, P. et al. (2019). High-temperature superconductivity in monolayer $Bi_2Sr_2CaCu_2O_8$+delta. *Nature* 575: 156–163. https://doi.org/10.1038/s41586-019-1718-x.

36 Desai, S.B., Madhvapathy, S.R., Amani, M. et al. (2016). Gold-mediated exfoliation of ultralarge optoelectronically-perfect monolayers. *Adv. Mater.* 28: 4053.

37 Huang, Y., Pan, Y.H., Yang, R. et al. (2020). Universal mechanical exfoliation of large-area 2D crystals. *Nat. Commun.* 11: 2453. https://doi.org/10.1038/s41467-020-16266-w.

38 Fu, Q., Dai, J.Q., Huang, X.Y. et al. (2022). One-step exfoliation method for plasmonic activation of large-area 2D crystals. *Adv. Sci.* 9: 2204247. https://doi.org/10.1002/advs.202204247.

39 Huang, Y., Wang, Y.K., Huang, X.Y. et al. (2021). An efficient route to prepare suspended monolayer for feasible optical and electronic characterizations of two-dimensional materials. *InfoMat* 13: https://doi.org/10.1002/inf2.12274.

40 Mounet, N., Gibertini, M., Schwaller, P. et al. (2018). Two-dimensional materials from high-throughput computational exfoliation of experimentally known compounds. *Nat. Nanotechnol.* 13: 246–252. https://doi.org/10.1038/s41565-017-0035-5.

41 Rhodes, D., Chae, S.H., Ribeiro-Palau, R., and Hone, J. (2019). Disorder in van der Waals heterostructures of 2D materials. *Nat. Mater.* 18: 541–549. https://doi.org/10.1038/s41563-019-0366-8.

42 Shin, B.G., Han, G.H., Yun, S.J. et al. (2016). Indirect bandgap puddles in monolayer MoS_2 by substrate-induced local strain. *Adv. Mater.* 28: 9378. https://doi.org/10.1002/adma.201602626.

43 Du, X., Skachko, I., Barker, A., and Andrei, E.Y. (2008). Approaching ballistic transport in suspended graphene. *Nat. Nanotechnol.* 3: 491–495. https://doi.org/10.1038/nnano.2008.199.

44 Du, X., Skachko, I., Duerr, F. et al. (2009). Fractional quantum Hall effect and insulating phase of Dirac electrons in graphene. *Nature* 462: 192–195. https://doi.org/10.1038/nature08522.

45 Bolotin, K.I., Ghahari, F., Shulman, M.D. et al. (2009). Observation of the fractional quantum Hall effect in graphene. *Nature* 462: 196–199. https://doi.org/10.1038/nature08582.

46 Dean, C.R., Young, A.F., Meric, I. et al. (2010). Boron nitride substrates for high-quality graphene electronics. *Nat. Nanotechnol.* 5: 722–726. https://doi.org/10.1038/nnano.2010.172.

47 Wang, L., Meric, I., Huang, P.Y. et al. (2013). One-dimensional electrical contact to a two-dimensional material. *Science* 342: 614–617. https://doi.org/10.1126/science.1244358.

48 Maher, P., Wang, L., Gao, Y. et al. (2014). Tunable fractional quantum Hall phases in bilayer graphene. *Science* 345: 61–64. https://doi.org/10.1126/science.1252875.

49 Wang, L., Gao, Y., Wen, B. et al. (2015). Evidence for a fractional fractal quantum Hall effect in graphene superlattices. *Science* 350: 1231–1234. https://doi.org/10.1126/science.aad2102.

50 Li, J.I.A., Tan, C., Chen, S. et al. (2017). Even-denominator fractional quantum Hall states in bilayer graphene. *Science* 358: 648–651. https://doi.org/10.1126/science.aao2521.

51 Koulakov, A.A., Fogler, M.M., and Shklovskii, B.I. (1996). Charge density wave in two-dimensional electron liquid in weak magnetic field. *Phys. Rev. Lett.* 76: 499–502. https://doi.org/10.1103/PhysRevLett.76.499.

52 Kaasbjerg, K., Thygesen, K.S., and Jacobsen, K.W. (2012). Phonon-limited mobility in n-type single-layer MoS_2 from first principles. *Phys. Rev. B* 85: 16. https://doi.org/10.1103/PhysRevB.85.115317.

53 Radisavljevic, B. and Kis, A. (2013). Mobility engineering and a metal-insulator transition in monolayer MoS_2. *Nat. Mater.* 12: 815–820. https://doi.org/10.1038/nmat3687.

54 Cui, X., Lee, G.H., Kim, Y.D. et al. (2015). Multi-terminal transport measurements of MoS_2 using a van der Waals heterostructure device platform. *Nat. Nanotechnol.* 10: 534–540. https://doi.org/10.1038/nnano.2015.70.

55 Allain, A., Kang, J.H., Banerjee, K., and Kis, A. (2015). Electrical contacts to two-dimensional semiconductors. *Nat. Mater.* 14: 1195–1205. https://doi.org/10.1038/nmat4452.

56 Liu, Y., Guo, J., Zhu, E. et al. (2018). Approaching the Schottky-Mott limit in van der Waals metal-semiconductor junctions. *Nature* 557: 696–700. https://doi.org/10.1038/s41586-018-0129-8.

57 Movva, H.C.P., Rai, A., Kang, S. et al. (2015). High-mobility holes in dual-gated WSe_2 field-effect transistors. *ACS Nano* 9: 10402–10410. https://doi.org/10.1021/acsnano.5b04611.

58 Fallahazad, B., Movva, H.C., Kim, K. et al. (2016). Shubnikov-de haas oscillations of high-mobility holes in monolayer and bilayer WSe_2: landau level degeneracy, effective mass, and negative compressibility. *Phys. Rev. Lett.* 116: 5. https://doi.org/10.1103/PhysRevLett.116.086601.

59 Pisoni, R., Kormányos, A., Brooks, M. et al. (2018). Interactions and magnetotransport through spin-valley coupled landau levels in monolayer MoS_2. *Phys. Rev. Lett.* 121: 6. https://doi.org/10.1103/PhysRevLett.121.247701.

60 Low, T. and Avouris, P. (2014). Graphene plasmonics for terahertz to mid-infrared applications. *ACS Nano* 8: 1086–1101.

61 Sundaram, R.S., Engel, M., Lombardo, A. et al. (2013). Electroluminescence in single layer MoS_2. *Nano Lett.* 13: 1416–1421. https://doi.org/10.1021/nl400516a.

62 Xia, F.N., Wang, H., Xiao, D. et al. (2014). Two-dimensional material nanophotonics. *Nat. Photonics* 8: 899–907. https://doi.org/10.1038/Nphoton.2010.271.

63 Xu, M., Liang, T., Shi, M., and Chen, H. (2013). Graphene-like two-dimensional materials. *Chem. Rev.* 113: 3766–3798. https://doi.org/10.1021/cr300263a.

64 Autere, A., Jussila, H., Dai, Y. et al. (2018). Nonlinear optics with 2D layered materials. *Adv. Mater.* 30: e1705963. https://doi.org/10.1002/adma.201705963.

65 Ferrari, A.C., Bonaccorso, F., Fal'Ko, V. et al. (2015). Science and technology roadmap for graphene, related two-dimensional crystals, and hybrid systems. *Nanoscale* 7: 4598–4810. https://doi.org/10.1039/c4nr01600a.

66 Sun, Z., Martinez, A. & Wang, F. Optical modulators with 2D layered materials. *Nat. Photonics* 10, 227 (2016). https://doi.org/10.1038/nphoton.2016.15. https://www.nature.com/articles/nphoton.2016.15#supplementary-information.

67 Sun, Z., Hasan, T., and Ferrari, A.C. (2012). Ultrafast lasers mode-locked by nanotubes and graphene. *Physica E* 44: 1082–1091.

68 Martinez, A. and Sun, Z. (2013). Nanotube and graphene saturable absorbers for fibre lasers. *Nat. Photonics* 7: 842–845. https://doi.org/10.1038/nphoton.2013.304.

69 Liu, X., Guo, Q., and Qiu, J. (2017). Emerging low-dimensional materials for nonlinear optics and ultrafast photonics. *Adv. Mater.* 29: 1605886. https://doi.org/10.1002/adma.201605886.

70 Yu, S., Wu, X., Wang, Y. et al. (2017). 2D materials for optical modulation: challenges and opportunities. *Adv. Mater.* 29: 1606128. https://doi.org/10.1002/adma.201606128.

71 Low, T., Chaves, A., Caldwell, J.D. et al. (2017). Polaritons in layered two-dimensional materials. *Nat. Mater.* 16: 182–194. https://doi.org/10.1038/NMAT4792.

72 Mak, K.F. and Shan, J. (2016). Photonics and optoelectronics of 2D semiconductor transition metal dichalcogenides. *Nat. Photonics* 10: 216–226. https://doi.org/10.1038/nphoton.2015.282.

73 Koppens, F.H., Mueller, T., Avouris, P. et al. (2014). Photodetectors based on graphene, other two-dimensional materials and hybrid systems. *Nat. Nanotechnol.* 9: 780–793. https://doi.org/10.1038/nnano.2014.215.

74 Manzeli, S., Ovchinnikov, D., Pasquier, D. et al. (2017). 2D transition metal dichalcogenides. *Nat. Rev. Mater.* 2: 17033. https://doi.org/10.1038/natrevmats.2017.33.

75 Wang, Q.H., Kalantar-Zadeh, K., Kis, A. et al. (2012). Electronics and optoelectronics of two-dimensional transition metal dichalcogenides. *Nat. Nanotechnol.* 7: 699–712. https://doi.org/10.1038/nnano.2012.193.

76 Kuzmenko, A.B., van Heumen, E., Carbone, F., and van der Marel, D. (2008). Universal optical conductance of graphite. *Phys. Rev. Lett.* 100: 117401. https://doi.org/10.1103/PhysRevLett.100.117401.

77 Nair, R.R., Blake, P., Grigorenko, A.N. et al. (2008). Fine structure constant defines visual transparency of graphene. *Science* 320: 1308. https://doi.org/10.1126/science.1156965.

78 Laxmi, V., Parveen, A., Tyagi, D. et al. (2022). Nanophotonic modulator based on Silicon-ITO heterojunction and slot waveguide with 2D-graphene sheet. *J. Opt.* 1-11.

79 Geim, A.K. and Novoselov, K.S. (2007). The rise of graphene. *Nat. Mater.* 6: 183–191. https://doi.org/10.1038/Nmat1849.

80 Novoselov, K.S., Falko, V.I., Colombo, L. et al. (2012). A roadmap for graphene. *Nature* 490: 192–200. https://doi.org/10.1038/nature11458.

81 Castro Neto, A.H., Guinea, F., Peres, N.M.R. et al. (2009). The electronic properties of graphene. *Rev. Mod. Phys.* 81: 109–162. https://doi.org/10.1103/RevModPhys.81.109.

82 Bonaccorso, F., Sun, Z., Hasan, T., and Ferrari, A.C. (2010). Graphene photonics and optoelectronics. *Nat. Photonics* 4: 611–622. https://doi.org/10.1038/nphoton.2010.186.

83 Mak, K.F., Sfeir, M.Y., Wu, Y. et al. (2008). Measurement of the optical conductivity of graphene. *Phys. Rev. Lett.* 101: 196405. https://doi.org/10.1103/PhysRevLett.101.196405.

84 Sun, Z.P., Popa, D., Hasan, T. et al. (2010). A stable, wideband tunable, near transform-limited, graphene-mode-locked, ultrafast laser. *Nano Res.* 3: 653–660. https://doi.org/10.1007/s12274-010-0026-4.

85 Fu, B., Hua, Y., Xiao, X. et al. (2014). Broadband graphene saturable absorber for pulsed fiber lasers at 1, 1.5, and 2 μm. *IEEE J. Select. Topics Quantum Electron.* 20: 411–415. https://doi.org/10.1109/jstqe.2014.2302361.

86 Splendiani, A., Sun, L., Zhang, Y. et al. (2010). Emerging photoluminescence in monolayer MoS2. *Nano Lett.* 10: 1271–1275. https://doi.org/10.1021/nl903868w.

87 Li, Y., Rao, Y., Mak, K.F. et al. (2013). Probing symmetry properties of few-layer MoS2 and h-BN by optical second-harmonic generation. *Nano Lett.* 13: 3329–3333. https://doi.org/10.1021/nl401561r.

88 Watanabe, K., Taniguchi, T., and Kanda, H. (2004). Direct-bandgap properties and evidence for ultraviolet lasing of hexagonal boron nitride single crystal. *Nat. Mater.* 3: 404. https://doi.org/10.1038/nmat1134.

89 Xue, H., Wang, Y., Dai, Y. et al. (2018). A $MoSe_2$/WSe_2 heterojunction-based photodetector at telecommunication wavelengths. *Adv. Funct. Mater.* 28: 1804388. https://doi.org/10.1002/adfm.201804388.

90 Lopez-Sanchez, O., Lembke, D., Kayci, M. et al. (2013). Ultrasensitive photodetectors based on monolayer MoS_2. *Nat. Nanotechnol.* 8: 497–501. https://doi.org/10.1038/nnano.2013.100.

91 Liu, Y., Shi, L., Schmitt, M.S. et al. (2017). Enhanced light-matter interactions in graphene-covered dielectric magnetic mirrors. *Opt. Express* 25: 30754–30763. https://doi.org/10.1364/OE.25.030754.

92 Echtermeyer, T.J., Britnell, L., Jasnos, P.K. et al. (2011). Strong plasmonic enhancement of photovoltage in graphene. *Nat. Commun.* 2: 458. https://doi.org/10.1038/ncomms1464.

93 Li, Y., Li, Z., Chi, C. et al. (2017). Plasmonics of 2D nanomaterials: properties and applications. *Adv. Sci.* 4: 1600430. https://doi.org/10.1002/advs.201600430.

94 Phare, C.T., Lee, Y.-H.D., Cardenas, J., and Lipson, M. (2015). Graphene electro-optic modulator with 30 GHz bandwidth. *Nat. Photonics* 9: 511–514. https://doi.org/10.1038/nphoton.2015.122.

95 Sensale-Rodriguez, B., Yan, R., Kelly, M.M. et al. (2012). Broadband graphene terahertz modulators enabled by intraband transitions. *Nat. Commun.* 3: 780. https://doi.org/10.1038/ncomms1787.

96 Zhang, H., Bao, Q., and Sun, Z. (2018). Introduction to two-dimensional layered materials for ultrafast lasers. *Photon. Res.* 6: TDL1–TDL2. https://doi.org/10.1364/prj.6.00tdl1.

97 Guo, X., Liu, R., Hu, D. et al. (2020). Efficient all-optical plasmonic modulators with atomically thin van der Waals heterostructures. *Adv. Mater.* 32: e1907105. https://doi.org/10.1002/adma.201907105.

98 Li, D., Xue, H., Qi, M. et al. (2017). Graphene actively Q-switched lasers. *2D Mater.* 4: https://doi.org/10.1088/2053-1583/aa6e6b.

99 Bogusławski, J., Wang, Y., Xue, H. et al. (2018). Graphene actively mode-locked lasers. *Adv. Funct. Mater.* 28: https://doi.org/10.1002/adfm.201801539.

100 Chen, H., Yin, J., Yang, J. et al. (2017). Transition-metal dichalcogenides heterostructure saturable absorbers for ultrafast photonics. *Opt. Lett.* 42: 4279–4282. https://doi.org/10.1364/ol.42.004279.

101 Luo, Z., Wu, D., Xu, B. et al. (2016). Two-dimensional material-based saturable absorbers: towards compact visible-wavelength all-fiber pulsed lasers. *Nanoscale* 8: 1066–1072. https://doi.org/10.1039/c5nr06981e.

102 Fu, K.-M.C., Clark, S.M., Santori, C. et al. (2008). Ultrafast control of donor-bound electron spins with single detuned optical pulses. *Nat. Phys.* 4: 780–784. https://doi.org/10.1038/nphys1052.

103 Aharonovich, I., Castelletto, S., Simpson, D.A. et al. (2011). Diamond-based single-photon emitters. *Rep. Prog. Phys.* 74: 076501. https://doi.org/10.1088/0034-4885/74/7/076501.

104 Lohrmann, A., Johnson, B.C., McCallum, J.C., and Castelletto, S. (2017). A review on single photon sources in silicon carbide. *Rep. Prog. Phys.* 80: 034502. https://doi.org/10.1088/1361-6633/aa5171.

105 Grosso, G., Moon, H., Lienhard, B. et al. (2017). Tunable and high-purity room temperature single-photon emission from atomic defects in hexagonal boron nitride. *Nat. Commun.* 8: 705. https://doi.org/10.1038/s41467-017-00810-2.

106 Tran, T.T., Wang, D., Xu, Z.Q. et al. (2017). Deterministic coupling of quantum emitters in 2D materials to plasmonic nanocavity arrays. *Nano Lett.* 17: 2634–2639. https://doi.org/10.1021/acs.nanolett.7b00444.

107 Kim, S., Fröch, J.E., Christian, J. et al. (2018). Photonic crystal cavities from hexagonal boron nitride. *Nat. Commun.* 9: 2623. https://doi.org/10.1038/s41467-018-05117-4.

108 Chakraborty, C., Kinnischtzke, L., Goodfellow, K.M. et al. (2015). Voltage-controlled quantum light from an atomically thin semiconductor. *Nat. Nanotechnol.* 10: 507. https://doi.org/10.1038/Nnano.2015.79.

109 He, Y.M., Clark, G., Schaibley, J.R. et al. (2015). Single quantum emitters in monolayer semiconductors. *Nat. Nanotechnol.* 10: 497–502. https://doi.org/10.1038/Nnano.2015.75.

110 Koperski, M., Nogajewski, K., Arora, A. et al. (2015). Single photon emitters in exfoliated WSe_2 structures. *Nat. Nanotechnol.* 10: 503. https://doi.org/10.1038/nnano.2015.67.

111 Srivastava, A., Sidler, M., Allain, A.V. et al. (2015). Optically active quantum dots in monolayer WSe_2. *Nat. Nanotechnol.* 10: 491.

112 Tonndorf, P., Schmidt, R., Schneider, R. et al. (2015). Single-photon emission from localized excitons in an atomically thin semiconductor. *Optica* 2: 347–352. https://doi.org/10.1364/Optica.2.000347.

113 Kumar, S., Kaczmarczyk, A., and Gerardot, B.D. (2015). Strain-induced spatial and spectral isolation of quantum emitters in mono-and bilayer WSe2. *Nano Lett.* 15: 7567–7573. https://doi.org/10.1021/acs.nanolett.5b03312.

114 Palacios-Berraquero, C. and Palacios-Berraquero, C. (2016). Atomically thin quantum light-emitting diodes. *Nat. Commun.* 7: 12978. https://doi.org/10.1038/ncomms12978.

115 Tongay, S., Suh, J., Ataca, C. et al. (2013). Defects activated photoluminescence in two-dimensional semiconductors: interplay between bound, charged, and free excitons. *Sci. Rep.* 3: 2657. https://doi.org/10.1038/Srep02657.

116 Tran, T.T., Zachreson, C., Berhane, A.M. et al. (2016). Quantum emission from defects in single-crystalline hexagonal boron nitride. *Phys. Rev. Appl.* 5: 034005. https://doi.org/10.1103/PhysRevApplied.5.034005.

117 Bourrellier, R., Meuret, S., Tararan, A. et al. (2016). Bright UV single photon emission at point defects in h-BN. *Nano Lett.* 16: 4317–4321. https://doi.org/10.1021/acs.nanolett.6b01368.

118 Tran, T.T., Bray, K., Ford, M.J. et al. (2016). Quantum emission from hexagonal boron nitride monolayers. *Nat. Nanotechnol.* 11: 37. https://doi.org/10.1038/Nnano.2015.242.

119 Choi, S., Tran, T.T., Elbadawi, C. et al. (2016). Engineering and localization of quantum emitters in large hexagonal boron nitride layers. *ACS Appl. Mater. Interfaces* 8: 29642–29648. https://doi.org/10.1021/acsami.6b09875.

120 Tran, T.T., Elbadawi, C., Totonjian, D. et al. (2016). Robust multicolor single photon emission from point defects in hexagonal boron nitride. *ACS Nano* 10: 7331–7338. https://doi.org/10.1021/acsnano.6b03602.

121 Hou, S., Birowosuto, M.D., Umar, S. et al. (2017). Localized emission from laser-irradiated defects in 2D hexagonal boron nitride. *2D Mater.* 5: 015010.

122 Tawfik, S.A., Ali, S., Fronzi, M. et al. (2017). First-principles investigation of quantum emission from hBN defects. *Nanoscale* 9: 13575–13582. https://doi.org/10.1039/c7nr04270a.

123 Geim, A.K. and Grigorieva, I.V. (2013). van der Waals heterostructures. *Nature* 499: 419. https://doi.org/10.1038/nature12385.

124 Hunt, B., Sanchez-Yamagishi, J.D., Young, A.F. et al. (2013). Massive Dirac fermions and hofstadter butterfly in a van der Waals heterostructure. *Science* https://doi.org/10.1126/science.1237240.

125 Dean, C.R., Wang, L., Maher, P. et al. (2013). Hofstadter's butterfly and the fractal quantum Hall effect in moiré superlattices. *Nature* 497: 598. https://doi.org/10.1038/nature12186.

126 Novoselov, K.S., Mishchenko, A., Carvalho, A., and Castro Neto, A.H. (2016). 2D materials and van der Waals heterostructures. *Science* 353: https://doi.org/10.1126/science.aac9439.

127 Carr, S., Fang, S., and Kaxiras, E. (2020). Electronic-structure methods for twisted moire layers. *Nat. Rev. Mater.* 5: 748–763. https://doi.org/10.1038/s41578-020-0214-0.

128 Yankowitz, M., Ma, Q., Jarillo-Herrero, P., and LeRoy, B.J. (2019). van der Waals heterostructures combining graphene and hexagonal boron nitride. *Nat. Rev. Phys.* 1: 112–125. https://doi.org/10.1038/s42254-018-0016-0.

129 Hofstadter, D.R. (1976). Energy-levels and wave-functions of bloch electrons in the rational and irrational magenetic fields. *Phys. Rev. B* 14: 2239–2249. https://doi.org/10.1103/PhysRevB.14.2239.

130 Bistritzer, R. and MacDonald, A.H. (2011). Moire bands in twisted double-layer graphene. *PNAS* 108: 12233–12237. https://doi.org/10.1073/pnas.1108174108.

131 Morell, E.S., Correa, J.D., Vargas, P. et al. (2010). Flat bands in slightly twisted bilayer graphene: tight-binding calculations. *Phys. Rev. B* 82: 4. https://doi.org/10.1103/PhysRevB.82.121407.

132 Cao, Y., Fetemi, V., Fang, S. et al. (2018). Unconventional superconductivity in magic-angle graphene superlattices. *Nature* 556: 43. https://doi.org/10.1038/nature26160.

133 Cao, Y. et al. (2018). Correlated insulator behaviour at half-filling in magic-angle graphene superlattices. *Nature* 556: 80. https://doi.org/10.1038/nature26154.

134 Chen, G.R. et al. (2020). Tunable correlated Chern insulator and ferromagnetism in a moire superlattice. *Nature* 579: 56. https://doi.org/10.1038/s41586-020-2049-7.

135 Chen, G.R. et al. (2019). Signatures of tunable superconductivity in a trilayer graphene moire superlattice. *Nature* 572: 215–219. https://doi.org/10.1038/s41586-019-1393-y.

136 Liu, X.M. et al. (2020). Tunable spin-polarized correlated states in twisted double bilayer graphene. *Nature* 583: 221–225. https://doi.org/10.1038/s41586-020-2458-7.

137 Park, J.M., Cao, Y., Watanabe, K. et al. (2021). Tunable strongly coupled superconductivity in magic-angle twisted trilayer graphene. *Nature* 590: 19. https://doi.org/10.1038/s41586-021-03192-0.

138 Serlin, M. et al. (2020). Intrinsic quantized anomalous Hall effect in a moire heterostructure. *Science* 367: 900–903. https://doi.org/10.1126/science.aay5533.

139 Balents, L., Dean, C.R., Efetov, D.K., and Young, A.F. (2020). Superconductivity and strong correlations in moire flat bands. *Nat. Phys.* 16: 725–733. https://doi.org/10.1038/s41567-020-0906-9.

140 Kennes, D.M. et al. (2021). Moire heterostructures as a condensed-matter quantum simulator. *Nat. Phys.* 17: 155–163. https://doi.org/10.1038/s41567-020-01154-3.

141 Wu, F.C., Lovorn, T., Tutuc, E., and MacDonald, A.H. (2018). Hubbard model physics in transition metal dichalcogenide moire bands. *Phys. Rev. Lett.* 121: 5. https://doi.org/10.1103/PhysRevLett.121.026402.

142 Wu, F.C., Lovorn, T., Tutuc, E. et al. (2019). Topological insulators in twisted transition metal dichalcogenide homobilayers. *Phys. Rev. Lett.* 122: 5. https://doi.org/10.1103/PhysRevLett.122.086402.

143 Wang, L. et al. (2020). Correlated electronic phases in twisted bilayer transition metal dichalcogenides. *Nat. Mater.* 19: 861–866. https://doi.org/10.1038/s41563-020-0708-6.

144 Gong, C. et al. (2017). Discovery of intrinsic ferromagnetism in two-dimensional van der Waals crystals. *Nature* 546: 265–269. https://doi.org/10.1038/nature22060.

145 Chittari, B.L. et al. (2016). Electronic and magnetic properties of single-layerMPX$_3$ metal phosphorous trichalcogenides. *Phys. Rev. B* 94: https://doi.org/10.1103/PhysRevB.94.184428.

146 Sachs, B., Wehling, T.O., Novoselov, K.S. et al. (2013). Ferromagnetic two-dimensional crystals: Single layers of K_2CuF_4. *Phys. Rev. B* 88: https://doi.org/10.1103/PhysRevB.88.201402.

147 Sivadas, N., Daniels, M.W., Swendsen, R.H. et al. (2015). Magnetic ground state of semiconducting transition metal trichalcogenide monolayers. *Phys. Rev. B* 91: https://doi.org/10.1103/PhysRevB.91.235425.

148 Han, M.G. et al. (2019). Topological magnetic-spin textures in two-dimensional van der Waals $Cr_2Ge_2Te_6$. *Nano Lett.* 19: 7859–7865. https://doi.org/10.1021/acs.nanolett.9b02849.

149 Huang, B. et al. (2017). Layer-dependent ferromagnetism in a van der Waals crystal down to the monolayer limit. *Nature* 546: 270–273. https://doi.org/10.1038/nature22391.

150 Deng, Y. et al. (2018). Gate-tunable room-temperature ferromagnetism in two-dimensional Fe_3GeTe_2. *Nature* 563: 94–99. https://doi.org/10.1038/s41586-018-0626-9.

151 Bonilla, M. et al. (2018). Strong room-temperature ferromagnetism in VSe_2 monolayers on van der Waals substrates. *Nat. Nanotechnol.* 13: 289–293. https://doi.org/10.1038/s41565-018-0063-9.

152 Liu, S. et al. (2017). Wafer-scale two-dimensional ferromagnetic Fe_3GeTe_2 thin films grown by molecular beam epitaxy. *NPJ 2D Mater. Appl.* 1: https://doi.org/10.1038/s41699-017-0033-3.

153 O'Hara, D.J. et al. (2018). Room temperature intrinsic ferromagnetism in epitaxial manganese selenide films in the monolayer limit. *Nano Lett.* 18: 3125–3131. https://doi.org/10.1021/acs.nanolett.8b00683.

154 Lin, M.-W. et al. (2016). Ultrathin nanosheets of $CrSiTe_3$: a semiconducting two-dimensional ferromagnetic material. *J. Mater. Chem. C* 4: 315–322. https://doi.org/10.1039/c5tc03463a.

155 Zhang, W.-B., Qu, Q., Zhu, P., and Lam, C.-H. (2015). Robust intrinsic ferromagnetism and half semiconductivity in stable two-dimensional single-layer chromium trihalides. *J. Mater. Chem. C* 3: 12457–12468. https://doi.org/10.1039/c5tc02840j.

156 Dzyaloshinsky, I. (1958). A thermodynamic theory of "Weak" ferromgntism of antiferromagnetics. *J. Phys. Chem. Solids* 4: 241–255.

157 Sander, D. (2004). The magnetic anisotropy and spin reorientation of nanostructures and nanoscale films. *J. Phys. Condens. Matter* 16: R603–R636. https://doi.org/10.1088/0953-8984/16/20/r01.

158 Gibertini, M., Koperski, M., Morpurgo, A.F., and Novoselov, K.S. (2019). Magnetic 2D materials and heterostructures. *Nat. Nanotechnol.* 14: 408–419. https://doi.org/10.1038/s41565-019-0438-6.

159 Moriya, T. (1960). Anisotropic superexchange interaction and weak ferromagnetism. *Physiol. Rev.* 120: 91–98. https://doi.org/10.1103/PhysRev.120.91.

160 Kitaev, A. (2006). Anyons in an exactly solved model and beyond. *Ann. Phys.* 321: 2–111. https://doi.org/10.1016/j.aop.2005.10.005.

161 Rau, J.G., Lee, E.K., and Kee, H.Y. (2014). Generic spin model for the honeycomb iridates beyond the Kitaev limit. *Phys. Rev. Lett.* 112: 077204. https://doi.org/10.1103/PhysRevLett.112.077204.

162 Lado, J.L. and Fernández-Rossier, J. (2017). On the origin of magnetic anisotropy in two dimensional CrI_3. *2D Mater.* 4: https://doi.org/10.1088/2053-1583/aa75ed.

163 Liu, J., Shi, M., Lu, J., and Anantram, M.P. (2018). Analysis of electrical-field-dependent Dzyaloshinskii-Moriya interaction and magnetocrystalline anisotropy in a two-dimensional ferromagnetic monolayer. *Phys. Rev. B* 97: https://doi.org/10.1103/PhysRevB.97.054416.

164 Friedt, J.M. and Sanchez, J.P. (1978). Origin of the magnetic hyperfine field transferred at iodine in ferromagnetic CrI_3. *J. Phys. C: Solid State Phys.* 11: 3731.

165 Kalvius, G.M., Oppliger, L.G., and Ruby, S.L. (1965). Mossbuer resonance of 127I in ferromagnetic CrI_3. *Phys. Lett. A* 18: 241.

166 Dillon, J.F. and Olson, C.E. (1965). Magnetization, resonance, and optical properties of the ferromagnet CrI_3. *J. Appl. Phys.* 36: 1259–1260. https://doi.org/10.1063/1.1714194.

167 Thiel, L. et al. (2019). Probing magnetism in 2D materials at the nanoscale with single-spin microscopy. *Science* 364: 973–976.

168 Soriano, D., Cardoso, C., and J., F. a.-R. (2019). Interplay between interlayer exchange and stacking in CrI_3 bilayers. *Solid State Commun.* 299: 113662.

169 Jiang, P. et al. (2019). Stacking tunable interlayer magnetism in bilayer CrI_3. *Phys. Rev. B* 99: 144401.

170 Jang, S.W., Jeong, M.Y., Yoon, H. et al. (2019). Microscopic understanding of magnetic interactions in bilayer CrI_3. *Phys. Rev. Mater.* 3: 031001(R).

171 May, A.F., Calder, S., Cantoni, C. et al. (2016). Magnetic structure and phase stability of the van der Waals bonded ferromagnet Fe_3−x$GeTe_2$. *Phys. Rev. B* 93: https://doi.org/10.1103/PhysRevB.93.014411.

172 Chen, B. et al. (2013). Magnetic properties of layered itinerant electron ferromagnet Fe_3GeTe_2. *J. Phys. Soc. Jpn.* 82: 124711. https://doi.org/10.7566/jpsj.82.124711.

173 Zhu, J.-X. et al. (2016). Electronic correlation and magnetism in the ferromagnetic metal Fe_3GeTe_2. *Phys. Rev. B* 93: https://doi.org/10.1103/PhysRevB.93.144404.

174 Bhatti, S. et al. (2017). Spintronics based random access memory: a review. *Mater. Today* 20: 530–548. https://doi.org/10.1016/j.mattod.2017.07.007.

175 Prange, R.E. and Korenman, V. (1979). Local-band theory of itinerant ferromagnetism. IV. Equivalent Heisenberg model. *Phys. Rev. B* 19: 4691–4697. https://doi.org/10.1103/PhysRevB.19.4691.

176 Long, G. et al. (2020). Persistence of magnetism in atomically thin $MnPS_3$ crystals. *Nano Lett.* 20: 2452–2459. https://doi.org/10.1021/acs.nanolett.9b05165.

177 Wildes, A.R., Rønnow, H.M., Roessli, B. et al. (2006). Static and dynamic critical properties of the quasi-two-dimensional antiferromagnetMnPS3. *Phys. Rev. B* 74: https://doi.org/10.1103/PhysRevB.74.094422.

178 Ressouche, E. et al. (2010). Magnetoelectric $MnPS_3$ as a candidate for ferrotoroidicity. *Phys. Rev. B* 82: https://doi.org/10.1103/PhysRevB.82.100408.

179 Okuda, K. et al. (1986). Magnetic properties of layered compound $MnPS_3$. *J. Phys. Soc. Jpn.* 55: 4456–4463.

180 Joy, P.A. and Vasudevan, S. (1992). Magnetism in the layered transition-metal thiophosphates MPS3 (M=Mn, Fe, and Ni). *Phys. Rev. B: Condens. Matter* 46: 5425–5433. https://doi.org/10.1103/physrevb.46.5425.

181 Pich, C. and Schwabl, F. (1995). Spin-wave dynamics of two-dimensional isotropic dipolar honeycomb antiferromagnets. *J. Magn. Magn. Mater.* 148: 30–31.

182 Mishra, R. et al. (2017). Anomalous current-induced spin torques in ferrimagnets near compensation. *Phys. Rev. Lett.* 118: 167201. https://doi.org/10.1103/PhysRevLett.118.167201.

183 Wadley, P. et al. (2016). Electrical switching of an antiferromagnet. *Science* 351: 587–590.

184 Malinowski, G. et al. (2008). Control of speed and efficiency of ultrafast demagnetization by direct transfer of spin angular momentum. *Nat. Phys.* 4: 855–858. https://doi.org/10.1038/nphys1092.

185 Gu, X., Wei, Y., Yin, X. et al. (2018). Colloquium: phononic thermal properties of two-dimensional materials. *Rev. Mod. Phys.* 90: 041002. https://doi.org/10.1103/RevModPhys.90.041002.

186 Zhang, Z. et al. (2020). Size-dependent phononic thermal transport in low-dimensional nanomaterials. *Phys. Rep.* 860: 1–26. https://doi.org/10.1016/j.physrep.2020.03.001.

187 Yang, N., Xu, X., Zhang, G., and Li, B. (2012). Thermal transport in nanostructures. *AIP Adv.* 2: 041410. https://doi.org/10.1063/1.4773462.

188 Liu, S., Xu, X., Xie, R. et al. (2012). Anomalous heat conduction and anomalous diffusion in low dimensional nanoscale systems. *Eur. Phys. J. B* 85: 337. https://doi.org/10.1140/epjb/e2012-30383-8.

189 Xu, X., Chen, J., and Li, B. (2016). Phonon thermal conduction in novel 2D materials. *J. Phys. Condens. Matter* 28: 483001. https://doi.org/10.1088/0953-8984/28/48/483001.

190 Gu, X. and Yang, R. (2016). Phonon transport and thermal conductivity in two-dimensional materials. *Ann. Rev. Heat Transfer* 19: 1.

191 Balandin, A.A. (2011). Thermal properties of graphene and nanostructured carbon materials. *Nat. Mater.* 10: 569. https://doi.org/10.1038/nmat3064.

192 Zhao, Y. et al. (2020). Thermal transport in 2D semiconductors—considerations for device applications. *Adv. Funct. Mater.* 30: 1903929. https://doi.org/10.1002/adfm.201903929.

193 Fu, Y. et al. (2020). Graphene related materials for thermal management. *2D Mater.* 7: 012001. https://doi.org/10.1088/2053-1583/ab48d9.

194 Wu, X., Tang, W., and Xu, X. (2020). Recent progresses of thermal conduction in two-dimensional materials. *Acta Phys. Sin.*, https://doi.org/10.7498/aps.7469.20200709 .

195 Balandin, A.A. et al. (2008). Superior thermal conductivity of single-layer graphene. *Nano Lett.* 8: 903.

196 Cai, W. et al. (2010). Thermal transport in suspended and supported monolayer graphene grown by chemical vapor deposition. *Nano Lett.* 10: 1645. https://doi.org/10.1021/nl9041966.

197 Faugeras, C. et al. (2010). Thermal conductivity of graphene in corbino membrane geometry. *ACS Nano* 4: 1889.

198 Ghosh, S. et al. (2008). Extremely high thermal conductivity of graphene: prospects for thermal management applications in nanoelectronic circuits. *Appl. Phys. Lett.* 92: 151911. https://doi.org/10.1063/1.2907977.

199 Lee, J.U., Yoon, D., Kim, H. et al. (2011). Thermal conductivity of suspended pristine graphene measured by Raman spectroscopy. *Phys. Rev. B* 83: 081419. https://doi.org/10.1103/PhysRevB.83.081419.

200 Li, Q. et al. (2017). Measurement of specific heat and thermal conductivity of supported and suspended graphene by a comprehensive Raman optothermal method. *Nanoscale* 9: 10784–10793. https://doi.org/10.1039/c7nr01695f.

201 Xu, X.F. et al. (2014). Length-dependent thermal conductivity in suspended single-layer graphene. *Nat. Commun.* 5: 3689. https://doi.org/10.1038/ncomms4689.

202 Pettes, M.T., Jo, I., Yao, Z., and Shi, L. (2011). Influence of polymeric residue on the thermal conductivity of suspended bilayer graphene. *Nano Lett.* 11: 1195–1200. https://doi.org/10.1021/nl104156y.

203 Jo, I. et al. (2013). Thermal conductivity and phonon transport in suspended few-layer hexagonal boron nitride. *Nano Lett.* 13: 550. https://doi.org/10.1021/nl304060g.

204 Bae, M.H. et al. (2013). Ballistic to diffusive crossover of heat flow in graphene ribbons. *Nat. Commun.* 4: 1734. https://doi.org/10.1038/ncomms2755.

205 Xu, X. *et al. Phonon Transport in Suspended Single Layer Graphene.* (2010), https://doi.org/10.48550/arXiv.1012.2937.

206 Munoz, E., Lu, J., and Yakobson, B.I. (2010). Ballistic thermal conductance of graphene ribbons. *Nano Lett.* 10: 1652–1656. https://doi.org/10.1021/nl904206d.

207 Mingo, N. and Broido, D.A. (2005). Carbon nanotube ballistic thermal conductance and its limits. *Phys. Rev. Lett.* 95: 096105.

208 Kim, P., Shi, L., Majumdar, A., and McEuen, P.L. (2001). Thermal transport measurements of individual multiwalled nanotubes. *Phys. Rev. Lett.* 87: 215502. https://doi.org/10.1103/PhysRevLett.87.215502.

209 Lindsay, L., Broido, D.A., and Mingo, N. (2010). Flexural phonons and thermal transport in graphene. *Phys. Rev. B* 82: 115427. https://doi.org/10.1103/PhysRevB.82.115427.

210 Seol, J.H. et al. (2010). Two-dimensional phonon transport in supported graphene. *Science* 328: 213.

211 Lindsay, L. and Broido, D.A. (2012). Theory of thermal transport in multilayer hexagonal boron nitride and nanotubes. *Phys. Rev. B* 85: 035436. https://doi.org/10.1103/PhysRevB.85.035436.

212 Alam, M.T., Bresnehan, M.S., Robinson, J.A., and Haque, M.A. (2014). Thermal conductivity of ultra-thin chemical vapor deposited hexagonal boron nitride films. *Appl. Phys. Lett.* 104: 013113. https://doi.org/10.1063/1.4861468.

213 Lin, Z., Liu, C., and Chai, Y. (2016). High thermally conductive and electrically insulating 2D boron nitride nanosheet for efficient heat dissipation of high-power transistors. *2D Mater.* 3: 041009. https://doi.org/10.1088/2053-1583/3/4/041009.

214 Zhou, H. et al. (2014). High thermal conductivity of suspended few-layer hexagonal boron nitride sheets. *Nano Res.* 7: 1232–1240. https://doi.org/10.1007/s12274-014-0486-z.

215 Wang, C. et al. (2016). Superior thermal conductivity in suspended bilayer hexagonal boron nitride. *Sci. Rep.* 6: 25334. https://doi.org/10.1038/srep25334.

216 Cai, Q. et al. (2019). High thermal conductivity of high-quality monolayer boron nitride and its thermal expansion. *Sci. Adv.* 5: eaav0129.

217 Ghosh, S. et al. (2010). Dimensional crossover of thermal transport in few-layer graphene. *Nat. Mater.* 9: 555. https://doi.org/10.1038/nmat2753.

218 Jang, W., Chen, Z., Bao, W. et al. (2010). Thickness-dependent thermal conductivity of encased graphene and ultrathin graphite. *Nano Lett.* 10: 3909–3913. https://doi.org/10.1021/nl101613u.

219 Yang, J. et al. (2014). Thermal conductance imaging of graphene contacts. *J. Appl. Phys.* 116: 023515. https://doi.org/10.1063/1.3212673.

220 Sadeghi, M.M., Jo, I., and Shi, L. (2013). Phonon-interface scattering in multilayer graphene on an amorphous support. *PNAS* 110: 16321–16326. https://doi.org/10.1073/pnas.1306175110.

221 Guo, J., Wang, C., Dong, L., and Xu, X. (2017). Thickness-dependent thermal conductivity of suspendede few-layer boron nitride. *J. Eng. Thermophys.* 38: 2220.

222 Zhao, Y. et al. (2018). Probing the physical origin of anisotropic thermal transport in black phosphorus nanoribbons. *Adv. Mater.* 30: 1804928. https://doi.org/10.1002/adma.201804928.

223 Guo, J. et al. (2019). Thickness-dependent in-plane thermal conductivity and enhanced thermoelectric performance in p-Type $ZrTe_5$ nanoribbons. *Phys. Status Solidi RRL* 13: 1800529. https://doi.org/10.1002/pssr.201800529.

224 Yang, L., Grassberger, P., and Hu, B. (2006). Dimensional crossover of heat conduction in low dimensions. *Phys. Rev. E* 74: 062101. https://doi.org/10.1103/PhysRevE.74.062101.

225 Nika, D.L., Ghosh, S., Pokatilov, E.P., and Balandin, A.A. (2009). Lattice thermal conductivity of graphene flakes: comparison with bulk graphite. *Appl. Phys. Lett.* 94: 203103. https://doi.org/10.1063/1.3136860.

226 Mei, S., Maurer, L.N., Aksamija, Z., and Knezevic, I. (2014). Full-dispersion Monte Carlo simulation of phonon transport in micron-sized graphene nanoribbons. *J. Appl. Phys.* 116: 164307. https://doi.org/10.1063/1.4899235.

227 Wang, H. et al. (2017). Experimental study of thermal rectification in suspended monolayer graphene. *Nat. Commun.* 8: 15843. https://doi.org/10.1038/ncomms15843.

228 Yu, Y., Minhaj, T., Huang, L. et al. (2020). In-plane and interfacial thermal conduction of two-dimensional transition-metal dichalcogenides. *Phys. Rev. Appl.* 13: 034059. https://doi.org/10.1103/PhysRevApplied.13.034059.

229 Radisavljevic, B., Radenovic, A., Brivio, J. et al. (2011). Single-layer MoS_2 transistors. *Nat. Nanotechnol.* 6: 147–150. https://doi.org/10.1038/nnano.2010.279.

230 Li, L., Yu, Y., Ye, G.J. et al. (2014). Black phosphorus field-effect transistors. *Nat. Nanotechnol.* 9: 372–377. https://doi.org/10.1038/nnano.2014.35.

231 Zhang, Y. et al. (2017). 2D black phosphorus for energy storage and thermoelectric applications. *Small* 13: 1700661. https://doi.org/10.1002/smll.201700661.

232 Wu, J., Chen, Y., Wu, J., and Hippalgaonkar, K. (2018). Perspectives on thermoelectricity in layered and 2D materials. *Adv. Electron. Mater.* 4: https://doi.org/10.1002/aelm.201800248.

233 Pop, E. (2010). Energy dissipation and transport in nanoscale devices. *Nano Res.* 3: 147–169. https://doi.org/10.1007/s12274-010-1019-z.

234 Ong, Z.Y. and Bae, M.H. (2019). Energy dissipation in van der Waals 2D devices. *2D Mater.* 6: 032005. https://doi.org/10.1088/2053-1583/ab20ea.

235 Yan, Z., Liu, G., Khan, J.M., and Balandin, A.A. (2012). Graphene quilts for thermal management of high-power GaN transistors. *Nat. Commun.* 3: 827. https://doi.org/10.1038/ncomms1828.

236 Gao, Z., Zhang, Y., Fu, Y. et al. (2013). Thermal chemical vapor deposition grown graphene heat spreader for thermal management of hot spots. *Carbon* 61: 342–348. https://doi.org/10.1016/j.carbon.2013.05.014.

237 Lee, J.H. et al. (2014). Wafer-scale growth of single-crystalmonolayer graphene on reusablehydrogen-terminated germanium. *Science* 344: 286.

238 Wu, T. et al. (2016). Fast growth of inch-sized single-crystalline graphene from a controlled single nucleus on Cu-Ni alloys. *Nat. Mater.* 15: 43–47. https://doi.org/10.1038/nmat4477.

239 Wang, N. et al. (2018). Tailoring the thermal and mechanical properties of graphene film by structural engineering. *Small* 14: e1801346. https://doi.org/10.1002/smll.201801346.

240 Shahil, K.M. and Balandin, A.A. (2012). Graphene-multilayer graphene nanocomposites as highly efficient thermal interface materials. *Nano Lett.* 12: 861–867. https://doi.org/10.1021/nl203906r.

241 Liu, D. et al. (2019). Conformal hexagonal-boron nitride dielectric interface for tungsten diselenide devices with improved mobility and thermal dissipation. *Nat. Commun.* 10: 1188. https://doi.org/10.1038/s41467-019-09016-0.

242 Soluyanov, A.A. et al. (2015). Type II Weyl semimetals. *Nature* 527: 495–498. https://doi.org/10.1038/nature15768.

243 Ali, M.N. et al. (2014). Large, non-saturating magnetoresistance in WTe2. *Nature* 514: 205–208. https://doi.org/10.1038/nature13763.

244 Fatemi, V. et al. (2018). Electrically tunable low-density superconductivity in a monolayer topological insulator. *Science* 10: 4642.

245 Shi, Y. et al. (2019). Imaging quantum spin Hall edges in monolayer WTe_2. *Sci. Adv.* 5: 8799.

246 Zhu, L. et al. (2018). Superconductivity in potassium-intercalated Td-WTe_2. *Nano Lett.* 18: 6585–6590. https://doi.org/10.1021/acs.nanolett.8b03180.

247 Falson, J. et al. (2020). Type-II Ising pairing in few-layer stanene. *Science* 10: 1126.

248 Ma, L. et al. (2019). Electric-field-controlled superconductor–ferromagnetic insulator transition. *Sci. Bull.* 64: 653–658. https://doi.org/10.1016/j.scib.2019.04.022.

249 Devarakonda, A. et al. (2020). Clean 2D superconductivity in a bulk van der Waals superlattice. *Science* 370: 231–236.

250 Naito, M. and Tanaka, a. S. (1982). Electrical transport properties in 2H-NbS_2, -$NbSe_2$, -TaS_2 and -$TaSe_2$. *J. Phys. Soc. Jpn.* 51: 219–227.

251 Wang, Y. et al. (2020). Strain-sensitive magnetization reversal of a van der Waals magnet. *Adv. Mater.* 32: e2004533. https://doi.org/10.1002/adma.202004533.

252 Li, Z. et al. (2020). Efficient strain modulation of 2D materials via polymer encapsulation. *Nat. Commun.* 11: https://doi.org/10.1038/s41467-020-15023-3.

253 Hui, Y.Y. et al. (2013). Exceptional tunability of band energy in a compressively strained trilayer MoS_2 sheet. *ACS Nano* 7: 7126.

254 Plechinger, G. et al. (2015). Control of biaxial strain in single-layer molybdenite using local thermal expansion of the substrate. *2D Mater.* 2: https://doi.org/10.1088/2053-1583/2/1/015006.

255 He, M. et al. (2017). Dichotomy between in-plane magnetic susceptibility and resistivity anisotropies in extremely strained $BaFe_2As_2$. *Nat. Commun.* 8: 504. https://doi.org/10.1038/s41467-017-00712-3.

256 Song, S. et al. (2016). Room temperature semiconductor-metal transition of $MoTe_2$ thin films engineered by strain. *Nano Lett.* 16: 188–193. https://doi.org/10.1021/acs.nanolett.5b03481.

7

Applications of Polymorphic Two-Dimensional Transition Metal Dichalcogenides in Electronics and Optoelectronics

Yao Yao[1], Siyuan Li[1], Jiajia Zha[2], Zhuangchai Lai[3], Qiyuan He[4], Chaoliang Tan[1,2,], and Hua Zhang[1,5]*

[1] *City University of Hong Kong, Department of Chemistry, 83 Tat Chee Avenue, Kowloon, Hong Kong, China*
[2] *City University of Hong Kong, Department of Electrical Engineering, 83 Tat Chee Avenue, Kowloon, Hong Kong, China*
[3] *The Hong Kong Polytechnic University, Department of Applied Physics, Hung Hom, Kowloon, Hong Kong, China*
[4] *City University of Hong Kong, Department of Materials Science and Engineering, 83 Tat Chee Avenue, Kowloon, Hong Kong, China*
[5] *City University of Hong Kong, Hong Kong Branch of National Precious Metals Material Engineering Research Center (NPMM), 83 Tat Chee Avenue, Kowloon, Hong Kong, China*

Ultrathin two-dimensional (2D) transition metal dichalcogenides (TMDs) are among the most studied 2D layered materials with great promises for future electronics and optoelectronics [1–3]. Compared to traditional bulk semiconductors, TMDs exhibit many unique and appealing features, such as the dangling-bond-free surface, efficient gate modulation, strong light–matter interaction, and superior mechanical flexibility. Moreover, they might also be compatible with the current complementary metal-oxide-semiconductor (CMOS) technology and may be used to fabricate advanced micro-/nanodevices and integrated circuits. Normally, TMDs present two kinds of crystal phases, i.e. the trigonal prismatic phase (2H) and the octahedral (1T) phase. According to structural differences, several distorted subtypes can be derived from 1T-TMDs, including 1T′, 1T″, and 1T‴ phases [4]. By virtue of the unique band structures determined by the atomic configurations, polymorphic TMDs present distinctive electronic properties and thus enable various applications. For example, the 2H-phase group VIB TMDs are typical semiconductors with bandgaps in the range of 1–2 eV, which have been extensively utilized in constructing electronic and optoelectronic devices. In contrast, (semi-)metallic TMDs with unconventional phases have been widely explored to form low-resistance ohmic contacts. The semiconducting and (semi-)metallic properties of TMDs with different phases can be explained by the d-orbital occupation of the transition metals. The unique electronic properties of polymorphic TMDs provide opportunities to design and fabricate novel devices, supplementing the traditional Si-based electronic

* Authors **Yao Yao** and **Siyuan Li** contributed equally to this work.

Two-Dimensional Transition-Metal Dichalcogenides: Phase Engineering and Applications in Electronics and Optoelectronics, First Edition. Edited by Chi Sin Tang, Xinmao Yin, and Andrew T. S. Wee.

industry. In this chapter, we will introduce the applications of polymorphic TMDs in electronics and optoelectronics. For electronic applications, polymorphic TMDs are introduced to lower contact resistance in TMD-based field-effect transistors (FETs), design novel memory devices and neuromorphic computing systems, and construct energy harvesting units. For optoelectronic applications, they are used in the fabrication of high-performance photodetectors and solar cells.

7.1 Field-Effect Transistors (FETs)

In modern electronic systems, FET is one of the most critical components for the basis of computing and memory units [5]. However, with the continuous scaling down of electronic devices, the short channel effect and quantum effect significantly impede further improvement of device performance [6]. In the past decade, ultrathin 2D layered materials, especially TMDs, have attracted increasing attention as potential candidates for next-generation nanoelectronics due to their unique chemical, physical, and electronic properties [7–11]. For example, ultrathin MoS_2-based FETs with back- and top-gate configurations were demonstrated in 2007 and 2011, respectively [12, 13]. For the top-gate FET based on single-layer MoS_2, a current on/off ratio up to 10^8 and a subthreshold swing (SS) of 74 mV dec^{-1} can be obtained [13]. Promising FET performance, such as decent field-effect electron mobility in the range of tens of cm^2 V^{-1} s^{-1}, high current on/off ratio, and small SS at the magnitude of tens of mV dec^{-1}, was obtained in these initial demonstrations [12–15]. Besides MoS_2, other TMDs, such as WS_2 and WSe_2, have also been extensively studied for building high-performance FETs [16–19].

Recently, based on the successful demonstration of TMD-based FETs, large-scale integrated circuits based on TMD thin films have also been developed [20–27]. For example, the first large-scale MoS_2-based integrated circuits designed by computer-aided design (CAD) methods were demonstrated in 2016 [26]. In 2017, a microprocessor consisting of 115 transistors was constructed based on a single-layer MoS_2 film, which could execute user-defined programs [27]. These successful attempts on single FETs and large-scale integrated circuits based on TMDs would greatly benefit the further development of novel TMD-based electronics.

However, the relatively high contact resistance in TMD-based FETs remains a critical challenge [28]. As a common method to reduce contact resistance in conventional bulk material-based transistors, the substitutional doping method would inevitably introduce extra defects into ultrathin TMDs, which would greatly lower carrier mobility. Therefore, it is not suitable to realize the ohmic contact in TMD-based devices by the substitutional doping method. On the other hand, a lower contact resistance could be achieved by tuning the work function of metal contacts to lower the Schottky barrier height. However, conventional metal deposition processes would also introduce extra defects in ultrathin TMDs, making the Schottky barrier height hardly tunable due to the Fermi-level pinning effect [29].

Hence, contact engineering is crucial to achieving high-performance TMD-based FETs [30]. In 2018, Liu et al. fabricated a van der Waals metal/semiconductor junction without the Fermi-level pinning effect at the metal/TMD interface [31].

Different from the conventional metal deposition method, pre-deposited metal contacts are transferred on TMD nanosheets. For devices with transferred metal contacts, the Schottky barrier height can be significantly reduced by aligning the work function of transferred metal contacts and the band edges of TMDs. In 2021, Shen et al. reported another effective method to suppress the Fermi-level pinning effect and obtain ultra-low contact resistance in MoS_2-based transistors [32]. In their work, the metal-induced gap states (MIGS) could be significantly suppressed by introducing semi-metallic Bi as metal contacts.

Recently, a study on TMDs with unconventional phases provided an alternative strategy to reduce the contact resistance of TMD-based FETs [33]. Because of the superior electrical conductivity, (semi-)metallic 1T-/1T′-TMDs can be utilized as low-resistance electrical contacts to construct homojunction-based FETs [33, 34]. Up to now, much progress has been made in this strategy. Normally, 1T-/1T′-TMDs can be introduced by two main methods, i.e. phase transition and direct synthesis. The following parts will introduce the related works in the two methods.

7.1.1 Homojunction-based FETs Formed by Phase Transition

In 2014, Kappera et al. realized partial 2H-to-1T phase transition of ultrathin MoS_2 with *n*-butyllithium treatment [34]. The defined regions on the MoS_2 nanosheet are transferred from 2H to 1T phase, owing to the electron donation from *n*-butyllithium. The metal electrodes (Au) are deposited on partially transferred 1T phase regions. With the transfer length method (TLM), the contact resistance of this Au/1T-/2H-MoS_2 junction is only 200–300 Ω μm at zero gate bias. In contrast, the contact resistance of the traditional Au/2H-MoS_2 junction is over 1000 Ω μm. Such improvement is attributed to the atomically sharp interface between the 1T and 2H phases and the perfect match of the work function of 1T-MoS_2 (~4.2 eV) with the conduction band edge of 2H-MoS_2. Besides, the SS value decreases significantly from 1.5 to 0.8 V dec^{-1} on introducing the 1T phase-based contact. In 2015, Ma et al. reported a similar method to improve the device performance of WSe_2-based transistors [35]. With the metal/1T-/2H-WSe_2 contacts, the WSe_2-based devices exhibited higher field-effect mobility (~66 cm^2 V^{-1} s^{-1}) and smaller SS values (~0.66 V dec^{-1}) than the devices with the metal/2H-WSe_2 contacts.

Besides the aforementioned chemical treatment method, some other phase transition approaches have been developed. For example, Cho et al. reported the laser-irradiation-induced local phase transition of $MoTe_2$ flakes from 2H to 1T′ phases in 2015 [36]. According to the surface-sensitive X-ray photoemission spectroscopy (XPS) results, the laser-irradiation-induced phase transition only occurs at the top layer at the initial stage. Then, due to the laser-thinning effect, the thickness of the $MoTe_2$ flakes decreases during the laser irradiation process, as shown in Figure 7.1a. From the *in-situ* scanning transmission electron microscopy (STEM) image, phase transition can be attributed to the atomic Te vacancies induced by the local heating effect from laser irradiation. In the fast Fourier transform (FFT) image, besides the hexagonal diffraction pattern of 2H-$MoTe_2$, extra spots with a rectangular symmetry are observed, which is the structural feature of the distorted octahedral structure (1T′ phase) [38, 39]. Using the laser-irradiation-induced 1T′-$MoTe_2$ as the contact material, the fabricated transistors exhibited superior performance to those with metal/2H-$MoTe_2$ contacts. As shown in Figure 7.1b, the Arrhenius plot shows a significant decrease in the energy barrier height, indicating the nearly ideal ohmic contacts of the metal/1T′-/2H-$MoTe_2$ junction. Furthermore, as shown in Figure 7.1c, the extracted field-effect mobility of the device with metal/1T′-/2H-$MoTe_2$ contact is about 50 times that of the device with metal/2H-$MoTe_2$ contacts.

7.1.2 Homojunction-based FETs Formed by Direct Synthesis

Although contact resistance can be reduced by phase transition methods as mentioned above, these methods would also induce undesired defects, limiting the device's performance [40]. Therefore, the direct synthesis strategy emerges as an attractive approach to obtaining high-quality TMD-based polymorphic homojunction for high-performance FETs.

In 2017, Sung et al. prepared a polymorphic homojunction of 1T′-/2H-$MoTe_2$ film by chemical vapor deposition (CVD). [37]. In this method, 1T′- and 2H-$MoTe_2$ were synthesized on the same substrate (SiO_2/Si) under different reaction temperatures (710 and 670 °C, respectively). Figure 7.1d shows the growth scheme for the coplanar epitaxy of the 1T′-/2H-$MoTe_2$ homojunction and the optical image of devices based on this junction. Benefiting from the epitaxial growth, 1T′- and 2H-$MoTe_2$ are connected by covalent bonds at the boundary, forming an ideal coplanar contact [37, 41–44]. The ideal 1T′-/2H-$MoTe_2$ interface endows the $MoTe_2$ transistors with much better performance than those devices with Au/2H-$MoTe_2$ contacts, as shown in Figure 7.1e,f. The measured Schottky barrier height of the device with Au/1T′-/2H-$MoTe_2$ contacts was ~22 meV, which is much lower than that of the device with Au/2H-$MoTe_2$ contacts (~150 meV). Following the strategy, some other works have also been reported in the past several years [45, 46].

In 2019, Zhang et al. reported a facile method to simultaneously synthesize 1T′- and 2H-$MoTe_2$ on the same substrate [47]. They systematically studied the effect of the precursors and reaction temperature on the phase of synthesized $MoTe_2$. At the same reaction temperature, 1T′- and 2H- $MoTe_2$ could be obtained by utilizing MoO_3 and $MoO_{2.5}$ as precursors, respectively. The precursors (MoO_3 and $MoO_{2.5}$) are first patterned according to the designed circuit structure. Then, after the one-step phase-controlled synthesis, the pre-patterned MoO_3 and $MoO_{2.5}$ can be converted into 1T′- and 2H- $MoTe_2$. The 2H regions act as the channels of FETs and the 1T′ regions serve as the electrical contacts and connections. The barrier height of the Pd/1T′-/2H-$MoTe_2$ contact is ~20 meV, which is only a tenth of the Pd/2H-$MoTe_2$ contact. The lower barrier height of the Pd/1T′-/2H-$MoTe_2$ contact would cause a much higher carrier injection efficiency and higher field-effect mobility. In addition, buried local gates and 12 nm thick high-k dielectrics (HfO_2) were introduced to improve gate controllability. The optimized FETs exhibited a high on/off ratio of $\sim10^5$, a high mobility of ~130 $cm^2\ V^{-1}\ s^{-1}$, and a small SS of ~69 mV dec^{-1}, approaching the theoretical limit of 60 mV dec^{-1}. By virtue of the

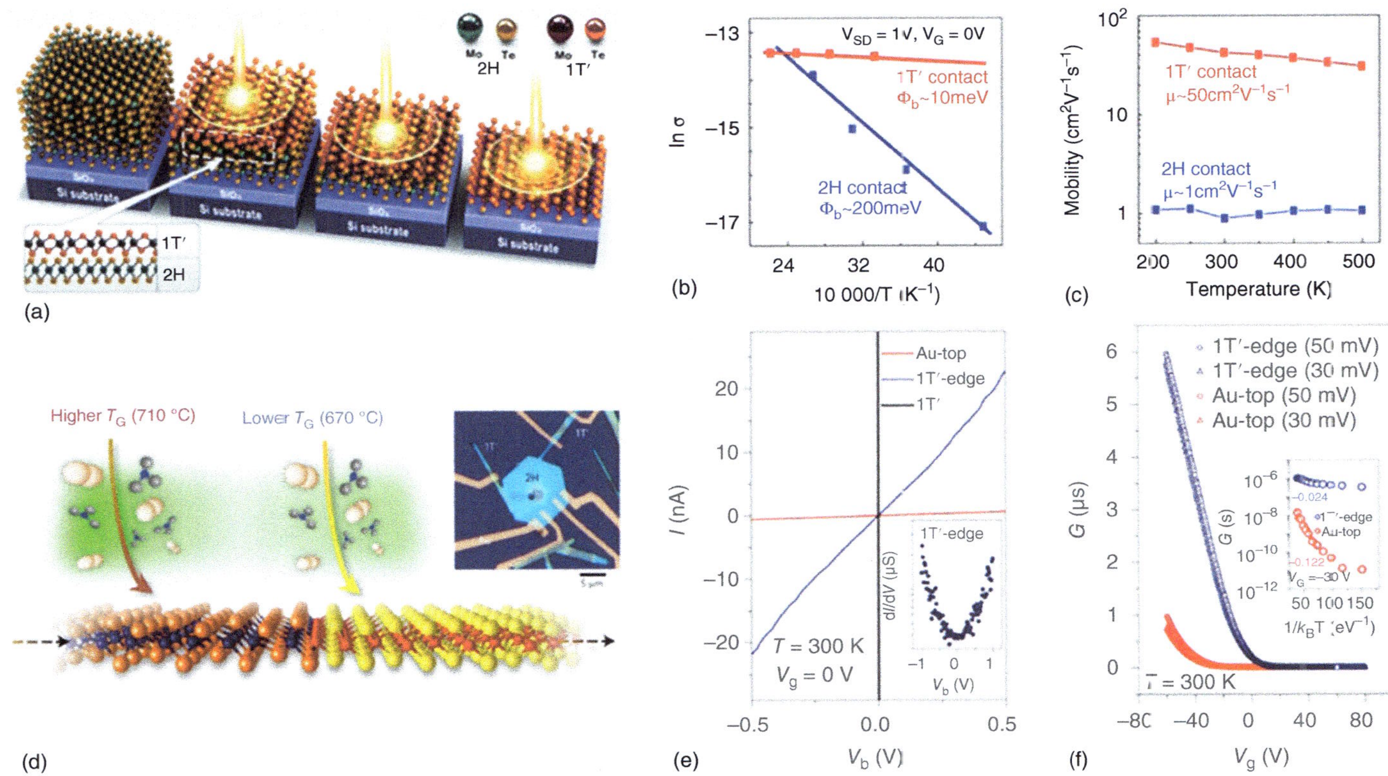

Figure 7.1 (a) Schematic illustration of the laser-irradiation-induced phase transition of $MoTe_2$. (b, c) Temperature-related conductance (b) and field-effect mobility (c) of $MoTe_2$-based FETs with Au/1T′-/2H-$MoTe_2$ contacts (red) and Au/2H-$MoTe_2$ contacts (blue), respectively. (d) Schematic illustration of the coplanar epitaxy growth of 1T′-/2H-$MoTe_2$ homojunction. Inset: optical image of devices based on 1T′-/2H-$MoTe_2$ homojunction. (e) I–V curves of 2H-$MoTe_2$ transistors with Au/1T′-/2H-$MoTe_2$ contacts (blue) and Au/2H-$MoTe_2$ contacts (red), and 1T′-$MoTe_2$ transistor with Au/1T′-$MoTe_2$ contacts (black). Inset: dI–dV curve of the 2H-$MoTe_2$ transistor with Au/1T′-/2H-$MoTe_2$ contacts. (f) Transfer characteristics of 2H-$MoTe_2$ transistors with Au/1T′-/2H-$MoTe_2$ contacts (blue) and Au/2H-$MoTe_2$ contacts (red), respectively. Inset: Arrhenius plots of the 2H channel conductance with the two kinds of contacts at $V_g = -30$ V. Source: (a–c) Cho et al. [36]/Reproduced with permission from American Association for the Advancement of Science.; (d–f) Sung et al. [37]/Reproduced with permission from Springer Nature.

excellent device performance and facile synthesis process, this method shows great potential in fabricating TMD-based circuits.

7.2 Memory and Neuromorphic Computing

It is well known that memory units play an indispensable role in modern electronics, in which bigger storage capacity and faster read/write speed are consistently being pursued [48]. Many emerging memory technologies, such as resistive random-access memories (RRAM), magnetic random-access memories (MRAM), phase-change memories (PCM), and ferroelectric random-access memories (FRAM), have been developed in the past several decades [49]. Besides advanced device structures, various emerging materials with modulable resistance also play a critical role in promoting future memory technologies. Therefore, TMDs have attracted tremendous interest in fabricating advanced memory devices due to many advantages, such as low-power switching capability, electrostatic gate tunability, and mechanical flexibility.

In 2015, Sangwan et al. reported a MoS_2-based lateral memristor, in which the resistance change originated from the defect migration toward grain boundaries [50]. In this work, they fabricated a lateral MoS_2-based memristor with an intersecting grain boundary. When applying a positive bias, sulfur vacancies move to the grain boundaries and induce a lower resistance of the MoS_2 channel, corresponding to the set process. On the other hand, when a negative bias was applied, sulfur vacancies moved away from the grain boundaries and resulted in a higher resistance of the MoS_2 channel, which corresponded to the reset process. In 2018, the same group reported a different lateral memristor based on the single-layer polycrystalline MoS_2 [51]. In this work, atomic vacancies in the MoS_2 channel could migrate to contact regions through the grain boundaries and modulate the Schottky barrier height at the contact interfaces, inducing memristive behavior.

Besides the aforementioned lateral structures, TMD-based vertical memristors have also been widely studied. In 2019, Yan et al. reported a low-power consumption vertical memristor with a sandwich structure of Pd/WS_2/Pt (Figure 7.2a) [52]. As shown in Figure 7.2b, the I–V characteristic exhibits typical switch behavior at a low set bias (0.6 V) and a low reset bias (−0.2 V). The resistance switching performance could come from the generation of vacancies and the electron hopping among vacancies. In addition, the memristor exhibits a short on/off switching time, a low on-state current, and low power consumption in switching processes. In 2019, Xu et al. reported another low-voltage vertical memristor with a Cu/MoS_2/Au structure [55]. The switching behavior could be attributed to the diffusion of Cu ions from the Cu electrode, forming the atomic-scale filaments. The cross-sectional area of the Cu filament can be modulated by applying different biases, causing a resistance switch between a low resistance state (LRS) and a high resistance state (HRS).

In the past few years, with the wide exploration of unconventional phase TMDs, the phase-dependent resistance switching performance has attracted considerable attention in fabricating novel memristors. Through various phase engineering

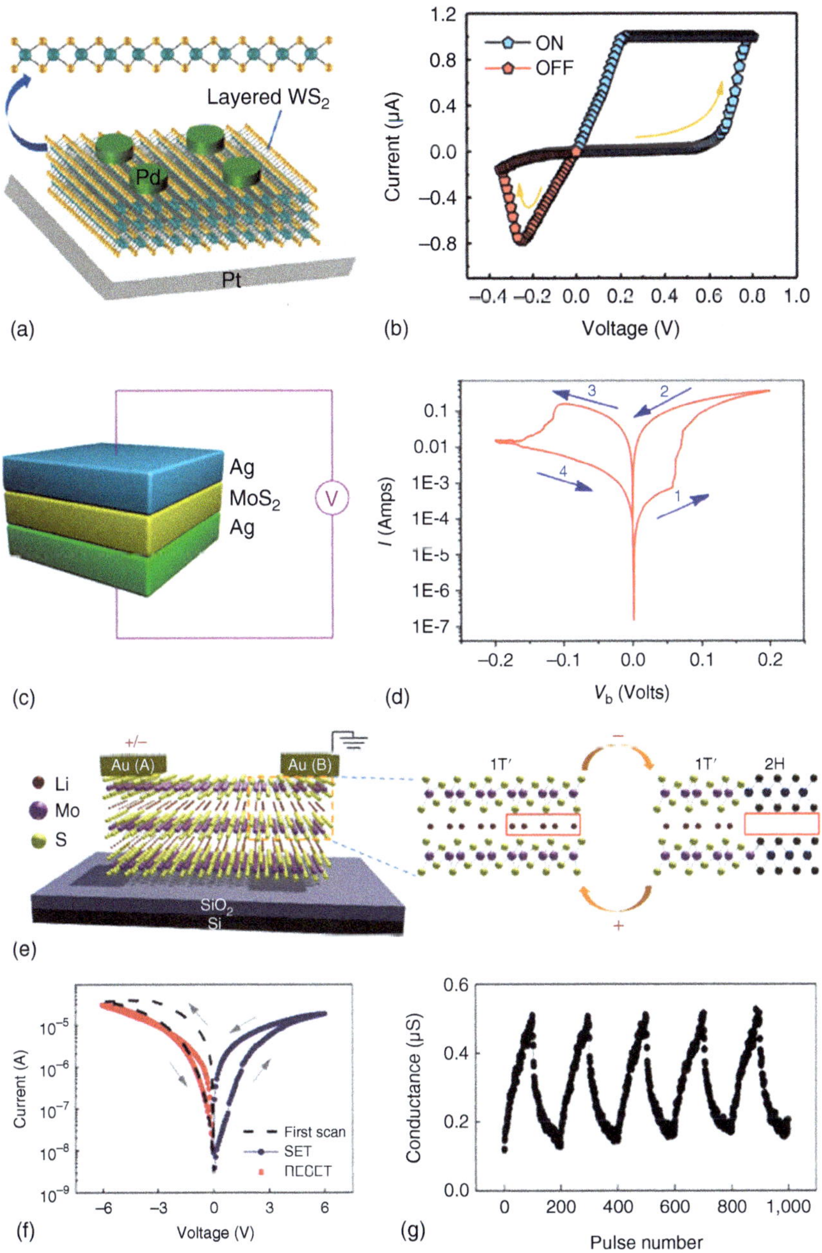

Figure 7.2 (a) Schematic illustration of the Pd/WS_2/Pt memristor. (b) $I-V$ characteristic of the Pd/WS_2/Pt memristor at sub-1 V range. (c) Schematic illustration of the Ag/1T-MoS_2/Ag memristor. (d) $I-V$ characteristic of Ag/1T-MoS_2/Ag memristor at room temperature. (e) Schematic illustration of the local 2H-1T′ phase transition in MoS_2 by the intercalation and migration of Li^+ ions. (f) $I-V$ characteristics of the Au/MoS_2/Au memristor. (g) Conductance changes induced by programmed stimulation. Source: (a, b) Yan et al. [52]/Reproduced with permission from John Wiley & Sons; (c, d) Reproduced from Cheng et al. [53]; (e–g) Reproduced from Zhu et al. [54].

strategies, TMDs could be modulated between (semi-)metallic and semiconducting phases, corresponding to the switching behavior between LRS and HRS [32]. For example, in 2016, Cheng et al. reported a 1T-MoS_2-based memristor with the typical crossbar structure, as shown in Figure 7.2c [53]. The active layer comprises 1T-MoS_2 nanosheets prepared by a chemical exfoliation method. Figure 7.2d shows the typical I–V characteristics of this memristor. The set voltage from HRS to LRS is 66 mV, and the reset voltage is −98 mV. The degradation of the I–V characteristics after 1000 cycles is negligible, indicating the excellent stability of the 1T-MoS_2-based memristor. Compared to the 2H-MoS_2, the octahedral coordination of 1T-MoS_2 would induce a strong interaction between the orbitals of Mo and S, leading to the delocalization of electrons [53]. When applying a strong forward electric field, the enhanced lattice distortion and electron delocalization could increase the conductivity of MoS_2 film. In contrast, when a strong reverse electric field is applied, the lattice distortion can be restored, and the conductivity is decreased.

In 2019, Zhang et al. reported their $MoTe_2$- and $Mo_{1-x}W_xTe_2$-based memristors, in which the multi-resistance states originate from the electric-field-induced phase transition from 2H to the distorted $2H_d$ and orthorhombic T_d phases [56]. According to the I–V characteristics of the $MoTe_2$- and $Mo_{1-x}W_xTe_2$-based memristors, the set/reset voltages and the width of hysteresis loops are positively related to the thickness of $MoTe_2$ and $Mo_{1-x}W_xTe_2$. Therefore, the set/reset voltage of memristors could be feasibly modulated by changing the thickness of the active layers. In addition, the $MoTe_2$-based memristor exhibits good performance stability and high switching speed. The switching between HRS and LRS could be accomplished within 10 ns. For practical applications, a tunneling barrier layer (Al_2O_3) was also inserted between the top electrode and the $MoTe_2$ film to realize the long-term maintenance and the high on/off ratio.

In the past decade, driven by the urgent requirement for high-speed calculation and high capacitive storage, the novel neuromorphic computing architecture has been widely studied to supplement the traditional von Neumann computing architecture [57]. For mimicking human information processing in neuromorphic electronics, multiple-state modulation of memristors' resistance is critical to memorize previous electrical inputs and affecting the response to upcoming electrical inputs [57]. As discussed previously, the resistance of TMDs can be modulated by phase transition, making TMDs good candidates for constructing artificial synaptic devices in neuromorphic electronics. For example, in 2019, Zhu et al. reported a representative ion-modulated MoS_2-based memristor. Figure 7.2e shows a schematic illustration of the local 2H-to-1T′ phase transition in MoS_2 [54]. By immersing the device in n-butyllithium solution, Li^+ ions can be intercalated into the MoS_2 film to partially form the 1T′ phase in MoS_2. Due to the highly anisotropic ionic transport properties, it is possible to guide the migration of the intercalated Li^+ ions in specific directions by controlling the lateral electric field. By applying an appropriate electric field to uniformly distribute Li^+ ions in the whole MoS_2 channel, 1T′ phase is dominant, exhibiting the lowest resistance. In contrast, when Li^+ ions are concentrated in a small region, the whole MoS_2 channel would be dominated by the 2H phase and the resistance would be the highest.

Therefore, the resistance of the MoS_2 channel could be modulated in multiple states by changing the distribution of Li^+ ions. Figure 7.2f shows the I–V characteristic of this MoS_2-based memristor. Figure 7.2g shows that the device conductance can be continuously increased/decreased by applying positive/negative voltage pulses and emulating the synaptic potentiation/depression behaviors. The synaptic interactions, such as synaptic cooperation [58], were also investigated in the multi-terminal device, which includes four independent electrodes (A, B, C, and D) and a common ground electrode (G). Initially, the Li^+ ions are concentrated under electrodes A, B, C, and D, and no ion is under the ground electrode G. When the positive electric pulses are applied to the electrodes A, B, and C in sequence, the channel conductance of A–G, B–G, and C–G increases with more Li^+ ions migrating from A, B, and C to electrode G. Interestingly, the conductance of D–G gradually increases without electrical pulse being applied to electrode D. This function is interesting as it successfully mimics synaptic modification among adjacent synapses.

7.3 Energy Harvesting

Recently, flexible electronics have attracted a lot of attention for constructing smart skins for soft robotics and health monitoring [59]. Due to their excellent electrical and mechanical properties, TMDs have been widely used to fabricate various flexible devices, such as transistors, sensors, and memristors. In addition, the self-powered system is of vital importance for practical applications of smart skins, in which TMD-based energy harvesting units also show great promise [60]. This section will introduce two typical TMD-based energy harvesting units whose energy sources are the heat of the human body and electromagnetic radiation, respectively.

In 2016, Oh et al. reported a TMD-based wearable thermoelectric generator [61]. As shown in Figure 7.3a, TMD nanosheets are chemically exfoliated from the bulk WS_2 and $NbSe_2$ with the n-butyllithium solution and purified using a 100 nm-pore-sized filter membrane. The purified TMD nanosheets were transferred onto a flexible polydimethylsiloxane (PDMS) substrate to fabricate the flexible thermoelectric generator. The power factors for the $NbSe_2$ film and the WS_2 film are experimentally confirmed as 26–34 and 5–7 $\mu W\,K^{-2}\,m^{-1}$, respectively. Figure 7.3b shows the overall device performance of a single thermoelectric generator. The output power of a single unit is calculated to be 0.63 nW at the temperature difference (ΔT) of 60 K. To further improve the output power, the generator units were connected in parallel. When 100 units were stacked in parallel, the output power could reach as high as 38 nW. As shown in Figure 7.3c, the output voltage of the wristband-type thermoelectric generator gradually increases to ~2.4 mV and is maintained stably during various human motions. Therefore, the TMD-based thermoelectric generator can be a potential power supply device for wearable electronics.

In 2019, Zhang et al. proposed another strategy to realize energy harvesting based on TMDs and a MoS_2-based flexible Schottky diode was fabricated to harvest the energy from Wi-Fi signals [62]. Figure 7.3d shows the schematic illustration of the lateral Schottky diode based on the 2H- and 1T-/1T′-MoS_2 heterophase junction.

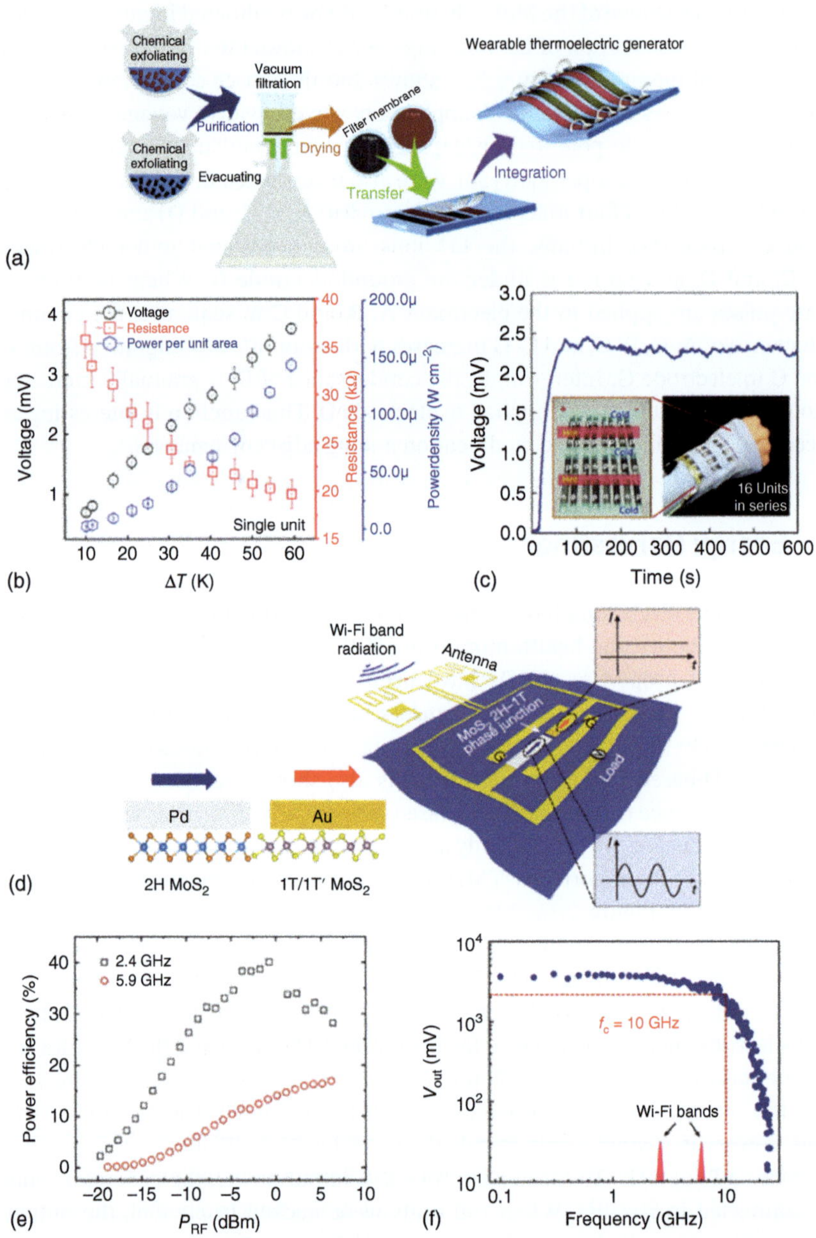

Figure 7.3 (a) Schematic illustration of the fabrication process of the wearable TMD-based thermoelectric generator. (b) The correlations between output voltage (black), resistance (red), output power per unit area (blue), and the temperature difference. (c) The output voltage of the wristband-type thermoelectric generator at a temperature difference of 3 K. Inset: images of the device configuration and the operating situation. (d) Schematic illustration of a flexible lateral Schottky diode based on the heterophase junction of 2H- and 1T-/1T′-MoS_2, which is integrated with a Wi-Fi band antenna for wireless energy harvesting. (e) The power efficiency of the MoS_2-based rectifier versus the input power with two frequencies in the Wi-Fi band. (f) The output voltage versus the input signal frequency (P_{input} = 5 mW). Source: (a–c) Reproduced from Oh et al. [61]; (d–f) Reproduced from Zhang et al. [62].

Pd electrode was deposited on a selected region of 2H-MoS_2 nanosheet to form a Schottky junction, serving as a rectifier. The exposed region without Pd contact was chemically transformed into the 1T/1T′ phase, forming an ohmic contact between 2H- and 1T-/1T′-MoS_2. Finally, the Au electrode was deposited on the transformed 1T-/1T′-MoS_2. From the current responsivity of the nonlinear rectifier, the rectified output current, at around zero bias, is 4.7 A W^{-1}, which is comparable to that of the traditional rigid silicon devices [63]. The ohmic contact between 2H- and 1T-/1T′-MoS_2 reduced the series resistance of this Schottky diode, which is critical to get a small RC (resistor-capacitor) time constant and a high cut-off frequency (f_c) [63, 64]. Meanwhile, due to the lateral configuration and the atomic thickness of the MoS_2 film, the device capacitance is substantially minimized [62]. The extracted junction capacitance (C_j) in the low bias region (<0.4 V) is in the sub-10-fF regime, one order of magnitude lower than that of the previously reported flexible rectifiers [65–67]. The small total capacitance endows the device with a cut-off frequency of around 10 GHz. Figure 7.3e shows the calculated power conversion efficiency in the Wi-Fi band. When the frequency is 2.4 GHz, the device delivers the highest power efficiency of 40.1% at an input power of −0.7 dBm. Moreover, as shown in Figure 7.3f, when the input power is 5 mW, the output voltage could exceed 2 V below the cut-off frequency of 10 GHz, covering two commonly used Wi-Fi channels (2.4 and 5.9 GHz). This work provides a universal energy-harvesting strategy, which holds great promise to be integrated into future flexible electronic systems.

7.4 Photodetectors

Photodetectors that convert incident light into electrical signals are the fundamental optoelectronic components in sensing, imaging, and communication systems. Conventional semiconducting TMDs, such as 2H-MoS_2 and 2H-WS_2, have demonstrated great success in high-performance photodetection [68–70]. Under light illumination, the photogenerated electron-hole pairs in the semiconducting TMD channels could be effectively separated by the built-in or external electric fields, leading to the obvious photo response. In 2012, photodetectors based on exfoliated single-layer 2H-MoS_2 nanosheets were first fabricated by our group [68]. The device exhibits a strong photo response in the visible-light range of 400–700 nm. The 2H-MoS_2-based phototransistor shows good photodetection performance, including a responsivity of 7.5 mA W^{-1} and a response time of about 50 ms. In addition, the photocurrent and responsivity of the single-layer MoS_2-based phototransistors can be tuned by adjusting the source-drain voltage and gate voltage. Lopez-Sanchez et al. subsequently demonstrated that single-layer MoS_2 phototransistors could reach a high responsivity of 880 A W^{-1} under 561 nm light at a large bias of $V_{DS} = 8$ V and $V_G = -70$ V [69].

Till now, various strategies have been put forward to improve the performance of photodetectors based on semiconducting TMDs. One of the widely used strategies is electrostatic gating, which can easily tune the Fermi-level of TMD channels and the Schottky barrier height of devices [71, 72]. Moreover, band-structure engineering makes it possible to control the carrier density, conduction type, and bandgap of semiconducting TMDs. There are many effective ways to precisely modulate the band structure of TMDs, such as doping, strain engineering, defect engineering, and alloying [73–77]. Recently, building heterojunctions emerged as another practical approach to improving the performance of TMD-based photodetectors. By combining the materials with different conduction types and work functions, p–n or Schottky junctions can be formed to effectively suppress dark current and separate the photogenerated carriers [76]. Nowadays, many techniques have been utilized to fabricate TMD-based heterojunctions, including stacking, epitaxial growth, chemical doping, and electrostatic doping [78–82]. In addition, when incorporated with metal nanostructures that have strong localized surface plasmon resonance, semiconducting TMD-based photodetectors show significantly improved light absorption and enhanced photocurrent [83], owing to the hot electron injection from the metal nanostructures into the semiconducting TMDs [84].

Despite the great progress, photodetectors based on semiconducting TMDs often suffer from the relatively narrow detection range limited by their bandgaps [85]. Moreover, the Fermi-level pinning effect of semiconducting TMDs with deposited metal contacts would result in high barrier height and large contact resistance, suppressing the extraction efficiency of photogenerated carriers in devices [86, 87]. In contrast, for photodetectors based on (semi-)metallic TMDs with unconventional phases, the detection wavelength will not be limited due to their gapless nature, and ohmic contact could be obtained [85]. Therefore, photodetectors with broadband photo response have been demonstrated based on (semi-)metallic TMDs. For example, Wang et al. investigated the photo response of 1T′-$MoTe_2$ synthesized by the CVD method [88]. Figure 7.4a shows the optical image of the fabricated device. The inset is the atomic force microscope (AFM) image of the 1T′-$MoTe_2$ channel. The photodetector based on the 1T′-$MoTe_2$ nanosheet exhibits a broadband photo response from visible to near-infrared (NIR) ranges (Figure 7.4b). Kim et al. obtained a good NIR responsivity in devices based on metal-organic chemical vapor deposition (MOCVD) produced large-scale 1T′-$MoTe_2$ films [90]. The device shows an ultrafast response time in the microsecond scale under 1000 nm pulsed laser irradiation, revealing the potential of (semi-)metallic TMDs in ultrafast optoelectronic applications. In addition, other TMDs with unconventional phases have also been developed for future optoelectronic applications. For example, Yu et al. fabricated a photodetector based on layered 1T-$SnSe_2$ with a fast response time down to ~2 ms at room temperature, as shown in Figure 7.4c,d [89].

Besides (semi-)metallic TMDs with a single composition, novel TMD alloys with unconventional phases have also been studied for photodetection. For example, Deng et al. synthesized $Mo_xRe_{1-x}S_2$ alloy (Figure 7.5a) with an adjustable band structure by introducing Mo into 1T′-ReS_2 or introducing Re into 2H-MoS_2 using the CVD method [91]. Figure 7.5b,c show the annular dark-field scanning transmission electron microscope (ADF-STEM) images of 1T′-$Mo_xRe_{1-x}S_2$ and 2H-$Mo_xRe_{1-x}S_2$, respectively. Compared with 1T′-ReS_2 or 2H-MoS_2, the $Mo_xRe_{1-x}S_2$ shows a noticeable bandgap narrowing (Figure 7.5d). Therefore, the photodetector based on the $Mo_xRe_{1-x}S_2$ alloy exhibits an obvious photo response in the NIR range, whereas photodetectors based on either 1T′-ReS_2 or 2H-MoS_2 show negligible

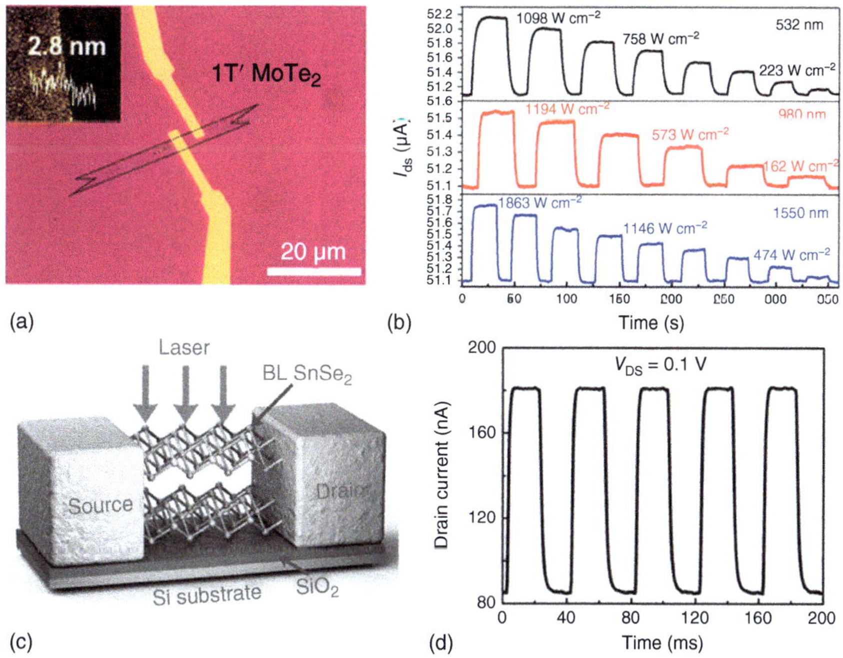

Figure 7.4 (a) Optical image of the 1T′-$MoTe_2$-based photodetector. Inset: AFM image of the 1T′-$MoTe_2$ nanosheet. (b) Photocurrent of the 1T′-$MoTe_2$ photodetector under light illumination with various wavelengths and power densities at V_{DS} = 5 V. (c) Schematic illustration of the photodetector based on bilayer (BL) 1T-$SnSe_2$ on SiO_2/Si substrate. (d) Time-resolved photocurrent of the 1T-$SnSe_2$ photodetector illuminated by 633 nm light (4 mW) at V_{DS} = 0.1 V. Source: (a, b) Reproduced from Wang et al. [88]; (c, d) Yu et al. [89]/Reproduced with permission from John Wiley & Sons.

photo response (Figure 7.5e). Thus, the synthesis of novel TMD alloys through phase engineering is an effective strategy to extend the photo response range and improve the photodetection performance.

In contrast to 2H-TMDs with trigonal-prismatic coordination, TMDs with anisotropic crystal structures are promising materials for polarized photodetection. Lai et al. demonstrated a broadband and polarization-angle-sensitive photodetector based on semi-metallic T_d-$MoTe_2$ [92]. Because of the broken inversion symmetry of the orthorhombic lattice structure of T_d-$MoTe_2$, the linear polarization dependence of photocurrent can be observed. It is worth noting that the anisotropic ratio of the photocurrent varies significantly with the incident photon energy. As shown in Figure 7.5f, the anisotropic ratio decreases from 2.72 to 1.19 with the light wavelength decreasing from 10.6 μm to 633 nm. A similar anisotropic photo response was observed in 1T′-WTe_2- and T_d-WTe_2-based photodetectors with an anisotropic ratio of 10.7 and 4.9, respectively [93, 94]. The anisotropic crystal structures of the unconventional phase TMDs offer the possibility for polarization-sensitive photodetection.

It is well known that semiconducting TMDs exhibit a variety of attractive features, such as the tunable bandgap, high absorption coefficient, and high on/off ratio, which can be applied in various optoelectronic application scenarios [77, 95].

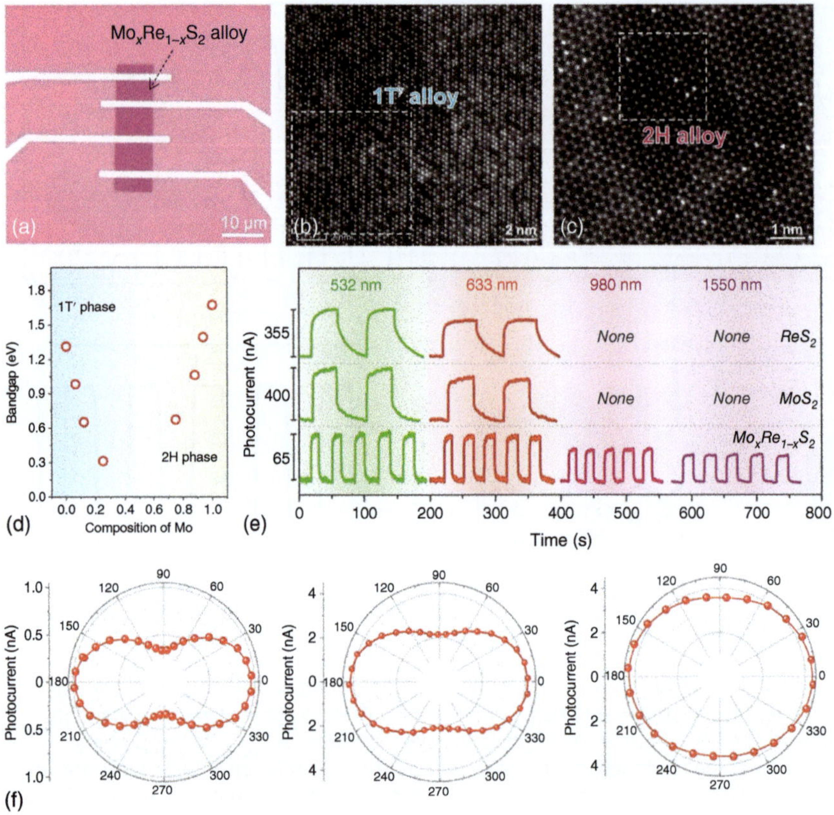

Figure 7.5 (a) Optical image of the $Mo_xRe_{1-x}S_2$-based device. (b, c) High-resolution ADF-STEM images of 1T′-$Mo_xRe_{1-x}S_2$ (b) and 2H-$Mo_xRe_{1-x}S_2$ (c). (d) The calculated bandgap of $Mo_xRe_{1-x}S_2$ as a function of the Mo composition x. (e) Time-resolved photocurrent of ReS_2- (upper), MoS_2- (middle), and $Mo_xRe_{1-x}S_2$-based (lower) photodetectors under light illumination with various wavelengths at $V_{DS} = 2$ V. (f) Anisotropic photocurrent of T_d-$MoTe_2$-based photodetector under linearly polarized light with wavelengths of 10.6 μm, 4 μm, and 633 nm, respectively. Source: (a–e) Deng et al. [91]/Reproduced with permission from John Wiley & Sons; (f) Lai et al. [92]/Reproduced with permission from John Wiley & Sons.

On the other hand, (semi-)metallic TMDs show outstanding charge transfer properties [40, 96]. Therefore, the integration of TMDs with different phases could lead to enhanced optoelectronic performance. The commonly used strategies for fabricating heterophase optoelectronic devices include phase transformation and stacking methods. Yamaguchi et al. reported the phase transformation in CVD-grown single-layer MoS_2 for high-performance photodetectors [97]. In this work, well-defined regions of single-layer 2H-MoS_2 were converted to 1T-MoS_2 by immersing in *n*-butyllithium solution. The metallic 1T-MoS_2 serves as heterophase contact with 2H-MoS_2. Under light illumination, the heterophase MoS_2 photodetector can exhibit a dramatically enhanced responsivity compared to the 2H-MoS_2 photodetector. The scanning photocurrent microscopy (SPCM) map (Figure 7.6a)

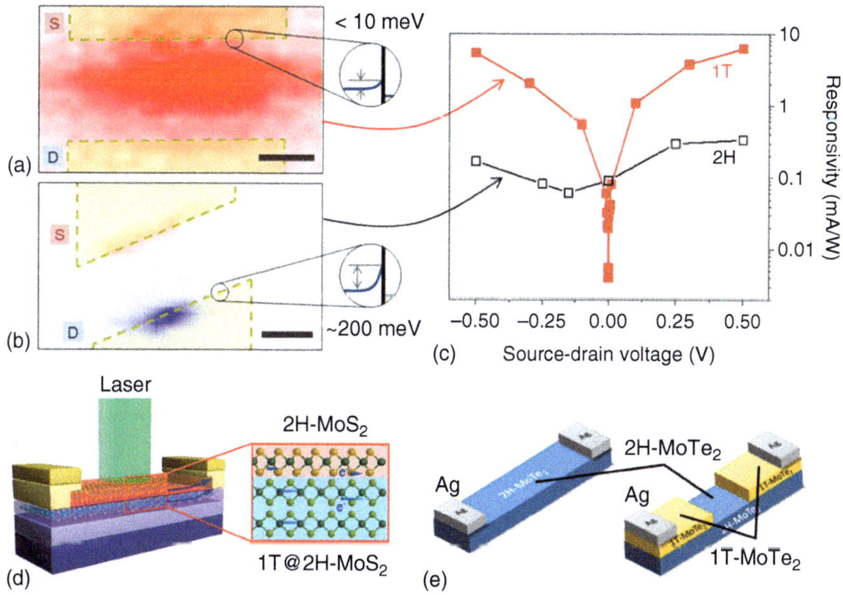

Figure 7.6 (a, b) SPCM maps of MoS_2-based photodetectors with metal/1T-/2H-MoS_2 contacts (a) and metal/2H-MoS_2 contacts (b). Scale bars: 3 µm. (c) Local responsivity of the two photodetectors extracted from the SPCM maps. (d) Schematic illustration of the vertically stacked 2H-MoS_2/1T@2H-MoS_2 photodetector. (e) Schematic illustration of $MoTe_2$-based photodetectors with Ag/2H-$MoTe_2$ and Ag/1T-/2H-$MoTe_2$ contacts. Source: (a–c) Reproduced from Yamaguchi et al. [97]; (d) Reproduced from Wang et al. [98]; (e) Lin et al. [99]/MDPI/Licensed under CC BY 4.0.

shows that the peak photocurrent in the heterophase MoS_2 device emerges at the center of the channel. In contrast, for the 2H-MoS_2 photodetector, the maximum photocurrent is observed at the edges of the contact regions, as shown in Figure 7.6b. The enhanced performance of the heterophase MoS_2 photodetector is revealed by the local responsivity extracted from the SPCM measurements (Figure 7.6c), which can be explained by the significantly decreased Schottky barrier height (<10 meV) at the 1T-/2H-MoS_2 lateral heterophase contacts. Besides, the 1T′/2H heterophase contact can also be achieved through multi-step growth strategies. For example, Xu et al. fabricated large-scale photodetector arrays based on 2H- and 1T′-$MoTe_2$ through a two-step CVD method [100]. In the synthetic process, the pre-patterned Mo films were converted into $MoTe_2$ areas with different phases by controlling the CVD growth conditions. The patterned areas of 2H-$MoTe_2$ and 1T′-$MoTe_2$ act as the channels and contacts in the device arrays, respectively. The low contact resistance of the 1T′-/2H-$MoTe_2$ heterophase structure results in a high responsivity of 1.02 A W^{-1} under NIR light.

Additionally, van der Waals integration is another facile and effective method to produce heterophase structures without considering the lattice mismatch. Wang et al. fabricated a high-performance MoS_2 photodetector consisting of a 1T@2H-MoS_2 mixed-phase layer and a thin layer of 2H-MoS_2 (Figure 7.6d) [98].

Under visible light illumination, photogenerated carriers induced in 2H-MoS_2 can be effectively transferred by the highly conductive 1T@2H-MoS_2 mixed-phase layer. Benefiting from high photon absorption and fast charge transfer, the hybrid photodetector shows an ultrahigh responsivity of 1917 A W^{-1} and an external quantum efficiency (EQE) of 4.5×10^5 % at $V_{GS} = 18$ V and $V_{DS} = 20$ V under 530 nm light illumination. The responsivity of the hybrid device is over one order of magnitude higher than that of the 2H-MoS_2-based device. Lin et al. studied the effect of contact engineering on $MoTe_2$-based photodetectors, as shown in Figure 7.6e [99]. After inserting a layer of 1T-$MoTe_2$ between 2H-$MoTe_2$ and the Ag contacts, the responsivity is increased by about one order of magnitude, and the response time is shortened significantly. This phenomenon can be attributed to the lower Schottky barrier height and the fewer trap states at the interface between 2H- and 1T-$MoTe_2$.

Apart from heterophase TMD-based photodetectors, phase-engineered TMDs have also found applications in constructing heterostructures with other materials. For instance, hybrid perovskites have emerged as rising stars in the field of optoelectronics which is mainly attributed to their high light absorption coefficient and long carrier diffusion length [101]. The efficient extraction of photogenerated carriers is of vital importance in achieving high-performance perovskite-based optoelectronic devices. In this case, metallic TMDs with outstanding charge transfer properties could be used to build heterostructures with high charge extraction efficiency. Wang et al. fabricated the 1T-MoS_2/$CH_3NH_3PbI_3$ hybrid photodetectors, in which 1T-MoS_2 served as the charge transport layer [102]. 1T-MoS_2-based hybrid photodetectors show low contact resistance and high-concentration carrier injection, contributing to a responsivity of over 3000 A W^{-1} and an EQE of 7.7×10^5 % (500 nm, 37.9 μW cm^{-2}), which are over 20 times the value of the hybrid photodetectors based on 2H-phase MoS_2.

7.5 Solar Cells

In addition to photodetectors, TMDs have found applications in photovoltaic devices which convert solar energy into electric power. For next-generation photovoltaic systems, dye-sensitized solar cells (DSSCs) are one of the most representative technological solutions. In 2011, 2H-MoS_2 and 2H-WS_2 were first used as counter-electrode catalysts in DSSCs by Wu et al. [103]. 2H-MoS_2 and 2H-WS_2 show low charge transfer resistance and outstanding catalytic activities for the redox couple I_3^-/I^-, leading to a high power conversion efficiency (PCE) of 7.59% and 7.73%, respectively. The achieved performance is comparable to that of DSSCs based on noble metal Pt counter electrodes, indicating the promising future of TMD-based catalysts in cost-effective DSSCs. The charge transport properties of TMD-based catalysts in DSSCs could be further enhanced by phase engineering [104]. Tan et al. synthesized $MoS_{2x}Se_{2(1-x)}$ and $Mo_xW_{1-x}S_2$ nanosheets with high 1T phase concentration by a lithium intercalation and exfoliation method (Figure 7.7a–c) [105]. The exfoliated $MoS_{2x}Se_{2(1-x)}$ nanosheets were then used as a cost-effective electrocatalyst on the counter electrode of DSSCs. The phase-engineered TMD-based DSSCs show comparable performance to the DSSC with the Pt catalyst.

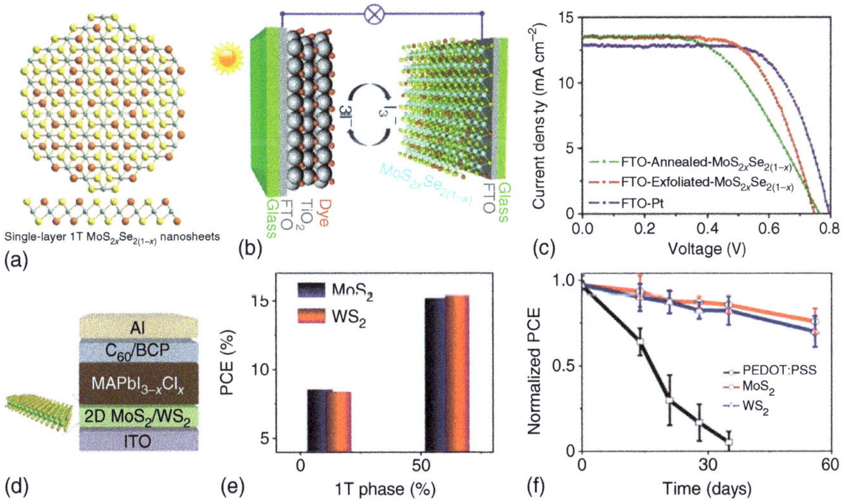

Figure 7.7 (a) Schematic illustration of single-layer 1T-$MoS_{2x}Se_{2(1-x)}$. (b) Schematic illustration of a DSSC device with $MoS_{2x}Se_{2(1-x)}$ catalyst on the counter electrode. (c) J–V curves of DSSCs based on FTO-exfoliated-$MoS_{2x}Se_{2(1-x)}$, FTO-annealed-$MoS_{2x}Se_{2(1-x)}$, and FTO-Pt counter electrodes under simulated sunlight with the power density of 100 mW cm^{-2} (AM 1.5). (d) Schematic illustration of a perovskite solar cell with the TMD-based hole transport layer. (e) PCE of perovskite solar cells with TMD hole transport layers. (f) Stability of perovskite solar cells with PEDOT:PSS and TMD hole transport layers. Source: (a–c) Tan et al. [105]/Reproduced with permission from John Wiley & Sons; (d–f) Reproduced from Huang et al. [106].

Recently, perovskite solar cells, with record-breaking PCE, are among the best candidates for next-generation photovoltaics. In perovskite solar cells, the electron transport layer (ETL) and hole transport layer (HTL) between the photoactive layer and the electrodes facilitate charge transfer and collection. However, many commonly used high-performance organic HTL materials are limited by high costs and low environmental stability [107]. To this end, polymorphic TMDs have been used as a cost-effective and stable substitute for organic HTL materials in multilayer perovskite solar cells. For example, Kim et al. replaced the organic HTL PEDOT:PSS with CVD-grown 2H-MoS_2 and 2H-WS_2 in an inverted planar $MAPbI_{3-x}Cl_x$ solar cell [108]. The suitable band alignment between TMDs and the perovskite layer promotes the charge transfer, resulting in a comparable PCE to that of devices based on organic HTL materials. Besides, the introduced TMD transport layers can passivate the surface of the perovskite layer, leading to a longer device lifetime [108, 109]. Furthermore, metallic TMDs are promising candidates as charge transport layers in solar cells because of their superior charge transport properties. For example, Huang et al. replaced the PEDOT:PSS with 1T and 2H mixed-phase MoS_2 and WS_2 in perovskite solar cells (Figure 7.7d) [106]. Effective photoluminescence quenching in solar cells based on mixed-phase TMDs suggests improved charge transfer and better band alignment at the TMD/perovskite interface. By increasing the 1T phase percentage of TMD transport layers, the solar cells exhibit higher PCE (Figure 7.7e). In addition, the mixed-phase TMDs can simultaneously improve the environmental stability of the solar cells owing to the higher moisture resistance (Figure 7.7f). Because of the intriguing properties and encouraging results mentioned above, polymorphic TMDs are attracting increasing attention in future high-performance photovoltaic applications.

7.6 Perspectives

In the past decades, polymorphic TMDs held great promise in fabricating novel electronic/optoelectronic devices to supplement traditional bulk material-based devices. As discussed in this chapter, unique properties such as the atomic thickness, the moderate bandgap (1–2 eV), and decent carrier mobility provide semiconducting TMDs with compelling advantages over bulk materials in fabricating high-performance electronic and optoelectronic devices. However, the performance of these devices still faces many limitations. For example, large contact resistance is often observed in devices with a traditional deposited metal contact. Therefore, the introduction of polymorphic TMDs provides many unique advantages that could help enhance their performance. In terms of electronic devices, by introducing (semi-)metallic TMDs to construct heterophase structures, contact resistance can be significantly decreased with suppressed Fermi-level pinning effect and reduced Schottky barrier height. For optoelectronic devices, the introduction of (semi-)metallic TMDs facilitates the separation of photogenerated carriers.

However, there is still a long way towards bringing polymorphic TMDs from research laboratories to industrial applications. Challenges remain in the synthetic methods of polymorphic TMDs and their integration in electronics and optoelectronics. For the synthesis of high-quality polymorphic TMD crystals, it is still challenging to prepare wafer-scale TMD films because of rigorous growth conditions. Nowadays, the fast development of substrate engineering and phase engineering provides opportunities for synthesizing large-scale TMDs with high phase purity.

For integrating polymorphic TMDs into current bulk material-based electronics and optoelectronics, developing a fabrication process compatible with the mature CMOS technology is a critical step. For metastable TMDs with unconventional phases, new encapsulation and surface modification strategies for better environmental stability are also highly demanded. Besides developing a mature fabrication process to ensure the uniform performance of polymorphic TMD-based devices, establishing standard device models is of great importance to future large-scale applications, such as very-large-scale integration (VLSI) circuits. Nowadays, novel flexible electronics and optoelectronics have attracted extensive attention from academic and industry communities. By virtue of their superior mechanical flexibility, ultrathin layered materials, including polymorphic TMDs, exhibit significant advantages for the preparation of novel flexible electronic systems. Therefore, it is appealing to develop a set of fabrication processes to integrate polymorphic TMDs and other layered materials into flexible electronic systems.

Besides the aforementioned applications in electronic and optoelectronics, polymorphic TMDs are being extensively studied in other fields. For instance, driven by the urgent requirement for high-speed calculation, high capacitive storage, and low energy consumption, future society calls for new technological routes to meet the increasing demands for computing power. However, in state-of-the-art micro/nanoelectronics, the continuous scaling down trend faces challenges because of the fundamental quantum limits of materials and the high production cost. To this end, neuromorphic computing and quantum computing have attracted much attention. As discussed in this chapter, phase transition in polymorphic TMDs results in the change of resistance in multiple states, which is essential for neuromorphic computing devices to store and process information. In addition, benefiting from the unique material properties, such as superconductivity, Weyl semi metallic feature, topological non-triviality, and ferromagnetism, polymorphic TMDs can serve as compelling candidates for quantum computing.

In summary, polymorphic TMDs are fascinating materials for building novel high-performance electronics/optoelectronics due to their unique properties over traditional bulk materials. Theoretical and experimental efforts are still ongoing to continually move this exciting field forward.

References

1. Zhu, K., Wen, C., Aljarb, A.A. et al. (2021). The development of integrated circuits based on two-dimensional materials. *Nat. Electron.* 4: 775–785.
2. Chang, C., Chen, W., Chen, Y. et al. (2021). Recent progress on two-dimensional materials. *Acta Phys. Chim. Sin.* 37: 2108017.
3. Tan, C., Cao, X., Wu, X. et al. (2017). Recent advances in ultrathin two-dimensional nanomaterials. *Chem. Rev.* 117: 6225–6331.
4. Sokolikova, M.S. and Mattevi, C. (2020). Direct synthesis of metastable phases of 2D transition metal dichalcogenides. *Chem. Soc. Rev.* 49: 3952–3980.
5. Liu, C., Chen, H., Wang, S. et al. (2020). Two-dimensional materials for next-generation computing technologies. *Nat. Nanotechnol.* 15: 545–557.
6. Schulman, D.S., Arnold, A.J., and Das, S. (2018). Contact engineering for 2D materials and devices. *Chem. Soc. Rev.* 47: 3037–3058.
7. Zeng, M., Xiao, Y., Liu, J. et al. (2018). Exploring two dimensional materials toward the next-generation circuits: from monomer design to assembly control. *Chem. Rev.* 118: 6236–6296.
8. Chhowalla, M., Jena, D., and Zhang, H. (2016). Two-dimensional semiconductors for transistors. *Nat. Rev. Mater.* 1: 16052.
9. Liu, H., Si, M., Deng, Y. et al. (2014). Switching mechanism in single-layer molybdenum disulfide transistors: an insight into current flow across Schottky barriers. *ACS Nano* 8: 1031–1038.
10. Radisavljevic, B. and Kis, A. (2013). Mobility engineering and a metal–insulator transition in monolayer MoS_2. *Nat. Mater.* 12: 815–820.

11 Radisavljevic, B., Whitwick, M.B., and Kis, A. (2011). Integrated circuits and logic operations based on single-layer MoS_2. *ACS Nano* 5: 9934–9938.

12 Ayari, A., Cobas, E., Ogundadegbe, O. et al. (2007). Realization and electrical characterization of ultrathin crystals of layered transition-metal dichalcogenides. *J. Appl. Phys.* 101: 014507.

13 Radisavljevic, B., Radenovic, A., Brivio, J. et al. (2011). Single-layer MoS_2 transistors. *Nat. Nanotechnol.* 6: 147–150.

14 Radisavljevic, B. and Kis, A. (2013). Measurement of mobility in dual-gated MoS_2 transistors. *Nat. Nanotechnol.* 8: 147–148.

15 Roy, T., Tosun, M., Kang, J.S. et al. (2014). Field-effect transistors built from all two-dimensional material components. *ACS Nano* 8: 6259–6264.

16 Tosun, M., Chuang, S., Fang, H. et al. (2014). High-gain inverters based on WSe_2 complementary field-effect transistors. *ACS Nano* 8: 4948–4953.

17 Hwang, W.S., Remskar, M., Yan, R. et al. (2012). Transistors with chemically synthesized layered semiconductor WS_2 exhibiting 10^5 room temperature modulation and ambipolar behavior. *Appl. Phys. Lett.* 101: 013107.

18 Wang, Q.H., Kalantar-Zadeh, K., Kis, A. et al. (2012). Electronics and optoelectronics of two-dimensional transition metal dichalcogenides. *Nat. Nanotechnol.* 7: 699–712.

19 Fang, H., Chuang, S., Chang, T.C. et al. (2012). High-performance single-layered WSe_2 p-FETs with chemically doped contacts. *Nano Lett.* 12: 3788–3792.

20 Ahn, C., Lee, J., Kim, H. et al. (2015). Low-temperature synthesis of large-scale molybdenum disulfide thin films directly on a plastic substrate using plasma-enhanced chemical vapor deposition. *Adv. Mater.* 27: 5223–5229.

21 Li, T., Guo, W., Ma, L. et al. (2021). Epitaxial growth of wafer-scale molybdenum disulfide semiconductor single crystals on sapphire. *Nat. Nanotechnol.* 16: 1201–1207.

22 Song, X., Guo, Z., Zhang, Q. et al. (2017). Progress of large-scale synthesis and electronic device application of two-dimensional transition metal dichalcogenides. *Small* 13: 1700098.

23 Yu, H., Liao, M., Zhao, W. et al. (2017). Wafer-scale growth and transfer of highly-oriented monolayer MoS_2 continuous films. *ACS Nano* 11: 12001–12007.

24 Tang, H., Zhang, H., Chen, X. et al. (2019). Recent progress in devices and circuits based on wafer-scale transition metal dichalcogenides. *Sci. China Inf. Sci.* 62: 220401.

25 Lin, Z.Y., Liu, Y., Halim, U. et al. (2018). Solution-processable 2D semiconductors for high-performance large-area electronics. *Nature* 562: 254–258.

26 Yu, L., El-Damak, D., Radhakrishna, U. et al. (2016). Design, modeling, and fabrication of chemical vapor deposition-grown MoS_2 circuits with E-mode FETs for large-area electronics. *Nano Lett.* 16: 6349–6356.

27 Wachter, S., Polyushkin, D.K., Bethge, O. et al. (2017). A microprocessor based on a two-dimensional semiconductor. *Nat. Commun.* 8: 14948.

28 Yoon, Y., Ganapathi, K., and Salahuddin, S. (2011). How good can monolayer MoS_2 transistors be? *Nano Lett.* 11: 3768–3773.

29 Tung, R.T. (2000). Chemical bonding and fermi level pinning at metal-semiconductor interfaces. *Phys. Rev. Lett.* 84: 6078–6081.

30 Yin, X., Tang, C.S., Zheng, Y. et al. (2021). Recent developments in 2D transition metal dichalcogenides: phase transition and applications of the (quasi-)metallic phases. *Chem. Soc. Rev.* 50: 10087–10115.

31 Liu, Y., Guo, J., Zhu, E. et al. (2018). Approaching the Schottky-Mott limit in van der Waals metal-semiconductor junctions. *Nature* 557: 696–700.

32 Shen, P.-C., Su, C., Lin, Y. et al. (2021). Ultralow contact resistance between semimetal and monolayer semiconductors. *Nature* 593: 211–217.

33 Chen, Y., Lai, Z., Zhang, X. et al. (2020). Phase engineering of nanomaterials. *Nat. Rev. Chem.* 4: 243–256.

34 Kappera, R., Voiry, D., Yalcin, S.E. et al. (2014). Phase-engineered low-resistance contacts for ultrathin MoS_2 transistors. *Nat. Mater.* 13: 1128–1134.

35 Ma, Y., Liu, B., Zhang, A. et al. (2015). Reversible semiconducting-to-metallic phase transition in chemical vapor deposition grown monolayer WSe_2 and applications for devices. *ACS Nano* 9: 7383–7391.

36 Cho, S., Kim, S., Kim, J.H. et al. (2015). Phase patterning for ohmic homojunction contact in $MoTe_2$. *Science* 349: 625–628.

37 Sung, J.H., Heo, H., Si, S. et al. (2017). Coplanar semiconductor–metal circuitry defined on few-layer $MoTe_2$ via polymorphic heteroepitaxy. *Nat. Nanotechnol.* 12: 1064–1070.

38 Heising, J. and Kanatzidis, M.G. (1999). Structure of restacked MoS_2 and WS_2 elucidated by electron crystallography. *J. Am. Chem. Soc.* 121: 638–643.

39 Keum, D.H., Cho, S., Kim, J.H. et al. (2015). Bandgap opening in few-layered monoclinic $MoTe_2$. *Nat. Phys.* 11: 482–486.

40 Yu, Y., Nam, G.-H., He, Q. et al. (2018). High phase-purity 1T′-MoS_2- and 1T′-$MoSe_2$-layered crystals. *Nat. Chem.* 10: 638–643.

41 Lin, Y.-C., Dumcenco, D.O., Huang, Y.-S. et al. (2014). Atomic mechanism of the semiconducting-to-metallic phase transition in single-layered MoS_2. *Nat. Nanotechnol.* 9: 391–396.

42 Allain, A., Kang, J., Banerjee, K. et al. (2015). Electrical contacts to two-dimensional semiconductors. *Nat. Mater.* 14: 1195–1205.

43 Wang, L., Meric, I., Huang, P.Y. et al. (2013). One-dimensional electrical contact to a two-dimensional material. *Science* 342: 614–617.

44 Cui, X., Lee, G.-H., Kim, Y.D. et al. (2015). Multi-terminal transport measurements of MoS_2 using a van der Waals heterostructure device platform. *Nat. Nanotechnol.* 10: 534–540.

45 Zhang, X., Jin, Z., Wang, L. et al. (2019). Low contact barrier in 2H/1T′ $MoTe_2$ in-plane heterostructure synthesized by chemical vapor deposition. *ACS Appl. Mater. Interfaces* 11: 12777–12785.

46 Xu, X., Chen, S., Liu, S. et al. (2019). Millimeter-scale single-crystalline semiconducting $MoTe_2$ via solid-to-solid phase transformation. *J. Am. Chem. Soc.* 141: 2128–2134.

47 Zhang, Q., Wang, X.-F., Shen, S.-H. et al. (2019). Simultaneous synthesis and integration of two-dimensional electronic components. *Nat. Electron.* 2: 164–170.

48 Chiang, C.-C., Ostwal, V., Wu, P. et al. (2021). Memory applications from 2D materials. *Appl. Phys. Rev.* 8: 021306.

49 Zhu, W., Low, T., Wang, H. et al. (2019). Nanoscale electronic devices based on transition metal dichalcogenides. *2D Mater* 6: 032004.

50 Sangwan, V.K., Jariwala, D., Kim, I.S. et al. (2015). Gate-tunable memristive phenomena mediated by grain boundaries in single-layer MoS_2. *Nat. Nanotechnol.* 10: 403–406.

51 Sangwan, V.K., Lee, H.-S., Bergeron, H. et al. (2018). Multi-terminal memtransistors from polycrystalline monolayer molybdenum disulfide. *Nature* 554: 500–504.

52 Yan, X., Zhao, Q., Chen, A.P. et al. (2019). Vacancy-induced synaptic behavior in 2D WS_2 nanosheet–based memristor for low-power neuromorphic computing. *Small* 15: 1901423.

53 Cheng, P., Sun, K., and Hu, Y.H. (2016). Memristive behavior and ideal memristor of 1T phase MoS_2 nanosheets. *Nano Lett.* 16: 572–576.

54 Zhu, X., Li, D., Liang, X. et al. (2019). Ionic modulation and ionic coupling effects in MoS_2 devices for neuromorphic computing. *Nat. Mater.* 18: 141–148.

55 Xu, R., Jang, H., Lee, M.-H. et al. (2019). Vertical MoS_2 double-layer memristor with electrochemical metallization as an atomic-scale synapse with switching thresholds approaching 100 mV. *Nano Lett.* 19: 2411–2417.

56 Zhang, F., Zhang, H., Krylyuk, S. et al. (2019). Electric-field induced structural transition in vertical $MoTe_2$- and $Mo_{1-x}W_xTe_2$-based resistive memories. *Nat. Mater.* 18: 55–61.

57 Huh, W., Lee, D., and Lee, C.-H. (2020). Memristors based on 2D materials as an artificial synapse for neuromorphic electronics. *Adv. Mater.* 32: 2002092.

58 Nishiyama, M., Hong, K., Mikoshiba, K. et al. (2000). Calcium stores regulate the polarity and input specificity of synaptic modification. *Nature* 408: 584–588.

59 Kim, J., Lee, Y., Kang, M. et al. (2021). 2D materials for skin-mountable electronic devices. *Adv. Mater.* 33: 2005858.

60 Kanahashi, K., Pu, J., and Takenobu, T. (2020). 2D materials for large-area flexible thermoelectric devices. *Adv. Energy Mater.* 10: 1902842.

61 Oh, J.Y., Lee, J.H., Han, S.W. et al. (2016). Chemically exfoliated transition metal dichalcogenide nanosheet-based wearable thermoelectric generators. *Energy Environ. Sci.* 9: 1696–1705.

62 Zhang, X., Grajal, J., Vazquez-Roy, J.L. et al. (2019). Two-dimensional MoS_2-enabled flexible rectenna for Wi-Fi-band wireless energy harvesting. *Nature* 566: 368–372.

63 Donchev, E., Pang, J.S., Gammon, P.M. et al. (2014). The rectenna device: from theory to practice (a review). *MRS Energy Sustain.* 1: E1.

64 Pozar, D.M. (2012). *Microwave Engineering*, 4e. Wiley.

65 Chasin, A., Steudel, S., Vanaverbeke, F. et al. (2012). UHF IGZO Schottky diode. *IEDM Tech. Dig.* 12: 287–290.

66 Chasin, A., Volskiy, V., Libois, M. et al. (2014). An integrated a-IGZO UHF energy harvester for passive RFID Tags. *IEEE Trans. Electron. Dev.* 61: 3289–3295.

67 Zhang, J., Li, Y., Zhang, B. et al. (2015). Flexible indium–gallium–zinc–oxide Schottky diode operating beyond 2.45 GHz. *Nat. Commun.* 6: 7561.

68 Yin, Z., Li, H., Li, H. et al. (2012). Single-layer MoS_2 phototransistors. *ACS Nano* 6: 74–80.

69 Lopez-Sanchez, O., Lembke, D., Kayci, M. et al. (2013). Ultrasensitive photodetectors based on monolayer MoS_2. *Nat. Nanotechnol.* 8: 497–501.

70 Xie, Y., Zhang, B., Wang, S. et al. (2017). Ultrabroadband MoS_2 photodetector with spectral response from 445 to 2717 nm. *Adv. Mater.* 29: 1605972.

71 Wang, X., Wang, P., Wang, J. et al. (2015). Ultrasensitive and broadband MoS_2 photodetector driven by ferroelectrics. *Adv. Mater.* 27: 6575–6581.

72 Xu, J., Luo, X., Hu, S. et al. (2021). Tunable linearity of high-performance vertical dual-gate vdW phototransistors. *Adv. Mater.* 33: 2008080.

73 Baugher, B.W.H., Churchill, H.O.H., Yang, Y. et al. (2014). Optoelectronic devices based on electrically tunable p–n diodes in a monolayer dichalcogenide. *Nat. Nanotechnol.* 9: 262–267.

74 Desai, S.B., Seol, G., Kang, J.S. et al. (2014). Strain-induced indirect to direct bandgap transition in multilayer WSe_2. *Nano Lett.* 14: 4592–4597.

75 Yao, J., Zheng, Z., and Yang, G. (2016). Promoting the performance of layered-material photodetectors by alloy engineering. *ACS Appl. Mater. Interfaces* 8: 12915–12924.

76 Wang, F., Pei, K., Li, Y. et al. (2021). 2D homojunctions for electronics and optoelectronics. *Adv. Mater.* 33: 2005303.

77 Mak, K.F. and Shan, J. (2016). Photonics and optoelectronics of 2D semiconductor transition metal dichalcogenides. *Nat. Photonics* 10: 216–226.

78 Duan, X., Wang, C., Shaw, J.C. et al. (2014). Lateral epitaxial growth of two-dimensional layered semiconductor heterojunctions. *Nat. Nanotechnol.* 9: 1024–1030.

79 Lee, C.-H., Lee, G.H., van der Zande, A.M. et al. (2014). Atomically thin p–n junctions with van der Waals heterointerfaces. *Nat. Nanotechnol.* 9: 676–681.

80 Li, J., Yang, X., Liu, Y. et al. (2020). General synthesis of two-dimensional van der Waals heterostructure arrays. *Nature* 579: 368–374.

81 Sun, J., Wang, Y., Guo, S. et al. (2020). Lateral 2D WSe_2 p–n homojunction formed by efficient charge-carrier-type modulation for high-performance optoelectronics. *Adv. Mater.* 32: 1906499.

82 Ghosh, S., Varghese, A., Thakar, K. et al. (2021). Enhanced responsivity and detectivity of fast WSe_2 phototransistor using electrostatically tunable in-plane lateral p–n homojunction. *Nat. Commun.* 12: 3336.

83 Miao, J., Hu, W., Jing, Y. et al. (2015). Surface plasmon-enhanced photodetection in few layer MoS_2 phototransistors with Au nanostructure arrays. *Small* 11: 2392–2398.

84 Hong, T., Chamlagain, B., Hu, S. et al. (2015). Plasmonic hot electron induced photocurrent response at MoS_2-metal junctions. *ACS Nano* 9: 5357–5363.

85 Liu, J., Xia, F., Xiao, D. et al. (2020). Semimetals for high-performance photodetection. *Nat. Mater.* 19: 830–837.

86 Kim, C., Moon, I., Lee, D. et al. (2017). Fermi level pinning at electrical metal contacts of monolayer molybdenum dichalcogenides. *ACS Nano* 11: 1588–1596.

87 Zhang, X., Liu, B., Gao, L. et al. (2021). Near-ideal van der Waals rectifiers based on all-two-dimensional Schottky junctions. *Nat. Commun.* 12: 1522.

88 Wang, X., Shang, J., Zhu, M. et al. (2020). Controlled growth of large-scale uniform 1T′ $MoTe_2$ crystals with tunable thickness and their photodetector applications. *Nanoscale Horiz.* 5: 954–959.

89 Yu, P., Yu, X., Lu, W. et al. (2016). Fast photoresponse from 1T tin diselenide atomic layers. *Adv. Funct. Mater.* 26: 137–145.

90 Kim, D., Lee, R., Kim, S. et al. (2019). Two-dimensional phase-engineered 1T′- and 2H-$MoTe_2$-based near-infrared photodetectors with ultra-fast response. *J. Alloys Compd.* 789: 960–965.

91 Deng, Q., Li, X., Si, H. et al. (2020). Strong band bowing effects and distinctive optoelectronic properties of 2H and 1T′ phase-tunable $Mo_xRe_{1-x}S_2$ alloys. *Adv. Funct. Mater.* 30: 2003264.

92 Lai, J., Liu, X., Ma, J. et al. (2018). Anisotropic broadband photoresponse of layered type-II Weyl semimetal $MoTe_2$. *Adv. Mater.* 30: 1707152.

93 Zhou, W., Chen, J., Gao, H. et al. (2019). Anomalous and polarization-sensitive photoresponse of T_d-WTe_2 from visible to infrared light. *Adv. Mater.* 31: 1804629.

94 Zhang, Q., Zhang, R., Chen, J. et al. (2019). Remarkable electronic and optical anisotropy of layered 1T′-WTe_2 2D materials. *Beilstein J. Nanotechnol.* 10: 1745–1753.

95 Kang, K., Xie, S., Huang, L. et al. (2015). High-mobility three-atom-thick semiconducting films with wafer-scale homogeneity. *Nature* 520: 656–660.

96 Lai, Z., He, Q., Tran, T.-H. et al. (2021). Metastable 1T′-phase group VIB transition metal dichalcogenide crystals. *Nat. Mater.* 20: 1113–1120.

97 Yamaguchi, H., Blancon, J.-C., Kappera, R. et al. (2015). Spatially resolved photoexcited charge-carrier dynamics in phase-engineered monolayer MoS_2. *ACS Nano* 9: 840–849.

98 Wang, W., Zeng, X., Warner, J.-H. et al. (2020). Photoresponse-bias modulation of a high-performance MoS_2 photodetector with a unique vertically stacked 2H-MoS_2/1T@2H-MoS_2 structure. *ACS Appl. Mater. Interfaces* 12: 33325–33335.

99 Lin, D.-Y., Hsu, H.-P., Liu, G.-H. et al. (2021). Enhanced photoresponsivity of 2H-$MoTe_2$ by inserting 1T-$MoTe_2$ interlayer contact for photodetector applications. *Crystals* 11: 964.

100 Xu, X., Liu, S., Han, B. et al. (2019). Scaling-up atomically thin coplanar semiconductor–metal circuitry via phase engineered chemical assembly. *Nano Lett.* 19: 6845–6852.

101 Wang, H.-P., Li, S., Liu, X. et al. (2021). Low-dimensional metal halide perovskite photodetectors. *Adv. Mater.* 33: 2003309.

102 Wang, Y., Fullon, R., Acerce, M. et al. (2017). Solution-processed MoS_2/organolead trihalide perovskite photodetectors. *Adv. Mater.* 29: 1603995.

103 Wu, M., Wang, Y., Lin, X. et al. (2011). Economical and effective sulfide catalysts for dye-sensitized solar cells as counter electrodes. *Phys. Chem. Chem. Phys.* 13: 19298–19301.

104 Wei, W., Sun, K., Hu, Y. et al. (2016). An efficient counter electrode material for dye-sensitized solar cells—flower-structured 1T metallic phase MoS_2. *J. Mater. Chem. A* 4: 12398–12401.

105 Tan, C., Zhao, W., Chaturvedi, A. et al. (2016). Preparation of single-layer $MoS_{2x}Se_{2(1-x)}$ and $Mo_xW_{1-x}S_2$ nanosheets with high-concentration metallic 1T phase. *Small* 12: 1866–1874.

106 Huang, P., Wang, Z., Liu, Y. et al. (2017). Water-soluble 2D transition metal dichalcogenides as the hole-transport layer for highly efficient and stable p–i–n perovskite solar cells. *ACS Appl. Mater. Interfaces* 9: 25323–25331.

107 Kakavelakis, G., Paradisanos, I., Paci, B. et al. (2018). Extending the continuous operating lifetime of perovskite solar cells with a molybdenum disulfide hole extraction interlayer. *Adv. Energy Mater.* 8: 1702287.

108 Kim, Y.G., Kwon, K.-C., Le, Q.-V. et al. (2016). Atomically thin two-dimensional materials as hole extraction layers in organolead halide perovskite photovoltaic cells. *J. Power Sources* 319: 1–8.

109 Capasso, A., Matteocci, F., Najafi, L. et al. (2016). Few-layer MoS_2 flakes as active buffer layer for stable perovskite solar cells. *Adv. Energy Mater.* 6: 1600920.

8

Polymorphic Two-dimensional Transition Metal Dichalcogenides: Modern Challenges and Opportunities

Chi Sin Tang[1,2], *Xinmao Yin*[2], *and Andrew T. S. Wee*[3,4]

[1] *National University of Singapore, Singapore Synchrotron Light Source (SSLS), 5 Research Link, Singapore, 117603, Singapore*
[2] *Shanghai University, Shanghai Key Laboratory of High Temperature Superconductors, Shanghai Frontiers Science Center of Quantum and Superconducting Matter States, Physics Department, 99 Shangda Road, Shanghai 200444, China*
[3] *National University of Singapore, Department of Physics, Faculty of Science, Science Drive 4, Singapore, 117542, Singapore*
[4] *National University of Singapore, Centre for Advanced 2D Materials, 6 Science Drive 2, Singapore, 117546, Singapore*

8.1 Summing up the Chapters

Low-dimensional semiconductors have, in recent years, emerged as a potential solution to address the issue of channel control in highly scaled transistors. While the practical limit for silicon channel thickness appears to be about 3 nm [1], few- and single-layered materials are promising alternatives for bulk semiconductors and semi-metallic materials in field-effect transistor (FET) and other novel functional devices.

Their atomic-scale features and robust properties render them ideal for device miniaturization and novel electronic applications that promise enhancement of energy efficiency. Besides, the technology to synthesize TMD materials such as MoS_2 is sufficiently matured, and these materials are very stable and robust even under harsh conditions.

The preceding chapters have provided an extensive treatment to the progress made in the treatment, phase engineering, and device fabrication processes of ultra-thin TMD systems in recent years. They provide comprehensive discussions on the current developments in the aspects of material synthesis, theoretical investigations, thermoelectric studies, magnetic properties, and other related applications where TMDs are and can potentially be utilized in multiple domains related to microdevices, computing systems, and catalytic applications.

We begin with a general overview of the properties of 2D-TMDs in Chapter 1. This chapter further outlines the general synthesis techniques of this class of materials and highlights several established characterization techniques to investigate the optical and electronic properties of both single and multi-layer TMD systems.

Two-Dimensional Transition-Metal Dichalcogenides: Phase Engineering and Applications in Electronics and Optoelectronics, First Edition. Edited by Chi Sin Tang, Xinmao Yin, and Andrew T. S. Wee.

Thereafter, Chapter 2 delves into the recent developments in phase engineering of 2D-TMDs. It discusses the crystal structures of 2D-TMDs in the 1H, 2H, 3R, 1T, and 1T′ phases and the common synthesis methods from the top-down and bottom-up approaches. It then focuses on various strategies to engineer the phases of 2D-TMDs which can be classified into direct synthesis with targeted phases, and external factor-induced phase transformation.

We then consider the specific thermoelectric properties of 2D-TMDs in Chapter 3 where the complex interplay between the fundamental electronic properties and phonons and heat transport features are closely examined. By considering the recent development of thermoelectrical properties in 2D-TMDs, readers can comprehend the advantages of utilizing 2D-TMDs for TE applications where their electrical and thermal transport can be effectively manipulated via parameters including thickness effect, structural defects, and mechanical strain. Most importantly, recent advances to the thermal and electrical transport in 2D-TMDs, the challenges, and how they could be effectively applied in diverse practical applications have also been discussed. An extensive discussion is then given in Chapter 4 on the electronic structures of 2D-TMDs and their resulting optical and other emerging properties. Through a computational approach, a series of comprehensive investigations which effectively combine both computational and experimental methods provide new insights into novel quantum effects including valley polarization, charge density wave, and magnetic properties of 2D-TMD systems. Apart from highlighting the unique electronic structures, transport, optical properties, and spin–valley physical features, the symmetry-breaking properties in the form of asymmetric Janus structures have also been considered. They serve as another unique strategy to tune the emerging properties of 2D-TMDs through their asymmetric and polarized structures.

With the theoretical and computational aspects of 2D-TMDs established, Chapter 5 focuses on the magnetism and spin structures of polymorphic 2D-TMDs; the extensively explored 2D Cr-based magnetic materials are introduced, followed by a brief overview of the recent experimental progress on other magnetic 2D-TMDs. Various approaches for fabricating few-layer or monolayer magnetic materials, together with the characterization of their magnetic properties are systematically surveyed. The modulation of magnetic properties and relevant device applications are discussed, concluding with perspectives on the existing challenges and opportunities in this field.

The aspects of sample preparation are extensively discussed in Chapter 6 where it focuses on the exfoliation techniques employed to derive large-area 2D-TMDs. Mechanical, electronic, optical, and mechanical properties unique to such exfoliated materials are also discussed. While mechanical exfoliation has been commonly regarded as a "mechanical craftsmanship," underlying physical and engineering properties often run profoundly into the domains of theoretical calculations, especially when it concerns more than 1500 cleavable layered materials. An improved understanding of exfoliation techniques allows one to produce large-area low-dimensional layers which in turn plays a significant role in device applications. Having established the theoretical and experimental methods of studying 2D-TMDs,

Chapter 7 introduces the applications of polymorphic TMDs in electronics and optoelectronics. For electronic applications, polymorphic TMDs are introduced to lower the contact resistance in TMD-based FETs, design novel memory devices and neuromorphic computing systems, and construct energy harvesting units. For optoelectronic applications, they are used in the fabrication of high-performance photodetectors and solar cells.

With a bird's-eye view of the preceding chapters, we can witness the sheer amount of research and extensive theoretical and experimental studies devoted to basic research and development for future applications in multiple disciplines and domains. The progress made has been remarkable over the past decades with considerable breakthroughs. Nevertheless, major challenges and bottlenecks persist over multiple fronts. One can regard the outlook of 2D-TMDs and their device applications with cautious confidence. While we witness significant progress in material growth and the fabrication of contact devices and structures, the ability to match state-of-the-art silicon technology is yet to be demonstrated. But as discussed in this chapter, there is a pressing urgency to catch up to the integration of 2D-TMD-based devices with conventional semiconductor technology. Crises such as climate change and the unprecedented surge in demands for computational and data capabilities are some underlying factors for 2D-TMDs as potential to some of these pressing problems confronting humanity today and in the near future.

8.2 Projecting the Future: Challenges and Opportunities

This final chapter attempts to project the scientific and technological trends in the development of this burgeoning discipline over the next few years and possibly decades. As the precursor to these novel technological applications, it is important to consider the diverse TMD phase engineering techniques crucial to current and future applications which have been dealt with in Chapters 2–7. While methods such as lattice strain engineering and irradiation techniques are effective phase engineering techniques, their potential for large-scale industrial implementation may be restricted due to the localized effects within the TMD materials. It is therefore necessary for one to overcome the immense difficulty of scaling up the preparation process and to increase the phase transition yield. In terms of techniques such as ion intercalation, it is certainly a well-established phase engineering technique effective for 2D-TMDs species such as MoS_2 with large 1H-1T′ energy differences. When applied to other 2D-TMD such as WTe_2 and $MoTe_2$ with considerably smaller 1H-1T′ energy differences, their unique Weyl semi-metallic behavior brings great interest to both fundamental science and practical applications [2]. With the strong influence of spin-orbit coupling, their attributes as topological insulators [3, 4] make 2D-TMDs a suitable class of materials for further intercalation studies.

As we consider the state-of-the-art, the scaling of semiconductor-based integrated circuits to even lower dimensions have been significantly hampered with the scale remaining stagnant for more than a decade. This is fundamentally attributed to the fundamental quantum limits of the material silicon. It is certain that atomic-scale and low-dimensional materials such as 2D-TMDs possess the potential to overcome the limits because of their atomic-level thickness. Nevertheless, a significant degree of technological development is still required to attain the current level of manufacturing standards found in conventional 3D silicon-based devices where their defect densities are measured in the scale of parts per billion. Meanwhile, TMD systems still lag far behind with the defect concentrations and structural phase yield at a high percentage scale. While synthetic techniques to derive high-quality 2D-TMD materials and TMD-based device fabrication techniques have made considerable progress over the past decades, a significant technological leap is still required for them to go beyond the level and quality currently restricted to the research laboratory. This is because of multiple manufacturability issues such as large-area growth and reproducible mechanical transfer of 2D materials.

At this juncture, experimental reports have been made of prototypical low-dimensional TMD-based devices. They display high-performance and low-power outcomes in a diverse range of setups such as FET systems, radio frequency (RF) and thin-film transistors, neuromorphic components, LED devices, solar cells, mechanical resonators, waveguides, and ultrasensitive device sensors. Such positive outcomes provide compelling reasons that they will become increasingly important over the next few decades as ease of production and device performance improves with technological advances. With the attractive mechanical, optical, and electronic properties of low-dimensional TMD systems, they can potentially complement or replace conventional silicon-based devices that pervade virtually all aspects of human lifestyle ranging from communication infrastructures to smart wearables, electronic entertainment to healthcare, where they promise to be more energy efficient, versatile, robust, and low-cost.

8.3 Global Challenges and Threats

The pressing problems that the world is currently facing come in multiple forms. Given the complication of these crises involving multiple stakeholders, there is no single or one-dimensional solution to address them effectively and to solve the problem. Instead, multi-pronged approaches ought to be employed.

The following threats confronting the human race in the next few years and decades are well-recognized and many have acknowledged the urgency to deal with them before we reach the tipping point – the detrimental effects of climate change, threats of future pandemics deadlier and more virulent than Covid-19, threats to global food security and unregulated development of new technologies, especially in areas related to artificial intelligence (AI).

While there are aspects of political will and cooperation that should be put in place at the national level and through extensive cooperation at the international level, the development of scientific and technological capabilities ought to be the driving force where more efficient infrastructures and environmentally cleaner technological devices replace existing technologies. These can potentially facilitate the shift to more efficient and greener environmental practices, while novel ways can be developed to address problems related to infectious diseases. In Sections 8.3 and 8.4, we discuss the global problems in greater detail and consider the specific roles that 2D-TMDs and their polymorphic behaviors can be capitalized on and manipulated in innovative ways to deal with the respective global problems. While the state of development is at different stages, we will also consider the viability of 2D-TMDs as solutions to some of these existing problems.

Climate change is perhaps the most pertinent subject posing a direct and imminent threat to the survival of the human race today. The rapidly expanding human population and the indiscriminate level of human activities have caused a major strain on natural resources. One prominent effect of such human activities would be the onset of global warming which is increasingly evident over the past few decades. The effects of global warming have been clearly manifested over multiple levels and facets ranging from extreme weather events, melting glaciers, to rising sea levels. These effects not only undermine the human standard of living, but also directly threaten the very existence of human beings.

The environmental effects of climate change are broad and far-reaching. The long-standing effects span the physical environment, ecosystems, and human societies as a result of excessive human activities. At the human and societal level, the effects of climate change are impacting humans globally on all continents and ocean regions [5], with low-latitude, less developed regions and countries facing the greatest existential threat [6]. Should this climatic trajectory continue unchecked, potentially severe, pervasive, and irreversible existential impacts for both the human race and the global ecosystem are expected [7], where inhabitants in developing countries are going to bear the greater brunt [7].

8.3.1 Clean and Renewable Energy Sources

Renewable energy is key to limiting the adverse effects of climate change [8]. Even though the usage of renewable sources of energy has been rapidly increasing over the past two decades, fossil fuels (coal, oil, and natural gas) still account for up to 80% of the global energy source in 2018 [9]. In order to achieve carbon neutrality by 2050, clean renewable energy must become the dominant form of electricity generation, amounting to 85% and above. Meanwhile, the use of coal must be eliminated and nearly phased out by then [10, 11].

With the urgent transition to clean and renewable energy sources, solar energy emerges as one of the most ubiquitous and viable renewable energy sources due to its large magnitude and it has been considered the cheapest source of electricity [12]. In fact, the effective tapping and utilization of solar energy has been regarded as an affordable, inexhaustible, and clean energy source that will yield huge long-term benefits. At the national level, the reliance on this inexhaustible, and virtually import-independent resource has its strategic benefits in terms of energy security. But internationally, it also benefits and enhances the push for sustainability, reduces pollution, and ultimately mitigates the effects of global warming that are mostly attributed to the burning of fossil fuels [13].

To tap into this unlimited energy source, photovoltaic systems (PVs) have been developed over the past decades for the generation of clean electricity. Since mass production of PV began at the turn of the millennium, the decreasing costs over the years due to economies of scale, improvements to manufacturing efficiency, and technology have made solar energy much more affordable and accessible [14].

While the development of PV technology has been notable, significantly greater efforts must be undertaken to ensure that solar energy replaces fossil fuels as the primary energy source. In this regard, the following ought to be considered [15]:

(1) enhancing the PV cell efficiency
(2) reducing the cost of the PV
(3) prolonging their operation lifetime.

When considered from another perspective, the challenge of energy storage due to the intermittent nature of solar radiation by weather and seasonal effects prevails. Hence, one considers alleviating this problem via the conversion of solar energy in the form of photoelectrochemical (PEC) systems into (clean) chemical fuels suitable for storage, transportation, and the eventual energy consumption [16–19].

Conventional solar cells have generally been manufactured using Group IV, III–V, or II–VI compounds and given the decades of massive improvement and development, they are generally of high quality and energy efficienct. However, the development has plateaued over the years and is unable to keep up with the exponential growth in energy demands. The difficulties involved in the implementation of epitaxial growth techniques and the limited supply of the elemental constituents have hampered the continuous growth of conventional solar cell technology. To circumvent this problem, the fabrication of PV and PEC devices in the ultrathin setup could possibly minimize material consumption whilst maintaining the device's robustness and efficiency [20–26].

In this regard, 2D-TMDs and other 2D materials fit the bill especially with their light weight and flexibility, whilst providing efficient charge transport and the storage of photo-generated carriers. With a bandgap typically in the range of ~1.0–2.0 eV, 2D-TMDs are excellent light absorbers as compared to other materials such as semi-metallic graphene. Hence, more superior light-harvesting property is expected for the TMD-based solar cells. Their mechanical and electronic versatility further allows them to be stacked into van der Waals (vdW) heterostructures in the form of 2D/3D or 2D/2D material configurations to create vdW heterojunctions for PV and PEC applications [27–31].

With regards to energy storage in the form of PEC, the use of conventional p-type semiconductors produces very low photovoltage due to the poor hydrogen evolution reaction (HER) at the semiconductor/liquid junction. 2D-TMD materials including MoS_2 and WS_2 have emerged as ideal candidates to circumvent this problem, thanks to the low permeability through the basal planes of their layered structure. This, in turn, allows for many catalytically active edge sites. Besides, the excellent stability in acidic electrolytes renders them ideal substitutes to replace noble metal catalysts. Furthermore, with 1H-phase MoS_2 as an example, it is a direct bandgap semiconductor of ~1.88 eV in the monolayer form and becomes an indirect bandgap semiconductor of ~1.29 eV in its bulk form. In terms of the energy band, the position of its conduction band maximum is situated at a level higher than the HER [32, 33]. Early experimental studies using MoS_2 crystals have also confirmed that PEC oxidation of water on the crystal surface occurs under visible light [34]. These are indications that TMDs in both their monolayer and multilayer form could be utilized as semiconducting absorbers and HER catalysts in particular for constructing a heterojunction for PEC water reduction processes.

Developments in photocatalytic applications involving the HER and CO_2 reduction based on TMD materials are expected to grow over the next decades. Nevertheless, major challenges prevail concerning the efficiency, reliability, and stability of the HER and CO_2 reduction processes especially under harsh experimental conditions for potential industrial applications. The ability to control the phase engineering process of TMDs with great precision is a fundamental issue that must be solved to optimize the material for photocatalytic applications. Even though the need to optimize the 1H-1T′ phase transition yield is important for applications involving PEC, the precise control of TMDs in their respective phases plays an essential role in optimizing the HER performances. A study conducted on $MoSe_2$ nanosheets has been an instructive case study where a phase content of 45% by 55% between the 1H and 1T components maximizes the HER photocatalytic performance [35]. Further investigative studies are essential to move this discipline forward with better and more efficient applications to advance the use of 2D-TMDs in PEC systems.

Even though 2D-TMD-based PV and PEC devices promise huge possibilities, current efficiencies of as-demonstrated devices are still far from satisfactory. It is therefore certain that further efforts are necessary to realize PV and PEC devices with state-of-the-art energy conversion efficiencies. Figure 8.1 provides a comprehensive summary of various prototypical vdW junction-based devices and compares their efficiency and performances with mainstream 3D/3D PV devices [38–41] and electrodes [42–47].

8.3.2 Water Treatment and Access to Clean Water

The resulting onset of extreme weather events is becoming increasingly commonplace [48], and crop failures are another worrying effect threatening massive starvation [49, 50], and multiple tropical infectious diseases including dengue fever and malaria are becoming easily transmissible in warmer climates [51].

The risk of water stress due to climate change has been an increasingly existential threat to many [52]. Warming climate has a radical effect on the nature and cycle of global rainfall, surface water evaporation, snowfall, stream flow, and other factors which, in turn, affects water supply and quality. The seepage of saltwater as a result of rising sea levels into the underground and freshwater sources also reduce the quantity of freshwater supplies for farming and human consumption. Furthermore, with about 50% of the world's freshwater consumption coming from glacial runoff [53], global consumption dependent on this water source including agriculture, power supply, and human health and well-being faces a grim prospect due to melting glaciers as a result of warming surface temperatures [54]. Meanwhile, the

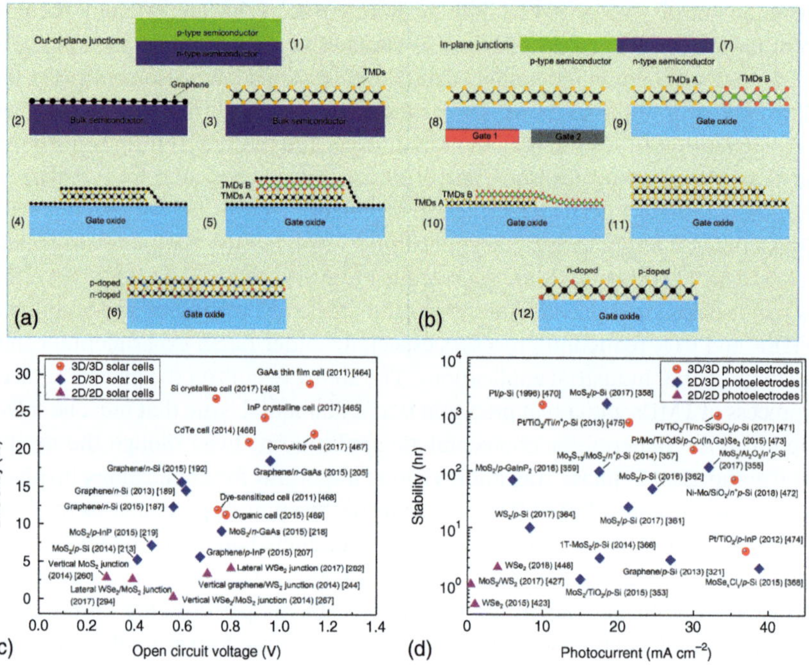

Figure 8.1 A list of schematics and charts displaying van der Waals junctions in both vertical and horizontal configurations along with their device performance and efficiency. (a) Vertically-stacked van der Waals systems with out-of-plane junctions [36, 37]. Panel (1): An out-of-plane junction that consists of a p- and n-type bulk semiconductors stacked on each other. Panels (2) and (3): Junctions comprising 2D materials such as graphene or TMDs on a bulk semiconductor. Panel (4): Multilayer heterojunction comprising a graphene/TMD/graphene heterostructure. Panel (5): 2D-TMD heterojunction with vertical graphene electrodes. Panel (6): An out-of-plane homojunction comprising a *p*-doped layer on top of an n-doped layer. (b) Lateral junction useful for the fabrication of PV devices [36, 37]. Panel (7): In-plane junction comprising a p- and an n-type bulk semiconductor. Panel (8): A planar homojunction configuration that is controlled by back gate electrodes 1 and 2. Panel (9): Lateral heterojunction fabricated by lateral hetero-epitaxy. Panel (10): A lateral contact in the form of a van der Waals heterojunction that consists of a p-type 2D-TMD stacked above an n-type 2D-TMD. Panel (11): 2D-TMD with variable number of layers at different part of the substrate. This configuration results in a functional PV device due to discontinuities in energy bands at the device interfaces. Panel (12): An in-plane junction formed by doping. (c) Power conversion efficiency of different device setups with respect to circuit voltage for the respective van der Waals junction solar cells and other conventional high-performance bulk solar cells. (d) Comparing the system stabilities with respect to photocurrent for the van der Waals junction photoelectrodes and other conventional high-performance bulk photoelectrodes. Source: Reproduced with permission Li et al. [15].

rapid pace of urbanization and industrialization has aggravated to a problem with the issue of water pollution. Pollutants in the form of industrial, human waste, etc. introduced into clean water supplies have given rise to significant threats to both the atmosphere and human health [55, 56]. Water contamination comprises both natural and artificial components ranging from biological contaminants (pathogens) [57], macroscopic contaminants (plastic, wood, metals) [58], organic materials (petroleum, pesticides, detergents, disinfectants, pharmaceutical products) [59–61], to thermal pollution (wastewater from industrial cooling processes) [62, 63].

In addressing the problem of water treatment, the utilization of 2D-TMDs nanotechnological applications come in as a viable solution. One-dimensional nanotubes and nanowires [64, 65] and three-dimensional structures [66] have been effective media for water purification due to their large surface area to volume ratio. Nevertheless, problems remain in the utilization of these materials including [67]:

1. Such purification microstructures tend to aggregate and/or re-stabilize in water media. With a significant reduction in surface area, they are no longer effective in the water purification process.
2. While water purification is duly carried out by the nanomaterials, it is important for the complete recovery of these deactivated or depleted nanomaterials which can be inherently harmful to both humans and the ecosystems. However, this recovery process is not an easy task.
3. Their applications are grossly limited by their dimension and size.

These problems can possibly be circumvented based on 2D-nanomaterials particularly due to their ease and efficiency in handling, and treating them, and that only a small quantity is required per unit volume of water [68]. Interestingly, the optoelectronic structure of 2D-TMDs such as monolayer-MoS_2 has enabled it to serve as a photocatalytic material for water purification based on solar energy. Figure 8.2 displays the schematic black phosphorus/MoS_2 (BP/MoS_2) photocatalytic system where the production of electron-hole pairs is essential for processes including hydrolysis, pollutant oxidation, and CO_2 reduction. Specifically, low-dimensional BP/MoS_2 heterostructures facilitate photo-generated charge separation. This, in turn, leads to a photocatalytic process of generating H_2 gas [70].

In general, with thermal stability, mechanical strength along with electronic transport properties of 2D-TMDs including MoS_2 [71–73] and WS_2 [74, 75], they are also ideal materials for water desalination processes. Particularly for the case of MoS_2, with the hydrophilic S-atoms exposed in its monolayer structure [76], water desalination processes based on nano-porous MoS_2 show that the membranes could obtain 70% higher water flux as compared to graphene nanopores whilst 88% of other ions remain rejected. This suggests the potential of 2D-TMD materials to be used in water desalination processes [77]. Likewise, the dangling chalcogen groups including the S and Se atoms on 2D-TMD monolayers can react readily with heavy metal ions such

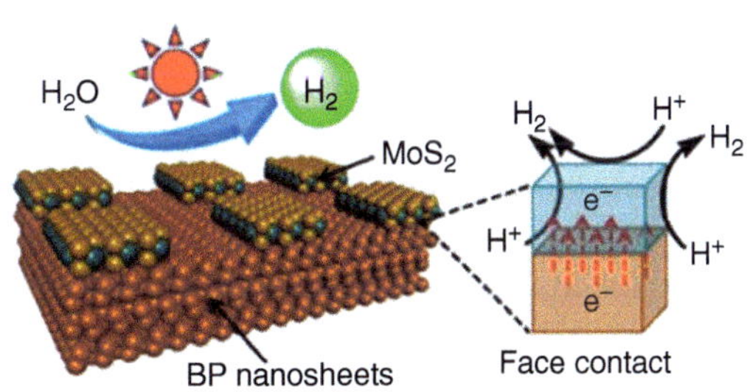

Figure 8.2 Schematic diagram of a photocatalyst comprising monolayer-MoS_2 on black phosphorus nanosheets. Source: Reproduced with permission from Fatima et al. [69] under the Creative Commons CC-BY license and adapted from Yuan et al. [70], Copyright Elsevier 2019.

as Hg^{2+} and Pb^{2+}. This allows them to be utilized as water purification media to remove heavy metal contaminants from polluted industrial water [78].

8.3.3 Healthcare and Pandemic Intervention

The distribution of arthropod-borne (mosquitos, ticks, etc.), food-borne, and water-borne diseases, as a result of climate change, is longstanding and widespread [79]. The warming surface temperature has facilitated the climate suitability for infectious disease transmission and vulnerability to mosquito-borne diseases previously typically only found in the tropics. Besides, the warming oceans have a lasting impact on extreme weather events which further promotes the spread of new and emerging infectious diseases, especially in less developed countries where the population is most vulnerable [54, 80].

The growing encroachment of human activities into natural habitats may inevitably result in an increasing frequency of similar global events within the century. Nevertheless, interspecies viral sharing can lead to viral spillovers. In particular, such biological events will increase due to geographic range shifts of mammalian species (most importantly bats) as a result of global climate change. Risk hotspots are mainly located in high-biodiversity regions where there are interactions with high human population density areas particularly in regions such as Asia and Africa [81].

Apart from national and international policies serving as overarching preventive measures to alleviate the possibilities of such major pandemic events taking place, improved medications and vaccine developments may serve as possible solutions in treating infectious diseases. Nevertheless, the combat against infectious diseases is multi-faceted involving diverse approaches at multiple levels. In this regard, the utilization of 2D materials, specifically 2D-TMDs, can play a part in facilitating material innovations and technological development processes in preventing and countering the detrimental effects of highly infectious diseases. Such technological and advanced material development can come in the form of high-performance, high-efficiency, and highly sensitive diagnostic devices, antimicrobial components, and essential personal protective equipment including face masks and shields, aprons, sterilizers, safe for human usage and other components and equipment essential for medical treatment.

Studies have been carried out in recent years concerning the utilization of 2D materials partially due to their suitability as carriers for antimicrobial materials. Their ability to interact with microorganisms physically and chemically has promising effects to reduce the possibility of developing drug resistance. MoS_2 has been reported to display antibacterial properties where their proposed mechanisms include the induction of membrane damage/oxidative stress [82–84] and ion (MoO_4^{2-}) release [85]. Given the larger surface area of monolayer-MoS_2, chemically exfoliated MoS_2 nanosheets are shown to have more potent antibacterial effects than their bulk counterpart [83]. Nanocomposite structures in the form of MoS_2-coated Ti substrates, graphene oxide-MoS_2 (GO/MoS_2), TiO_2 nano-particles anchored on MoS_2 sheets provide them with a variety of rich surface chemical properties that have potent effects on pathogens such as methicillin-resistant *Staphylococcus aureus* (MRSA) and *Escherichia coli* [82, 85, 86]. In the case of the antibiotic-resistant MRSA, MoS_2 nanosheets effectively kill the superbugs and destroy the biofilms through a combination of oxidative stress and depolarization of the bacterial membrane. While silver in both its ionic (Ag^+) and nanoparticle (Ag) form is toxic to humans at elevated concentrations, the former could not be released or metabolized and results in its accumulation in the body. The amount of Ag^+ is thus minimized when loaded onto Cysteine (Cys)-modified MoS_2 nanosheets and then coated with positively charged polyelectrolyte polydiallyldimethylammonium chloride (PDDA) for the effective elimination of bacteria *in-vitro* and speeds up wound healing *in-vivo* [87, 88]. In direct relation to water treatment and decontamination, a combination of MoS_2 in both the 1H and 1T′ phases can facilitate the generation of reactive-oxygen-species (ROS) under exposure to visible-light illumination. This in turn enhances the oxidative stress on the contiguous bacteria. Based on this concept, a few-layered vertically aligned MoS_2 (FLV-MoS_2) films have been fabricated for rapid high-efficiency water disinfection using visible light [89]. Another example involves the fabrication of a binary composite heterojunction photocatalyst that comes in the form of a MoS_2/TiO_2 nanotube arrays to carry out photocatalytic sterilization based on anodic oxidation and hydrothermal method (Figure 8.3). Based on visible light irradiation, the water disinfection efficiency against pathogens could go up to 98.5% and it remains high even after multiple recovery and recycling processes [90].

The tungsten-based (WX_2) 2D-TMDs have also been investigated for their bactericidal properties [91–93]. While W- and Mo-based 2D-TMD nanosheets are

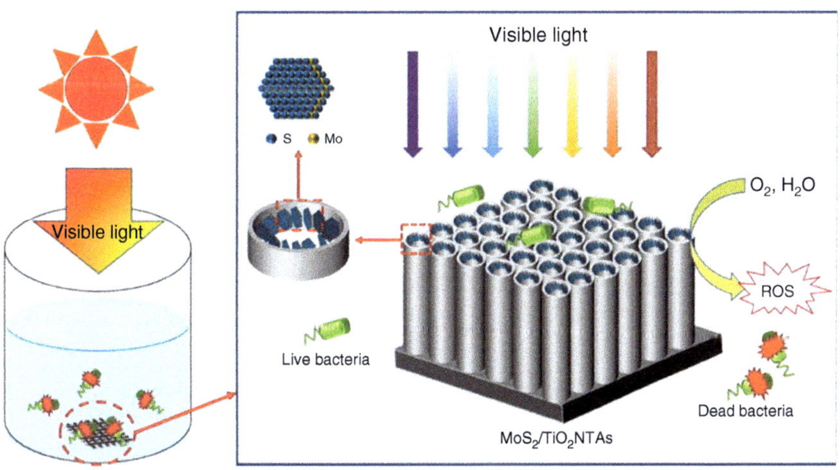

Figure 8.3 The optoelectronic properties of monolayer-MoS_2 in its respective structural phases can be utilized for Photothermal and photocatalytic bactericidal applications. Schematic depicting the binary composite heterojunction photocatalyst comprising MoS_2/TiO_2 nanotube arrays to carry out photocatalytic sterilization based on anodic oxidation and hydrothermal method. Source: Reproduced with permission from Yan et al. [90].

expected to exhibit comparable antibacterial characteristics and mechanisms, i.e. membrane destruction and induction ofoxidative stress. It has been reported that the anti-bacterial activity of WS_2 monolayers is mainly attributed to the induction of structural damage instead of the generation of ROS [92]. Nevertheless, it is still inconclusive whether the membrane disruption is attributed to a physical interaction or oxidative damage that is ROS-independent. Studies have also been conducted to compare the efficacy of bacterial inactivation between WS_2 and WSe_2. Using single-strand DNA (ssDNA), the antibacterial activity of WS_2-ssDNA and WSe_2-ssDNA is compared [93]. Results indicated that the latter had a stronger bacterial inactivation ability as WSe_2 possesses inherent p-type semiconducting properties, thus allowing the ROS-independent oxidation of glutathione. By comparing the antibacterial properties of monolayer-WS_2 and other related derivatives such as WS_2-reduced graphene oxide (WS_2-rGO) nanocomposites [91], it is been reported that while these materials can induce oxidative stress and membrane rupture on bacterial cells, WS_2-rGO nanocomposites have greater efficacy due to their ability of simultaneously generating O_2^-.

At a general level, 2D materials can also be utilized in a new paradigm known as nanomedicines where they are applied extensively in treating diseases such as cancer, bacterial infection, tissue engineering, biological protection, neurodegenerative diseases, and cardiovascular disease [94].

Apart from their mechanical and electronic properties, the diversity of synthesis methods for 2D-TMDs also renders them readily producible at superior few-layered or monolayer quality and performance. These are crucial properties ideal for applications in areas related to bio-sensing, imaging, optical therapy, etc. For instance, in an integrative anti-tumor/bone repairing treatment, a composite MoS_2-(Poly(lactic-*co*-glycolic acid)) film has been constructed on the surface of 3D-printed borosilicate bioactive glass (BG) (Figure 8.4) [95]. Apart from being able to elevate the temperature rapidly and effectively, this BG scaffold exhibits

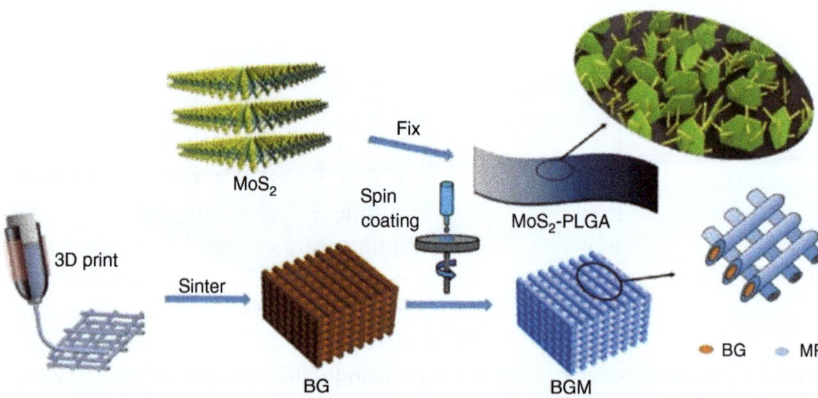

Figure 8.4 Schematic depicting the synthetic procedures of MoS_2-integrated composite bioactive glass scaffolds for the integrative functionalities of anti-tumour and bone repair. Source: Reproduced with permission from Wang et al. [95].

excellent photothermal stability. In their study on rats, besides being able to kill osteosarcoma cells and suppress the growth of malignant bone tumor growth through photothermal therapy, this system could further activate the proliferation and differentiation of the rat bone mesenchymal stem cells to facilitate bone repair.

Of course, the case studies presented in this chapter are non-exhaustive, especially with the diverse applications available for use in healthcare and biomedical applications. Other emerging applications in areas such as wearable technologies, neural interfacing, optogenetics, and biomimicry systems can also be considered [96]. Moreover, the unique electronic, electro-chemical, and optical properties of 2D-TMDs further allow them to be developed in the domains of *in-vitro* detection of target analytes, *in-vivo* stimulation/interrogation of cellular systems, drug delivery media, and even *in-situ* monitoring of biomarkers [96].

But amidst all these promises, one ought to consider the potential toxicity and adverse effects they may have on humans before the actualization of these technological advances. In general, 2D materials have been reported to affect (human) cells, thus rendering them incompatible and unsuitable for safe utilization [97]. Likewise, it is essential to assess the bio-compatibility of 2D TMDs before utilizing them in devices and assays [98].

A comprehensive understanding of the material toxicity and their interaction mechanisms would help in the design of nanomaterials which suit specific applications by making them safer for certain cell types or more toxic for others such as cancer and pathogenic cells. While there remains numerous questions pertaining to the roles and cytotoxic properties of 2D TMDs and other 2D materials, it is essential to assess the biocompatibility of these materials extensively and perhaps on a case-by-case basis to be tailored for specific biological applications.

8.3.4 Food Safety and Security

Food security has been a major theme over the course of human civilization. Food supplies have become significantly more accessible with the advent of globalization and with the previous Green Revolutions. The intricate global supply chain has been the backbone to make sure many have access to the food supply. According to the United Nations Committee on World Food Security, **food security** is defined as a state in which all people, at all times, have physical, social, and economic access to sufficient, safe, and nutritious food that meets their food preferences and dietary needs for an active and healthy life. While the global supply chain has been well-established for the well-being of a significant proportion of the global population, it is a delicate one that can easily be disrupted and undermined by major events such as war [99, 100] and pandemic [100] with the poorest and least developed countries bearing the brunt of its effects [101]. Meanwhile, natural disasters and the effects of climate change have also severely threatened food production and undermined global food security, especially with the growing population amongst the developing countries [102, 103].

The world population has breached the 8 billion mark [104] and is projected to grow (albeit at a slower rate) to around 8.5 billion in 2030 and to 9.7 billion in 2050.

Eventually, it is projected to peak at about 10.4 billion in the 2080s [105]. Despite the decelerating growth, the pressures brought by this global population size continues to put a strain on the environment, energy resources, and global food supplies [106]. Hence, more has to be done and committed to sustainably feed the massive global population [107].

Similar to the threats of climate change, multi-faceted approaches, and management systems must be put in place to address this massive challenge.

From the technological front, multiple approaches can also be carried out in alleviating the threats of global hunger and nutritional deficiency. They can come in the form of crop yield enhancements and productivity and preservation [108], more efficient food preservation measures through improved packaging [109], and food packing sensors and food spoilage detection [110].

8.3.4.1 Agricultural Production, Sustainability, Productivity, and Protection

Strategies involving the application of 2D-based nanomaterials have been implemented in agricultural activities in the form of pest- and weed control, improvements to soil quality, and food nutritional value [111]. Low-dimensional graphene materials are a case in point where they have been utilized in enhancing the contents of agro-active materials. Through their unique low-dimensional and large surface area, they can effectively deliver and regulate the release of nutrients and other substances which can promote crop growth or increase the shelf-life/active time of the agro-active components without undermining soil quality and, at the same time, keep soil pollution to a minimum [112–114].

With their unique low-dimensional mechanical and electronic properties, CVD-grown 2D-TMDs have found themselves useful and have been utilized in biological applications related to biomedical and bio-sensory systems and even in disciplines related to DNA sequencing and personalized medication. Specifically, the unique anti-bacterial and highly-sensitive sensory properties of MoS_2-based nano-composites have rendered it a useful component in combating microbial growth and detecting multi-component pesticide residues on vegetable and fruit produce [115], while a range of 2D-TMDs have been demonstrated to be a favorable platform for the indirect electrochemical detection of fenitrothion, a class of organophosphate pesticide [116]. First-principles study has indicated that engineered 2D-TMD nanoribbons can optimize the efficiency of nitrogen fixation processes due to the preference of nitrogen molecules to stay at the metal edges of 2D-TMD-based nanoribbons [117].

8.3.4.2 Roles of 2D-TMDs in Food Packaging and Preservation

Food packaging has played a critical role and its increasing importance is notable, especially in the global food supply chain where one has to ensure the preservation of food quality from the manufacturing stage right to the hands of the consumers [118]. Global demands for high-quality, safe, and rapid access to food supplies have been on a rising trajectory with globalization and urbanization. With such growing demands, economical multi-functional food packaging techniques, and methodologies that retain food quality and promote the propagation and advertising of food products have been the focus of food scientists and researchers [119]. The rise of nanotechnology and maturing expertise in low-dimensional materials have therefore offered new opportunities in the design and development of food packaging and preservation techniques.

Studies have been conducted to infuse MoS_2 into polymers such as poly(vinyl alcohol) [120] and polyacrylonitrile [121]. This is particularly useful due to the robust mechanical properties of monolayer-MoS_2 which boasts an in-plane stiffness of $180 \pm 60\,N\,m^{-1}$ – equivalent to a Young's modulus of $\sim 270 \pm 100\,GPa$ [122]. Such infusion of MoS_2 into the polymer material can significantly enhance the thermal stability, barrier property, and mechanical property of the packaging material due to the formation of hydrogen bonds between the MoS_2 and polymer molecules.

The optoelectronic and catalytic properties of 2D-TMDs can also be readily utilized in the fabrication of sensory systems and smart labels as part of the functionality of intelligent packaging. For instance, a MoS_2/Cu_2S hybrid has been synthesized on disposable cellulose paper by a hydrothermal method that is capable of sensing humidity, temperature, breath, and ethanol adulteration in the environment [123]. The development of intelligent labels and printable sensors can also be realized as a form of intelligent packaging [124] through the utilization of 2D-TMDs. MoS_2 and WS_2 are reported to fuse with graphene-based inks which are both water-based and biocompatible for the fabrication of multi-functional identification tags [125]. Such "smart" inks can be imprinted on common packaging materials such as polyethylene terephthalate (PET) films or papers and they serve as a large-area platform where photosensors and programmable logic memory devices could be constructed.

Despite the extensive studies that have affirmed the versatility and promising potential of 2D-TMDs and their applications towards food security, this discipline is still considered to be in the experimental phase with multiple gaps to be duly addressed. In particular, the eco-toxicity and toxicity of these materials have to be studied in greater detail before the mass implementation of these technological advances. Nevertheless, the utilization of 2D-TMD materials and for that matter, other nanostructures should not be overlooked given the immense potential they promise in the global food chain.

8.4 Exponential Growth in Demands for Modern Computation

8.4.1 Deep Learning and Artificial Intelligence

Having discussed the major threat facing humanity in the near future, we now turn our direction toward the challenges and the prospects of future computing. Catchphrases such as "Deep Learning," "Internet of Things," "Cloud-based Computing," etc. occupy our headlines and are the preoccupation of data scientists and computer scientists alike. Some look forward to it with bated breath while others are tackling the current problems and those they foresee in the very near future.

Modern computing systems consume far too much energy. Should this trajectory of development continue, it will not be a sustainable platform for the complex AI applications that are becoming an integral part of our daily lives. While many are preoccupied with the quest for speed, accuracy, and efficiency, few are aware that at the global level, data centers are using a whopping 200-terawatt hours of energy per year and are forecast to increase by another order of magnitude by 2030 [126, 127].

8.4.2 Internet of Things and Data Overload

With the rise of smart technologies, smart devices are no longer restricted to smartphones and computers. Instead, they are now a ubiquitous feature in our daily lives with about 20 billion devices ranging from basic home appliances to air and water quality sensors, trackers, and the list goes on. This quantity of devices implies the explosion of data and that explains the prevalence of IoT.

With the development of artificial neural network architectures, atomic layer materials especially 2D-TMDs are of significant interest due to their favorable electronic and mechanical properties that include interfacial charge trapping, ferroelectricity, ferromagnetism, polymorphism, and diverse interchangeable structural phases [128, 129].

In essence, they fulfill the following conditions in neuromorphic devices [130–132]:

1. Fast switching process
2. High-density integration
3. Low energy consumption

The low-dimensionality of 2D-TMDs allows for a highly integrated device architecture which enables fast switching processes at low power consumption (Figure 8.5a). Neuromorphic operations of 2D-TMD-based devices include memristive switching, phase transformation, and ferroelectronics [134, 135] and their structure further allows for reliable operative functionalities [136]. Furthermore, their mechanical robustness and flexibility enable flexible device applications for the development of wearable and highly flexible optical and electrical synaptic devices (Figure 8.5b) [132, 137].

2D-TMDs can act as basic building blocks by bypassing the limitations of single-layer systems to create more complex and elaborate heterostructure systems to cater to different types of electronic, optoelectronic memristors, and synaptic devices and structures (Figure 8.5c). Such customized systems can imitate complex biological systems that integrate optical perception, memory, and computation functions [138, 139]. Furthermore, the heterostructure systems create vdW contact between multiple layers without current leakage or failure [140, 141]. Such vdW heterostructures can be utilized to achieve desired synaptic characteristics similar to biological systems through gate tuning and adjustments to energy band configurations [142, 143].

Defect engineering is another strategy in the development of ion-based resistive devices (Figure 8.5d). This can be regulated through processes including H_2 etching, surface plasma treatment, and thermal annealing to generate site vacancies which are efficient in facilitating charge transfer processes [144, 145]. Given the

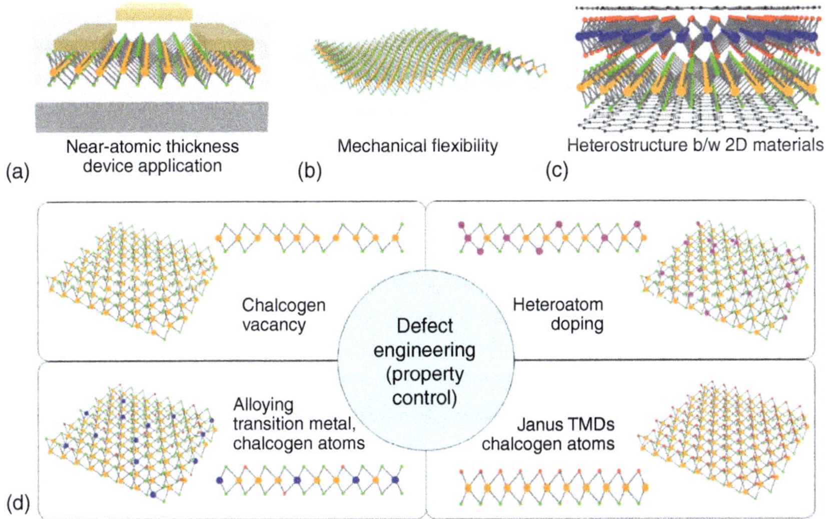

Figure 8.5 Different ways of manipulating 2D-TMDs to be utilized for memristor-based neuromorphic devices. (a) Device fabrication based on their atomic-level thickness, (b) mechanical flexibility, (c) formation of heterostructures to reduced contact resistance. (d) Defect-engineering techniques such as the introduction of chalcogen vacancies, hetero-atom doping, alloying, and Janus structures. Source: Kwon et al. [133]/Springer Nature/Licensed under CC BY 4.0.

polymorphic properties of 2D-TMDs, phase engineering serves as another strategy to tune their electronic properties [146, 147]. Heteroatom doping is a strategy to stimulate such phase transition processes to extend the material capabilities. A case in point is the introduction of Re-atoms to MoS_2 to trigger a 1H-1T phase transition [148] where such Re-doped MoS_2-based FETs are shown to be degenerated n-type devices [149]. Furthermore, structural distortion and photon scattering can be induced by such heteroatom processes which, in turn, is able to control the optical properties and their recombination dynamics. Hence, methods including heteroatom doping and metal or chalcogen atom alloying further extend the application of 2D-TMDs to be active materials in memristive and neuromorphic applications. Meanwhile, the breaking of out-of-plane mirror symmetry – Janus structuring is another means of defect-engineering of 2D-TMDs where they demonstrate piezo and ferroelectric functionalities [150]. This is a useful property which in turn can be extended to memristive and neuromorphic applications.

When capitalizing the polymorphic properties of 2D-TMDs in memristive applications, the semiconducting 1H-phase and the quasi-metallic 1T′-phase of monolayer-MoS_2 comes into consideration due to its robust properties, easy accessibility, and reversibility between the respective structural phases [151, 152]. Multiple techniques including the lithiation process can be easily carried out for the phase transformation of the MoS_2 monolayers. Based on these straightforward experimental processes, MoS_2-based devices exhibit very reliable memristive properties with the respective structural phases regulated by the local low-high Li^+

concentration driven by an external electrical field. The control of Li^+ intercalation leads to variation between high-resistance (HRS) and low-resistance (LRS) states which is also evident through a contraction and extension of the lattice between the respective states. This reversible transition between the respective states induced by the migration of Li^+ allows the MoS_2 system to behave like an artificial system with synaptic plasticity of weight potentiation and depression. Moreover, these prototypical devices can be directly connected between adjacent synapses that achieve similar (different) learning outcomes through signal inputs from multiple terminals similar to the framework of synaptic cooperation (competition) in biological systems [153].

Nevertheless, it is acknowledged that the exposure of 2D-TMDs to the environment can undermine their performance due to chemically active bonds at the grain boundaries and vacancies [154]. While the high surface-to-volume ratio of 2D materials has been advantageous in many aspects, this feature renders it vulnerable to external elements. A notable example is where MoS_2 readily decomposes in biofluids and is easily etched by annealing under conditions including oxidizing atmosphere and plasma treatment [155–158]. Encapsulation of the material serves as an effective way to minimize any forms of degradation of the 2D-material-based devices due to effects from external elements. In this case, MoS_2 memristor device is encapsulated by applying an amorphous fluoropolymer layer by barrier layer deposition (Figure 8.6) [159]. CYTOP has been chosen as the encapsulated material due to its excellent electrical insulation, and chemical resistance, and thermal stability [160]. This encapsulated device has an enhanced switching endurance of 100 potentiation and depression cycles and exhibited resistive switching behavior for 25 conductance states. This is in contrast to a ~50% degradation for bare devices exposed to the environment (Figure 8.6) [159].

Indeed, the development of quantum computing and neuromorphic devices hold the future development of computing. The aforementioned example is merely the tip of the iceberg amidst its burgeoning development. They are considered the basis

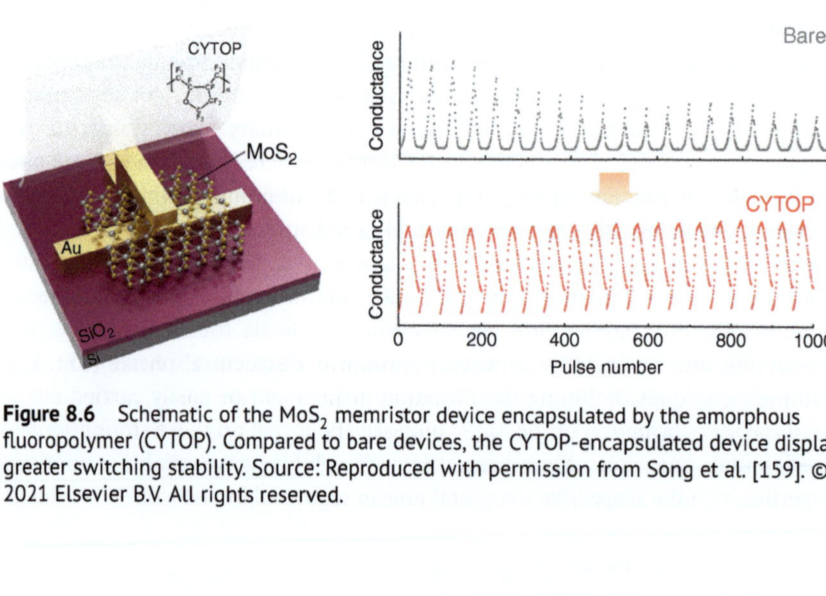

Figure 8.6 Schematic of the MoS_2 memristor device encapsulated by the amorphous fluoropolymer (CYTOP). Compared to bare devices, the CYTOP-encapsulated device displays greater switching stability. Source: Reproduced with permission from Song et al. [159].

of edge computing and the IoT, which requires an all-rounded development of novel chip architectures and device technologies. The utilization of 2D materials, specifically 2D-TMDs, as the basis of synaptic devices can potentially lead to the effective integration of memristors and sensors into a single entity. This is an important quantum leap for the eventual materialization of multifunctional neuromorphic computing. Nevertheless, major challenges prevail for the eventual scaling of the relevant qubit system. Problems associated with material synthesis, defect management, and material characterization are stumbling blocks to be overcome. Nevertheless, extensive experimental and theoretical studies have continued over the years to improve the reliability and overcome the challenges in alleviating the impact of Fermi-level pinning and Schottky barrier injection in both monolayer heterostructure systems, while expanding the scalability of current device fabrication and sample growth techniques [161].

Herein, a few technical issues must be addressed for the eventual implementation of 2D-TMDs at a substantial chip-level scale [133, 162].

1. **Large-scale synthesis of high-quality defect-free 2D-TMDs**: As mentioned in Section 8.1, it is a very critical requirement for the successful integration of 2D-TMDs to be integrated into circuit design and applications. The ability to grow large-area material with uniform physical properties is an essential prerequisite. To achieve this, a detailed understanding of synthesis parameters such as temperature, pressure, and ambience are essential to maximize monolayer qualities including stoichiometry, thickness, defects, and grain size.
2. **Effective control of physical properties and doping strategy:** The ability to fine-tune the properties of 2D-TMDs lies at the core of designing high-precision and high-performance devices tailored for neuromorphic computing. Precise doping techniques, therefore, come into play to regulate their electronic, optical, and structural properties effectively and efficiently. At this juncture, conventional doping processes based on ion implementation remain unsuitable to perform such precise modulation at the atomic level. Instead, stable doping processes have to be developed [163].
3. **Low-energy consumption high-stability:** As highlighted previously, efficient energy consumption is a hallmark of memristive devices. While the parallel, compact, and energy-efficient structure of two-terminal synaptic devices are desirable, the high contact resistance is a critical stumbling block and bottleneck to the charge-transfer capability in 2D-TMD devices. Furthermore, the issues of Fermi-level pinning and the formation of Schottky contact [164, 165], along with problems such as material degradation and oxidation can hamper device performance. The implementation of a vdW heterostructure system is a possible means to circumvent this problem [166].
4. **Development of multi-functional synaptic devices:** External stimuli to activate and control the artificial synaptic behaviors can come in the form of electric field, optical stimulation, magnetic field, pressure, chemical adsorption/desorption, and temperature, etc. These would require an elaborate multi-gated neural architecture that is able to send and process multiple complex

functions that may take place simultaneously. The effective modulation of the 2D-TMD properties by the introduction of external stimuli can take place by adding a functional layer to the 2D-TMD. The actualization of this device can bring the development of synaptic systems a step closer to simulating the advanced properties of biological neuro-systems in the form of vision and motion recognition/detection sensors.

8.5 Conclusion

As seen from all the preceding chapters from 1 to 7, the mechanical and optoelectronic versatility of 2D-TMDs is immensely interesting and their potential for applications in diverse disciplines is certainly appealing.

Nevertheless, virtually all these developments are still at the research and developmental stage. While we anticipate 2D-TMDs to be commonplace in memristive devices, or to be the staple material in highly efficient PVs, it is hard to predict the future trajectory of such technological developments. As mentioned in Sections 8.3 and 8.4, the urgency of the climate crisis calls for the need to rapidly develop the relevant technologies to meet these needs. It is certainly possible to see 2D materials, or 2D-TMDs in particular, as the future of technological advances which may eventually be an integral part of human life, existence, and environmental sustainability. We hope that these ambitions may soon become reality, where energy sources are cleaner, the quality of life significantly improved, computer technologies outstrip current limits, yet with minimal effects on the global environment.

References

1 Derbyshire, K. (2021). Thinner Channels With 2D Semiconductors. Semiconductor Engineering. https://semiengineering.com/thinner-channels-with-2d-semiconductors/.

2 Qian, X., Liu, J., Fu, L., and Li, J. (2014). Quantum spin Hall effect in two-dimensional transition metal dichalcogenides. *Science* 346: 1344–1347. https://doi.org:10.1126/science.1256815.

3 Sakai, H., K. Ikeura, M. S. Bahramy et al. Critical enhancement of thermopower in a chemically tuned polar semimetal $MoTe_2$. *Sci. Adv.* 2, e1601378. https://doi.org:10.1126/sciadv.1601378.

4 Takahashi, H., Hasegawa, K., Akiba, T. et al. (2019). Giant enhancement of cryogenic thermopower by polar structural instability in the pressurized semimetal $MoTe_2$. *Phys. Rev. B* 100: 195130. https://doi.org:10.1103/PhysRevB.100.195130.

5 Smith, K.R., Woodward, A., Campbell-Lendrum, D. et al. (2014). Human health: impacts, adaptation, and co-benefits. In: *Climate Change 2014: Impacts, Adaptation, and Vulnerability. Part A: Global and Sectoral Aspects. Contribution of Working Group II to theFifth Assessment Report of the Intergovernmental Panel*

on *Climate Change*, 709–754. Cambridge, United Kingdom and New York, NY, USA: Cambridge University Press.

6 Gleditsch, N.P. and Nordås, R. (2014). Conflicting messages? The IPCC on conflict and human security. *Polit. Geogr.* 43: 82–90. https://doi.org/10.1016/j.polgeo.2014.08.007.

7 De Pryck, K. (2021). Intergovernmental expert consensus in the making: the case of the summary for policy makers of the IPCC 2014 synthesis report. *Glob. Environ. Politics* 21: 108–129. https://doi.org:10.1162/glep_a_00574.

8 Sepulveda, N.A., Jenkins, J.D., de Sisternes, F.J., and Lester, R.K. (2018). The role of firm low-carbon electricity resources in deep decarbonization of power generation. *Joule* 2: 2403–2420. https://doi.org/10.1016/j.joule.2018.08.006.

9 Friedlingstein, P., Jones, M.W., O'Sullivan, M. et al. (2019). Global carbon budget 2019. *Earth Syst. Sci. Data* 11: 1783–1838. https://doi.org/10.5194/essd-11-1783-2019.

10 Hermansen, E.A.T., Lahn, B., Sundqvist, G., and Øye, E. (2021). Post-Paris policy relevance: lessons from the IPCC SR15 process. *Clim. Change* 169: 7. https://doi.org/10.1007/s10584-021-03210-0.

11 Teske, S. (2019). Trajectories for a just transition of the fossil fuel industry. In: *Achieving the Paris Climate Agreement Goals: Global and Regional 100% Renewable Energy Scenarios with Non-energy GHG Pathways for +1.5°C and +2°C* (ed. S. Teske), 403–411. Switzerland AG: Springer Open.

12 *Renewable Power Generation Costs in 2020.* (2021), International Renewable Energy Agency, ISBN: 978-92-9260-348-9.

13 'What Is Renewable Energy?' United Nations, https://www.un.org/en/climatechange/what-is-renewable-energy. Accessed 1 Aug. 2023.

14 Renewable Power Generation Costs in 2020 (2021), International Renewable Energy Agency, ISBN: 978-92-9260-348-9.

15 Li, C., Cao, Q., Wang, F. et al. (2018). Engineering graphene and TMDs based van der Waals heterostructures for photovoltaic and photoelectrochemical solar energy conversion. *Chem. Soc. Rev.* 47: 4981–5037. https://doi.org:10.1039/C8CS00067K.

16 Walter, M.G., Warren, E.L., and McKone, J.R. (2010). Solar water splitting cells. *Chem. Rev.* 110: 6446–6473. https://doi.org:10.1021/cr1002326.

17 Kumar, B., Llorente, M., Froehlich, J. et al. (2012). Photochemical and photoelectrochemical reduction of CO_2. *Annu. Rev. Phys. Chem.* 63: 541–569. https://doi.org:10.1146/annurev-physchem-032511-143759.

18 White, J.L., Baruch, M.F., Pander, J.E. III, et al. (2015). Light-driven heterogeneous reduction of carbon dioxide: photocatalysts and photoelectrodes. *Chem. Rev.* 115: 12888–12935. https://doi.org:10.1021/acs.chemrev.5b00370.

19 Sivula, K. (2015). Toward economically feasible direct solar-to-fuel energy conversion. *J. Phys. Chem. Lett.* 6: 975–976. https://doi.org:10.1021/acs.jpclett.5b00406.

20 Mathis, A., Courvoisier, F., Froehly, L. et al. (2012). Micromachining along a curve: Femtosecond laser micromachining of curved profiles in diamond and silicon using accelerating beams. *Appl. Phys. Lett.* 101: 071110. https://doi.org:10.1063/1.4745925.

21 Wang, Z., White, T.P., and Catchpole, K.R. (2013). Plasmonic near-field enhancement for planar ultra-thin photovoltaics. *IEEE Photon. J.* 5: 8400608–8400608. https://doi.org:10.1109/JPHOT.2013.2280518.

22 Kang, G., Park, H., Shin, D. et al. (2013). Broadband light-trapping enhancement in an ultrathin film a-si absorber using whispering gallery modes and guided wave modes with dielectric surface-textured structures. *Adv. Mater.* 25: 2617–2623. https://doi.org/10.1002/adma.201204596.

23 Kats, M.A., Blanchard, R., Genevet, P., and Capasso, F. (2013). Nanometre optical coatings based on strong interference effects in highly absorbing media. *Nat. Mater.* 12: 20–24. https://doi.org:10.1038/nmat3443.

24 Massiot, I., Vandamme, N., Bardou, N. et al. (2014). Metal nanogrid for broadband multiresonant light-harvesting in ultrathin GaAs layers. *ACS Photonics* 1: 878–884. https://doi.org:10.1021/ph500168b.

25 Aeberhard, U. (2016). Simulation of ultrathin solar cells beyond the limits of the semiclassical bulk picture. *IEEE J. Photovolt.* 6: 654–660. https://doi.org:10.1109/JPHOTOV.2016.2528405.

26 Aeberhard, U. and Rau, U. (2017). Microscopic perspective on photovoltaic reciprocity in ultrathin solar cells. *Phys. Rev. Lett.* 118: 247702. https://doi.org:10.1103/PhysRevLett.118.247702.

27 Yin, Z., Zhu, J., He, Q. et al. (2014). Graphene-based materials for solar cell applications. *Adv. Energy Mater.* 4: 1300574. https://doi.org/10.1002/aenm.201300574.

28 Song, Y., Fang, W., Brenes, R., and Kong, J. (2015). Challenges and opportunities for graphene as transparent conductors in optoelectronics. *Nano Today* 10: 681–700. https://doi.org/10.1016/j.nantod.2015.11.005.

29 Tan, C., Cao, X., Wu, X.J. et al. (2017). Recent advances in ultrathin two-dimensional nanomaterials. *Chem. Rev.* 117: 6225–6331. https://doi.org:10.1021/acs.chemrev.6b00558.

30 Cao, X., Tan, C., Zhang, X. et al. (2016). Solution-processed two-dimensional metal dichalcogenide-based nanomaterials for energy storage and conversion. *Adv. Mater.* 28: 6167–6196. https://doi.org/10.1002/adma.201504833.

31 Huang, X., Qi, X., Boey, F., and Zhang, H. (2012). Graphene-based composites. *Chem. Soc. Rev.* 41: 666–686. https://doi.org:10.1039/C1CS15078B.

32 Kang, J., Tongay, S., Zhou, J. et al. (2013). Band offsets and heterostructures of two-dimensional semiconductors. *Appl. Phys. Lett.* 102: 012111. https://doi.org:10.1063/1.4774090.

33 Gong, C., Zhang, H., Wang, W. et al. (2013). Band alignment of two-dimensional transition metal dichalcogenides: application in tunnel field effect transistors. *Appl. Phys. Lett.* 103: 053513. https://doi.org:10.1063/1.4817409.

34 Tributsch, H. & Bennett, J. C. Electrochemistry and photochemistry of MoS_2 layer crystals. I. *J. Electroanal. Chem. Interfacial Electrochem.* 81, 97-111 (1977). https://doi.org/10.1016/S0022-0728(77)80363-X.

35 Yin, Y., Zhang, Y., Gao, T. et al. (2017). Synergistic phase and disorder engineering in 1T-$MoSe_2$ nanosheets for enhanced hydrogen-evolution reaction. *Adv. Mater.* 29: 1700311. https://doi.org/10.1002/adma.201700311.

36 Liu, Y., Weiss, N.O., Duan, X. et al. (2016). Van der Waals heterostructures and devices. *Nat. Rev. Mater.* 1: 16042. https://doi.org:10.1038/natrevmats.2016.42.

37 Jariwala, D., Davoyan, A.R., Wong, J., and Atwater, H.A. (2017). Van der Waals materials for atomically-thin photovoltaics: promise and outlook. *ACS Photonics* 4: 2962–2970. https://doi.org:10.1021/acsphotonics.7b01103.

38 Yoshikawa, K., Kawasaki, H., Yoshida, W. et al. (2017). Silicon heterojunction solar cell with interdigitated back contacts for a photoconversion efficiency over 26%. *Nat. Energy* 2: 17032. https://doi.org:10.1038/nenergy.2017.32.

39 Green, M.A., Hishikawa, Y., Warta, W. et al. (2017). Solar cell efficiency tables (version 50). *Prog. Photovoltaics Res. Appl.* 25: 668–676. https://doi.org/10.1002/pip.2909.

40 Yang, W. S., B. W. Park, E. H. Jung et al. Iodide management in formamidinium-lead-halide–based perovskite layers for efficient solar cells. *Science* 356, 1376-1379 (2017). https://doi.org:10.1126/science.aan2301

41 Mori, S., Oh-oka, H., Nakao, H. et al. (2014). Organic photovoltaic module development with inverted device structure. *MRS Online Proc. Lib.* 1737: 26–31. https://doi.org:10.1557/opl.2015.540.

42 Maier, C.U., Specht, M., and Bilger, G. (1996). Hydrogen evolution on platinum-coated p-silicon photocathodes. *Int. J. Hydrogen Energy* 21: 859–864. https://doi.org/10.1016/0360-3199(96)00023-7.

43 Bae, D., Pedersen, T., Seger, B. et al. (2017). Carrier-selective p- and n-contacts for efficient and stable photocatalytic water reduction. *Catal. Today* 290: 59–64. https://doi.org/10.1016/j.cattod.2016.11.028.

44 Vijselaar, W., Westerik, P., Veerbeek, J. et al. (2018). Spatial decoupling of light absorption and catalytic activity of Ni–Mo-loaded high-aspect-ratio silicon microwire photocathodes. *Nat. Energy* 3: 185–192. https://doi.org:10.1038/s41560-017-0068-x.

45 Kumagai, H., Minegishi, T., Sato, N. et al. (2015). Efficient solar hydrogen production from neutral electrolytes using surface-modified Cu(In,Ga)Se_2 photocathodes. *J. Mater. Chem. A* 3: 8300–8307. https://doi.org:10.1039/C5TA01058F.

46 Lee, M.H., Takei, K., Zhang, J. et al. (2012). p-type InP nanopillar photocathodes for efficient solar-driven hydrogen production. *Angew. Chem. Int. Ed.* 51: 10760–10764. https://doi.org/10.1002/anie.201203174.

47 Seger, B., Tilley, D.S., Pedersen, T. et al. (2013). Silicon protected with atomic layer deposited TiO_2: durability studies of photocathodic H_2 evolution. *RSC Adv.* 3: 25902–25907. https://doi.org:10.1039/C3RA45966G.

48 IPCC, 2014: Climate Change 2014: Synthesis Report. Contribution of Working Groups I, II and III to the Fifth Assessment Report of the Intergovernmental Panel on Climate Change [Core Writing Team, R.K. Pachauri and L.A. Meyer (eds.)]. IPCC, Geneva, Switzerland, 151

49 Costello, A., Abbas, M., Allen, A. et al. (2009). Managing the health effects of climate change. *Lancet* 373: 1693–1733. https://doi.org:10.1016/S0140-6736(09)60935-1.

50 Watts, N., Adger, W.N., Agnolucci, P. et al. (2015). Health and climate change: policy responses to protect public health. *Lancet* 386: 1861–1914. https://doi.org:10.1016/S0140-6736(15)60854-6.

51 Watts, N., Amann, M., Arnell, N. et al. (2019). The 2019 report of The Lancet Countdown on health and climate change: ensuring that the health of a child born today is not defined by a changing climate. *Lancet* 394: 1836–1878. https://doi.org:10.1016/S0140-6736(19)32596-6.

52 Holding, S., Allen, D.M., Foster, S. et al. (2016). Groundwater vulnerability on small islands. *Nat. Clim. Change* 6: 1100–1103. https://doi.org:10.1038/nclimate3128.

53 Hall, M.H.P. and Fagre, D.B. (2003). Modeled climate-induced glacier change in Glacier National Park, 1850–2100. *BioScience* 53: 131–140. https://doi.org:10.1641/0006-3568(2003)053[0131:MCIGCI]2.0.CO;2.

54 Epstein, P.R. and Ferber, D. (2011). *Changing Planet, Changing Health: How the Climate Crisis Threatens Our Health and What We can do About It*. University of California Press.

55 Wu, B., Wang, G., Wang, Z. et al. (2017). Integrated hydrologic and hydrodynamic modeling to assess water exchange in a data-scarce reservoir. *J. Hydrol.* 555: 15–30. https://doi.org/10.1016/j.jhydrol.2017.09.057.

56 Wang, G., Li, J., Sun, W. et al. (2019). Non-point source pollution risks in a drinking water protection zone based on remote sensing data embedded within a nutrient budget model. *Water Res.* 157: 238–246. https://doi.org/10.1016/j.watres.2019.03.070.

57 Douterelo, I., Jackson, M., Solomon, C., and Boxall, J. (2017). Spatial and temporal analogies in microbial communities in natural drinking water biofilms. *Sci. Total Environ.* 581, 582: 277–288. https://doi.org/10.1016/j.scitotenv.2016.12.118.

58 Yu, L., Ruan, S., Xu, X. et al. (2017). One-dimensional nanomaterial-assembled macroscopic membranes for water treatment. *Nano Today* 17: 79–95. https://doi.org/10.1016/j.nantod.2017.10.012.

59 Lipczynska-Kochany, E. (2018). Humic substances, their microbial interactions and effects on biological transformations of organic pollutants in water and soil: a review. *Chemosphere* 202: 420–437. https://doi.org/10.1016/j.chemosphere.2018.03.104.

60 Syafrudin, M., Kristanti, R.A., Yuniarto, A. et al. (2021). Pesticides in drinking water—a review. *Int. J. Environ. Res. Public Health* 18 (2): 468. https://doi.org/10.3390/ijerph18020468.

61 Jureczko, M. and Kalka, J. (2020). Cytostatic pharmaceuticals as water contaminants. *Eur. J. Pharmacol.* 866: 172816. https://doi.org/10.1016/j.ejphar.2019.172816.

62 Raptis, C.E., van Vliet, M.T.H., and Pfister, S. (2016). Global thermal pollution of rivers from thermoelectric power plants. *Environ. Res. Lett.* 11: 104011. https://doi.org:10.1088/1748-9326/11/10/104011.

63 Råman Vinnå, L., Wüest, A. & Bouffard, D. Physical effects of thermal pollution in lakes. *Water Resour. Res.* 53, 3968-3987 (2017). https://doi.org/10.1002/2016WR019686

64 Dehghani, M.H., Mostofi, M., Alimohammadi, M. et al. (2016). High-performance removal of toxic phenol by single-walled and multi-walled carbon nanotubes: Kinetics, adsorption, mechanism and optimization studies. *J. Ind. Eng. Chem.* 35: 63–74. https://doi.org/10.1016/j.jiec.2015.12.010.

65 Akinpelu, A.A., Ali, M.E., Johan, M.R. et al. (2019). Polycyclic aromatic hydrocarbons extraction and removal from wastewater by carbon nanotubes: a review of the current technologies, challenges and prospects. *Process Saf. Environ. Prot.* 122: 68–82. https://doi.org/10.1016/j.psep.2018.11.006.

66 Shehzad, K., Xu, Y., Gao, C., and Duan, X. (2016). Three-dimensional macro-structures of two-dimensional nanomaterials. *Chem. Soc. Rev.* 45: 5541–5588. https://doi.org:10.1039/C6CS00218H.

67 Zeng, M., Chen, M., Huang, D. et al. (2021). Engineered two-dimensional nanomaterials: an emerging paradigm for water purification and monitoring. *Mater. Horiz.* 8: 758–802. https://doi.org:10.1039/D0MH01358G.

68 Anichini, C., Czepa, W., Pakulski, D. et al. (2018). Chemical sensing with 2D materials. *Chem. Soc. Rev.* 47: 4860–4908. https://doi.org:10.1039/C8CS00417J.

69 Fatima, J., Shen, Z., Wu, S. et al. (2022). Tunable 2D nanomaterials; their key roles and mechanisms in water purification and monitoring. *Front. Environ. Sci.* 10: https://doi.org:10.3389/fenvs.2022.766743.

70 Yuan, Y.-J., Shen, Z., Wu, S. et al. (2019). Liquid exfoliation of g-C_3N_4 nanosheets to construct 2D-2D MoS_2/g-C_3N_4 photocatalyst for enhanced photocatalytic H_2 production activity. *Appl. Catal., B* 246: 120–128. https://doi.org/10.1016/j.apcatb.2019.01.043.

71 Gao, J., Zhang, M., Wang, J. et al. (2019). Bioinspired modification of layer-stacked molybdenum disulfide (MoS_2) membranes for enhanced nanofiltration performance. *ACS Omega* 4: 4012–4022. https://doi.org:10.1021/acsomega.9b00155.

72 Wang, Z. and Mi, B. (2017). Environmental applications of 2D molybdenum disulfide (MoS_2) nanosheets. *Environ. Sci. Technol.* 51: 8229–8244. https://doi.org:10.1021/acs.est.7b01466.

73 Wu, M.-h., Li, L., Liu, N. et al. (2018). Molybdenum disulfide (MoS_2) as a co-catalyst for photocatalytic degradation of organic contaminants: a review. *Process Saf. Environ. Prot.* 118: 40–58. https://doi.org/10.1016/j.psep.2018.06.025.

74 Eftekhari, A. (2017). Tungsten dichalcogenides (WS_2, WSe_2, and WTe_2): materials chemistry and applications. *J. Mater. Chem. A* 5: 18299–18325. https://doi.org:10.1039/C7TA04268J.

75 Rahmanian, E., Mayorga-Martinez, C.C., Malekfar, R. et al. (2018). 1T-phase tungsten chalcogenides (WS_2, WSe_2, WTe_2) decorated with TiO_2 nanoplatelets

with enhanced electron transfer activity for biosensing applications. *ACS Appl. Nano Mater.* 1: 7006–7015. https://doi.org:10.1021/acsanm.8b01796.

76 Zheng, Z., Grünker, R., and Feng, X. (2016). Synthetic two-dimensional materials: a new paradigm of membranes for ultimate separation. *Adv. Mater.* 28: 6529–6545. https://doi.org/10.1002/adma.201506237.

77 Heiranian, M., Farimani, A.B., and Aluru, N.R. (2015). Water desalination with a single-layer MoS_2 nanopore. *Nat. Commun.* 6: 8616. https://doi.org:10.1038/ncomms9616.

78 Li, X., Ma, W., Li, H. et al. (2020). Sulfur-functionalized metal-organic frameworks: synthesis and applications as advanced adsorbents. *Coord. Chem. Rev.* 408: 213191. https://doi.org/10.1016/j.ccr.2020.213191.

79 Romanello, M., McGushin, A., Napoli, C.D. et al. (2021). The 2021 report of the Lancet Countdown on health and climate change: code red for a healthy future. *Lancet* 398: 1619–1662. https://doi.org:10.1016/S0140-6736(21)01787-6.

80 Epstein, P.R. (2001). Climate change and emerging infectious diseases. *Microbes Infect.* 3: 747–754. https://doi.org/10.1016/S1286-4579(01)01429-0.

81 Carlson, C.J., Albery, G.F., Merow, C. et al. (2022). Climate change increases cross-species viral transmission risk. *Nature* 607: 555–562. https://doi.org:10.1038/s41586-022-04788-w.

82 Pal, A., Jana, T.K., Roy, T. et al. (2018). MoS_2-TiO_2 nanocomposite with excellent adsorption performance and high antibacterial activity. *ChemistrySelect* 3: 81–90. https://doi.org/10.1002/slct.201702618.

83 Yang, X., Li, J., Liang, T. et al. (2014). Antibacterial activity of two-dimensional MoS_2 sheets. *Nanoscale* 6: 10126–10133. https://doi.org:10.1039/C4NR01965B.

84 Wu, N., Y. Yu, T. Li et al. Investigating the Influence of MoS_2 nanosheets on *E. coli* from metabolomics level. *PLoS One* 11, e0167245 (2016). https://doi.org:10.1371/journal.pone.0167245

85 Tang, K., Wang, L., Geng, H. et al. (2020). Molybdenum disulfide (MoS_2) nanosheets vertically coated on titanium for disinfection in the dark. *Arabian J. Chem.* 13: 1612–1623. https://doi.org/10.1016/j.arabjc.2017.12.013.

86 Pandit, S., Karunakaran, S., Boda, S.K. et al. (2016). High antibacterial activity of functionalized chemically exfoliated MoS_2. *ACS Appl. Mater. Interfaces* 8: 31567–31573. https://doi.org:10.1021/acsami.6b10916.

87 Cao, F., Ju, E., Zhang, Y. et al. (2017). An efficient and benign antimicrobial depot based on silver-infused MoS_2. *ACS Nano* 11: 4651–4659. https://doi.org:10.1021/acsnano.7b00343.

88 Zhang, X., Zhang, W., Liu, L. et al. (2017). Antibiotic-loaded MoS_2 nanosheets to combat bacterial resistance via biofilm inhibition. *Nanotechnology* 28: 225101. https://doi.org:10.1088/1361-6528/aa6c9b.

89 Liu, C., Kong, D., Hsu, P.C. et al. (2016). Rapid water disinfection using vertically aligned MoS_2 nanofilms and visible light. *Nat. Nanotechnol.* 11: 1098–1104. https://doi.org:10.1038/nnano.2016.138.

90 Yan, H., Liu, L., Wang, R. et al. (2020). Binary composite MoS_2/TiO_2 nanotube arrays as a recyclable and efficient photocatalyst for solar water disinfection. *Chem. Eng. J.* 401: 126052. https://doi.org/10.1016/j.cej.2020.126052.

91 Navale, G.R., Rout, C.S., Gohil, K.N. et al. (2015). Oxidative and membrane stress-mediated antibacterial activity of WS_2 and rGO-WS_2 nanosheets. *RSC Adv.* 5: 74726–74733. https://doi.org:10.1039/C5RA15652A.

92 Liu, X., Duan, G., Li, W. et al. (2017). Membrane destruction-mediated antibacterial activity of tungsten disulfide (WS_2). *RSC Adv.* 7: 37873–37880. https://doi.org:10.1039/C7RA06442J.

93 Bang, G.S., Cho, S., Son, N. et al. (2016). DNA-assisted exfoliation of tungsten dichalcogenides and their antibacterial effect. *ACS Appl. Mater. Interfaces* 8: 1943–1950. https://doi.org:10.1021/acsami.5b10136.

94 Ouyang, J., Rao, S., Liu, R. et al. (2022). 2D materials-based nanomedicine: from discovery to applications. *Adv. Drug Delivery Rev.* 185: 114268. https://doi.org/10.1016/j.addr.2022.114268.

95 Wang, H., Zeng, X., Pang, L. et al. (2020). Integrative treatment of anti-tumor/bone repair by combination of MoS_2 nanosheets with 3D printed bioactive borosilicate glass scaffolds. *Chem. Eng. J.* 396: 125081. https://doi.org/10.1016/j.cej.2020.125081.

96 Bolotsky, A., Butler, D., Dong, C. et al. (2019). Two-dimensional materials in biosensing and healthcare: from in vitro diagnostics to optogenetics and beyond. *ACS Nano* 13: 9781–9810. https://doi.org:10.1021/acsnano.9b03632.

97 Huang, C.-C., Aronstam, R.S., Chen, D.-R., and Huang, Y.-W. (2010). Oxidative stress, calcium homeostasis, and altered gene expression in human lung epithelial cells exposed to ZnO nanoparticles. *Toxicol. in Vitro* 24: 45–55. https://doi.org/10.1016/j.tiv.2009.09.007.

98 Yan, L., Zhao, F., Li, S. et al. (2011). Low-toxic and safe nanomaterials by surface-chemical design, carbon nanotubes, fullerenes, metallofullerenes, and graphenes. *Nanoscale* 3: 362–382. https://doi.org:10.1039/C0NR00647E.

99 Asian Development Bank. 'ADB Plans $14 Billion to Ease Food Crisis, Promote Long-Term Food Security in Asia and the Pacific'. Asian Development Bank, 27 Sept. 2022, https://www.adb.org/news/adb-plans-14-billion-ease-food-crisis-promote-long-term-food-security-asia-and-pacific.

100 'The War in Ukraine Triggered a Global Food Shortage'. Brookings, https://www.brookings.edu/articles/the-war-in-ukraine-triggered-a-global-food-shortage/ (accessed 1 August 2023).

101 FAO, IFAD, UNICEF, WFP and WHO. 2020. *The State of Food Security and Nutrition in the World 2020. Transforming food systems for affordable healthy diets.* Rome, FAO. https://doi.org/10.4060/ca9692en.

102 Solaraju-Murali, B., N Gonzalez-Reviriego, L. Caron et al. Multi-annual prediction of drought and heat stress to support decision making in the wheat sector *npj Clim. Atmos. Sci.* 4, 34 (2021). https://doi.org:10.1038/s41612-021-00189-4

103 Grote, U., Fasse, A., Nguyen, T.T., and Erenstein, O. (2021). Food security and the dynamics of wheat and maize value chains in Africa and Asia. *Front. Sustain. Food Syst.* 4: https://doi.org:10.3389/fsufs.2020.617009.

104 Nations, U (2022). *World Population to Reach 8 billion on 15 November 2022.* United Nations.

105 World Population Prospects 2022: Summary of Results, Department of Economic and Social Affairs, the United Nations, ISBN: 978-92-1-148373-4.

106 Weber, H. and Sciubba, J.D. (2019). The effect of population growth on the environment: evidence from European regions. *Eur. J. Popul.* 35: 379–402. https://doi.org:10.1007/s10680-018-9486-0.

107 Ranganathan, J., Waite, R., Searchinger, T. & Hanson, C. *How to Sustainably Feed 10 Billion People by 2050, in 21 Charts.* (2018).

108 Singhal, J., Verma, S., and Kumar, S. (2022). The physio-chemical properties and applications of 2D nanomaterials in agricultural and environmental sustainability. *Sci. Total Environ.* 837: 155669. https://doi.org/10.1016/j.scitotenv.2022.155669.

109 Yu, Y., Zheng, J., Li, J. et al. (2021). Applications of two-dimensional materials in food packaging. *Trends Food Sci. Technol.* 110: 443–457. https://doi.org/10.1016/j.tifs.2021.02.021.

110 Joshi, N., Pransu, G. & Adam Conte-Junior, C. Critical review and recent advances of 2D materials-based gas sensors for food spoilage detection. *Crit. Rev. Food Sci. Nutr.*, 1-24 (2022). https://doi.org:10.1080/10408398.2022.2078950

111 Myers, S.S., Smith, M.R., Guth, S. et al. (2017). Climate change and global food systems: potential impacts on food security and undernutrition. *Annu. Rev. Public Health* 38: 259–277. https://doi.org:10.1146/annurev-publhealth-031816-044356.

112 Kabiri, S., Degryse, F., Tran, D.N.H. et al. (2017). Graphene oxide: a new carrier for slow release of plant micronutrients. *ACS Appl. Mater. Interfaces* 9: 43325–43335. https://doi.org:10.1021/acsami.7b07890.

113 Tong, Y., Shao, L., Li, X. et al. (2018). Adhesive and stimulus-responsive polydopamine-coated graphene oxide system for pesticide-loss control. *J. Agric. Food. Chem.* 66: 2616–2622. https://doi.org:10.1021/acs.jafc.7b05500.

114 Thirumalraj, B., Krishnapandi, A., Chen, S.-M., Sudhakaran M. S. P., and Choe, H. Rational design and interlayer effect of dysprosium-stannate nanoplatelets incorporated graphene oxide: a versatile and competent electrocatalyst for toxic carbamate pesticide detection in vegetables. *ACS Sustainable Chem. Eng.* 8, 17882-17892 (2020). https://doi.org:10.1021/acssuschemeng.0c07973

115 Chen, Y., Liu, H., Tian, Y. et al. (2020). In situ recyclable surface-enhanced Raman scattering-based detection of multicomponent pesticide residues on fruits and vegetables by the flower-like MoS_2@Ag hybrid substrate. *ACS Appl. Mater. Interfaces* 12: 14386–14399. https://doi.org:10.1021/acsami.9b22725.

116 Nasir, M.Z.M., Mayorga-Martinez, C.C., Sofer, Z., and Pumera, M. (2017). Two-dimensional 1T-phase transition metal dichalcogenides as nanocarriers to enhance and stabilize enzyme activity for electrochemical pesticide detection. *ACS Nano* 11: 5774–5784. https://doi.org:10.1021/acsnano.7b01364.

117 Li, F., Chen, L., Liu, H. et al. (2019). Enhanced N_2-fixation by engineering the edges of two-dimensional transition-metal disulfides. *J. Phys. Chem. C* 123: 22221–22227. https://doi.org:10.1021/acs.jpcc.9b04730.

118 Risch, S.J. (2009). Food packaging history and innovations. *J. Agric. Food. Chem.* 57: 8089–8092. https://doi.org:10.1021/jf900040r.

119 Ketelsen, M., Janssen, M., and Hamm, U. (2020). Consumers' response to environmentally-friendly food packaging - a systematic review. *J. Cleaner Prod.* 254: 120123. https://doi.org/10.1016/j.jclepro.2020.120123.

120 Yamaguchi, M., Terao, T., Ohira, A. et al. (2019). Size and shape of Nafion particles in water after high-temperature treatment. *J. Polym. Sci., Part B: Polym. Phys.* 57: 813–818. https://doi.org/10.1002/polb.24833.

121 Peng, H., Wang, D., and Fu, S. (2020). Tannic acid-assisted green exfoliation and functionalization of MoS_2 nanosheets: significantly improve the mechanical and flame-retardant properties of polyacrylonitrile composite fibers. *Chem. Eng. J.* 384: 123288. https://doi.org/10.1016/j.cej.2019.123288.

122 Bertolazzi, S., Brivio, J., and Kis, A. (2011). Stretching and breaking of ultrathin MoS_2. *ACS Nano* 5: 9703–9709. https://doi.org:10.1021/nn203879f.

123 Sahatiya, P., Kadu, A., Gupta, H., Thanga Gomathi, P. & Badhulika, S. Flexible, Disposable cellulose-paper-based MoS_2/Cu_2S hybrid for wireless environmental monitoring and multifunctional sensing of chemical stimuli. *ACS Appl. Mater. Interfaces* 10, 9048-9059 (2018). https://doi.org:10.1021/acsami.8b00245

124 Liao, Y., Zhang, R., and Qian, J. (2019). Printed electronics based on inorganic conductive nanomaterials and their applications in intelligent food packaging. *RSC Adv.* 9: 29154–29172. https://doi.org:10.1039/C9RA05954G.

125 McManus, D., Vranic, S., Withers, F. et al. (2017). Water-based and biocompatible 2D crystal inks for all-inkjet-printed heterostructures. *Nat. Nanotechnol.* 12: 343–350. https://doi.org:10.1038/nnano.2016.281.

126 Mehonic, A. and Kenyon, A.J. (2022). Brain-inspired computing needs a master plan. *Nature* 604: 255–260. https://doi.org:10.1038/s41586-021-04362-w.

127 Jones, N. (2018). How to stop data centres from gobbling up the world's electricity. *Nature* 561: 163–166. https://doi.org:10.1038/d41586-018-06610-y.

128 Arnold, A.J., Razavieh, A., Nasr, J.R. et al. (2017). Mimicking neurotransmitter release in chemical synapses via hysteresis engineering in MoS_2 transistors *ACS Nano* 11: 3110–3118. https://doi.org:10.1021/acsnano.7b00113.

129 Bao, L., Zhu, J., Yu, Z. et al. (2019). Dual-gated MoS_2 neuristor for neuromorphic computing. *ACS Appl. Mater. Interfaces* 11: 41482–41489. https://doi.org:10.1021/acsami.9b10072.

130 Shi, Y., Liang, X., Yuan, B. et al. (2018). Electronic synapses made of layered two-dimensional materials. *Nat. Electron.* 1: 458–465. https://doi.org:10.1038/s41928-018-0118-9.

131 Wang, M., Cai, S., Pan, C. et al. (2018). Robust memristors based on layered two-dimensional materials. *Nat. Electron.* 1: 130–136. https://doi.org:10.1038/s41928-018-0021-4.

132 Ge, R., Wu, X., Kim, M. et al. (2018). Atomristor: nonvolatile resistance switching in atomic sheets of transition metal dichalcogenides. *Nano Lett.* 18: 434–441. https://doi.org:10.1021/acs.nanolett.7b04342.

133 Kwon, K.C., Baek, J.H., Hong, K. et al. (2022). Memristive devices based on two-dimensional transition metal chalcogenides for neuromorphic computing. *Nano-Micro Lett.* 14: 58. https://doi.org:10.1007/s40820-021-00784-3.

134 Zhao, Q., Xie, Z., Peng, Y.-P. et al. (2020). Current status and prospects of memristors based on novel 2D materials. *Mater. Horiz.* 7: 1495–1518. https://doi.org:10.1039/C9MH02033K.

135 Ge, R., Wu, X., Liang, L. et al. (2021). A library of atomically thin 2D materials featuring the conductive-point resistive switching phenomenon. *Adv. Mater.* 33: 2007792. https://doi.org/10.1002/adma.202007792.

136 He, H.-K., Yang, F.-F., and Yang, R. (2020). Flexible full two-dimensional memristive synapses of graphene/$WSe_{2-x}O_y$/graphene. *Phys. Chem. Chem. Phys.* 22: 20658–20664. https://doi.org:10.1039/D0CP03822A.

137 Meng, J.-L., Wang, T.-Y., Chen, L. et al. (2021). Energy-efficient flexible photoelectric device with 2D/0D hybrid structure for bio-inspired artificial heterosynapse application. *Nano Energy* 83: 105815. https://doi.org/10.1016/j.nanoen.2021.105815.

138 Mennel, L., Symonowicz, J., Wachter, S. et al. (2020). Ultrafast machine vision with 2D material neural network image sensors. *Nature* 579: 62–66. https://doi.org:10.1038/s41586-020-2038-x.

139 Hou, X., Liu, C., Ding, Y. et al. (2020). A logic-memory transistor with the integration of visible information sensing-memory-processing. *Adv. Sci.* 7: 2002072. https://doi.org/10.1002/advs.202002072.

140 Ajayan, P., Kim, P., and Banerjee, K. (2016). Two-dimensional van der Waals materials. *Phys. Today* 69: 38–44. https://doi.org:10.1063/PT.3.3297.

141 Sun, L., Zhang, Y., Han, G. et al. (2019). Self-selective van der Waals heterostructures for large scale memory array. *Nat. Commun.* 10: 3161. https://doi.org:10.1038/s41467-019-11187-9.

142 Ren, Y., Hu, L., Mao, J.-Y. et al. (2018). Phosphorene nano-heterostructure based memristors with broadband response synaptic plasticity. *J. Mater. Chem. C* 6: 9383–9393. https://doi.org:10.1039/C8TC03089H.

143 Zhang, W., Gao, H., Deng, C. et al. (2021). An ultrathin memristor based on a two-dimensional WS_2/MoS_2 heterojunction. *Nanoscale* 13: 11497–11504. https://doi.org:10.1039/D1NR01683K.

144 Lin, Z., Bruno R Carvalho, Ethan Kahn et al. Defect engineering of two-dimensional transition metal dichalcogenides. *2D Mater.* 3, 022002 (2016). https://doi.org:10.1088/2053-1583/3/2/022002

145 Liang, Q., Zhang, Q., Zhao, X. et al. (2021). Defect engineering of two-dimensional transition-metal dichalcogenides: applications, challenges, and opportunities. *ACS Nano* 15: 2165–2181. https://doi.org:10.1021/acsnano.0c09666.

146 Cheng, Y., Nie, A., Zhang, Q. et al. (2014). Origin of the phase transition in lithiated molybdenum disulfide. *ACS Nano* 8: 11447–11453. https://doi.org:10.1021/nn505668c.

147 Huang, J., Pan, X., Liao, X. et al. (2020). In situ monitoring of the electrochemically induced phase transition of thermodynamically metastable 1T-MoS_2 at nanoscale. *Nanoscale* 12: 9246–9254. https://doi.org:10.1039/D0NR02161J.

148 Lin, Y.-C., Dumcenco, D.O., Huang, Y.-S., and Suenaga, K. (2014). Atomic mechanism of the semiconducting-to-metallic phase transition in single-layered MoS_2. *Nat. Nanotechnol.* 9: 391–396. https://doi.org:10.1038/nnano.2014.64.

149 Zhang, K. et al. (2018). Tuning the electronic and photonic properties of monolayer MoS_2 via in situ rhenium substitutional doping. *Adv. Funct. Mater.* 28: 1706950. https://doi.org/10.1002/adfm.201706950.

150 Lu, A.-Y. et al. (2017). Janus monolayers of transition metal dichalcogenides. *Nat. Nanotechnol.* 12: 744–749. https://doi.org:10.1038/nnano.2017.100.

151 Zhu, X., Li, D., Liang, X., and Lu, W.D. (2019). Ionic modulation and ionic coupling effects in MoS_2 devices for neuromorphic computing. *Nat. Mater.* 18: 141–148. https://doi.org:10.1038/s41563-018-0248-5.

152 Chhowalla, M. et al. (2013). The chemistry of two-dimensional layered transition metal dichalcogenide nanosheets. *Nat. Chem.* 5: 263–275. https://doi.org:10.1038/nchem.1589.

153 Miller, K.D. (1996). Synaptic economics: competition and cooperation in synaptic plasticity. *Neuron* 17: 371–374. https://doi.org:10.1016/S0896-6273(00)80169-5.

154 Ye, G. et al. (2016). Defects engineered monolayer MoS_2 for improved hydrogen evolution reaction. *Nano Lett.* 16: 1097–1103. https://doi.org:10.1021/acs.nanolett.5b04331.

155 Chen, X., Shinde, S.M., Dhakal, K.P. et al. (2018). Degradation behaviors and mechanisms of MoS_2 crystals relevant to bioabsorbable electronics. *NPG Asia Mater.* 10: 810–820. https://doi.org:10.1038/s41427-018-0078-6.

156 Lu, X., Utama, M.I.B., Zhang, J. et al. (2013). Layer-by-layer thinning of MoS_2 by thermal annealing. *Nanoscale* 5: 8904–8908. https://doi.org:10.1039/C3NR03101B.

157 Yang, X., Tang, S., Ding, G. et al. (2012). Layer-by-layer thinning of graphene by plasma irradiation and post-annealing. *Nanotechnology* 23: 025704. https://doi.org:10.1088/0957-4484/23/2/025704.

158 Wu, J., Li, H., Yin, Z. et al. (2013). Layer thinning and etching of mechanically exfoliated MoS_2 nanosheets by thermal annealing in air. *Small* 9: 3314–3319. https://doi.org/10.1002/smll.201301542.

159 Song, Y.-W., Song, M.-K., Choi, D., and Kwon, J.-Y. (2021). Encapsulation-enhanced switching stability of MoS_2 memristors. *J. Alloys Compd.* 885: 161016. https://doi.org/10.1016/j.jallcom.2021.161016.

160 Roh, J., Cho, I.-T., Shin, H. et al. (2015). Fluorinated CYTOP passivation effects on the electrical reliability of multilayer MoS_2 field-effect transistors. *Nanotechnology* 26: 455201. https://doi.org:10.1088/0957-4484/26/45/455201.

161 Yin, X., Tang, C.S., Zheng, Y. et al. (2021). Recent developments in 2D transition metal dichalcogenides: phase transition and applications of the (quasi-)metallic phases. *Chem. Soc. Rev.* 50: 10087–10115. https://doi.org:10.1039/D1CS00236H.

162 Liu, C., Chen, H., Wang, S. et al. (2020). Two-dimensional materials for next-generation computing technologies. *Nat. Nanotechnol.* 15: 545–557. https://doi.org:10.1038/s41565-020-0724-3.

163 Chhowalla, M., Jena, D., and Zhang, H. (2016). Two-dimensional semiconductors for transistors. *Nat. Rev. Mater.* 1: 16052. https://doi.org:10.1038/natrevmats.2016.52.

164 Cui, X., Lee, G.-H., Kim, Y.D. et al. (2015). Multi-terminal transport measurements of MoS_2 using a van der Waals heterostructure device platform. *Nat. Nanotechnol.* 10: 534–540. https://doi.org:10.1038/nnano.2015.70.

165 Zhu, W., Perebeinos, V., Freitag, M., and Avouris, P. (2009). Carrier scattering, mobilities, and electrostatic potential in monolayer, bilayer, and trilayer graphene. *Phys. Rev. B* 80: 235402. https://doi.org:10.1103/PhysRevB.80.235402.

166 Liu, C., Chen, H., Hou, X. et al. (2019). Small footprint transistor architecture for photoswitching logic and in situ memory. *Nat. Nanotechnol.* 14: 662–667. https://doi.org:10.1038/s41565-019-0462-6.

Index

Two-Dimensional Transition-Metal Dichalcogenides: Phase Engineering and Applications in Electronics and Optoelectronics, First Edition. Edited by Chi Sin Tang, Xinmao Yin, and Andrew T. S. Wee.

n

o

p